NEUROPEPTIDES IN DEVELOPMENT AND AGING

ANNALS OF THE NEW YORK ACADEMY OF SCIENCES
Volume 814

NEUROPEPTIDES IN DEVELOPMENT AND AGING

Edited by Bill E. Beckwith, Alois Saria, Bibie M. Chronwall, Curt A. Sandman, and Fleur L. Strand

The New York Academy of Sciences
New York, New York
1997

Library of Congress Cataloging-in-Publication Data

Neuropeptides in development and aging/edited by Bill E. Beckwith
    . . . [et al.].
        p.   cm.—(Annals of the New York Academy of Sciences, ISSN
0077-8923; v. 814)
    Proceedings of the first joint meeting of the European Neuropeptide
Club and the 17th Winter Neuropeptide Conference, held Feb. 3–6, 1996
in Breckenridge, Colorado.
    Includes bibliographical references and index.
    ISBN 1-57331-066-2 (cloth: alk. paper).—ISBN 1-57331-067-0
(pbk.: alk. paper)
    1. Neuropeptides—Pathophysiology—Congresses.   2. Developmental
neurobiology—Congresses.   3. Brain—Pathophysiology.   4. Senile
dementia—Pathophysiology.   I. Beckwith, Bill E.   II. Winter
Neuropeptide Conference (17th: 1996: Breckenridge, Colo.)
III. Series.
    [DNLM: 1. Neuropeptides—metabolism—congresses. 2. Aging—
metabolism—congresses. 3. Brain—growth & development—
congresses.   4. Brain—physiopathology—congresses.   W1 AN626YL
v.814 1997/WL 104 N49424 1997]
    Q11.N5 vol. 814
    [QP552.N48]
    500 s—dc21
    [612.6]
    DNLM/DLC
for Library of Congress   97-5944
                                                                            CIP

BiC/PCP
Printed in the United States of America
ISBN 1-57331-066-2 (cloth)
ISBN 1-57331-067-0 (paper)
ISSN 0077-8923

ANNALS OF THE NEW YORK ACADEMY OF SCIENCES

Volume 814
April 24, 1997

NEUROPEPTIDES IN DEVELOPMENT AND AGING[a]

Editors and Conference Organizers
BILL E. BECKWITH, ALOIS SARIA, BIBIE M. CHRONWALL,
CURT A. SANDMAN, AND FLEUR L. STRAND

CONTENTS

[a] This volume comprises the proceedings of a conference entitled Neuropeptides in Development and Aging, the first joint meeting of the European Neuropeptide Club and the 17th Winter Neuropeptide Conference, held February 3–6, 1996 in Breckenridge, Colorado.

Financial assistance was received from:

- BIOMEASURE, INC.
- BURROUGHS WELLCOME
- CEPHALON, INC.
- MERCK
- PFIZER, INC.
- SMITH KLINE BEECHAM PHARMACEUTICALS
- THE COUNCIL FOR TOBACCO RESEARCH
- THE UPJOHN COMPANY

Preface

The Seventeenth Annual Winter Neuropeptide Conference (WNPC) once again was held in Breckenridge, Colorado. This year's conference was unique because it marked the first joint meeting of the WNPC and the European Neuropeptide Club, both affiliated with the International Neuropeptide Society. This venture provided an opportunity to share concepts and directions in neuropeptide research and theory in a broader forum than in the past. The proceedings of two previous WNPCs were published as *Annals of the New York Academy of Sciences* (Vol. 579, 1990 and Vol. 739, 1994). We are now pleased to provide a third *Annals* on conference proceedings.

The focus of this year's meeting was the influence of neuropeptides on development throughout the life span. It has become increasingly clear that the central nervous system (CNS) is not a passive entity determined by either environment or heredity. Rather, the brain appears to interact with its environment at critical periods throughout early and late development and thereby contributes to its own structural and functional systems. The sessions in this year's meeting were aimed at describing the role neuropeptides play in this process.

Part I, on neuropeptides in the pathology of aging, opens with the plenary address by Carl Cotman. Dr. Cotman reviews current concepts in peptidergic alterations underlying dementias of the Alzheimer's type and their relation to pathology and course. He also presents novel concepts aimed at untangling factors that modulate manifestations and rate of progression of the disorder, including a discussion of possible neuroprotective factors that may be induced by activities such as exercise. These processes may not only relate to the development of pathological processes in aging but may also explain normal variation in aging. For example, Albert *et al.* recently published interesting data suggesting that education, strenuous exercise, peak pulmonary expiratory flow rate, and self-efficacy modulate cognitive functioning in aging (Psychol. & Aging. 1995. **10:** 578–589). These relationships may be a result of yet-to-be identified neuroprotective factors. Dr. Cotman's paper is followed by descriptions of peptidergic changes and peptide processing enzymes related to Alzheimer's-like disease. For example, neuropeptides may hold promise as biochemical markers for Alzheimer's disease and specific neuropeptidases appear to be correlated with degradation of peptidergic and related neurotransmitter systems.

In the second section, an overview is presented of the family of insect neuropeptides which controls development. One of the first demonstrations that the brain and its neurosecretions are essential for normal metamorphosis was presented through studies of the gypsy moth, *Lymantria dispar,* by the Polish biologist Stephan Kopec at the turn of the century. Subsequent demonstrations have proven that all physiological processes in insects, including development and aging, are influenced by neuropeptides. Papers in this section provide details regarding the role of intracellular messengers such as eclosion hormone and its action on the nervous system, the role of allatostatins (which may be similar in function and distribution to somatostatin in vertebrates), and pyrokinin/PBAN in modulating pupariation. This section ends with a discussion of structure-activity relationships underlying the biological activity of insect neuropeptides.

The biosynthesis and processing of neuropeptides during development are reviewed in part III. For example, secretoneurin is associated with development of the amygdala that links hypothalamic, limbic, and striatal functions. Additionally, the unique role of proenkephalin processing as it contributes to brain development is discussed. Cytokines broaden the array of factors guiding CNS maturation and appear to have important functional roles during brain development extending the role of cytokines beyond mediation of peripheral immune function. Establishing the role of neuropeptides in development depends on the presence of the peptide precursors and the presence of specific cleavage enzymes. The use of metal-ion sites and nonpeptide antagonists is discussed to clarify the process of signal transduction with neurotensin. These areas of research offer important implications for understanding the development and treatment of neuropsychiatric disorders.

Part IV discusses neuropeptidergic regulation of mitogenesis, cell survival, and embryonic growth. Among the peptides discussed is vasoactive intestinal peptide (VIP), which has a dramatic influence on the regulation of embryonic growth and maturation related to brain ontogenesis. It has become clear that ontogenetic cell death is an essential component determining the structure of functional systems in the brain. VIP appears to contribute to the determination of neuronal circuitry that underlies many biological processes ranging from exact distribution of functional modules within the CNS to regulation of behavioral processes. The papers on VIP present a logical description of a strategy for understanding and describing these processes. At the other end of the developmental spectrum is the discussion of the possible role of a VIP analogue as a neuroprotective agent which may guard against neuronal death that is a part of Alzheimer's-like neurodegenerative processes. This further underscores the need to gain an understanding of neuroprotective agents discussed by Dr. Cotman in part I.

In the fifth section, the role of target tissues in the developmental expression of neuropeptide expression is described. Two papers in this section review models for the role that target organs have in neuropeptide expression and the relative distribution of VIP, galanin, neuropeptide Y, and substance P in the autonomic nervous system. It has been known for a long time that acetylcholine and norepinephrine are essential classical neurotransmitters in the regulation of sympathetic and parasympathetic actions. More recently, it has become clear that neuropeptides also regulate autonomic balance. Given the important role of the autonomic nervous system in modulating such processes as learning, disease states, and stress-related functions, this important area of neuropeptide research has implications for describing individual differences in the development of vulnerabilities in our response to stress. These papers remind us of the interactive nature of nervous system development. The pattern of development of functional units within the nervous system is determined by cells of origin, local supportive structures, target organs, and the environment in which the organism develops and interacts.

In part VI the role of neuropeptides in the ontogeny of the brain systems involved in the stress response is reviewed. The roles of corticotropin-releasing hormone (CRH) and adrenocorticotropic hormone (ACTH) are discussed as they influence the development of neurocircuitry. The interaction of these neuropeptides (and neurosteroids) during differentiation of the CNS has an important role in determining the dynamics of regulation of the hypothalamic-pituitary axis (HPA) and the hypothalamic-pituitary-gonadal (HPG) axis—systems important in both the determination of coping with stress and the regulation of reproduction and possibly sexual identity. Finally, this section concludes with a discussion of the role of neuropeptides in the interaction between mother and fetus during development. Research with human mothers is discussed and an important question is posed which remains unresolved to date: What effect does maternal stress have on the development of the biological machinery underlying the ontogeny of our adaptation to stress?

The papers presented at this meeting underscore the fact that "there is no mental function without brain and social context" (Eisenberg, L. 1995. Am. J. Psychiatry **152:** 1563–1575). The cytoarchitectonics of the nervous system are complexly determined by the interaction of the organism with its environment throughout the life span—beginning with the maternal environment and continuing into old age. The papers from this conference present a critical discussion of how neuropeptides may play a central role in the ontogeny of neurosystems from conception through death. These discussions have important implications for gaining an understanding of the biological underlay of adaptation and coping mechanisms and have far-reaching implications for model development and treatment.

—BILL E. BECKWITH
—ALOIS SARIA
—BIBIE M. CHRONWALL
—CURT A. SANDMAN
—FLEUR L. STRAND

The β-Amyloid Peptide, Peptide Self-Assembly, and the Emergence of Biological Activities

A New Principle in Peptide Function and the Induction of Neuropathology

CARL W. COTMAN

Institute for Brain Aging and Dementia
University of California, Irvine
Irvine, California 92697

INTRODUCTION

In the classic definition, neuropeptides are a class of molecules that modulate synaptic transmission, neuroendocrine responses, and in general participate in various adaptive responses of the central nervous system. Although most peptides regulate such supportive functions, some peptides in the brain are associated with diseases. In Alzheimer's disease (AD), a peptide known as β-amyloid accumulates in the extracellular space in the form of small deposits commonly referred to as senile plaques.

Until recently, β-amyloid was thought to be metabolically inactive. It is now clear, however, that when this peptide self-assembles, it alters its conformation and can drive signal transduction processes that, over time, contribute to brain dysfunction. Indeed, our research and that of others show that β-amyloid destroys neuronal processes and causes neurons to enter into programmed cell death (PCD). This new role makes this peptide unusual in the context of most other peptides in the nervous system. However, it appears to be only one example of a peptide that participates in a newly recognized class of peptide/protein conformation-dependent diseases. Others include prion protein (spongiform encephalopathies), transthyretin (senile systemic amyloidosis), islet amyloid polypeptide (type II diabetes), fibrogen (hereditary renal amyloidosis), and any of at least 16 other related disorders.[1] The common feature of these peptides is the intrinsic property of self-assembling into β-sheet structures and the emergence of various types of pathology. In the case of β-amyloid, understanding the mechanism for its production, assembly, and how it causes neuronal damage is extremely important in view of the increasing population of the elderly and the growing prevalence of AD.

This article is a brief review of recent studies demonstrating that when amyloid self-assembles and forms β-sheet or related structures, a new biological activity emerges which impacts neurons and other brain cells. β-Amyloid is normally produced by many different cells, but in the elderly human brain, and to an even greater extent in AD, it changes conformation, self-assembles, resists degradation, and accumulates in the form of senile plaques. In cell culture, β-amyloid transiently stimulates neuronal growth, but then causes them to degenerate. β-Amyloid initiates a series of signal transduction processes that activates PCD. Recent research

1

has begun to define the properties of the stimulus as embodied by self-assembled β-amyloid that drives cells into PCD. Specifically, we suggest that protein assemblies cross-link select membrane receptors and initiate signal transduction events that cause neurons to enter PCD. These cell culture experiments predicted the existence of a novel mechanism for neuronal loss in AD and indeed we and others have now found evidence for PCD in the brain of those with AD.

Overproduction of amyloid, its self-assembly, persistence, and interactions with the cell surface converge with other stimuli, such as oxidative injury, to accelerate decline. The significance of these observations is that they emphasize the need to maintain the health of neurons and keep them from the threshold of entering into the program to degenerate. Strategies are needed to prevent decline. In the final section of this article, we explore the implications of this research for the development of strategies for preserving the health and resilience of the brain.

Whereas the activity of amyloid *in vitro* is now generally recognized, the relationship with overall cognitive function has a long and debated history. Accordingly, we will begin this discussion with recent data that crtically reexamines the possible relationship of amyloid accumulation with the decline in cognitive function and demonstrates that as amyloid accumulates, function declines.[2] This highlights the relevance of understanding the fundamental properties of this peptide's activities and the possibility that there may be a general principle that places neurons and other cells at risk for degeneration.

β-AMYLOID DEPOSITION IS RELATED TO GENERAL COGNITIVE STATUS

Although at one level it is obvious that the presence of various forms of pathology or structural loss will correlate with functional decline, the nature of the correlation can support or render less likely various competing theories as to the origin or development of the pathology. Accordingly, it is important to evaluate the relationship between cellular and molecular changes in the AD brain and progressive cognitive decline to identify the most critical changes that drive brain dysfunction.

The nature of the neuropathological changes that occur in the brain of patients who have developed AD were described nearly 90 years ago. In 1907, Alzheimer reported the case history of a 51-year-old woman suffering from memory impairments, disorientation, and changes in affect. Upon autopsy, Alzheimer described what we now recognize as the neuropathological hallmarks of AD: neurofibrillary tangles and senile plaques. Neurofibrillary tangles develop within neurons, contribute to their degeneration, and persist as the skeletal remains of the cell long after it has died.

Senile plaques are extracellular deposits of various proteins (primarily β-amyloid) that accumulate as the disease progresses. They are an invariant feature of the disease. β-Amyloid is a 40–42 amino acid peptide that is derived from a large precursor protein, the amyloid precursor protein. It is normally produced by many different cells, but in the elderly human brain and to an even greater extent in AD, it accumulates because it self-assembles and forms senile plaques.

Some studies suggest that plaque number correlates with dementia severity;[3,4] others suggest that neurofibrillary tangles and neuropil threads best predict decline.[5] On the other hand, it has been argued that synapse number (synaptophysin immunoreactivity) shows a strong correlation with dementia severity and that plaque number is at best a weak index of the extent of dementia.[6] Indeed, the lack of a correlation between plaques and cognitive function has been used to argue that amyloid accumulation is unlikely to play a central role in decline. In view of our *in vitro* data and other studies on the biological activities of amyloid, we have recently reexamined the potential relationship between amyloid and general cognitive function. Amyloid can cause a series of transformations on neurons as well as on glial cells and smooth muscle cells in the brain,[7,8] and, as such, it is quite possible that its accumulation correlates with changes in general cognitive function.[2]

To rigorously evaluate a possible relationship between neuropathological and clinical variables, it is essential to design a study that takes into account several factors:

- The sample should include a range of cases from normal to severely demented. This should include mildly impaired individuals because uneven weighting of cases will bias the possible relationship. Some earlier studies have been criticized for overweighting of severe cases.[9]
- The cases should have a short time interval between the last neuropsychological evaluation and autopsy to avoid outdated clinical data.
- The postmortem delay should be sufficiently short so as to not influence the structural or molecular marker.
- The histological technique should be sensitive and able to specifically measure the parameter. Many of the early studies have used classical silver stains that can be highly variable and generally lack sensitivity. Immunocytochemistry and the use of specific antibodies are an improved alternative.
- The analytical technique must be quantitative and objective. It is now feasible to use computer-assisted image analysis on coded samples.

Recently, we addressed and controlled each parameter to examine the relationship between several commonly measured tests of global cognitive function and β-amyloid load. β-Amyloid load is defined as the area occupied by β-amyloid immunoreactivity. This measure is better able to take into account tiny neuropil deposits of amyloid too small to be considered as plaques and the wide difference in size between plaques. We analyzed the entorhinal cortex because AD clearly has an impact upon this brain region and because lesion experiments have demonstrated that damage to this area is known to cause loss in recall and cognitive functions.

We found that β-amyloid load showed a strong correlation with scores on three commonly used clinical measures of dementia severity, the Blessed Information, Memory, and Concentration test ($r = -0.93$), the Clinical Dementia Rating Scale ($r = 0.89$), and the Mini Mental State Exam ($r = -0.90$) (FIG. 1). The correlation was maintained after correction for tissue shrinkage and the interval between testing and death, and was relatively insensitive to weighting by the control cases.[2] In a follow-up study on these same cases, we found that other measures of neuropathology also correlated with these clinical tests but the strongest correlation was with β-amyloid load.[10] These findings along with other data provide growing support for a central role for β-amyloid in the progression of AD pathogenesis and brain dysfunction. Indeed, as discussed below, β-amyloid can exert a profound effect on neurons and other cells in the brain.

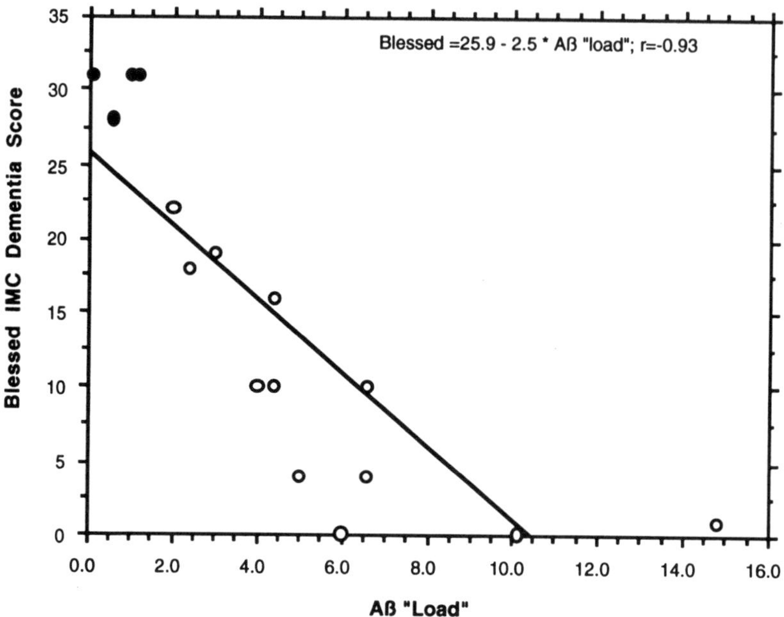

FIGURE 1. β-Amyloid (Aβ) ''load'' is correlated with a decline in global brain function in 20 individuals. Aβ ''load'' was measured using the computer image analysis in five randomly imaged fields from entorhinal cortex and compared to performance on the Blessed IMC test. A score of 31 on the Blessed indicates normal cognitive performance, whereas a score of less than 14 suggests severe cognitive impairment ($r = -0.93$, $p = 0.0003$). Estimated scores are denoted by *filled circles* (with estimated scores excluded, $r = -0.86$, $p = 0.0017$). (Cummings & Cotman.[2] Reprinted with permission, from *Lancet*.)

SELF-ASSEMBLED β-AMYLOID CAUSES NEURONS TO DEGENERATE

Many plaques in the AD brain show an extensive involvement of neurites, which appear to sprout into the plaque, become dystrophic, and degenerate.[11] This raised the possibility that β-amyloid is not an inactive substance as was assumed, but possesses a biological activity.[12] Initial investigations revealed that β-amyloid stimulated process outgrowth and enhanced survival over short time intervals in cultured hippocampal neurons.[12,13] These peptides, however, could also enhance neuronal death in response to excitotoxins[14] and induce neurodegeneration in culture.[13,15–19] Thus, β-amyloid peptides were paradoxically capable of either enhancing growth or inducing toxicity. Further studies have clarified this paradox and demonstrated that the *in vitro* activity of β-amyloid peptides is dependent on the assembly (or polymerization) state of these peptides.[16,17,20–22]

As β-amyloid is aged, it spontaneously aggregates and self-assembles into higher-order structures, similar to those in the AD brain. This conformation change transforms the peptide into a stimulus that initiates neonatal cell death (FIG. 2). After incubation of synthetic β-amyloid$_{1–42}$ peptides for several days *in vitro*, insoluble aggregates form.[17,23] These β-amyloid peptide assemblies demonstrate positive Congo red and thioflavine S staining similar to that observed in the

AD brain.[23,24] Critically, β-amyloid peptides that exhibit aggregation demonstrate toxicity in cultured neurons, whereas β-amyloid peptides that do not exhibit an aggregated state do not exhibit toxicity.[16,25]

SELF-ASSEMBLED β-AMYLOID INDUCES PROGRAMMED CELL DEATH

In our studies on β-amyloid-induced cell death, we noted that a cytoplasmic marker of cell lysis and death (lactic acid dehydrogenase) did not accurately reflect cell death and that the cells appeared to shrink over time. This profile of cell death was quite distinct from our previous studies and others on excitotoxic-induced neuronal death.[14]

Several lines of evidence suggest that cell death can occur by one of two general pathways, necrosis or apoptosis (also called PCD). Necrosis is associated with nonphysiological conditions that disrupt cellular homeostasis (e.g., hypoxia, ischemia, and excitotoxicity). In contrast, apoptosis or PCD is a type of regulated

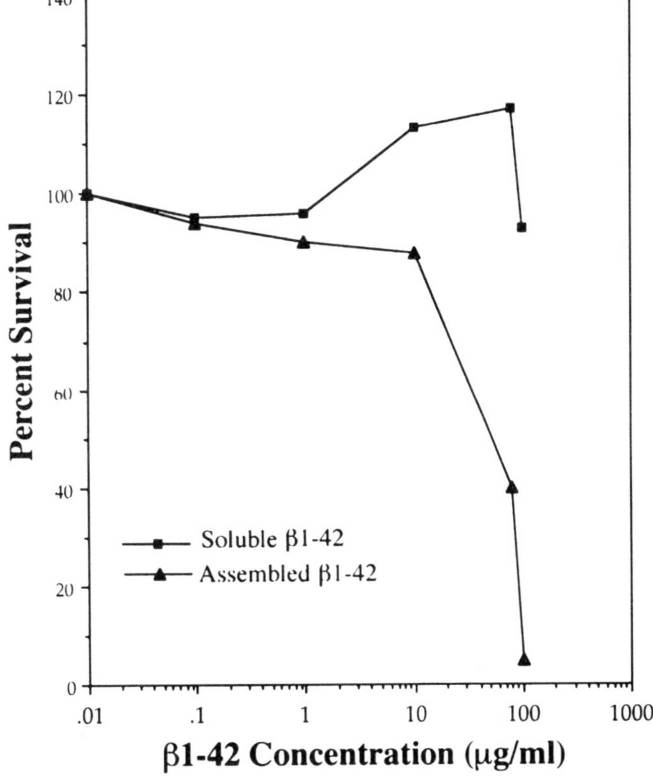

FIGURE 2. As β-amyloid (β1–42) is aged, it self-assembles into higher-order structures and fibrils, similar to those in the Alzheimer's disease brain, and transforms from a stimulus that is neurotrophic-like (soluble), into a stimulus that initiates neuronal cell death (assembled). (Modified from Cotman *et al.*[55])

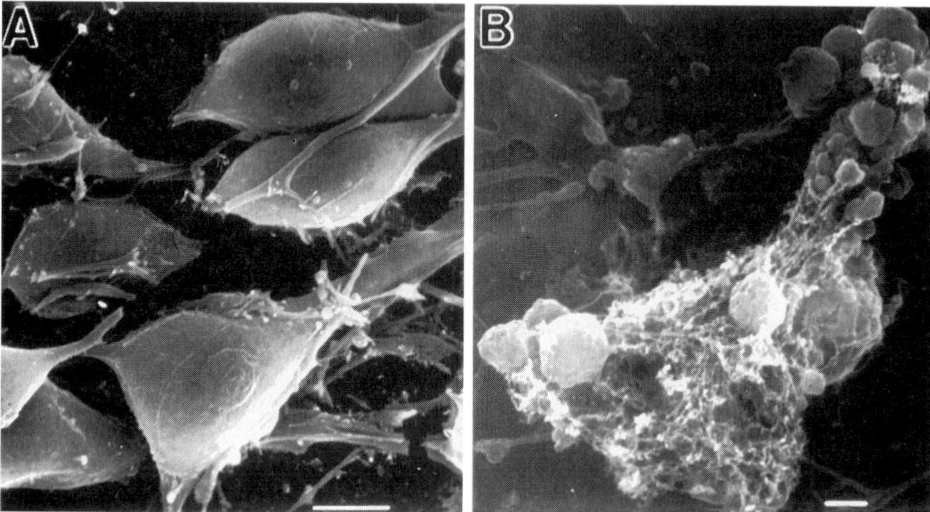

FIGURE 3. In β-amyloid–treated cultures, scanning electron microscopic analysis shows that the surface of some neurons develops an extensive fibrillar matrix that appears to represent amyloid accumulation. **(A)** Untreated control neurons are shown. **(B)** Neurons treated with β-amyloid are shown. Magnification bar in A = 5 μm; B = 1 μm. (Modified from Watt *et al.*[27])

cellular self-destruction that functions in the normal control of development and tissue homeostasis. β-Amyloid may serve as a type of stimulus that can initiate PCD and, if so, PCD may be one of the fundamental mechanisms driving neuronal degeneration in the AD brain.

In a variety of systems, it has been shown that cells undergoing PCD exhibit focal cell surface protrusions (or blebs), nuclear and cytoplasmic condensation, and dense aggregation of chromatin that abuts the nuclear membrane. A biochemical hallmark of cells undergoing apoptosis is internucleosomal cleavage of DNA into oligonucleosome-length fragments. Additionally, cells undergoing apoptosis preserve their membrane integrity until late in degeneration. These features would be consistent with the delayed release of cytoplasm and cell shrinkage. Although several events are known to initiate PCD in nonneural cells, only a few specific insults have been associated with PCD of cultured neurons.

In a series of studies, we discovered that synthetic β-amyloid triggers the degeneration of cultured neurons through activation of PCD. Neurons treated with amyloid exhibit morphological and biochemical characteristics of PCD, including membrane blebbing, compaction of nuclear chromatin, and internucleosomal DNA fragmentation. Aurintricarboxylic acid, an inhibitor of nucleases, prevented DNA fragmentation and delayed cell death. Our *in vitro* results suggest that PCD may have a role in neuronal loss associated with AD.[26,27] Interestingly, cultured neurons in the process of PCD commonly display a network of assembled amyloid on their surface.[27] As revealed by scanning electron microscopy, some neurons in these cultures can develop an extensive fibrillar matrix on their surface (FIG. 3).

IMPORTANCE OF β-STRUCTURE AND THE LACK OF STEREOSPECIFICITY IN NEURONAL PCD

In order to understand the emergent biological activity and the mechanism driving degeneration induced by β-amyloid, it is essential to define the characteristics of β-amyloid salient to its function as a neurotoxic stimulus. We synthesized a series of β-amyloid peptides with progressively truncated carboxy-termini to demonstrate that the length of this hydrophobic region is a crucial determinant of peptide ability to both aggregate and induce neurotoxicity *in vitro*.[20] We also synthesized a series of truncated β-amyloid peptides (β4, β8, β12, β17) to examine the effects of amino-terminal heterogeneity, which occurs *in vivo* and *in vitro*, on the assembly and biological activity of β-amyloid. The amino-terminal truncated isoforms enhance aggregation into neurotoxic β-sheet fibrils, which suggests that these truncated peptides may initiate the pathological neurodegeneration in AD by acting as a nucleation site for β-amyloid deposition.[28] Thus far, we have observed that assembled, bioactive β-amyloid peptides exhibit β-sheet structure, and that amino acid substitutions that disrupt β-amyloid assembly also prevent β-sheet structure and abolish toxicity (for review see ref. 29).

According to classic receptor pharmacology, the D-stereoisomer of β-amyloid would not be predicted to exhibit bioactivity comparable to the native L-peptide, if β-amyloid acts via a specific receptor similar to a classic neurotransmitter receptor. For example, glutamate receptors readily discriminate L- versus D-agonists or antagonists.[30] To determine whether the neuronal surface has a stereospecific requirement for amyloid, we have examined the all-D- versus all-L-stereoisomer of β-amyloid. Interestingly, we have found that the all-D-stereoisomer of β-amyloid$_{1-42}$ not only forms aggregates indistinguishable from the all-L-form, but also induces neurotoxicity with a nearly identical dose-response curve (FIG. 4).[31] As we[32] and others[33,34] have reported, it appears that some aspect of β-sheet structure or a related higher-order assembly is necessary, but there is a range of tolerance in the exact stereospecific requirements of the peptide assembly. We

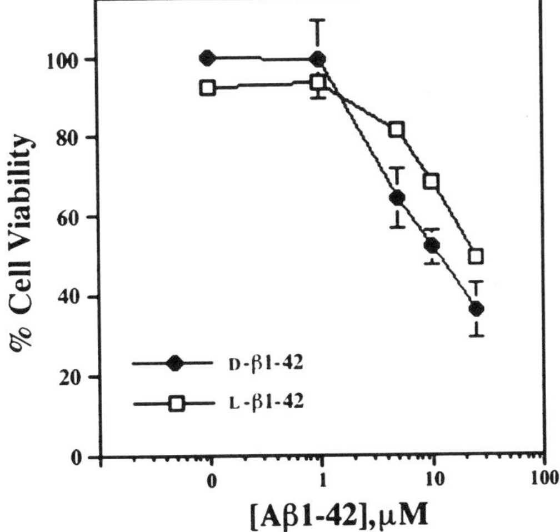

FIGURE 4. The all-D-stereoisomer of β1–42 forms aggregates indistinguishable from the all-L-β1–42 and induces neurotoxicity with a nearly identical dose-response curve. (Modified from Cribbs *et al.*[31])

would further suggest that β-amyloid activates PCD via a mechanism distinct from a classic ligand-receptor interaction.

CROSS-LINKING OF MEMBRANE RECEPTORS CAN INITIATE PROGRAMMED CELL DEATH

A number of well-characterized extracellular agents that bind to the cell surface have been reported to initiate PCD through the cross-linking of membrane receptors. These agents include bacterial superantigens,[35,36] the Fas ligand,[37] and certain viruses.[38] This form of PCD is called activation-induced PCD. In several cases, the receptor involved has been identified, but it is possible that the same mechanism can operate through several receptor types.

In several ways, β-amyloid-initiated PCD may be another example of activation-induced PCD. β-Amyloid induces PCD only when bound to neuronal surfaces and only when in a polymerized or fibrillar β-sheet assembly state.[26,39] This suggests that β-amyloid may initiate activation-induced PCD in a manner analogous to superantigens and Fas receptor cross-linking. To begin to address this hypothesis we have identified a reagent that can cross-link neuronal membrane receptors and examined the cellular response.

Although specific activation-induced PCD-linked receptors have not yet been identified in neurons, lectins are capable of binding and cross-linking many different glycosylated membrane receptors on cells, activating the receptors and causing a wide variety of cellular responses, including the induction of cell death. We have examined the response of neurons to lectin concanavalin A (Con A) because neurons contain a high density of receptors for this lectin[40] and it is possible to examine binding with cross-linking and without cross-linking by chemically modifying the Con A.

Con A is a powerful stimulus that initiates cell death in neurons. It produced measurable cell death and morphological changes at concentrations greater than 10 nM, and at 100 nM, induced massive cell death in neuronal cultures. The Con A-induced neuronal death exhibited many of the hallmarks associated with PCD, such as membrane blebbing, nuclear condensation and margination, and internucleosomal DNA cleavage.[41]

To examine the importance of receptor cross-linking in the Con A-initiated PCD, the effect of succinylated Con A was compared to that of native Con A. Succinylated Con A is a stable dimer and is much less effective at forming membrane aggregates of Con A receptors than is native Con A. Native Con A is a tetramer and binds with other tetramers to induce receptor aggregation. Native Con A caused a clustering of Con A receptors in agreement with previous findings.[41] In contrast, succinylated Con A bound to the cell surface but did not produce receptor clusters, nor did it produce significant cell death even at micromolar concentrations (FIG. 5). Thus, when the cross-linking of membrane receptors to form receptor clusters was blocked, the subsequent downstream intracellular signaling that initiates PCD was also blocked. It appears that receptor cross-linking is a necessary part of the signal that initiates Con A-induced PCD.

The Con A model shares certain features with β-amyloid-induced PCD. The time courses for both the Con A- and β-amyloid-induced morphological changes and the subsequent PCD are comparable.[20] Although most protocols for β-amyloid-induced cell death now employ preassembled peptide, over a much extended time course, β-amyloid in the soluble form will also cause cell death. Shortly prior to cell death, the surface of the neurons exhibits clusters of the peptide[25,42]

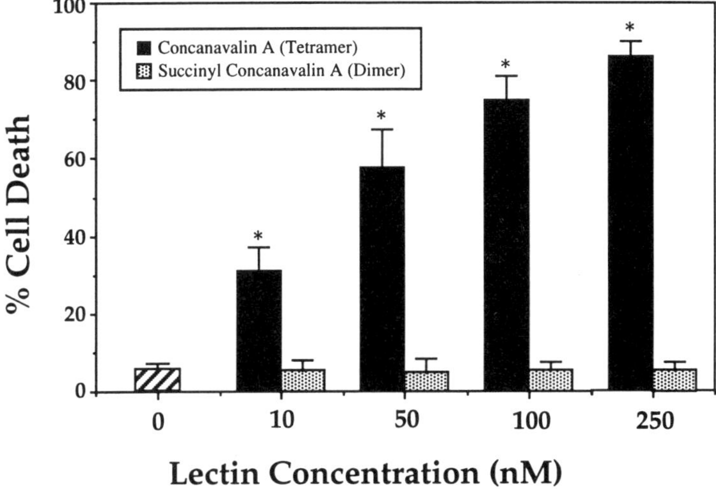

FIGURE 5. Receptor cross-linking by Con A initiates programmed cell death (PCD) in primary neurons. This graph provides a comparison of the ability of Con A and succinyl Con A to induce PCD in neuronal cultures. Lectin-induced cell death dose-response curves for Con A and succinyl Con A were performed to measure their relative activities. The extent of neuronal cell death was determined after 24 h by morphological cell counts. *$p <$ 0.0001 relative to untreated controls. (From Cribbs et al.[41] Reprinted, with permission, from *Neuroscience*.)

(and unpublished observations), not unlike those observed with Con A. It appears as if neurons are extremely sensitive to the cross-linking of the membrane receptors and that this stimulus will cause them to activate a program to degenerate via PCD.

PROGRAMMED CELL DEATH IN THE AD BRAIN

We suggest, therefore, that as the brain ages and protein assemblies develop, neurons are placed at increasing risk of dysfunctioning, losing their processes, and, over time, degenerating. This is consistent with our recent report that the amount of amyloid that accumulates in the aged and AD brain appears to correlate with the decline in cognitive function.[2] Further, it is becoming increasingly clear that PCD in most brain neurons can be initiated by a variety of stimuli and that most neurons are extremely vulnerable to PCD stimuli. For example, oxidative insults readily initiate PCD and oxidative damage is known to occur in the AD brain. Significantly, however, GABA neurons are relatively resistant to β-amyloid,[43] Con A,[41] and oxidative insults such as hydrogen peroxide;[44] in the AD brain these neurons are also relatively resistant to cell death in support of predictions from culture data.

From these conditions and others, it can be predicted that PCD may be one of the mechanisms driving neuronal death in AD. In order to examine this hypothesis, our strategy has been to employ multiple markers of PCD because it is unlikely

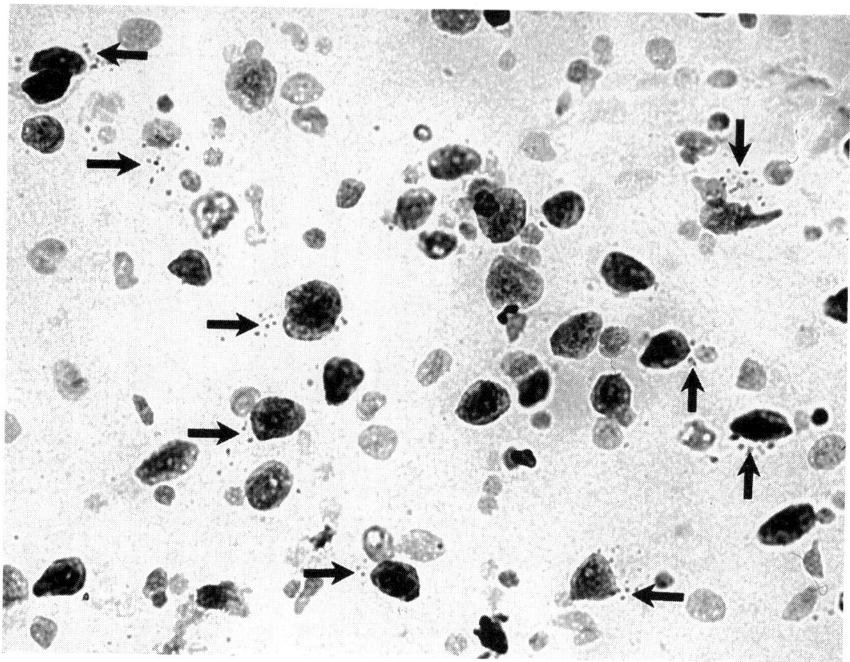

FIGURE 6. Nuclei in the Alzheimer's disease (AD) brain show DNA damage and the presence of nuclear apoptotic bodies indicating the presence of PCD. Area CA1 of hippocampus of AD brain labeled for DNA damage using the *in situ* nick translation technique. *Arrows* show neuronal nuclei expressing DNA damage and the presence of apparent nuclear apoptotic bodies. This field contains an unusually high density of nuclear apoptotic bodies.

that one marker by itself is sufficient and multiple indices are necessary to dissect the relevant specific pathways. We have used cell culture experiments to identify possible markers and then examined the presence or absence of these in postmortem tissues. For example, in addition to morphological changes and DNA damage, amyloid-induced PCD will induce the expression of the immediate early gene c-Jun.[45]

As predicted from culture experiments, the AD brain shows extensive DNA damage in some neurons (as detected by the TUNEL method which labels cut 3′ ends of DNA), nuclear morphology consistent with apoptosis, and the presence and coexpression in some cells of c-Jun as predicted from culture experiments.[45,46] FIGURE 6 shows the appearance of neurons that display DNA damage and the presence of nuclear apoptotic bodies. The appearance of these neurons is very similar to that described in rodent neurons following an ischemic episode.[47] However, although many of the cells labeled by the TdT method in the AD brain exhibit clear morphological characteristics of apoptosis, studies in AD tissue reveal a surprisingly large number of TdT-labeled nuclei in the entorhinal cortex/hippocampal formation. This raised the possibility that an artifact exists, which is associated with postmortem tissues, or that the AD brain shows extensive and previously undetected DNA damage. As discussed in detail elsewhere, the existence of artifacts does not appear to account for the extent of DNA damage observed.[45]

The accumulation of DNA damage could be a result of deficient DNA repair, an unexpectantly slow course of PCD such as delayed clearing, or the induction of a set of compensatory mechanisms to arrest the program. In view of the extent of DNA damage, it is unlikely that the clearing response is the only factor. Recently, we have demonstrated that Bcl-2 is up-regulated in most DNA-damaged neurons, and we have suggested that this up-regulation may protect neurons or arrest the program at some stage.[48] Irrespective of the cause, DNA damage is known to drive most cell types into PCD and thus supports the contention that at least some neurons degenerate via PCD in the AD brain.

POSSIBLE BEHAVIORAL STRATEGIES FOR PROTECTING NEURONS

Although many therapeutic agents will ultimately be available to aid the aging and AD brain, it may be that the brain has its own strategies for self-maintenance and repair with age. For example, it would be surprising if the brain did not have a way to transduce certain types of behaviors into protective programs for the involved circuits.

PCD is a process regulated by the cells themselves and thus is amenable to certain types of interventions. Neurons possess certain positive feedback mechanisms that can enhance their vulnerability. For example, in the case of amyloid production, once neurons enter into a state of stress, they appear in some cases to create more of the product that contributes to their dysfunction unless the problem is arrested. Thus, neurons exposed to amyloid induce more amyloid precursor protein and more amyloidogenic peptoids.[49] If amyloid self-assembles and resists degradation it will add to the amyloid load and brain dysfunction. Thus, once initiated, the process can cascade and further the probability of functional decline and degeneration. In fact, amyloid can organize and drive many cascades in parallel.[29] Clearly, it is essential to protect neurons and render them resistant to adverse events.

Recent research in two generally unrelated areas of research suggests a strategy for increasing the margin of protection to the brain. On the one hand, basic science research has clearly shown that some proteins, such as neurotrophic factors, can raise the threshold for neuronal stress and death and provide a margin of protection for neurons. It is important, therefore, to develop new strategies for regulating these molecules. On the other hand, several recent epidemiology studies have shown that certain forms and amounts of physical activity can translate into better health for the individual and reduced rates of decline in cognitive performance and physical health with age[50] (see also reviews cited in ref. 51). The mechanism for this protective effect is unknown.

Perhaps certain types of behavioral activity such as exercise can regulate the expression of neurotrophic factors that may improve neuronal function and help protect these cells from damage. To evaluate this hypothesis, we investigated the effect of physical activity on brain neurotrophic factors. Recently, we found in animal experiments that physical exercise (voluntary running) can increase select brain neurotrophic factors.[52] Specifically, rats were provided access to a running wheel and after various periods of running, the expression of the neurotrophic factor, brain-derived neurotrophic factor (BDNF), was measured. BDNF will

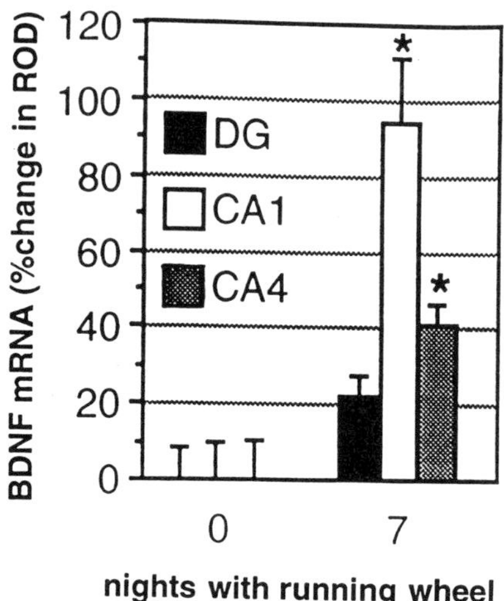

FIGURE 7. Voluntary wheel running induces a rapid increase in BDNF mRNA in the dentate gyrus (DG), CA1, and CA4 fields of the hippocampus. BDNF has been shown to be neuroprotective and also to enhance synaptic transmission and certain forms of stimulus-dependent plasticity. Relative optical density (ROD) values for each target area minus background ROD level (corpus callosum), expressed as percent change ($\pm$ SEM) from control values. *$p < 0.05$. (Modified from Neeper *et al.*[52])

protect neurons from PCD[53] and restore long-term potentiation in animals in which the BDNF gene has been knocked out.[54]

Rats appear to find some level of physical activity rewarding and will voluntarily run several kilometers per night if given access to a running wheel. Exercise induced significant increases in BDNF mRNA in the hippocampus and caudal cortex after only a few nights of running. The increase in the hippocampus was significant after only two nights of running and was sustained at similar levels after four and seven nights (FIG. 7). The response of the hippocampus was unexpected because this brain area is not directly associated with motor activity. Evidently a mechanism exists to translate a simple, widely practiced activity into increased expression of a neuroprotective molecule.

Thus it is possible that some types of behavior practiced late in life may impact on the health of neurons and the preservation of cognitive and other functions. Clearly, there are many exciting new advances that will help define ways to promote the quality of life in the later years.

CONCLUSIONS

Amyloid is a peptide that self-assembles into higher-order structures, and when it does, a new biological activity evolves that can access key signal transduction pathways in a variety of cells. In neurons, β-amyloid initiates a transient growth response followed by cellular degeneration. We suggest that β-amyloid in neurons cross-links select cell surface receptors and that this is interpreted by neurons as a signal to degenerate. This may be similar to developmental and injury signals which participate in PCD, but in this case amyloid mimics these mechanisms and recruits cell responses into pathological degeneration. β-Amyloid not only initiates degeneration in neurons but also in smooth muscle cells and microglial cells; other cells, however, such as astrocytes, fibroblasts, and even subpopulations of neurons resist degeneration. β-Amyloid can also initiate and organize molecular cascades because β-amyloid resists proteolysis and has adhesive properties for other molecules. This can, for example, contribute to the development of inflammatory cascades.[29] Taken together, it is clear that amyloid, once self-assembled, initiates a constellation of activities which redirect normal functions. As such, it is not surprising that the accumulation of β-amyloid correlates with the decline in cognitive status.

It is suggested that amyloid is generated by cells as part of a natural stress response, and once produced it may self-assemble, accumulate, and develop its own activity and self-perpetuation capacity. Thus it is essential to develop means to protect cells, arrest amyloid accumulation, and protect neurons. Our own data and that of others suggest that simple behavioral strategies such as exercise may be significant interventions. Exercise can induce growth factors and other gene products that have the capacity to protect neurons by increasing their resilience to adverse stimuli.

Amyloid represents a growing class of peptides and proteins that have an innate capacity to self-assemble into β-sheet or related structures that, once formed, can direct new biological activities and promote pathologies. Thus, it would appear that β-amyloid accumulation in the brain is not unique, but is part of a broader, more fundamental, class of defects in tissues as they age or are subject to other pathological processes. Research into these issues will return significant benefits in terms of health care and improving the quality of life.

REFERENCES

1. KELLY, J. W. 1996. Alternative conformations of amyloidogenic proteins govern their behavior. Curr. Opin. Struct. Biol. **6:** 11–17.
2. CUMMINGS, B. J. & C. W. COTMAN. 1995. Image analysis of beta-amyloid load in Alzheimer's disease and relation to dementia severity. Lancet **346:** 1524–1528.
3. BLESSED, G., B. E. TOMLINSON & M. ROTH. 1968. The association between quantitative measures of dementia and of senile change in the cerebral gray matter of elderly subjects. Br. J. Psychiatry **114:** 797–811.
4. HYMAN, B. T., K. MARZLOFF & P. V. ARRIAGADA. 1993. The lack of accumulation of senile plaques or amyloid burden in Alzheimer's disease suggests a dynamic balance between amyloid deposition and resolution. J. Neuropathol. Exp. Neurol. **52:** 594–600.
5. MULLER, W. E., S. KOCH, A. ECKERT, H. HARTMANN & K. SCHEUER. 1995. Beta-amyloid peptide decreases membrane fluidity. Brain Res. **674:** 133–136.
6. ZHAN, S. S., K. BEYREUTHER & H. P. SCHMITT. 1993. Quantitative assessment of the synaptophysin immuno-reactivity of the cortical neuropil in various neurodegenerative disorders with dementia. Dementia **4:** 66–74.
7. PIKE, C. J., B. J. CUMMINGS, R. MONZAVI & C. W. COTMAN. 1994. β-Amyloid-induced changes in cultured astrocytes parallel reactive astrocytosis associated with senile plaques in Alzheimer's disease. Neuroscience **63:** 517–531.
8. DAVIS-SALINAS, J., S. M. SAPORITO-IRWIN, C. W. COTMAN & W. E. VAN NOSTRAND. 1995. Amyloid β-protein induces its own production in cultured degenerating cerebrovascular smooth muscle cells. J. Neurochem. **65:** 931–934.

9. TERRY, R., E. MASLIAH, D. SALMON, N. BUTTERS, R. DETERESA, R. HILL, L. A. HANSEN & R. KATZMAN. 1991. Physical basis of cognitive alterations in Alzheimer's disease: Synapse loss is the major correlate of cognitive impairment. Ann. Neurol. **30:** 572–580.

10. CUMMINGS, B. J., C. J. PIKE, R. SHANKLE & C. W. COTMAN. β-Amyloid deposition and other measures of neuropathology predict cognitive status in Alzheimer's disease. Neurobiol. Aging. In press.

11. Geddes, J. W., D. T. Monaghan, C. W. Cotman, I. T. Lott, R. C. Kim & H. C. Chui. 1985. Plasticity of hippocampal circuitry in Alzheimer's disease. Science **230:** 1179–1181.

12. WHITSON, J. S., D. J. SELKOE & C. W. COTMAN. 1989. Amyloid β protein enhances the survival of hippocampal neurons in vitro. Science **243:** 1488–1490.

13. YANKNER, B. A., L. K. DUFFY & D. A. KIRSCHNER. 1990. Neurotrophic and neurotoxic effects of amyloid β protein: Reversal by tachykinin neuropeptides. Science **250:** 279–282.

14. KOH, J. Y., L. L. YANG & C. W. COTMAN. 1990. β-Amyloid protein increases the vulnerability of cultured cortical neurons to excitotoxic damage. Brain Res. **533:** 315–320.

15. YANKNER, B. A., L. R. DAWES, S. FISHER, L. VILLA-KOMAROFF, M. L. OSTER-GRANITE & R. L. NEVE. 1989. Neurotoxicity of a fragment of the amyloid precursor associated with Alzheimer's disease. Science **245:** 417–420.

16. PIKE, C. J., A. J. WALENCEWICZ, C. G. GLABE & C. W. COTMAN. 1991. Aggregation-related toxicity of synthetic β-amyloid protein in hippocampal cultures. Eur. J. Pharmacol. **207:** 367–368.

17. PIKE, C. J., A. J. WALENCEWICZ, C. G. GLABE & C. W. COTMAN. 1991. In vitro aging of β-amyloid protein causes peptide aggregation and neurotoxicity. Brain Res. **563:** 311–314.

18. BEHL, C., J. DAVIS, G. M. COLE & D. SCHUBERT. 1992. Vitamin E protects nerve cells from amyloid beta protein toxicity. Biochem. Biophys. Res. Commun. **186:** 944–950.

19. TAKADERA, T., N. SAKURA, T. MOHRI & T. HASHIMOTO. 1993. Toxic effect of a beta-amyloid peptide (beta 22-35) on the hippocampal neuron and its prevention. Neurosci. Lett. **161:** 41–44.

20. PIKE, C. J., D. BURDICK, A. WALENCEWICZ, C. G. GLABE & C. W. COTMAN. 1993. Neurodegeneration induced by β-amyloid peptides in vitro: The role of peptide assembly state. J. Neurosci. **13:** 1676–1687.

21. MATTSON, M. P., K. J. TOMASELLI & R. E. RYDEL. 1993. Calcium-destabilizing and neurodegenerative effects of aggregated β-amyloid peptide are attenuated by basic FGF. Brain Res. **621:** 35–49.

22. BUSCIGLIO, J., D. H. GABUZDA, P. MATSUDAIRA & B. A. YANKNER. 1993. Generation of beta-amyloid in the secretory pathway in neuronal and nonneuronal cells. Proc. Natl. Acad. Sci. USA **90:** 2092–2096.

23. BURDICK, D., B. SOREGHAN, M. KWON, J. KOSMOSKI, M. KNAUER, A. HENSCHEN. J. YATES, C. COTMAN & C. GLABE. 1992. Assembly and aggregation properties of synthetic Alzheimer's A4/β amyloid peptide analogs. J. Biol. Chem. **267:** 546–554.

24. HILBICH, C., B. KISTERS-WOIKE, J. REED, C. L. MASTERS & K. BEYREUTHER. 1991. Aggregation and secondary structure of synthetic amyloid βA4 peptides of Alzheimer's disease. J. Mol. Biol. **218:** 149–163.

25. PIKE, C. J., B. J. CUMMINGS & C. W. COTMAN. 1992. β-Amyloid induces neuritic dystrophy *in vitro*: Similarities with Alzheimer pathology. NeuroReport **3:** 769–772.

26. LOO, D. T., A. G. COPANI, C. J. PIKE, E. R. WHITTEMORE, A. J. WALENCEWICZ & C. W. COTMAN. 1993. Apoptosis is induced by beta-amyloid in cultured central nervous system neurons. Proc. Natl. Acad. Sci. USA **90:** 7951–7955.

27. WATT, J. A., C. J. PIKE, A. J. WALENCEWICZ-WASSERMAN & C. W. COTMAN. 1994. Ultrastrucutral analysis of β-amyloid-induced apopotsis in cultured hippocampal neurons. Brain Res. **661:** 147–156.

28. PIKE, C. J., M. J. OVERMAN & C. W. COTMAN. 1995. Amino-terminal deletions enhance aggregation of beta-amyloid peptides in vitro. J. Biol. Chem. **270:** 23895–23898.

29. COTMAN, C. W., A. J. TENNER & B. J. CUMMINGS. 1996. β-Amyloid converts an acute phase injury response to chronic injury responses. Neurobiol. Aging **17(5):** 723–732.

30. MONAGHAN, D. T., R. J. BRIDGES & C. W. COTMAN. 1989. The excitatory amino acid receptors: Their classes, pharmacology, and distinct properties in the function of the central nervous system. Annu. Rev. Pharmacol. Toxicol. **29:** 365–402.

31. CRIBBS, D. H., C. J. PIKE, S. L. WEINSTEIN & C. W. COTMAN. β-Amyloid stereoisomers exhibit similar structural and biological properties: Implications for mechanisms of toxicity. J. Biol. Chem. In press.

32. PIKE, C. J., A. J. WALENCEWICZ-WASSERMAN, J. KOSMOSKI, D. H. CRIBBS, C. G. GLABE & C. W. COTMAN. 1995. Structure-activity analyses of beta-amyloid peptides: Contributions of the beta 25-35 region to aggregation and neurotoxicity. J. Neurochem. **64:** 253–265.

33. HOWLETT, D. R., K. H. JENNINGS, D. C. LEE, M. S. CLARK, F. BROWN, R. WETZEL, S. J. WOOD, P. CAMILLERI & G. W. ROBERTS. 1995. Aggregation state and neurotoxic properties of Alzheimer beta-amyloid peptide. Neurodegeneration **4:** 23–32.

34. SIMMONS, L. K., P. C. MAY, K. J. TOMASELLI, R. E. RYDEL, K. S. FUSON, E. F. BRIGHAM, S. WRIGHT, I. LIEBERBURG, G. W. BECKER, D. N. BREMS & W. Y. LI. 1994. Secondary structure of amyloid beta peptide correlates with neurotoxic activity in vitro. Mol. Pharmacol. **45:** 373–379.

35. MARRACK, P. & J. KAPPLER. 1990. The staphylococcal enterotoxins and their relatives [published erratum appears in Science 1990 Jun 1; **248(4959):** 1066, see comments]. Science **248:** 705–711.

36. DELLABONA, P., J. PECCOUD, J. KAPPLER, P. MARRACK, C. BENOIST & D. MATHIS. 1990. Superantigens interact with MHC class II molecules outside of the antigen groove. Cell **62:** 1115–1121.

37. NAGATA, S. & P. GOLSTEIN. 1995. The Fas death factor. Science **267:** 1449–1456.

38. BANDA, N. K., J. BERNIER, D. K. KURAHARA, R. KURRLE, N. HAIGWOOD, R.-P. SEKALY & T. H. FINKEL. 1992. Crosslinking CD4 by human immunodeficiency virus gp120 primes T cells for activation-induced apopotsis. J. Exp. Med. **176:** 1099–1106.

39. COTMAN, C. W. & A. J. ANDERSON. 1995. A potential rule for apoptosis in neurodegeneration and Alzheimer's disease. Mol. Neurobiol. **10:** 19–45.

40. COTMAN, C. W. & D. TAYLOR. 1974. Localization and characterization of concanavalin A receptors in the synaptic cleft. J. Cell Biol. **62:** 236–242.

41. CRIBBS, D. H., V. M. KRENG, A. J. ANDERSON & C. W. COTMAN. 1996. Crosslinking of membrane glycoproteins by concanavalin A induces apoptosis in cortical neurons. Neuroscience. **75:** 173–185.

42. BUSCIGLIO, J., A. LORENZO & B. A. YANKNER. 1992. Methodological variables in the assessment of beta amyloid neurotoxicity. Neurobiol. Aging **13:** 609–612.

43. PIKE, C. J. & C. W. COTMAN. 1993. Cultured GABA-immunoreactive neurons are resistant to toxicity induced by β-amyloid. Neuroscience **56:** 269–274.

44. WHITTEMORE, E. R., D. T. LOO & C. W. COTMAN. 1994. Exposure to hydrogen peroxide induces cell death via apoptosis in cultured rat cortical neurons. NeuroReport **5:** 1585–1588.

45. ANDERSON, A. J., J. H. SU & C. W. COTMAN. 1996. DNA damage and apoptosis in Alzheimer's disease: Colocalization with c-Jun immunoreactivity, relationship to brain area, and effect of postmortem delay. J. Neurosci. **16:** 1710–1719.

46. SU, J. H., A. J. ANDERSON, B. J. CUMMINGS & C. W. COTMAN. 1995. Immunohistochemical evidence for apoptosis in Alzheimer's disease. NeuroReport **5:** 2529–2533.

47. CHARRIAUT-MARLANGUE, C. & Y. BEN-ARI. 1995. A cautionary note on the use of the TUNEL stain to determine apoptosis. NeuroReport **7:** 61–64.

48. SU, J. H., T. SATOU, A. J. ANDERSON & C. W. COTMAN. 1996. Up-regulation of Bcl-2 is associated with neuronal DNA damage in Alzheimer's disease. NeuroReport **7:** 437–440.

49. CRIBBS, D. H., J. DAVIS-SALIANAS, C. W. COTMAN & W. E. VAN NOSTRAND. 1995. β-Amyloid induces increased expression and processing of amyloid precursor protein in cortical neurons. Alzheimer's Res. **1:** 197–200.

50. BERKMAN, L. F., T. E. SEEMAN, M. ALBERT, D. BLAZER, R. KAHN, R. MOHS, C. FINCH, E. SCHNEIDER, C. COTMAN, G. MCCLEARN, J. NESSELROADE, D. FEATHERMAN, N. GARMEZY, G. MCKHANN, G. BRIM, D. PRAGER & J. ROWE. 1993. High, usual and impaired functioning in community-dwelling older men and women: Findings from the MacArthur Foundation Research Network on Successful Aging. J. Clin. Epidemiol. **46:** 1129–1140.

51. COTMAN, C. W. & S. NEEPER. 1996. Activity-dependent plasticity and the aging brain. *In* Handbook of the Biology of Aging. 4th edit. E. L. Schneider & J. W. Rowe, Eds.: 284–293. Academic Press. San Diego, CA.

52. NEEPER, S. A., F. GÓMEZ-PINILLA, J. CHOI & C. COTMAN. 1995. Exercise and brain neurotrophins [letter]. Nature **373:** 109.

53. KOH, J.-Y., G. J. GWAG, D. LOBNER & D. W. CHOI. 1995. Potentiated necrosis of cultured cortical neurons by neurotrophins. Science **268:** 573–575.

54. PATTERSON, S. L., T. ABEL, T. A. S. DEUEL, K. C. MARTIN, J. C. ROSE & E. R. KANDEL. 1996. Recombinant BDNF rescues deficits in basal synaptic transmission and hippocampal LTP in BDNF knockout mice. Neuron **7:** 695–702.

55. COTMAN, C. W., D. H. CRIBBS & A. J. ANDERSON. 1997. The β-amyloid model of Alzheimer's disease. *In* Molecular Mechanisms of Dementia. W. Wasco & R. E. Tanzi, Eds.: 73–90. Humana Press Inc. Totowa, NJ.

Neuropeptides and Alzheimer's Disease Pathology

GARTH BISSETTE[a]

Laboratory of Psychoneuroendocrinology
Department of Psychiatry and Human Behavior
University of Mississippi Medical Center
2500 North State Street
Jackson, Mississippi 39216-4505

INTRODUCTION

Alzheimer's disease (AD) is a progressive neurodegenerative disease that produces disturbances in memory and learning eventually leading to dementia. Responsible for two-thirds of the total cases of dementia in the United States, AD will claim between 3 and 4 million victims by the year 2030 at the present rates of incidence and prevalence. The disease presents with a variable course that ranges from a few years to decades, but which invariably ends in death. Current estimates indicate that over half of the available nursing home beds in this country are now occupied by patients with AD, and the associated economic costs of this disease are staggering. These costs do not consider the emotional toll of the disease on the survivors and caregivers who must witness the slowly vanishing personality and intellect of a family member or close friend. No truly effective treatment exists for this disease although one drug has been approved by the Food and Drug Administration and several more are in clinical trials. Many years of effort and several hundred millions of dollars have been spent researching the causes of AD, and much has been learned in the last 20 years; however, the specific etiology of even one of the many possible forms of AD remains unknown. Approximately 10–15% of clinical cases of AD are of the familial form which has an onset as early as the fourth decade and often has a more rapid course than the non-familial or sporadic type that represents the majority of AD cases and usually has a later age of onset in the sixth and seventh decades. Several genetic loci have now been identified for variants of the familial form of the disease, and the e4 allele of the gene encoding the apolipoprotein E molecule has now been shown to be associated with the late-onset form of AD. What remains unknown is whether and how these genetically different forms of the disease have the same profile of neurochemical deficits during the course of the disease. This is important because until the molecular precipitant of the disease process(es) is identified and a strategy developed to combat it, a rational treatment of the existing population of patients should be directed toward normalization of the existing neurochemical alterations. Unfortunately, the available research materials do not give a clear or consistent picture of the course of AD in terms of the neurochemical changes at different stages of the disease.

These research materials usually take two forms: (1) the postmortem brain from a patient with a clinical history of dementia and neuropathologically confirmed

[a] E-mail: gbiss@umsmed.edu

elevations in the numbers of senile plaques and neurofibrillary tangles relative to unaffected individuals of the same age, or (2) a cerebrospinal fluid (CSF) sample from a patient suspected of having AD due to the presence of progressive dementia that cannot be ascribed to some other medical cause. Only a very few studies have used biopsied brain tissue from living patients[1] because of the relatively high risk of morbidity when no truly effective treatment is available. Both postmortem brain and CSF have inherent limitations in the type of information one can extract regarding the neurochemical alterations induced by the disease. The postmortem brain obviously represents the endstage of the disease process and is confounded by possible changes in the neurochemical milieu during and after the agonal state leading to death. Also, as a single time point, the postmortem brain cannot be effectively used to address the neurochemical changes occurring relative to the deteriorating mental status during the course of the disease. Although changes in CSF parameters could conceivably be used to track alterations in a particular neurochemical indicator over the course of the disease, in practice it is extremely rare to obtain more than one or two samples from the same patient and even more unlikely that psychometric testing would be performed in a temporally congruent period. This problem is further compounded by the existing uncertainty in the diagnosis of AD in living patients without the corroboration of increased numbers of senile plaques and neurofibrillary tangles and in the absence of any reliable biochemical marker for the disease. Of course, CSF itself represents at best an average of the various compartments of the brain and spinal cord contributing to the ultimate concentration of the substance of interest sought in the sample. Both types of material are subject to possible complications induced by drug treatment prior to obtaining the sample. These caveats notwithstanding, the current state of research into the neurochemistry of AD has identified several potential targets of pharmacotherapy from among the myriad of neurotransmitter systems used by the human brain.

CLASSICAL NONPEPTIDE TRANSMITTER SYSTEMS IN ALZHEIMER'S DISEASE

Space limitations prevent a comprehensive review of the various classical neurotransmitter systems investigated in AD over the past 20 years. At death and after many years of disease presence, it is likely that many neurochemical systems that may not be involved in the specific mechanisms of AD pathology responsible for the associated dementia may be altered by secondary processes. Thus alterations in dopamine, serotonin, and norepinephrine have all been reported in brain tissues of some populations of postmortem AD patients, but have not been found in the great majority of cases that is usually indicative of a primary pathological process. However, a convincing case for exceptions to the above statement may be made for certain changes such as the noradrenergic neuron loss in the locus coeruleus as a function of aging.[2] Almost all AD researchers would now recognize the involvement of the acetylcholinergic system as of paramount importance to the underlying pathological process in AD (see ref. 3 for review). The degeneration of the acetylcholine (ACh)-containing cells of the nucleus basalis of Meynert[4] and the associated loss of choline acetyltransferase (CAT) in the neuronal projections from this nucleus to the entire cortex was the evidence responsible for the first neurochemical system recognized as specifically altered by AD pathology and was obtained using postmortem brain. Recent work from Duke University Medical Center[5] has shown that the rate-limiting step in the production of ACh, which is the transport of choline across the neuronal membrane by the high-affinity choline transporter, is highly up-regulated in brain tissue from patients with AD obtained within two hours of death, when this energy-dependent membrane process remains viable and active. Evidence for ACh receptor changes in AD have been published,[6,7] but are conflicting in determining whether the postsynaptic neurons receiving ACh input are up-regulating their receptor population to augment the reduced ACh signal during the disease process. There is also some controversy about the response of acetylcholinesterase (AChE), the enzyme that terminates the action of ACh after release into the synapse, to the degeneration of ACh-containing neurons.[8] Its concentration and activity have been reported to be increased and decreased in AD tissue and CSF, respectively. The therapeutic intent of the currently approved drug, Tacrine®, is to block the action of AChE and thereby increase the synaptic availability of ACh that is still being produced.[9] The actual effect of these various strategies in increasing ACh synaptic availability remains unknown because postmortem degradation of ACh by AChE prevents direct assessment of ACh concentrations unless AChE activity is denatured within seconds of death by some process such as microwave irradiation, which is possible with experimental animals. Some evidence exists for AChE inhibitors acting at non-neuronal sites.[8] The entire picture looks as if the surviving cholinergic neurons are attempting to compensate for the degenerating ACh neurons with an increase in the synthetic rate of ACh. It is therefore disconcerting to examine the evidence for therapeutic effects on dementia by amelioration of this reduction in ACh synaptic availability through the use of precursor loading strategies and AChE blockade.[10] Once the cholinergic deficit in AD was recognized, the obvious therapeutic approach was to increase ACh synaptic availability by increasing dietary sources of choline or administration of cholinergic receptor agonists; these have been tried at often heroic doses, and even with direct intraventricular delivery,[11] without notable success in reducing the AD-induced dementia.[12] Now that it is known that the high-affinity choline uptake protein activity is up-regulated enough to make such a precursor loading scheme successful in elevating ACh concentrations, it must be questioned whether the cholinergic deficit is responsible for the associated dementia.[13] This is underscored by the apparent inability of Tacrine® to effectively reverse the dementia associated with some cases of AD, although it may slow future progression of dementia somewhat.[12]

NEUROPEPTIDES AND ALZHEIMER'S DISEASE

A neuropeptide is by definition any chain of more than one amino acid connected by the characteristic peptide bond. In practice, most neuropeptide neurotransmitters range between 2 and around 50 amino acids with chains larger than 90 residues being considered proper proteins. Of the 80 or more known neuropeptides that are thought to function as neurotransmitters, only about 20 have been investigated as being involved in AD pathology, and of these only a few have been convincingly demonstrated to be altered in the majority of cases studied. For many peptides only one or two studies have been conducted, and these either have found no change or have failed to replicate a previous finding of change. This failure to replicate ranges from failures to reproduce the specific region of change, the absolute magnitude of change or even the direction of change. Typically, one group of researchers will publish the preliminary report of an alteration in peptide concentration—which is usually found to be a relative decrease when studying neurodegenerative diseases—in AD CSF or in one or more postmortem brain regions, and other groups will attempt to reproduce and extend this finding. Only when independent confirmation is obtained can the finding be considered reproducible and therefore valid. One advantage that peptide researchers enjoy over those working with the more classical neurotransmitter systems is the relative resistance to peptidase degradation of the major pool of active peptides while contained within their synaptic vesicles[14] in postmortem brain. This greatly reduces the potential for confusion due to postmortem degradation when investigating neuropeptides. Because there have only recently been practical treatment strategies for AD that address the known neuropeptide deficits, partly because of the logistical difficulties in getting active analogues of neuropeptides that can cross the blood–brain barrier, it is of great interest to determine whether such approaches could effectively reverse the dementia associated with AD.

Somatostatin is a neuropeptide that was originally isolated from the hypothalamus of sheep and pigs as a factor controlling the release of pituitary hormones (see ref. 15 for review). Somatostatin (somatotropin-release inhibiting factor, SRIF) was shown to physiologically regulate the release of growth hormone and many other pituitary hormones by providing the inhibitory component of the dual regulation experienced by most of the anterior pituitary hormones. Upon further investigation, it was quickly realized that the distribution of SRIF outside of the hypothalamus was more indicative of a neurotransmitter role in addition to that of a hypothalamic releasing factor, and application of SRIF to neurons consistently inhibited their firing rate. It was subsequently found to often be co-localized within neurons containing the inhibitory transmitter, gamma amino butyric acid (GABA) in the cortex, and two active forms have been identified which are cleaved from a larger precursor molecule and contain either 14 or 28 amino acids. Five molecular subtypes of SRIF receptors have now been cloned, and the specific peptidases contributing to the degradation of active SRIF have also been identified.

Somatostatin concentration decreases in the cortex of the postmortem brain of the AD patient were the second neurotransmitter system to be implicated in AD pathology after the discovery of the cholinergic deficits (see refs. 16 and 17 for review). Of the peptides found co-localized with GABA in cortical neurons, SRIF is the second most abundant after substance P and is more pervasive than either vasoactive intestinal peptide or cholecytokinin, which do not appear to be principal targets of AD pathology.[18,19] The major population of SRIF neurons at risk in AD are cortical interneurons that do not contain other peptides such as neuropeptide Y,[20,21] with most subcortical regions such as the nucleus basalis and hypothalamus largely unaffected, although a recent report indicates SRIF deficits in subcortical white matter do exist[22] in AD. A recent paper[23] has implicated SRIF 28 as the species most involved in AD pathology, and *in situ* hybridization surveys of messenger RNA for the pre-pro-SRIF precursor molecule have shown that residual synthetic capacity for SRIF exists in many regions where absolute SRIF concentrations are decreased.[24] This would indicate that the presynaptic neuronal machinery necessary to produce SRIF is capable of continued synthesis and could be a potential target for therapeutic intervention. Further support for a disregulation of SRIF metabolism is found in reports that the amount of SRIF precursor is increased relative to the active form of the peptide in CSF from AD patients[25] and that SRIF-28 is the species reduced in CSF from patients with AD.[26] However, the presumably postsynaptic SRIF receptors are not up-regulated in an attempt to compensate for the reduced synaptic availability of SRIF; in fact, the opposite is the case. Several research groups[27–29] have now observed SRIF receptor populations in postmortem cortex to be decreased in number, but not ligand affinity, in AD. In addition to the decreases in endogenous ligand concentration and receptor disregulation, the peptidases that degrade the active forms of SRIF are also altered inappropriately,[30] degrading SRIF more efficiently in post-mortem AD cortex than peptidases from non-AD brain.

Attempts to address the SRIF deficit with long-acting SRIF analogues have been unsuccessul in reversing AD dementia,[31] partly because of the decreased numbers of receptors and increased peptidase activity. The SRIF concentration deficit has been widely observed in CSF (see ref. 16) as well as in postmortem brain tissue from AD patients. Several papers have now described treatment-induced amelioration of these CSF deficits in SRIF concentrations that may correlate with cognitive improvement.[32–34] Unfortunately, any disease that produces significant cognitive impairment seems to reduce SRIF concentrations in CSF, as indicated by the temporary and reversible deficits in CSF SRIF during bouts of delirium.[35] Although this lack of specificity prevents SRIF in CSF from being useful as a diagnostic indicator for the presence of AD, these data indicate the concentration of SRIF in CSF may be a useful biological marker for treatments that successfully improve cognitive state.

Corticotropin-releasing factor (CRF) is a 41 amino acid neuropeptide that was isolated after a long search for the principal regulatory agent controlling release of the anterior pituitary hormone, adrenocorticotropin (ACTH). Outside of the hypothalamus, CRF is located in limbic system regions and within a population of cortical interneurons. A plethora of behavioral studies have implicated CRF systems as major mediators of behavioral responses to stimuli producing fear and anxiety, and CRF hypersecretion is suspected as a mediator of some of the endocrine changes and behavioral symptoms of major depressive disorder in humans.[36] Two major subtypes of CRF receptors are now known to exist,[37,38] and a portion of the bioavailable CRF is sequestered from receptors and peptidase degradation by the action of an endogenous CRF binding protein.

Using tissue obtained from the Medical Research Centre Brain Bank in Cambridge, U.K., our group[39] was the first to report CRF concentration decreases in frontal and temporal cortex of AD postmortem brain compared to non-AD control tissues. The concentration of CRF in the AD group was not significantly different from the controls in subcortical regions such as the nucleus basalis or hypothalamus, although in this patient population the caudate nucleus had significantly diminished concentrations of CRF in the AD group. The deficits in cortical CRF concentrations in AD were quickly confirmed by DeSouza's group,[40] who extended this finding to other cortical regions and observed that the postsynaptic neurons containing CRF receptors were attempting to compensate for the reduced synaptic availability of CRF by inversely proportional, regionally specific increases in CRF receptor number relative to decrease in the concentration of the CRF endogenous ligand. Based upon the regional distribution of the CRF receptor population that was found to be up-regulated in AD, it is likely that the CRF_1 receptor is the subtype involved (D. Chalmers, personal communication), and previous work has shown that the increase in CRF binding in AD tissue is due to functional CRF receptors.[41] Immunohistochemical staining confirmed that the cortical CRF concentration decreases in AD were due to degeneration of the interneurons of the cortex containing CRF.[42] Other areas of the brain such as the hippocampus, hypothalamus, and cerebellum have been shown to retain substantial levels of CRF immunoreactive fibers and cells.[43] Most of the CRF-related changes reported in AD have involved the cortical interneuron CRF circuits, and few groups have reported CRF concentration changes outside of the cortex or within the hypothalamus. However, evidence of hypothalamic disregulation in AD has been reported for the response of the hypothalamic-pituitary-adrenal (HPA) axis to challenge with exogenous synthetic CRF.[44] Others have reported elevations in the level of CRF mRNA in the postmortem paraventricular nucleus of the hypothalamus in patients with AD and even greater elevations in patients with a clinical history of major depression during life.[45] As elevations of CRF are reported in CSF of depressed patients relative to controls[36] along with other indicators of increased secretion of hypothalamic CRF, it would seem possible, based upon the CRF mRNA elevations in the hypothalamus of patients with AD, that CSF concentrations of CRF may also be elevated in AD relative to non-AD patients if hypothalamic CRF contributes significantly to CSF concentrations. One research group has reported such an increase,[46] and demented patients with depression[51] have also exhibited such increases of CRF in CSF concentrations. However, CSF concentrations of CRF in simple AD have also been reported to be either unchanged[36,47–51] or decreased[52–56] depending upon the control and AD patient populations being compared and the laboratory reporting the data. These disparities are almost certainly due to the uncertainty of the diagnosis of AD in patients during life. Our group has seen some populations of AD patients with a nonsignificant increase in CRF in the postmortem hypothalamus[39] and other AD groups with relative decreases in hypothalamic CRF and SRIF that did achieve statistical significance (Bissette *et al.*, in press). Thus a definitive picture of CSF and hypothalamic CRF in AD remains elusive.

Recently, a strategy[57] was proposed whereby the apparently reduced synaptic availability of CRF might be restored to near-normal levels in patients with AD. With the use of shortened analogues of CRF to displace endogenous CRF from the CRF-binding protein, the amount of free CRF in postmortem AD brain was able to be increased to control levels. The amount of CRF-binding protein was similar in the AD cortical tissue and in the normal control brain in the frontal, temporal, parietal, and occipital cortices. The ligand that displaces native CRF from the binding protein does not produce anxiety in laboratory animals at doses that effectively improve memory and learning as assessed with the Morris water maze task. Once such compounds are developed that can cross the blood–brain barrier in sufficient quantities to similarly affect the CRF-binding proteins of living patients, it will be possible to determine which symptoms of AD are mediated by the CRF concentration deficits seen at the time of death.

Because the SRIF and CRF deficits in AD are most frequently confined to cortical interneurons, the possibility that the same element of the pathological process in AD may affect both neuropeptide neuronal populations must be considered. Correlation between the magnitude of regional decreases in each of these distinct systems has been sought in the four existing studies where both SRIF and CRF were measured in the same postmortem samples of individual AD patients compared to controls (see TABLE 1). The temporal and frontal, but not parietal, cortex of the AD patients in our first study had similar levels of depletions in both CRF and SRIF (>50%).[39,58] In contrast, the caudate nucleus deficits in CRF were not reproduced for SRIF in the AD cases. Leake *et al.*[59] found temporal cortex SRIF depletions to be similar in magnitude to CRF deficits in this region, but saw only CRF changes in the occipital cortex, the only other cortical region examined. Gabriel *et al.*[18] recently described SRIF and CRF depletions in four of six cortical regions that were of similar magnitude in the AD cases compared to the controls. We have compared over 20 cortical regions and 13 subcortical regions for similarities in the amount of CRF and SRIF among 16 neuropathologically confirmed cases of AD compared to nine non-AD controls with and without dementia (Bissette *et al.*, in press). Five cortical regions and the hypothalamus had similar levels of pathological involvement between the two peptide systems as indicated by their

TABLE 1. Cortical CRF and SRIF Measured in the Same Postmortem Samples[a]

Cortical Region	Bissette (in press)[b] CRF	SRIF	Gabriel[18,c] CRF	SRIF	Leake[59,d] CRF	SRIF	Bissette[39]/ Nemeroff[58,e] CRF	SRIF
Temporal								
BA20	26	84						
BA21	28	52						
BA22	47	83						
BA23	51	77						
BA36			54	40				
BA38	80	90	45	25	50	63	60	60
BA39	32	59						
BA40	43	69						
Frontal								
BA6	75	78						
BA8			46	80				
BA9	68	71						
BA10	87	97					42	63
BA11	48	81						
BA12	42	67						
BA24	92	86	57	47				
BA32	61	83						
BA33	46	88						
BA44	50	74	59	47				
BA44	39	87						
Parietal								
BA4	63	84						
BA7	74	107					93	91
Occipital								
BA17	62	53	60	60	42	67		

[a] All CRF and SRIF values are in units of % control.
[b] Maximum $n = 9$ for control; $n = 16$ for AD.
[c] Maximum $n = 8$ for control; $n = 13$ for AD.
[d] $n = 9$ for control; $n = 10$ for AD.
[e] $n = 10$ for control; $n = 16$ for AD.

relative decreases in concentration in the AD group, and absolute concentrations were significantly correlated between the two peptides within individual patients for 16 cortical regions and five subcortical regions, including the hypothalamus. When both peptides were measured in the same CSF sample (see TABLE 2), their levels were significantly correlated,[47,52,60] although some reports have not attempted to specifically make correlations between SRIF and CRF concentrations in CSF.[50,51] Thus, these two anatomically distinct neuropeptide neuronal systems are linked to one another with or without AD pathology in several brain regions and possibly in CSF.

The contribution of these neuropeptide deficits to the symptoms of dementia remains unknown. Both SRIF- and CRF-staining nerve terminals are found in neuritic plaques, but so are peptides without major involvement in the disease process. Correlation of severity of dementia using cognitive test results to measure degree of dementia have more consistently identified SRIF as statistically associated with degree of dementia than CRF,[61] although CRF in CSF has been correlated

TABLE 2. CRF and SRIF Measured in the Same CSF Samples

Study	Subjects (n)	CRF (pM/L ± SD)	SRIF (pM/L ± SD)
Jolkkonen et al.[49]	Control (14)	8.7 ± 2.6	45.7 ± 8.9
	AD (17)	8.3 ± 1.1	39.4 ± 10.8
Banki et al.[51]	Control (17)	9.0 ± 2.33	17.5 ± 6.8
	AD (77)	12.5 ± 5.4	20.7 ± 10.9
Molchan et al.[47]	Control (13)	18.3 ± 3.3	38.1 ± 8.12
	AD (49)	16.1 ± 5.5	22.5 ± 9.3
Edvinsson et al.[50]	Control (11)	12.5 ± 3.3	62.0 ± 32.0
	AD (39)	9.0 ± 6.0	28.0 ± 15.0
Heilig et al.[52]	Control (30)	13.0 ± 3.0	34.0 ± 5.0
	AD (36)	8.0 ± 1.5	24.0 ± 6.0

with dementia severity in some AD populations.[48] A recent study using CSF found no correlation with duration or severity of dementia in the CRF and SRIF deficits of the AD group relative to the controls,[52] and two of the postmortem tissue studies[18] (and Bissette et al., in press) also found no major correlation between duration of dementia and SRIF or CRF regional concentrations. However, the cholinergic deficits were recently reported to be more predictive of degree of dementia than either peptide in postmortem tissue[62] and were found to be more widespread than the CRF or SRIF deficits, although significant correlations between ChAT and SRIF concentrations were obtained in frontal and temporal cortex.[63] Attempts to elucidate the relationship between the cholinergic input to the cortex from the nucleus basalis with the CRF and SRIF cortical interneurons have been generally disappointing. Lesions of the nucleus basalis in experimental animals cause compensatory increases in the concentrations of SRIF[64–66] and CRF[67] in the cortex, an effect opposite to that seen in AD, although recently one group observed ipsilateral deficits in SRIF and its receptor three months after unilateral lesions of the nucleus basalis.[68] Various studies have reported the ability of drugs that interact with cholinergic receptors to release CRF or SRIF from neurons containing them, but no one has convincingly demonstrated direct synaptic contact between the neuropeptide cortical interneurons and the cholinergic cortical projections, although changes in AChE and SRIF in CSF have been correlated in some populations of patients with AD,[69,70] as has CRF.[60]

This state of affairs has rendered it difficult to determine whether the changes in the neuropeptide system in AD precede, are subsequent to or are contemporary with the cholinergic degeneration. This information is crucial to designing a therapy that addresses the neurochemical changes in the beginning of the disease when intervention is likely to be most effective. Now that imaging techniques are available that can measure decreases in perfusion of cortical regions in as-yet-undemented patients with greatly increased genetic risk for AD,[71] it may be possible to address this issue by pairing cognitive testing to CSF samples drawn over the course of the disease in a high-risk population of patients. By using proxy measures of ACh systems such as AChE and measuring concentrations of SRIF and CRF in serial samples of CSF from the same individual before and after the onset of dementia, it should be possible to determine which pathological target(s) should receive priority in therapeutic attempts to slow or reverse the course of the disease. It may even be necessary to obtain a reference CSF sample from all otherwise healthy people between the ages of 20 and 30 years as part of a normal, routine physical check-up to use as a baseline for furture comparison. This sample would then provide a reference to that individual's personal neurochemical profile when they are healthy should major neurologic or psychiatric disease manifest itself later in life. Such samples would allow researchers to determine the most likely targets of neuropathology at the different stages of progressive neurologic and psychiatric diseases with a much more empirical approach than in the current situation where single CSF samples are used to make group comparisons or postmortem tissue is used to try to elucidate the disease process during the patient's life.

At present it is not possible to assign the dementia resulting from AD to a particular neurochemical deficit. That more than the cholinergic system is involved is apparent from the modest improvements seen with choline augmentation therapies and acetylcholinesterase blockade. The SRIF and CRF deficits in AD are promising potential therapeutic targets, which in the next few years will certainly be addressed pharmacologically. As synaptic loss seems to be the best physiological correlate for the degree of dementia,[72] it is imperative that such therapies be instituted as early in the course of the disease as possible. Thus, the development of a reliable biochemical marker for the disease remains a paramount necessity and neuropeptides may prove useful in this regard.

REFERENCES

1. FRANCIS, P. T., D. M. BOWEN, S. L. LOWE, D. NEARY, D. M. A. MANN & J. S. SNOWDEN. 1987. Somatostatin content and release measured in cerebral biopsies from demented patients. J. Neurol. Sci. **78:** 1–16.
2. MANAYE, K. F., D. D. MCINTIRE, D. M. A. MANN & D. C. GERMAN. 1995. Locus coeruleus cell loss in the aging human brain: A non-random process. J. Comp. Neurol. **358:** 79–87.
3. GEULA, C. & M.-M. MESULAM. 1994. Cholinergic systems and related neuropathological predilection patterns in Alzheimer disease. *In* Alzheimer Disease. R. D. Terry, R. Katzman & K. L. Bick, Eds.: 263–291. Raven Press. New York.
4. DAVIES, P. & A. J. F. MALONEY. 1976. Selective loss of central cholinergic neurons in Alzheimer's disease. Lancet **2:** 1403.
5. SLOTKIN, T. A., F. J. SEIDLER, B. J. CRAIN, J. M. BELL, G. BISSETTE & C. B. NEMEROFF. 1990. Regulatory changes in presynaptic cholinergic function assessed in rapid autopsy material from patients with Alzheimers disease: Implications for etiology and therapy. Proc. Natl. Acad. Sci. USA **87:** 2452–2455.
6. AUBERT, I., D. M. ARAUJO, D. CÉCYRE, Y. ROBITAILLE, S. GAUTHIER & R. QUIRION. 1992. Comparative alterations of nicotinic and muscarinic binding sites in Alzheimer's and Parkinson's disease. J. Neurochem. **58:** 529–541.
7. SVENSSON, A.-L., I. ALAFUZOFF & A. NORDBERG. 1992. Characterization of muscarinic receptor subtypes in Alzheimer and control brain cortices by selective muscarinic antagonists. Brain Res. **596:** 142–148.
8. MESULAM, M.-M., C. GEULA & M. A. MORÁN. 1987. Anatomy of cholinesterase inhibition in Alzheimer's disease: Effect of physostigmine and tetrahydroaminoacridine on plaques and tangles. Ann. Neurol. **22:** 683–691.
9. SOARES, J. C. & S. GERSHON. 1995. THA—Historical aspects, review of pharmacological properties and therapeutic effects. Dementia **6:** 225–234.
10. ALLEN, N. H. P. & A. BURNS. 1995. The treatment of Alzheimer's disease. J. Psychopharmacol. **9:** 43–56.
11. WHITEHOUSE, P. J. 1988. Intraventricular bethanecol in Alzheimer's disease: A continuing controversy. Neurology **38:** 307–308.
12. GIACOBINI, E. 1994. Cholinomimetic therapy of Alzheimer disease: Does it slow down deterioration? *In* Recent Advances in the Treatment of Neurodegenerative Disorders

and Cognitive Dysfunction. G. Racagni, N. Brunello & S. Z. Langer, Eds.: 51–57. Karger. Basel.

13. TRAUB, M. & S. B. FREEDMAN. 1992. The implication of current therapeutic approaches for the cholinergic hypothesis of dementia. Dementia **3:** 189–193.

14. LEE, C.-M., P. C. EMSON & L. L. IVERSEN. 1981. Chromatographic behaviour and post-mortem stability of somatostatin in the rat and mouse brain. Brain Res. **220:** 159–166.

15. EPELBAUM, J., P. DOURNAUD, M. FODOR & C. VIOLLET. 1994. The neurobiology of somatostatin. Crit. Rev. Neurobiol. **8:** 25–44.

16. BISSETTE, G. & B. MYERS. 1992. Somatostatin in Alzheimer's disease and depression. Life Sci. **51:** 1389–1410.

17. VÉCSAI, L. & P. KLIVÉNYI. 1995. Somatostatin and Alzheimer's disease. Arch. Gerontol. Geriatr. **21:** 35–41.

18. GABRIEL, S. M., M. DAVIDSON, V. HAROUTUNIAN, P. POWCHIK, L. M. BIERER, D. P. PUROHIT, D. P. PERL & K. L. DAVIS. 1996. Neuropeptide deficits in schizophrenia vs. Alzheimer's disease cerebral cortex. Biol. Psychiatry **39:** 82–91.

19. MAZUREK, M. F. & M. F. BEAL. 1991. Cholecystokinin and somatostatin in Alzheimer's disease postmortem cerebral cortex. Neurology **41:** 716–719.

20. DAVIES, C. A., D. R. MORROLL, D. PRINJA, D. M. A. MANN & A. GIBBS. 1990. A quantitative assessment of somatostatin-like and neuropeptide Y-like immunostained cells in the frontal and temporal cortices of patients with Alzheimer's disease. J. Neurol. Sci. **96:** 59–73.

21. GABRIEL, S. M., L. M. BIERER, V. M. HAROUTUNIAN, D. P. PUROHIT, D. P. PERL & K. L. DAVIS. 1993. Widespread deficits in somatostatin but not neuropeptide Y concentrations in Alzheimer's disease cerebral cortex. Neurosci. Lett. **155:** 116–120.

22. ANG, L.-C. & D. D. SHUL. 1995. Peptidergic neurons of subcortical white matter in aging and Alzheimer's brain. Brain Res. **674:** 329–335.

23. GASPAR, P., C. DUYCKAERTS, A. FEBVRET, R. BENOIT, B. BEEK & B. BERGER. 1989. Subpopulations of somatostatin 28-immunoreactive neurons display different vulnerability in senile dementia of the Alzheimer type. Brain Res. **490:** 1–13.

24. DOURNAUD, P., P. CERVERA-PIEROT, E. HIRSCH, F. JAVOY-AGID, C. KORDON, Y. AGID & J. EPELBAUM. 1994. Somatostatin messenger RNA-containing neurons in Alzheimer's disease: An *in situ* hybridization study in hippocampus, parahippocampal cortex and frontal cortex. Neuroscience **61:** 755–764.

25. GOMEZ, S., J. PUYMIRAT, P. VALADE, P. DAVOUS, P. RONDOT & P. COHEN. 1986. Patients with Alzheimer's disease show an increased content of 15K-dalton somatostatin precursor and a lowered level of tetradecapeptide in their cerebrospinal fluid. Life Sci. **39:** 623–627.

26. YASUDA, M., K. MAEDA, T. KAKIGI, N. MINAMITANI, T. KAWAGUCHI & C. TANAKA. 1995. Low cerebrospinal fluid concentrations of peptide histidine valine and somatostatin-28 in Alzheimer's disease: Altered processing of prepro-vasoactive intestinal peptide and prepro-somatostatin. Neuropeptides **29:** 325–330.

27. BEAL, M. F., M. F. MAZUREK, V. T. TRAN, G. CHATTA, E. D. BIRD & J. B. MARTIN. 1985. Reduced numbers of somatostatin receptors in the cerebral cortex in Alzheimer's disease. Science **229:** 289–291.

28. KRANTIC, S., Y. ROBITAILLE & R. QUIRION. 1992. Deficits in the somatostatin SS_1 receptor sub-type in frontal and temporal cortices in Alzheimer's disease. Brain Res. **573:** 299–304.

29. BERGSTRÖM, L., A. GARLIND, L. NILSSON, I. ALAFUZOFF, C. J. FOWLER, B. WINGBLAD & R. F. COWBURN. 1991. Regional distribution of somatostatin receptor binding and modulation of adenylyl cyclase activity in Alzheimer's disease brain. J. Neurol. Sci. **105:** 225–233.

30. WEBER, S. J., R. B. LOUIS, L. TROMBLEY, G. BISSETTE, P. DAVIES & T. P. DAVIS. 1992. Metabolic half-life of somatostatin and peptidase activities are altered in Alzheimer's disease. J. Gerontol. **47:** B18–25.

31. MOURADIAN, M. M., J. BLIN, M. GIUFFRA, I. J. HEUSER, F. BARONTINI, J. OWNBY & T. N. CHASE. 1991. Somatostatin replacement therapy for Alzheimer dementia. Ann. Neurol. **30:** 610–613.

32. KARLSSON, I., E. WIDERLOV, E. V. MELIN, A.-L. NYTH, G. A. M. BRANE, E. RYBO, J. F. REHFELD, G. BISSETTE & C. B. NEMEROFF. 1985. Changes in CSF neuropeptides after environmental stimulation in dementia. Nord. Psychiatr. J. **39:** 75–81.
33. WIDERLÖV, E., G. BRÅNE, R. EKMAN, M. KIHLGREN, A. NORBERG & I. KARLSSON. 1989. Elevated CSF somatostatin concentrations in demented patients parallel improved psychomotor functions induced by integrity-promoting care. Acta Psychiatr. Scand. **79:** 41–47.
34. ALHAINEN, K., J. SIRVIÖ, E.-L. HELKALA, K. REINIKAINEN & P. REIKKINEN. 1991. Somatostatin and cognitive functions in Alzheimer's disease—The relationship of cerebrospinal fluid somatostatin increase with clinical response to tetrahydroaminoacridine. Neurosci. Lett. **130:** 46–48.
35. KOPONEN, H., K. REINIKAINEN & P. J. REIKKINEN. 1990. Cerebrospinal fluid somatostatin in delirium. II. Changes at the acute stage and at one year follow-up. Psychol. Med. **20:** 501–505.
36. NEMEROFF, C. B., E. WIDERLOV, G. BISSETTE, H. WALLEUS, I. KARLSSON, K. EKLUND, C. D. KILTS, P. T. LOOSEN & W. W. VALE. 1984. Elevated concentrations of CSF corticotropin-releasing factor-like immunoreactivity in depressed patients. Science **226:** 1342–1344.
37. CHALMERS, D. T., T. W. LOVENBERG & E. B. DESOUZA. 1995. Localization of novel corticotropin-releasing factor receptor (CRF$_2$) mRNA expression to specific subcortical nuclei in rat brain: Comparison with CRF$_1$ receptor mRNA expression. J. Neurosci. **15:** 6340–6350.
38. LOVENBERG, T. W., C. W. LIAW, D. E. GRIGORIADIS, W. CLEVENGER, D. T. CHALMERS, E. B. DESOUZA & T. OLTERSDORF. 1995. Cloning and characterization of a functionally distinct corticotropin-releasing factor receptor subtype from rat brain. Proc. Natl. Acad. Sci. USA **92:** 836–840.
39. BISSETTE, G., G. P. REYNOLDS, C. D. KILTS, E. WIDERLÖV & C. B. NEMEROFF. 1985. Corticotropin-releasing factor-like immunoreactivity in senile dementia of the Alzheimer type. J. Am. Med. Assoc. **254:** 3067–3069.
40. DESOUZA, E. B., P. J. WHITEHOUSE, M. J. KUHAR, D. L. PRICE & W. W. VALE. 1986. Reciprocal changes in corticotropin-releasing factor (CRF)-like immunoreactivity and CRF receptors in cerebral cortex of Alzheimer's disease. Nature **319:** 593–595.
41. GRIGORIADIS, D. E., R. G. STRUBLE, D. L. PRICE & E. B. DESOUZA. 1989. Normal pattern of labeling of cerebral cortical corticotropin-releasing factor (CRF) receptors in Alzheimer's disease: Evidence from chemical crosslinking studies. Neuropharmacology **28:** 761–764.
42. POWERS, R. E., L. C. WALKER, E. B. DESOUZA, W. W. VALE, R. G. STRUBLE, P. J. WHITEHOUSE & D. J. PRICE. 1987. Immunohistochemical study of neurons containing corticotropin-releasing factor in Alzheimer's disease. Synapse **1:** 405–410.
43. KELLEY, M. & N. KOWALL. 1989. Corticotropin-releasing factor immunoreactive neurons persist throughout the brain in Alzheimers disease. Brain Res. **501:** 392–396.
44. HARTZINGER, M., A. Z'BRUN, U. HEMMETER, E. SEIFRITZ, F. BAUMANN, E. HOLSBOER-TRACHSLER & I. J. HEUSER. 1995. Hypothalamic-pituitary-adrenal system function in patients with Alzheimer's disease. Neurobiol. Aging **16:** 205–209.
45. RAADSHEER, F. C., J. J. VAN HEERIKHUIZE, P. J. LUCASSEN, W. J. G. HOOGENDIJK, F. J. H. TILDERS & D. F. SWAAB. 1995. Corticotropin-releasing hormone mRNA levels in the paraventricular nucleus of patients with Alzheimer's disease and depression. Am. J. Psychiatry **152:** 1372–1376.
46. MARTIGNONI, E., F. PETRAGLIA, A. COSTA, G. BONO, A. R. GENAZZANI & G. NAPPI. 1990. Dementia of the Alzheimer type and the hypothalamus-pituitary-adrenocortical axis: Changes in cerebrospinal fluid corticotropin-releasing factor and plasma cortisol levels. Acta Neurol. Scand. **81:** 452–456.
47. MOLCHAN, S. E., J. L. HILL, R. A. MARTINEZ, B. A. LAWLOR, A. M. MELLOW, D. R. RUBINOW, G. BISSETTE, C. B. NEMEROFF & T. SUNDERLAND. 1993. CSF somatostatin in Alzheimer's disease and major depression: Relationship to hypothalamic-pituitary-adrenal axis and clinical measures. Psychoneuroendocrinology **18:** 509–519.

48. POMARA, N., R. R. SINGH, D. DEPTULA, P. A. LEWITT, G. BISSETTE, M. STANLEY & C. B. NEMEROFF. 1989. CSF corticotropin-releasing factor (CRF) in Alzheimer's disease: Its relationship to severity of dementia and monoamine metabolites. Biol. Psychiatry 26: 500–504.

49. JOLKKONEN, J., R. SOIKKELI, P. HARTIKAINEN, G. BISSETTE & P. REIKKINEN. 1990. CSF neuropeptides in Alzheimer's disease and Parkinson's disease. Anu. Psiquiatrico 1: 251–257.

50. EDVINSSON, L., L. MINTHON, R. EKMAN & L. GUSTAFSON. 1993. Neuropeptides in cerebrospinal fluid of patients with Alzheimer's disease and dementia with frontotemporal lobe degeneration. Dementia 4: 167–171.

51. BANKI, C., L. KARMACSI, G. BISSETTE & C. B. NEMEROFF. 1992. Cerebrospinal neuropeptides in dementia. Biol. Psychiatry 32: 452–456.

52. HEILIG, M., M. SJÖGREN, K. BLENNOW, R. EKMAN & A. WALLIN. 1995. Cerebrospinal fluid neuropeptides in Alzheimer's disease and vascular dementia. Biol. Psychiatry 38: 210–216.

53. MAY, C., S. I. RAPOPORT, T. P. TOMAI, G. P. CHROUSOS & P. W. GOLD. 1987. Cerebrospinal fluid concentrations of corticotropin-releasing hormone (CRH) and corticotropin (ACTH) are reduced in Alzheimer's disease. Neurology 37: 535–538.

54. MOURADIAN, M. M., J. M. FARAH, E. MOHR, G. FABBRINI, T. L. O'DONOHUE & T. N. CHASE. 1986. Spinal fluid CRF reduction in Alzheimer's disease. Neuropeptides 8: 393–400.

55. SUEMARU, S., K. HASHIMOTO, T. OGASA, R. HIRASAWA, S. MAKINO, Z. OTA, J. KAGEYAMA & K. SUEMARU. 1991. Cerebrospinal fluid and plasma corticotropin-releasing hormone in senile dementia. Life Sci. 48: 1871–1879.

56. SUEMARU, S., K. SUEMARU, K. KAWAI, S. MIYATA, K. NOBUKINI, Y. THARA, R. NAMBA, K. URAKAMI & K. HASHIMOTO. 1995. Cerebrospinal fluid corticotropin-releasing hormone in neurodegenerative diseases: Reduction in spinocerebellar degeneration. Life Sci. 57: 2231–2235.

57. BEHAN, D. P., S. C. HEINRICHS, J. C. TRONCOSO, X-J. LIU, C. H. KAWAS, N. LING & E. B. DESOUZA. 1995. Displacement of corticotropin-releasing factor from its binding protein as a possible treatment for Alzheimer's disease. Nature 378: 284–287.

58. NEMEROFF, C. B., J. S. KIZER, G. P. REYNOLDS & G. BISSETTE. 1989. Neuropeptides in Alzheimer's disease: A postmortem study. Regul. Pept. 25: 123–130.

59. LEAKE, A., E. K. PERRY, R. H. PERRY, S. JABEEN, A. F. FAIRBAIRN, I. G. MCKEITH & I. N. FERRIER. 1991. Neocortical concentrations of neuropeptides in senile dementia of the Alzheimer and Lewy body type. Biol. Psychiatry 29: 357–364.

60. JOLKONNEN, J., P. HARTIKAINEN, R. SOIKKELI, G. BISSETTE, C. B. NEMEROFF & P. REIKKINEN. 1991. A correlation study of CSF neuropeptides in Alzheimer's and Parkinson's disease. Neuropeptides 19: 97–102.

61. TAMMINGA, C. A., N. L. FOSTER, P. FEDIO, E. D. BIRD & T. N. CHASE. 1987. Alzheimer's disease: Low cerebral somatostatin levels correlate with impaired cognitive function and cortical metabolism. Neurology 37: 161–165.

62. BIERER, L. M., V. HAROUTUNIAN, S. M. GABRIEL, P. J. KNOTT, L. S. CARLIN, D. P. PUROHIT, D. P. PERL, J. SCHMEIDLER, P. KANOF & K. L. DAVIS. 1995. Neurochemical correlates of dementia severity in Alzheimer's disease: Relative importance of the cholinergic deficits. J. Neurochem. 64: 749–760.

63. DORNAUD, P., P. DELAERE, J. J. HAUW & J. EPELBAUM. 1995. Differential correlation between neurochemical deficits, neuropathology and cognitive status in Alzheimer's disease. Neurobiol. Aging 16: 817–823.

64. ARENDASH, G. W., W. J. MILLARD, R. DAWSON, A. J. DUNN & E. M. MEYER. 1990. Different long-term effects of bilateral and unilateral nucleus basalis lesions on rat cerebral cortical neurotransmitter content. Neurochem. Res. 14: 1113–1118.

65. GAYKEMA, R. P. A., J. C. COMPAAN, C. NYAKAS, E. HORVATH & P. G. M. LUITEN. 1989. Long-term effects of cholinergic basal forebrain lesions on neuropeptide Y and somatostatin immunoreactivity in rat neocortex. Brain Res. 489: 392–396.

66. UNGER, J. W. & Y. SCHMIDT. 1994. Neuropeptide Y and somatostatin in the neocortex of young and aging rats: Response to nucleus basalis lesions. J. Clin. Neuroanat. 7: 25–34.

67. MILLARD, W. J., G. W. ARENDASH, A. J. DUNN & E. M. MEYER. 1990. Effects of nucleus basalis lesions on cerebral cortical concentrations of corticotropin-releasing hormone (CRH)-like immunoreactivity in the rat. Neurosci. Lett. **113:** 233–239.
68. MOYSE, E., E. SZIGETHY, J. M. DANGER, H. VAUDRY, G. L. WENK, A. BEAUDET & J. EPELBAUM. 1993. Short- and long-term effects of nucleus basalis magnocellularis lesions on cortical levels of somatostatin and its receptors in the rat. Brain Res. **607:** 154–160.
69. GOMEZ, S., P. DAVOUS, P. RONDOT, A. FAIVRE-BAUMAN, D. VALADE & J. PUYMIRAT. 1986. Somatostatin immunoreactivity and acetylcholinesterase activities in cerebrospinal fluid of patients with Alzheimer disease and senile dementia of the Alzheimer type. Psychoneuroendocrinology **11:** 69–73.
70. URAKAMI, K., Y. ADACHI, E. AWAKI & K. TAKAHASHI. 1989. Characterization of the course of senile dementia of the Alzheimer type using cerebrospinal fluid levels of acetylcholinesterase and somatostatin. Acta Psychiatr. Scand. **80:** 232–237.
71. REIMAN, E. M., R. J. CASSELL, L. S. YUN, K. CHEN, D. BANDY, S. MINOSHIMA, S. N. THIBODEAU & D. OSBORN. 1996. Preclinical evidence of Alzheimer's disease in persons homozygous for the e4 allele for apolipoprotein E. N. Engl. J. Med. **334:** 752–758.
72. TERRY, R. D., E. MASLIAH, D. P. SALMON, N. BUTTERS, R. DETERESA, R. HILL, L. A. HANSEN & J. R. KATZMAN. 1991. Physical basis of cognitive alterations in Alzheimer's disease: Synapse loss is the major correlate of cognitive impairment. Ann. Neurol. **30:** 572–580.

Alterations of Peptide Metabolism and Neuropeptidase Activity in Senile Dementia of the Alzheimer's Type[a]

STEPHEN M. WATERS[b] AND THOMAS P. DAVIS[c]

Department of Pharmacology, College of Medicine
University of Arizona Health Sciences Center
Tucson, Arizona 85724

INTRODUCTION

Senile dementia of the Alzheimer's type (SDAT) is characterized by a progressive loss of memory and associated deficits of language and behavior.[1] Extracellular plaques and intracellular neurofibrillary tangles present throughout the cerebral cortex and hippocampus confirm a diagnosis of SDAT upon postmortem analysis.[2] Although deficits of cholinergic neurons (i.e., decreased choline acetyltransferase activity) have been demonstrated and associated with loss of memory and function in SDAT,[3-5] cholinergic enhancement or replacement therapy has been largely unsuccessful for treatment of these patients. The limited benefit of these current therapies suggests the involvement of other neurotransmitter/neuromodulator systems in SDAT progression and pathology. Neuropeptides are one of the systems known to be affected in SDAT brain.

The level of several neuropeptides including somatostatin[6-8] and substance P[9,10] are decreased in human postmortem SDAT brain and cerebrospinal fluid. Decrease of somatostatin levels in cerebral cortex is one of the most consistent neuropeptide abnormalities in SDAT brain.[11-13] Interestingly, a correlation exists between alterations in choline acetyltransferase activity and somatostatin levels in cortical brain regions,[6] suggesting a role for somatostatin in SDAT pathology.

Substance P also is thought to play a role in the progression of SDAT. The sequence of substance P is highly homologous to the major proteinaceous component of the senile plaques observed in SDAT brain, β-amyloid. Of the 40–42 amino acids of β-amyloid, a 10 amino acid portion (25–35) has been shown to possess the neurotoxic activity of this protein.[14] This protein fragment is 56–73% homologous to the tachykinin neuropeptides (i.e., substance P). Interestingly, the neurotoxic action of β-amyloid 25–35 is mimicked by tachykinin receptor antagonists and reversed by substance P *in vivo*[18] and *in vitro*.[14] These results suggest a protective role for substance P against β-amyloid toxicity.

Because no reuptake system for the termination of peptide activity has been demonstrated, the action of extracellularly oriented neuropeptidases has been implicated as the mechanism of peptide inactivation.[16] By degrading peptide structure, these enzymes control the duration of activity of neuropeptides. Alterations

[a] This work was supported by a basic Public Health Service research grant, MH-42600, from the National Institute on Mental Health and a fellowship granted by The American Foundation for Pharmaceutical Education to S.M.W.

[b] Present address: Department of Anatomy and Neurobiology, Box 8108, Washington University School of Medicine, St. Louis, MO 63110.

[c] Corresponding author.

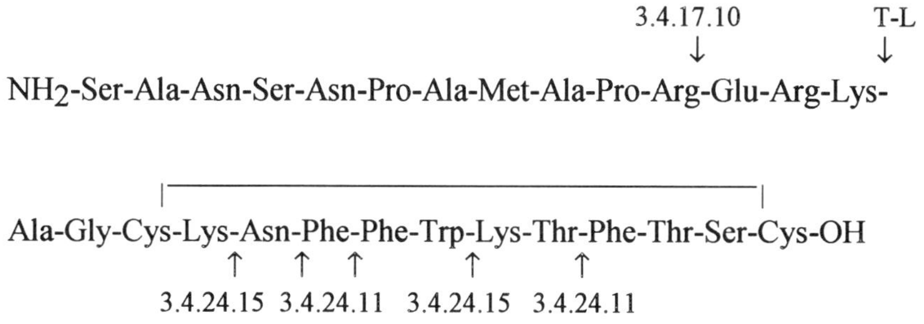

$$3.4.17.10 \qquad \text{T-L}$$
$$\downarrow \qquad\qquad \downarrow$$

NH$_2$-Ser-Ala-Asn-Ser-Asn-Pro-Ala-Met-Ala-Pro-Arg-Glu-Arg-Lys-

Ala-Gly-Cys-Lys-Asn-Phe-Phe-Trp-Lys-Thr-Phe-Thr-Ser-Cys-OH
$$\uparrow \quad \uparrow \quad \uparrow \qquad \uparrow \qquad \uparrow$$
$$3.4.24.15 \quad 3.4.24.11 \quad 3.4.24.15 \quad 3.4.24.11$$

FIGURE 1. Enzymatic cleavage sites of somatostatin-28. Shown are cleavage sites for enzymes trypsin-like serine protease (T-L), carboxypeptidase E (EC 3.4.17.10), neutral endopeptidase 24.11 (EC 3.4.24.11), and metalloendopeptidase 24.15 (EC 3.4.24.15).

in neuropeptidase activity can affect peptide half-life and therefore the action of specific peptides. Our laboratory has demonstrated that neuropeptidase activity is influenced by the administration of centrally-acting drugs[17–20] and disease states such as SDAT[21,22] and schizophrenia.[23]

In this chapter, we discuss reported alterations in somatostatin and substance P degradation in SDAT brain and couple these results to alterations in the activity of specific neuropeptidases implicated in the metabolism of these peptides. These changes in neuropeptidase activity can alter the metabolism of several peptide substrates. As a consequence, the interaction of neuropeptides with their respective physiological targets can be affected.

SOMATOSTATIN DEGRADATION IN SDAT BRAIN REGIONS

The neuropeptide somatostatin (FIG. 1) exists in three main forms in the brain, somatostatin-28 (SS-28), somatostatin-28 (1–12), and somatostatin-14 (SS-14). So-matostatin is widely distributed throughout the central nervous system with high concentrations throughout the cortex.[24] The carboxy-terminal tetradecapeptide SS-14 is cleaved from SS-28 by the action of a trypsin-like serine protease.[25] Somatostatin-14 is the form of somatostatin shown to be reduced in the cortex and hippocampus of SDAT brain.[11] Depletion of SS-14 has also been associated with memory impairment in the rat,[26,27] and therefore alterations in the metabolism of this peptide can affect the half-life of released somatostatin and its putative role in SDAT pathology and progression.

To assess the degradation of SS-28 and SS-14 in SDAT brain, the metabolic half-life of these peptides was examined in specific regions of postmortem SDAT and sex- and age-matched control brain tissues.[21] By using twice-washed, regional brain homogenates (2% w/v),[28] the degradation of SS-28 (50 μM) and SS-14 (50 μM) was determined. After time-course incubations with exogenous somatostatin peptides and high-performance liquid chromatography (HPLC) analysis of peptide fragments, regional differences in peptide half-life were observed (TABLES 1 and 2).

In the temporal cortex region Brodmann area 22, the half-life of SS-28 was decreased by 40% in SDAT brain. However, in another temporal cortex region, Brodmann area 21, the half-lives of SS-28 and SS-14 were only slightly decreased

TABLE 1. Metabolic Half-life of Somatostatin-28[a]

Brain Region	Control	SDAT
Brodmann area 21[b]	248 ± 91	228 ± 77
Brodmann area 22[c]	428 ± 108	258 ± 86[d]
Medial hippocampus[b]	178 ± 40	260 ± 130
Posterior hippocampus[c]	340 ± 40	336 ± 101

[a] Results are expressed in minutes as means ± SD of five to six samples.
[b] Samples were a generous gift from the Alzheimer's Disease Brain Bank, Bronx, NY.
[c] Samples were a generous gift from the Medical Research Council Brain Bank, Cambridge, U.K.
[d] $p < 0.01$ by a Student's unpaired t test.

in SDAT samples (10 and 15%, respectively). The half-lives of SS-28 and SS-14 were also slightly decreased in SDAT posterior hippocampus and caudate, respectively. These data suggest an increase in degradation of somatostatin peptides in SDAT temporal cortex, medial hippocampus, and caudate. In contrast to these brain regions, the half-life of SS-28 and SS-14 were slightly increased in SDAT medial hippocampus suggesting a decrease in somatostatin degradation in this region.

Interestingly, Ichai and co-workers[29] have reported significant decreases in SS-14 degradation in total brain homogenates of SDAT frontal and parietal cortex. These findings are in contrast to the increases in somatostatin degradation in the temporal cortex. This disparity underlies the regional specificity of disease-induced effects on peptide degradation together with the regional diversity of neuropeptidases capable of degrading somatostatin and other peptides. The regional localization of somatostatin-degrading neuropeptidases will be discussed in greater detail in a later section.

SUBSTANCE P DEGRADATION IN SDAT BRAIN REGIONS

Substance P, an undecapeptide of the tachykinin family (FIG. 2), is ubiquitously distributed throughout the brain with highest levels in the substantia nigra, caudate, and globus pallidus and with moderate levels in hippocampus and cortex.[30,31] Like somatostatin, substance P levels are decreased in several cortical regions of SDAT brain.[9] Because substance P is a putative protectant against the neurotoxicity of the senile plaque component β-amyloid,[14,15] alterations in the metabolism of substance P may affect this activity.

Assessment of the degradation of substance P in SDAT and sex- and age-

TABLE 2. Metabolic Half-Life of Somatostatin-14[a]

Brain Region	Control	SDAT
Brodmann area 21	164 ± 14	140 ± 37
Medial hippocampus	117 ± 32	130 ± 53
Caudate	84 ± 18	80 ± 11

[a] Results are expressed in minutes as means ± SD of five to six samples.
(Samples were a generous gift from the Alzheimer's Disease Brain Bank, Bronx, NY.)

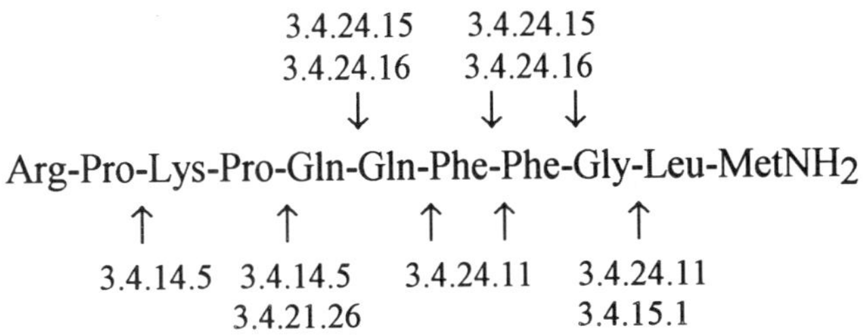

$$3.4.24.15 \quad 3.4.24.15$$
$$3.4.24.16 \quad 3.4.24.16$$
$$\downarrow \qquad \downarrow \quad \downarrow$$

$$\text{Arg-Pro-Lys-Pro-Gln-Gln-Phe-Phe-Gly-Leu-MetNH}_2$$

$$\uparrow \qquad \uparrow \qquad \uparrow \quad \uparrow \qquad \uparrow$$

$$3.4.14.5 \quad 3.4.14.5 \qquad 3.4.24.11 \qquad 3.4.24.11$$
$$3.4.21.26 \qquad\qquad 3.4.15.1$$

FIGURE 2. Enzymatic cleavage sites of substance P. Shown are cleavage sites for enzymes: dipeptidyl aminopeptidase IV (3.4.14.5), prolyl oligopeptidase (EC 3.4.21.26), neutral endopeptidase 24.11 (EC 3.4.24.11), metalloendopeptidase 24.15 (EC 3.4.24.15), metalloendopeptidase 24.16 (EC 3.4.24.16), and angiotensin converting enzyme (3.4.15.1).

matched control brain regions was performed on twice-washed membrane homogenates as previously described (2% w/v).[18] Exogenous substance P (100 μM) was incubated over a time course from 0 to 180 min, and substance P fragments were detected by HPLC analysis of collected supernatants. Regional half-life measurements were determined by the percentage of intact substance P remaining after each time point (TABLE 3).

In temporal cortex samples, the metabolic half-life of substance P was significantly ($p < 0.01$) increased in SDAT samples (50%) compared to matched controls. This finding correlates to decreased degradation of substance P in temporal cortex as a result of SDAT. In SDAT caudate, there was a slight incerase (10%) in the half-life of substance P, in contrast to a decrease (20%) in SDAT hippocampus. Therefore, substance P was degraded at a faster rate in SDAT hippocampus versus control. Interestingly, the degradation of both substance P and somatostatin is affected in temporal cortex regions where SDAT pathology is prevalent. However, somatostatin and substance P degradation in the hippocampus and caudate is not significantly altered because these regions are somewhat less affected by SDAT plaques and tangles.

These data also allude to the regional variability of somatostatin and substance P degradation. In the SDAT and control caudate, somatostatin and substance P were degraded at a much faster rate than in temporal cortex and hippocampus. These differences in peptide half-life support the fact that somatostatin and substance P-degrading neuropeptidases are at higher concentrations in the caudate compared to the cortex and hippocampus.[32]

TABLE 3. Metabolic Half-Life of Substance P[a]

Brain Region	Control	SDAT
Temporal Cortex	183 ± 16	277 ± 25[b]
Hippocampus	168 ± 16	132 ± 11
Caudate	83 ± 9	91 ± 6

[a] Results are expressed in minutes as means ± SEM of six to eight samples.
[b] $p < 0.01$ by a Student's unpaired t test.
(Samples were a generous gift from the Alzheimer's Disease Brain Bank, Bronx, NY.)

NEUROPEPTIDASE ACTIVITIES IN SDAT BRAIN REGIONS

As shown in FIGURE 1 for somatostatin and FIGURE 2 for substance P, several neuropeptidases are capable of degrading these peptides. Somatostatin-28 is cleaved to SS-14 by a trypsin-like serine protease.[25] Additionally, somatostatin is degraded by neutral endopeptidase 24.11 (NEP 24.11, neprilysin, EC 3.4.24.11), metalloendopeptidase 24.15 (MEP 24.15, EC 3.4.24.15), and carboxypeptidase E (CPE, EC 3.4.17.10). Substance P is also cleaved by NEP 24.11[33] and MEP 24.15.[34] Moreover, substance P is a substrate for angiotensin-converting enzyme (ACE, EC 3.4.15.1),[35] dipeptidyl aminopeptidase IV,[36] prolyl oligopeptidase, and metalloendopeptidase 24.16.[34]

The neuropeptidases likely responsible for alterations in neuropeptide metabolism in SDAT brain are membrane-associated ectoenzymes. These enzymes degrade released neuropeptides at the synapse and extrasynaptic sites. Neutral endopeptidase 24.11 is a completely membrane-bound enzyme with no measurable soluble activity.[37] Carboxypeptidase E, trypsin-like serine protease, and MEP 24.15 have activity in soluble preparations but have also appreciable membrane-associated activity.[38] Determination of the activity of these enzymes in SDAT brain samples allows for correlation of peptide degradation and activity changes of specific neuropeptidases.

In addition to degrading somatostatin and substance P, NEP 24.11 also is known to cleave met-enkephalin, leu-enkephalin, and angiotensin.[39] This exclusively membrane-associated enzyme has highest activity in the substantia nigra and caudate, with lower levels in cortical regions.[32] Neutral endopeptidase 24.11 activity in SDAT brain samples,[21,22] measured by the method of Bateman and Hersh,[40] is shown in TABLE 4. The activity of NEP 24.11 is highest in the caudate of both control and SDAT samples and with lower levels in hippocampus and the two temporal cortex regions (Brodmann areas 21 and 22). SDAT samples had higher NEP 24.11 activity than control in all regions with the largest contrast being in the caudate (15%). However, because significant changes are not present in the cortex, alterations in somatostatin and substance P degradation are likely not due to changes in NEP 24.11 activity.

Metalloendopeptidase 24.15 activity is more ubiquitously distributed in the brain than NEP 24.11[32] and can also degrade other neuropeptides such as neurotensin.[39] The activity of MEP 24.15, measured by the method of Orlowski *et al.*[41] in SDAT and matched control brain regions,[21,22] is shown in TABLE 4. In contrast to NEP 24.11, MEP 24.15 activity is generally lower in SDAT samples. In Brodmann area 21, MEP 24.15 activity was decreased by 50%, which correlates to a decreased degradation of substance P in the temporal cortex. Because SS-28 and SS-14 degradation was not affected in Brodmann area 21 (TABLE 1 and 2), somatostatin may not be a substrate for MEP 24.15 in this region of human brain.

Carboxypeptidase E (CPE) degrades peptides by sequentially removing basic amino acids from the carboxy terminus. Therefore, after trypsin-like serine protease cleaves SS-28 to SS-14, CPE can further degrade the produced peptide fragments. By using a method modified from Stack *et al.*,[42] CPE activity was determined[21] and is shown in TABLE 4. The activity of CPE was quite low in all human brain regions studied. SDAT and control postmortem brain had similar CPE activity suggesting that changes in somatostatin degradation are not due to changes in this peptidase.

Because changes in NEP 24.11, MEP 24.15, and CPE do not correlate to changes in SS-28 degradation, alterations in the activity of the trypsin-like serine protease that cleaves SS-28 to SS-14 may account for this effect. As shown in

TABLE 4. Neuropeptidase Activities of Postmortem Human Brain[a]

Peptidase	Brodmann Area 21[b]		Brodmann Area 22[c]		Medial Hippocampus[b]		Caudate[c]	
	Control	SDAT	Control	SDAT	Control	SDAT	Control	SDAT
Neutral endopeptidase 24.11	61.9 ± 16.6	64.7 ± 13.3	36.4 ± 12.5	38.6 ± 16.9	17.0 ± 11.1	17.9 ± 22.5	1387 ± 365	1586 ± 419
Metalloendopeptidase 24.15	1279 ± 146	600 ± 320^d	579 ± 81	727 ± 237	755 ± 165	735 ± 347	575 ± 174	210 ± 189^d
Carboxypeptidase E	3.5 ± 0.4	0.7 ± 0.4	2.4 ± 1.0	3.5 ± 0.4	3.7 ± 0.6	3.7 ± 0.6	1.8 ± 0.4	1.5 ± 0.5
Trypsin-like serine protease	0.4 ± 0.2	0.7 ± 0.4	1.8 ± 1.2	10.1 ± 6.8^e	0.3 ± 0.1	0.4 ± 0.1	0.3 ± 0.1	1.0 ± 1.0

[a] Results are expressed in pmoles/mg protein/min as means $\pm$ SD of four to six samples.
[b] Samples were a generous gift from the Alzheimer's Disease Brain Bank, Bronx, NY.
[c] Samples were a generous gift from the Kathleen Price Bryan Brain Bank, Durham, NC.
[d] $p < 0.01$ by Student's t test.
[e] $p < 0.05$ by Student's t test.

TABLE 4, using the method of Lindberg *et al.*,[25] the activity of the trypsin-like serine protease is increased in SDAT Brodmann area 22.[21] This finding correlates to the increased degradation (i.e., shorter half-life) of SS-28 in this brain region. The activity of trypsin-like serine protease was relatively low in other brain regions studied with no difference observed between SDAT and control Brodmann area 22, medial hippocampus, and caudate.

Other neuropeptidases are also known to be affected in SDAT brain regions. Ichai *et al.*[29] observed decreases in aminopeptidase M and post-proline dipeptidyl aminopeptidase in frontal cortex. In parietal cortex, angiotensin-converting enzyme and metalloendopeptidase 24.15 were shown to be affected in SDAT. Interestingly, these altered neuropeptidases degrade somatostatin and substance P. Indeed, somatostatin degradation was decreased in the frontal cortex and parietal cortex in parallel with these reported neuropeptidase alterations.[39] One would expect that substance P degradation would also be affected in these two cortical regions as a result of activity changes in these neuropeptidases.

Angiotensin-converting enzyme, although shown by Ichai *et al.*,[29] as being decreased is not always decreased as a result of SDAT. Arregui *et al.*[48] reported significant increases in the activity of this neuropeptidase in the caudate nucleus and frontal cortex (Brodmann area 10). This discrepancy may be due to several factors. First, different Brodmann areas of the same brain region can be differentially affected by SDAT pathology. As presented in this report, in two distinct temporal cortex regions (Brodmann areas 21 and 22) neuropeptidases were differentially affected in SDAT brain. Second, based on the tissue preparation used in an assay, a homogenate may contain varying levels of soluble and/or nuclear enzymes which can affect enzymatic activity measurements.[44] Third, the drug histories of control and SDAT patients must be considered as centrally-acting drugs have been shown to affect the activity of the neuropeptidases examined in these studies.[17,20] Additionally, although brain samples with a postmortem interval under 24 h have been shown to have stable neuropeptidase activities, samples collected after 24 h postmortem may have altered enzyme activity as a result of brain tissue degradation.[21]

SUMMARY

Work in our laboratory has shown that in addition to previously characterized changes in the level of neuropeptides in SDAT brain, the activity of degradative enzymes responsible for peptide metabolism is also affected. In addition to other reported alterations in peptide metabolism, we have observed that SS-28 degradation is increased in Brodmann area 22 whereas substance P degradation is increased in temporal cortex. Changes in the degradation of these neuropeptides known to be affected in SDAT correlate well with alterations in the activity of specific neuropeptidases. Trypsin-like serine protease activity is increased in SDAT Brodmann area 22 which parallels the increased degradation of SS-28. The activity of MEP 24.15 is decreased in temporal cortex which corresponds to the decreased degradation of substance P. Changes in the activity of these degradative enzymes in SDAT brain can potentially affect the action of other neuropeptide substrates because the neuropeptidases discussed here terminate the action of several neuropeptides. As more neuropeptide and degradative peptidase alterations are discovered in SDAT, greater emphasis may be placed on the role that peptides and neuropeptidases play in the progression of SDAT.

ACKNOWLEDGMENTS

The authors thank the Medical Research Council Brain Bank (Addenbrooke's Hospital, Cambridge, U.K.), the Kathleen Price Bryan Brain Bank (Duke University, Durham, NC), and Dr. Peter Davies at the Alzheimer's Disease Brain Bank (Albert Einstein College of Medicine, Bronx, NY) which is supported by the Alzheimer's Association, for the postmortem brain tissue samples.

REFERENCES

1. PRICE, D. L., E. H. KOO & A. UNTERBACK. 1989. Cellular and molecular biology of Alzheimer's disease. Bioessays 10: 69–74.
2. TERRY, R. D. & P. DAVIES. 1980. Dementia of the Alzheimer's type. Annu. Rev. Neurosci. 3: 77–95.
3. DAVIES, P. 1979. Neurotransmitter related enzymes in senile dementia of the Alzheimer's type. Brain Res. 171: 319–327.
4. ROSSOR, M. N., N. J. GARRET, A. L. JOHNSON, C. Q. MOUNTJOY, M. ROTH & L. L. IVERSON. 1982. A post-mortem study of the cholinergic and GABA systems in senile dementia. Brain 105: 313–330.
5. PERRY, E. K. 1986. The cholinergic hypothesis: Ten years on. Br. Med. Bull. 42: 63–69.
6. FERRIER, I. N., A. J. CROSS, J. A. JOHNSON, B. W. ROBERTS, T. J. CROW, S. A. N. CORSELLIS, Y. C. LEE, D. O'SHAUGHNESSY, T. E. ADRIAN, G. P. MCGREGOR, A. J. BARACESE-HAMILTON & S. R. BLOOM. 1983. Neuropeptides in Alzheimer's type dementia. J. Neurol. Sci. 62: 159–170.
7. JOLKKONEN, J., P. HARTIKAINEN, R. SOIKKELI, G. BISSETTE, C. B. NEMEROFF & P. RIEKKINEN. 1991. A correlation study of CSF neuropeptides in Alzheimer's and Parkinson's disease. Neuropeptides 19: 97–102.
8. GABRIEL, S. M., L. M. BIERER, V. HAROTUNIAN, D. P. PUROHIT, D. P. PERL & K. L. DAVIS. 1993. Widespread deficits in somatostatin but not neuropeptide Y concentrations in Alzheimer's disease cerebral cortex. Neurosci. Lett. 155: 116–120.
9. CRYSTAL, H. A. & P. DAVIES. 1982. Cortical substance P-like immunoreactivity in cases of Alzheimer's disease and senile dementia of the Alzheimer's type. J. Neurochem. 38: 1781–1784.
10. QUIGLEY, B. J. & N. W. KOWALL. 1991. Substance P-like immunoreactive neurons are depleted in Alzheimer's disease cerebral cortex. Neuroscience 41: 41–60.
11. DAVIES, P., R. KATZMAN & R. D. TERRY. 1980. Reduced somatostatin-like immunoreactivity in cerebral cortex from cases of Alzheimer's disease and Alzheimer's senile dementia. Nature 288: 279–280.
12. BEAL, M. F., R. BENOIT, M. F. MAZUREK, E. D. BIRD & J. B. MARTIN. 1986. Somatostatin-28 (1-12)-like immunoreactivity is reduced in Alzheimer's disease cerebral cortex. Brain Res. 368: 380–383.
13. HUSAIN, M. M. & C. B. NEMEROFF. 1990. Neuropeptides and Alzheimer's disease. J. Am. Geriatr. Soc. 38: 918–925.
14. YANKNER, B. A., L. K. DUFFY & D. A. KIRSCHNER. 1990. Neurotrophic and neurotoxic effects of amyloid β protein: Reversal by tachykinin neuropeptides. Science 250: 279–282.
15. KOWALL, N. W., M. F. BEAL, J. BUSCIGLIO, L. K. DUFFY & B. A. YANKNER. 1991. An in vivo model for the neurodegenerative effects of β amyloid and protection by substance P. Proc. Natl. Acad. Sci. USA 88: 7247–7251.
16. MCKELVY, J. F. & S. BLUMBERG. 1986. Inactivation and metabolism of neuropeptides. Annu. Rev. Neurosci. 9: 415–434.
17. KONKOY, C. S., S. M. WATERS & T. P. DAVIS. 1994. Acute administration of neuroleptics decreases neurotensin metabolism on intact, regional rat brain slices. J. Pharmacol. Exp. Ther. 269: 555–563.

18. WATERS, S. M., C. S. KONKOY & T. P. DAVIS. 1995. Neuropeptide metabolism on intact, regional brain slices: Effect of dopaminergic agents on substance P, cholecystokinin and met-enkephalin degradation. J. Pharmacol. Exp. Ther. **274:** 783–789.

19. KONKOY, C. S., S. M. WATERS & T. P. DAVIS. 1996. Subchronic haloperidol administration decreases aminopeptidase N activity and [Met^5]enkephalin metabolism in rat striatum and cortex. Eur. J. Pharmacol. **297:** 47–51.

20. WATERS, S. M., C. S. KONKOY & T. P. DAVIS. 1996. Haloperidol and apomorphine differentially affect neuropeptidase activity. J. Pharmacol. Exp. Ther. **277:** 113–120.

21. WEBER, S. J., R. B. LOUIS, L. TROMBLEY, G. BISSETTE, P. DAVIES & T. P. DAVIS. 1992. Metabolic half-life of somatostatin and peptidase activities are altered in Alzheimer's disease. J. Gerontol. **47:** B18–25.

22. WATERS, S. M. & T. P. DAVIS. 1995. Alterations of substance P metabolism and neuropeptidases in Alzheimer's disease. J. Gerontol. **50A:** B315–B319.

23. SCHOEMAKER, H. & T. P. DAVIS. 1984. Differential in vitro metabolism of β-endorphin in schizophrenia. Peptides **5:** 1049–1054.

24. DELFS, J., R. ROBBINS, J. L. CONNOLLY, M. DICHTER & S. REICHLIN. 1980. Somatostatin production by rat cerebral neurones in dissociated cell culture. Nature **283:** 676–677.

25. LINDBERG, I., H.-Y.T. YANG & E. COSTA. 1984. Further characterization of an enkephalin-generating enzyme from adrenal medullary chromaffin granules. J. Neurochem. **42:** 1411–1419.

26. VECSEI, L., I. BOLLOK & G. TELEGDY. 1984. Phenoxybenzamine antagonizes somatostatin-induced anti-amnesia in rats. Eur. J. Pharmacol. **99:** 325–328.

27. SCHETTINI, G., T. FLORIO, G. MAGRI, M. GRIMALDI, O. MENCCI, E. LANDOLFI & A. MANIRCO. 1988. Somatostatin and SMS 201-995 reverse the impairment of cognitive functions induced by cysteamine depletion of brain somatostatin. Eur. J. Pharmacol. **151:** 399–407.

28. DAVIS, T. P., A. J. CULLING-BERGLUND, H. SCHOEMAKER. 1986. Specific regional differences of in vitro β-endorphin metabolism in schizophrenia. Life Sci. **39:** 2601–2609.

29. ICHAI, C., N. CHEVALLIER, P. DELAERE, P. DOURNAUD, J. EPELBAUM, J.-J. HAUW, J.-P. VINCENT & F. CHECLER. 1994. Influence of region-specific alterations of neuropeptidase content on the catabolic fates of neuropeptides in Alzheimer's disease. J. Neurochem. **62:** 645–655.

30. EMSON, P. C., A. ARREGUI, V. CLEMENT-JONES, B. E. B. SANDBERG & M. ROSSOR. 1980. Regional distribution of methionine-enkephalin and substance P-like immunoreactivity in normal human brain and in Huntington's disease. Brain Res. **199:** 147–160.

31. COOPER, P. E., M. H. FERNSTROM, O. P. RORSTAD, S. E. LEEMAN & J. B. MARTIN. 1981. The regional distribution of somatostatin, substance P and neurotensin in human brain. Brain Res. **218:** 219–232.

32. DAUCH, P., Y. MASUO, J.-P. VINCENT & F. CHECLER. 1993. A survey of the cerebral regionalization and ontogeny of eight exo- and endopeptidases in murines. Peptides **14:** 593–599.

33. LEE, C.-M., B. E. B. SANDBERG, M. R. HANLEY & L. L. IVERSEN. 1981. Purification and characterization of a membrane-bound substance P-degrading enzyme from human brain. Eur. J. Biochem. **114:** 315–327.

34. MENTLEIN, R. & P. DAHMS. 1994. Endopeptidases 24.16 and 24.15 are responsible for the degradation of somatostatin, neurotensin and other neuropeptides by cultivated rat cortical astrocytes. J. Neurochem. **62:** 27–36.

35. THIELE, E. A., S. M. STRITTMATTER & S. H. SNYDER. 1985. Substance K and substance P as possible endogenous substrates of angiotensin converting enzyme in the brain. Biochem. Biophys. Res. Commun. **128:** 317–324.

36. WANG, L., S. AHMAD, I. F. BENTER, A. CHOW, S. MIZUTANI & P. WARD. 1991. Differential processing of substance P and neurokinin A by plasma dipeptidyl(amino)peptidase IV, aminopeptidase M and angiotensin converting enzyme. Peptides **12:** 1357–1364.

37. ALMENOFF, J., S. WILK & M. ORLOWSKI. 1981. Membrane bound pituitary metalloendopeptidase: Apparent identity to enkephalinase. Biochem. Biophys. Res. Commun. **102:** 206–214.

38. ACKER, G. R., C. MOLINEAUX & M. ORLOWSKI. 1987. Synaptosomal membrane-bound form of endopeptidase-24.15 generates Leu-enkephalin from dynorphin$_{1-8}$, α- and β-neoendorphin, and Met-enkephalin from Met-enkephalin-Arg^6-Gly^7-Leu^8. J. Neurochem. **48:** 284–292.

39. CHECLER, F. 1993. Neuropeptide-degrading peptidases. *In* Methods in Neurotransmitter and Neuropeptide Research. S. H. Parvez, M. Naoi, T. Nagatsu & S. Parvez, Eds.: 375–418. Elsevier Science Publishers. New York.

40. BATEMAN, R. C. & L. B. HERSH. 1987. Evidence for an essential histidine in neutral endopeptidase 24.11. Biochemistry **26:** 4237–4242.

41. ORLOWSKI, M., C. MICHAUD & T. G. CHU. 1983. A soluble metalloendopeptidase from rat brain. Eur. J. Biochem. **135:** 81–88.

42. STACK, G., L. D. FRICKER & S. H. SNYDER. 1984. A sensitive radiometric assay for enkephalin convertase and other carboxypeptidase B-like enzymes. Life Sci. **34:** 113–121.

43. ARREGUI, A., E. K. PERRY, M. ROSSOR, B. E. TOMLINSON. 1982. Angiotensin converting enzyme in Alzheimer's disease: Increased activity in caudate nucleus and cortical areas. J. Neurochem. **38:** 1490–1492.

44. GRAY, E. G. & V. P. WHITTAKER. 1962. Isolation of nerve endings from brain: An electron microscopic study of cell fragments derived by homogenization and centrifugation. J. Anat. **96:** 79–87.

Eclosion Hormone Action on the Nervous System

Intracellular Messengers and Sites of Action[a]

DAVID B. MORTON[b]

ARL Division of Neurobiology and Department of Biochemistry
The University of Arizona
Tucson, Arizona 85721

INTRODUCTION

One of the most dramatic morphological and functional transformations in biology is the metamorphosis of a caterpillar into a butterfly or moth. This process takes several weeks to complete and the final event is when the fully formed butterfly breaks free of its cuticle through a series of stereotyped movements known as eclosion or ecdysis. A similar behavioral sequence takes place at the end of each larval molt (larval ecdysis) and at the end of the larval to pupal transition (pupal ecdysis). In the early 1970s, Jim Truman and Lynn Riddiford showed that adult ecdysis was triggered by a hormone circulating in the blood,[1] which was subsequently named eclosion hormone (EH).[2] In the tobacco hornworm, *Manduca sexta*, EH has been shown to trigger ecdysis behavior at each larval molt[3] and the pupal molt,[4] as well as at adult ecdysis. Furthermore, it is thought that EH mediates ecdysis in all insects.[5] There is also evidence that in addition to triggering ecdysis behavior, EH has other physiological roles at the end of the molt such as triggering post-ecdysial muscle death and the release of other neuropeptides.[6]

Over the 15 years or so following the discovery of EH, a great deal of effort was invested in its purification and characterization. EH is a peptide hormone and it has been sequenced in *Manduca*[7-9] and the silkmoth, *Bombyx mori*,[10] and the gene for EH has been cloned in *Manduca*, *Bombyx*, and *Drosophilia melanogaster*.[11-13] EH shows a high degree of sequence conservation between these species. It is a 62 amino acid peptide with conserved cystine residues, and in *Manduca* these cystines have been shown to form three disulfide bonds that are necessary for biological activity.[8,14] In addition to the sequence conservation, there is a high degree of biological cross-reactivity across different orders of insects. This has been shown by the ability of extracts of the nervous system from a wide variety of insects to trigger premature ecdysis behavior when injected into *Manduca*.[5]

EH is found in, and released from, the central nervous system (CNS) of *Manduca*. Immunocytochemistry and *in situ* hybridization studies have localized

[a] This work was supported by the Alfred P. Sloan Foundation and National Institutes of Health Grant No. NS29740.

[b] Address correspondence to David B. Morton, ARL Division of Neurobiology, 611 Gould Simpson Building, The University of Arizona, 1040 E. 4th Street, Tucson, AZ 85721. E-mail: drdan@manduca.neurobio.arizona.edu

40

EH to two pairs of cells in the ventral midline of the brain in *Manduca*, known as the VM cells.[15,16] These cells send axons along the entire length of the ventral nervous system and project to the proctodeal nerve where they form neurohemal release sites.[16] By bioassaying the amount of EH present in the proctodeal nerve, both before and after ecdysis, it has been shown that EH is released from this site at each ecdysis.[17] During adult development, the VM cells sprout processes which project to the corpora allata/corpora cardiaca complex[17,18] to form an additional release site at adult ecdysis.[5]

Although injections of EH into the blood of *Manduca* will trigger ecdysis, there is evidence to suggest that blood-borne EH is not necessary for ecdysis. For example, when the proctodeal release sites were removed from animals prior to pupal ecdysis, they still showed ecdysis behavior at the appropriate time.[19] Blood from these animals contained no detectable EH, and dermal glands on the body wall, which appear to require EH to release their contents, showed no release when the proctodeal nerve was removed.[19] Furthermore, in intact animals a decrease in EH immunoreactivity in the VM cell axons within the CNS was seen after ecdysis, suggesting that in addition to its release into the blood, EH was released centrally where it could also trigger ecdysis.[19]

Recently, a new layer of complexity has been added to the hormonal control of ecdysis with the discovery of a new peptide hormone that can trigger ecdysis behavior.[20] This hormone, named Mas-ETH (*Manduca sexta* ecdysis-triggering hormone), is a 26 amino acid peptide that is found in a newly described endocrine system, the epitracheal glands, located on tracheal trunks near the spiracles.[20] Mas-ETH appears to act directly on the nervous system of *Manduca* to trigger ecdysis, and evidence suggests that EH acts on the epitracheal glands to trigger Mas-ETH release.[20] This, however, requires that EH acts as a blood-borne hormone and cannot explain the finding that removal of peripheral release sites for EH does not prevent ecdysis.[19] Clearly more work is required before the relationship of these two peptides to ecdysis behavior is fully understood.

MECHANISMS OF ECLOSION HORMONE ACTION

Regardless of the site of EH action, several lines of evidence indicate that EH triggers ecdysis behavior via the intracellular messenger, guanosine 3'5' cyclic monophosphate (cGMP).[21,22] The levels of cGMP in the ventral nervous system of *Manduca* are elevated after animals are injected with EH and ecdysis can be triggered by injecting animals with either phosphodiesterase inhibitors or cGMP itself.[22] In addition, levels of cGMP in the nervous system are elevated just prior to, and during, naturally occurring ecdysis.[22] Other actions of EH also appear to be mediated by cGMP, such as the post-ecdysial degeneration of certain abdominal muscles in the silk moth, *Antheraea polyphemus*.[23]

The majority of the actions of cyclic nucleotides are believed to be mediated by a stimulation of protein phosphorylation. We have shown that EH and cGMP stimulate the phosphorylation of two proteins in the nervous system of *Manduca*, named the EGPs (EH- and cGMP-regulated phosphoproteins).[24,25] Phosphorylation of the EGPs occurs as a result of EH action at larval, pupal, and adult ecdysis in *Manduca*, and phosphorylation of these proteins can also be stimulated by incubating isolated nervous tissue with EH, thus demonstrating a direct effect of EH on the CNS.[25] One particularly intriguing aspect of the EGPs is their developmental regulation. EH can only trigger ecdysis during a narrow window of time immediately preceding the normal time for ecdysis. For pupal ecdysis this period of sensitivity lasts for about 8 h.[26] Furthermore, the lack of sensitivity prior to this window of time is not due to a lack of EH receptors, because EH will still stimulate an increase in cGMP in the CNS at least 24 h before ecdysis.[22] For larval and pupal ecdysis, this sensitivity could be due to the presence of the EGPs because they are only detectable in the CNS at times when EH is capable of triggering ecdysis.[24,25] The development of the ability of EH to trigger ecdysis is regulated by the steroid hormones, the ecdysteroids,[26] and requires protein synthesis.[27] Similarly, the developmental appearance of the EGPs is also regulated by the ecdysteroids and requires protein synthesis at the same time in development.[27,28]

Recently, we have been using isolated ventral nervous systems to focus on the mechanisms by which EH stimulates an increase in cGMP in the CNS.[29] Such isolated nervous systems respond to EH with a dose-dependent increase in cGMP which has a similar time course to that seen *in vivo*.[29] The synthesis of cGMP from GTP is catalyzed by the enzyme guanylyl cyclase (GC). This enzyme exists in two isoforms, a receptor GC (rGC), which is an integral membrane protein in cells, and a cytoplasmic or soluble GC (sGC).[30] The breakdown of cGMP is catalyzed by phosphodiesterases (PDE). Hormonally stimulated increases in cGMP could be achieved by activation of either type of GC or by inhibition of PDE. Although we have been able to measure large increases in cGMP in intact nervous tissue in response to EH, when nervous tissue is homogenized and incubated with EH, no increase in cGMP is seen.[29] In addition, the EH-stimulated increase in cGMP in intact nervous tissue requires extracellular calcium and is potentiated by the PDE inhibitor isobutyl methylxanthine.[29] These characteristics indicate that EH stimulates an sGC.[31]

We have also carried out experiments to determine how EH stimulates the sGC in the CNS of *Manduca*. The most commonly accepted mechanism for activation of an sGC is through production of the gaseous messenger, nitric oxide (NO).[32] NO is generated by the calcium-dependent activation of NO synthase (NOS), which converts arginine to citrulline with the stoichiometric production of NO.[32] A variety of inhibitors, such as arginine analogues, have been shown to block NOS in vertebrates and hence prevent agonist-stimulated increases in cGMP. We have tried a variety of these inhibitors and have shown that they have no effect on the EH-stimulated increase in cGMP in isolated *Manduca* nervous tissue.[29,33] In addition, we have shown that by monitoring the conversion of [³H]arginine to [³H]citrulline, EH does not stimulate NOS.[29,33] Although all the evidence that we have gathered suggests that EH does not increase cGMP levels through the production of NO in the CNS of *Manduca*,[29,33] experiments with *Bombyx mori* tissue indicate that NO might be involved in the EH-stimulated increase in cGMP in that insect.[34] The reason for this discrepancy is not clear because a high degree of sequence similarity exists between EH from *Manduca* and *Bombyx*,[11] and *Bombyx* EH can trigger ecdysis in *Manduca*[35] suggesting that the receptors from both species are similar. There are, however, a number of differences between these studies. The time course of the increase in cGMP was more rapid in *Bombyx*[36] than in *Manduca*, and the experiments in *Bombyx* used nervous tissue solubilized with detergents before EH was applied. When these detergents were used in *Manduca*, the tissue lost hormonal responsiveness.[33]

We have considered a number of other possibilities for the activation of the sGC in *Manduca*. Another gaseous messenger, carbon monoxide (CO), has also been proposed to activate sGC in mammalian tissues.[37] CO is believed to be generated by the action of heme-oxygenase, which can be blocked by zinc protoporphyrine-IX; we have also shown that this inhibitor has no effect on the EH-stimulated increase in cGMP in *Manduca*.[33] In addition, before the discovery of NO as an endogenous activator of sGC, it was generally believed that free fatty acids or their lipoxygenase metabolites were intermediates in the activation of sGC.[31] We have some evidence to support the idea that this might be the pathway for EH activation of sGC in *Manduca*. A wide variety of lipase inhibitors are effective in blocking the EH-stimulated cGMP increase in the intact nervous system, and arachidonic acid will stimulate the sGC in extracts of *Manduca* CNS.[29,33] The inhibitors used include blockers of phopholipase C (PLC), diacylglycerol (DAG) lipase, phospholipase A_2, and lipoxygenase, supporting the idea that a lipid messenger, possibly a free fatty acid, is involved in the EH-stimulated cGMP increase.[33] A common mechanism for generating a lipid messenger is the activation of PLC, which generates inositol (1,4,5) trisphosphate ($InsP_3$), and DAG. In both *Manduca* and *Bombyx*, EH stimulates the production of $InsP_3$ with a time course consistent with its production preceding that of cGMP.[33,36]

Although we believe that the first step in the cascade of EH action is the activation of PLC, the next step in the cascade and the identity of the direct activator of the GC are still unknown. The ability of a wide variety of lipase inhibitors to block the EH-stimulated increase in cGMP suggests that EH stimulates the production of a lipid messenger. At the present time, however, we have not been able to detect effects of EH on the levels of any lipid messengers in the CNS.[38] Although we can measure a small but significant increase in $InsP_3$ in response to EH,[33] we have not been able to detect a concomitant increase in DAG.[38] Similarly, when nervous tissue was prelabeled with [^{3}H]arachidonic acid and then stimulated with EH, we did not detect release of any labeled compounds.[38] In addition, incubating the intact nervous system with a wide variety of fatty acids or DAGs failed to stimulate an increase in cGMP[33] although arachidonic acid is a good stimulator of sGC in extracts made from *Manduca* CNS.[29]

Another possible mediator for the EH activation of sGC is calcium. Experiments with isolated nervous tissue showed that extracellular calcium was required for the EH-stimulated increase in cGMP.[29] The addition of calcium ionophores stimulated an increase in cGMP in intact nervous tissue, although this increase was significantly smaller than that seen with the addition of EH.[33] It is possible that calcium could either act alone, or in combination with another messenger, to directly stimulate the sGC. Preliminary experiments with CNS homogenates showed that concentrations of calcium above 1 μM moderately stimulated GC activity.[38] We have not, however, been able to potentiate the calcium ionophore-stimulated increase in cGMP with either free fatty acids or other potential lipid messengers.[33] Another possibility is that calcium does not activate the sGC directly, but rather stimulates protein kinase C (PKC), which has been shown to phosphorylate and activate sGC.[39] We have also shown that the PKC inhibitor tamoxifen inhibited the EH-stimulated increase in cGMP by 71 $\pm$ 5%[38] suggesting that PKC may also play a role in the elevation of cGMP. These potential pathways do not, however, explain the requirement for extracellular calcium unless the intracellular stores are rapidly depleted and require refilling from outside the CNS.

Putting all these findings together has allowed us to generate a model for the action of EH, although this model still contains unanswered questions (FIG. 1). We believe that EH acts through cell surface receptors. Although we have no information on the nature of these receptors, we do know that they act via a pertussis toxin-insensitive mechanism.[33] Activation of the receptor by EH leads to the stimulation of PLC, which generates $InsP_3$ and, we assume, DAG. At this point at least three possibilities exist for the activation of the sGC in the CNS.

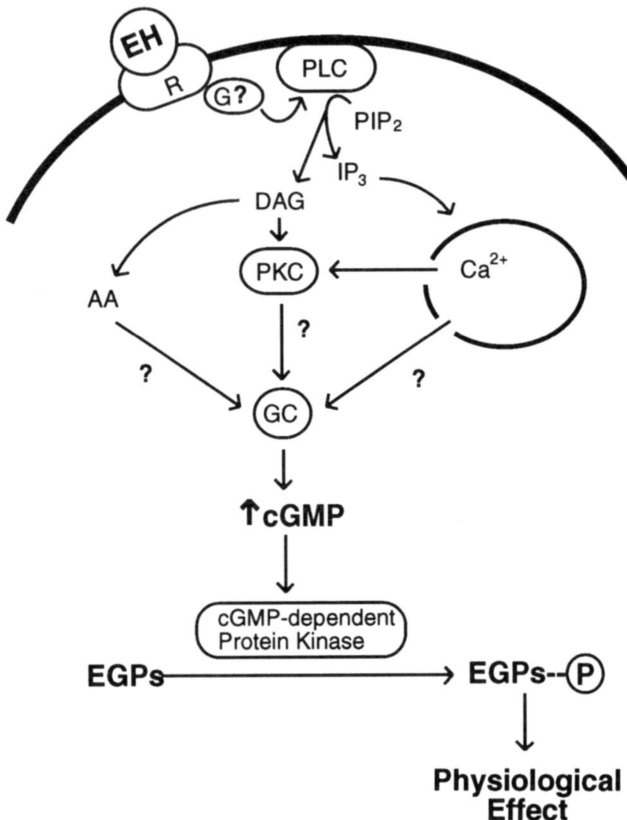

FIGURE 1. Model for the action of EH on the CNS of *Manduca sexta*. EH acts on cell-surface receptors and its initial action is to activate a PLC (phospholipase C), although it is unknown whether this is via a G protein–coupled pathway. Activation of the PLC leads to an increase in InsP$_3$ [inositol (1,4,5) trisphosphate] and a predicted increase in DAG (diacylglycerol). This then leads to one of, or a combination of, the following steps. The InsP$_3$ should stimulate the release of intracellular calcium which could directly stimulate a GC (guanylyl cyclase). The calcium, in combination with DAG, could activate a PKC (protein kinase C), which could activate a GC via phosphorylation. An alternative pathway would involve the generation of free fatty acids such as AA (arachidonic acid), which would either directly or after further metabolism activate a GC. The activation of the GC leads to an increase in cGMP within the target cells leading to an activation of cGMP-dependent protein kinase which stimulates the phosphorylation of the EGPs (EH and cGMP-regulated phospho-proteins). The phosphorylated EGPs then mediate a change in the physiological properties of the target cells in an unknown manner.

(1) The DAG could be hydrolyzed by DAG lipase to produce a free fatty acid, such as arachidonic acid (AA), which either directly or indirectly stimulates the GC. (2) The InsP$_3$ generated by the activation of PLC releases calcium from intracellular stores which directly activates the GC. (3) The increase in intracellular calcium acts in concert with DAG to activate PKC and phosphorylation of GC by PKC, which leads to an increase in cGMP. Regardless of the mechanism of cGMP increase, we believe that it results in the activation of cGMP-dependent protein kinase, which phosphorylates the EGPs.

Until recently, we had thought that the phosphorylated EGPs would then lead to the activation of the ecdysis central pattern generator resulting in ecdysis behavior. In light of the recent identification of Mas-ETH and the epitracheal glands, which appear to be directly responsible for triggering ecdysis,[20] it would appear that EH does not act directly on the CNS to trigger ecdysis. EH does, however, act directly on the CNS to increase cGMP levels and phosphorylate the EGPs. The physiological outcome of these biochemical changes can only be speculative at this point, but might involve the regulation of neuropeptide release from neurohemal sites in the CNS (see below).

SITES OF ECLOSION HORMONE ACTION

The original isolation of EH was as a neuropeptide that triggered a stereotyped behavioral motor program and thus led to the hypothesis that EH acted directly on sites in the CNS to activate a central pattern generator.[2] When EH was applied to isolated abdominal nervous systems of the silkmoth, *Hyalophora cecropia*, electrical activity recorded from the nerve roots had a similar pattern to normal ecdysis behavior.[40] Interestingly, these preparations only responded to EH if the tracheal system was left attached to the nervous system and was ventilated by forcing air through the tracheae. This observation was interpreted to mean that the nervous system was the direct target for EH, but that it was sensitive to good aeration.[40] To achieve similar ecdysis motor patterns in *Manduca*, it was found to be necessary to inject intact animals with EH before nervous system isolation. This requirement was interpreted to mean that targets for EH action were in the anterior part of the CNS, the brain or subesophageal ganglion,[41] but it was still thought that the CNS was the target for EH to trigger ecdysis behavior.

This concept has now been called into question with the finding of a new neuropeptide, Mas-ETH, which is capable of triggering ecdysis behavior.[20] Mas-ETH is found in the epitracheal glands situated on tracheae in *Manduca*, and this location might explain the need for the presence of tracheae for EH to trigger ecdysis in *Hyalophora* if EH is stimulating release of Mas-ETH.[20] Isolated abdominal ganglia from *Manduca* show the ecdysis motor program when exposed to Mas-ETH in the absence of the tracheal system, whereas EH (corpora cardiaca/corpora allata complex extracts) has no effect.[20] Interestingly, EH is capable of triggering the ecdysis motor program from isolated abdominal ganglia if the nervous tissue is incubated in the presence of newly dissected epitracheal glands, which contain Mas-ETH.[20] These findings suggest that the CNS is not a direct target for EH, at least in triggering ecdysis, but that EH acts on the epitracheal glands as a releasing hormone to stimulate the release of Mas-ETH.[20]

Evidence exists, however, that EH does indeed have a direct effect on the

CNS. EH will act on isolated abdominal nervous systems of *Manduca* to increase cGMP levels[29,33] and to stimulate the phosphorylation of the EPGs.[25] In these experiments, although abdominal nervous systems were removed along with substantial tracheal branches in an effort to keep the tissues as healthy as possible, insufficient lengths of tracheae were left attached to have inadvertently included the epitracheal gland. The major tracheal branch leaving the abdominal ganglia extends laterally from the CNS and passes under the intersegmental muscles (ISMs) before re-emerging and joining the spiracle. The epitracheal glands are located on this ventral tracheal branch distal to the ISMs.[20] In the experiments investigating the action of EH on isolated ventral nervous systems,[25,29,33] the tracheae were cut proximal to the ISMs; thus the observed effects cannot be attributed to the stimulated release and action of Mas-ETH.

To understand the physiological effects of EH on the CNS we need to know where the increase in cGMP takes place. This has now become possible with the development of an antibody that recognizes cGMP in paraformaldehyde-fixed tissue.[42] When animals were injected with EH prior to larval ecdysis and the nervous tissue removed, fixed, and stained with this antibody, a population of 50 neurons were revealed which are distributed throughout the CNS.[42] The increase in cGMP in these neurons peaked before ecdysis, at about 30 min after injection, and in some cells remained elevated for a few hours after ecdysis.[42] The identity of some of the cells which showed the EH-stimulated increase in cGMP is known: they consist of five pairs of cells in the subesophageal ganglion and two pairs of cells in the thoracic and abdominal ganglia. The neurons in the abdominal ganglia are also immunoreactive for crustacean cardioactive peptide (CCAP) and can be identified as a neurosecretory neuron, cell 27 or NS-L$_1$,[43] and an interneuron, cell 704.[42]

Because the increase in cGMP in these cells was triggered by injecting animals with EH, it is not clear whether their activation was due to the action of EH or Mas-ETH on the CNS. It is known, however, that the effect was not due to the direct action of EH on the cGMP-immunoreactive (cGMP-IR) cells themselves. If the connectives between the first and second abdominal ganglion were cut before EH injection, only those cells anterior to the transection showed an increase in cGMP.[42] This suggests that EH or Mas-ETH acts on sites in the anterior portion of the CNS (possibly brain or subesophageal ganglion) and activates descending neurons that synaptically drive the cGMP-IR neurons.[42] The identity of these anterior targets of EH or Mas-ETH is, as yet, unknown.

Since no cGMP-IR cells were detected when the CNS was transected between the thoracic and abdominal ganglia, it is an apparent paradox that isolated abdominal nervous tissue showed a large and robust increase in cGMP when incubated with EH.[29,33] In an effort to resolve this issue, I have used the same anti-cGMP antiserum to stain isolated abdominal nervous systems after exposure to EH.[44] Such nervous tissue removed from animals prior to pupal ecdysis, exposed to EH, and then stained for cGMP showed dense immunoreactivity in a network of fibers in the transverse nerve of each abdominal ganglion.[44] An example of this staining is shown in FIGURE 2. The transverse nerve of insects is a well-known neurohemal organ which contains the terminals of a variety of neurosecretory neurons, motor axons, and glial cells.[45] In an effort to get better resolution of the immunoreactivity, these preparations were viewed with a confocal microscope. This revealed that the immunoreactivity was primarily located at the surface of the transverse nerve and was not present in glial cells. This is consistent with the hypothesis that EH stimulates an increase in cGMP in the neurosecretory terminals of the transverse nerve.[44]

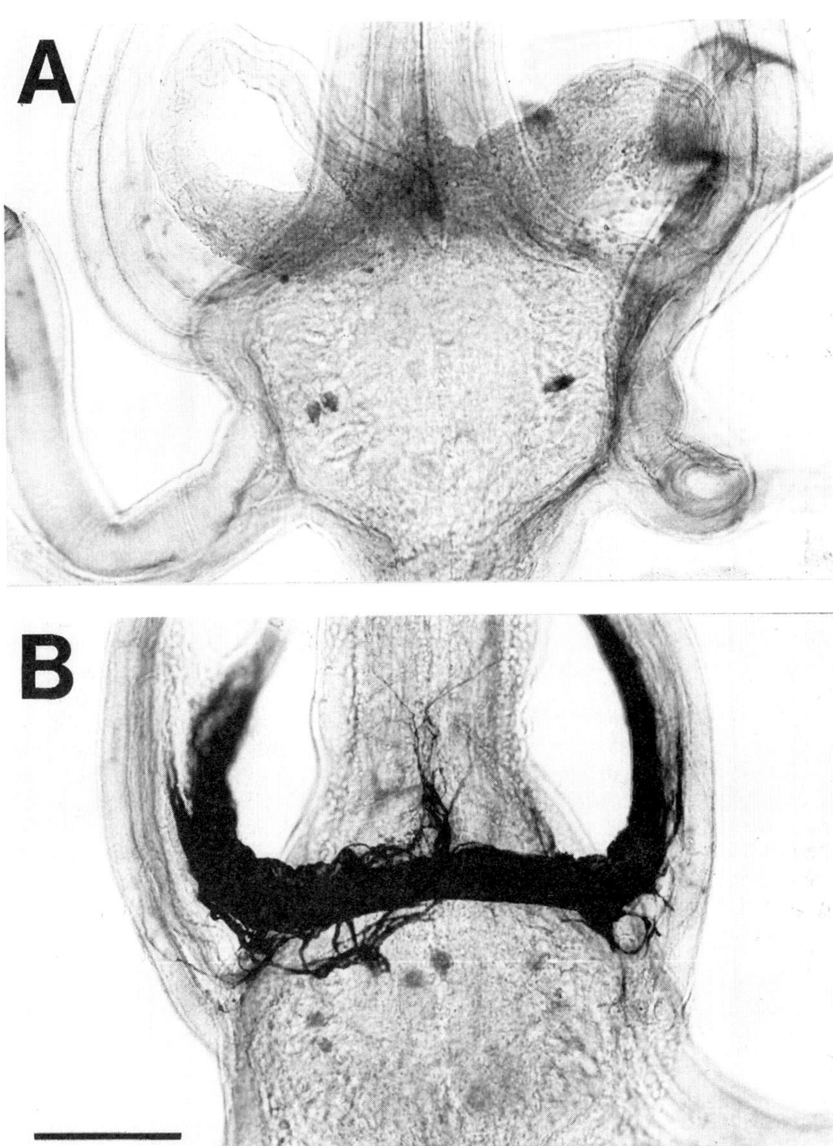

FIGURE 2. Example of the EH-stimulated increase in cGMP immunoreactivity in the transverse nerve of an abdominal ganglion of *Manduca sexta*. (**A**) Abdominal ganglion incubated in the absence of EH and then stained with a cGMP antibody. (**B**) Abdominal ganglion incubated with 1 nM EH for 15 min and stained with a cGMP antibody. Scale bar = 200 μM.

A number of observations and manipulations suggest that it is the immunoreactivity in the transverse nerve that accounts for most, if not all, of the increases in cGMP we had previously measured biochemically.[29,33] Removal of the transverse nerve prior to treatment with EH significantly reduced the EH-stimulated increase in cGMP.[44] The time course and sensitivity to EH of the biochemically measured increase in cGMP and the immunocytochemically measured increase in cGMP were similar, although the biochemical measurements were more sensitive.[44] In addition, the EH-stimulated cGMP-immunoreactivity in the transverse nerve was blocked by 4-bromophenacyl bromide, one of the same lipase inhibitors that blocked the biochemically measured EH-stimulated increase in cGMP.[44]

Although this finding explains why isolated abdominal nervous systems showed an increase in cGMP, it does not explain why cGMP-immunoreactivity in the transverse nerve was not detected previously.[42] This earlier study[42] used animals just prior to larval ecdysis whereas the transverse nerve study used animals prior to pupal ecdysis.[44] When isolated abdominal nervous systems from animals prior to larval ecdysis were exposed to EH, no cGMP-immunoreactivity in the transverse nerve was stimulated, and this was also true of animals just prior to adult ecdysis.[44] Thus, the EH-stimulated increase in cGMP in the transverse nerve is stage-specific, being associated only with pupal ecdysis.[44]

It was also important to determine whether the increase in cGMP in the transverse nerve occurred in the intact animals. The EH-stimulated increase in cGMP measured in isolated nervous systems showed a maximum response when exposed to about 1 nM EH.[29] When animals were injected with 5 pmoles EH, an amount calculated to give a maximum concentration of 2 nM, cGMP-immunoreactivity was seen in the transverse nerve 15, 30, and 45 min after injection.[44] Ecdysis occurred about 60 min after injection and at this time immunoreactivity was still seen in the transverse nerve, although it was considerably weaker.[44] The amount of EH used in these experiments, however, is considerably more than is required to trigger ecdysis, which at pupal ecdysis shows a half-maximal response of 10 fmoles, giving a calculated concentration of 3.5 pM.[7] Another discrepancy is that when the nervous system was removed from animals undergoing natural ecdysis and stained for cGMP, no cGMP-immunoreactivity was seen in the transverse nerve.[44] The amount of EH that is released prior to pupal ecdysis can be calculated by measuring the amount of EH present in the proctodeal nerve before and after ecdysis.[17] This has been estimated to be about 0.75 pmoles, which gives a calculated maximal concentration of 288 pM (see ref. 44 for calculations). This concentration is at or below the level necessary to detect EH-stimulated cGMP-immunoreactivity in the transverse nerve,[44] but it is at about the EC_{50} value for a biochemically measured EH-stimulated increase in cGMP.[29] Thus, it would appear that more EH is released into the hemolymph than is needed to trigger ecdysis, but there should be enough EH released to stimulate a significant increase in cGMP in the transverse nerve.

At this time, the physiological effect of an increase in cGMP in the neurosecretory endings of the transverse nerve is unknown. One possible function, however, would be to modulate the release of neuropeptides from these neurosecretory neurons. There are a number of examples in the mammalian nervous system where increases in cGMP in neurosecretory terminals have been shown to modulate release of neuropeptides.[46,47] In addition, the idea that EH stimulates the release of Mas-ETH from the epitracheal glands[20] also supports the notion that EH could act as a releasing hormone. It has long been thought that EH coordinates a variety of physiological and behavioral processes that occur at ecdysis.[6] One way it could achieve this coordination is by regulating the release of a wide variety of different peptides, each of which would then have specific roles to play.

CONCLUDING REMARKS

The increase in cGMP in the transverse nerve has also provided us with further information on the mechanisms underlying cGMP regulation. Earlier I summarized information which led us to believe that EH increases cGMP levels by activating an sGC in an NO-independent manner. All known sGCs are sensitive to NO; therefore it should be possible to bypass the EH receptor and increase cGMP in the transverse nerve by incubating nervous systems with NO. Indeed, when intact abdominal nervous systems were incubated with sodium nitroprusside (SNP), an NO donor, a six- to sevenfold increase in cGMP was measured.[29] When abdominal nervous systems were incubated in SNP and then stained for cGMP-immunoreactivity, however, the transverse nerve did not show an increase in cGMP.[44] What was revealed instead was a large number of neuronal cell bodies and central processes containing cGMP immunoreactivity, but no staining was seen in the transverse nerve.[44] Interestingly, some of these NO-sensitive neurons appeared to show cGMP-immunoreactivity in a segment-specific manner, with some cells stained only in one particular abdominal ganglion.[44] This finding seems to make the mechanism by which EH stimulates an increase in cGMP even more unclear, yet more interesting. The two most likely possibilities are that either EH activates an NO-insensitive sGC or that it activates an rGC which is inhibited by a variety of lipase inhibitors and requires extracellular calcium to be present as well as some other unknown factor for activation which is lost, or diluted, when the nervous system is homogenized. This latter requirement is needed to explain why we could not detect any EH-stimulated GC activity in CNS homogenates.[29] Neither of these two possibilities has any precedent in the literature, and the full elucidation of the pathway of EH-stimulated cGMP increases should yield novel information about cGMP regulation.

One way to try to understand this signaling system is to clone the different types of GCs present in *Manduca* CNS, determine where they are expressed, and then examine the biochemical characteristics of those that are expressed in cells which project to the transverse nerve. We have started such efforts and at this time have cloned PCR fragments of eight different cyclases which appear to fall into two classes. Five of the cyclases show highest sequence similarity to rCGs, whereas the other three appear to be sGCs. Of the sGCs, two show the highest similarity to β subunits and the third to α subunits.[48] Interestingly, the 50 neurons that show EH-stimulated cGMP-immunoreactivity in intact animals are also NO-insensitive suggesting that the increase in cGMP is mediated by an rGC.[42] If we can identify the rGC expressed in these cells, we might be able to identify its ligand and then the location and nature of the descending inputs postulated to activate these cells.

EH was first identified 25 years ago and has provided a fertile ground for physiological, biochemical, and behavioral studies. Each new line of investigation seems to turn up surprises and open new avenues for research. This review has focused on the mechanisms of cGMP increases stimulated by EH and on the transverse nerve as a newly identified target for EH action. Both of these systems have, true to form, yielded unexpected results and novel aspects of neuropeptide signaling.

ACKNOWLEDGMENTS

I wish to thank Sharon Hesterlee, Alan Nighorn, David Rivers, and Jeanette Simpson for reading this manuscript and many helpful comments on my use of the English language. I also wish to thank Sharon Hesterlee for providing FIGURE 1, and I am indebted to Dr. Dan Vimoto for providing the inspiration for much of this work.

REFERENCES

1. TRUMAN, J. W. & L. M. RIDDIFORD. 1970. Neuroendocrine control of ecdysis in silkmoths. Science **167**: 1624–1626.
2. TRUMAN, J. W. 1971. Physiology of insect ecdysis. I. The eclosion behavior of saturniid moths and its hormonal release. J. Exp. Biol. **54**: 805–814.
3. COPENHAVER, P. F. & J. W. TRUMAN. 1982. The role of eclosion hormone in the larval ecdyses of *Manduca sexta*. J. Insect Physiol. **28**: 695–701.
4. TRUMAN, J. W., P. H. TAGHERT & S. E. REYNOLDS. 1980. Physiology of pupal ecdysis in the tobacco hornworm, *Manduca sexta*. I. Evidence for control by eclosion hormone. J. Exp. Biol. **88**: 327–337.
5. TRUMAN, J. W., P. H. TAGHERT, P. F. COPENHAVER, N. J. TUBLITZ & L. M. SCHWARTZ. 1981. Eclosion hormone may control all ecdysis in insects. Nature **291**: 70–71.
6. TRUMAN, J. W. 1992. The eclosion hormone system of insects. *In* Progress in Brain Research. J. Joose, R. M. Buijs & F. J. H. Tilders, Eds. Vol. 92: 361–374. Elsevier. Amsterdam.
7. TERZI, G., J. W. TRUMAN & S. E. REYNOLDS. 1988. Purification and characterization of eclosion hormone from the moth, *Manduca sexta*. Insect Biochem. **18**: 701–707.
8. MARTI, T., K. TAKIO, K. A. WALSH, G. TERZI & J. W. TRUMAN. 1987. Microanalysis of the amino acid sequence of the eclosion hormone from the tobacco hornworm, *Manduca sexta*. FEBS Lett. **219**: 415–418.
9. KATAOKA, H., R. G. TROESTSCHLER, S. J. KRAMER, B. J. CESARIN & D. A. SCHOOLEY. 1987. Isolation and primary structure of the eclosion hormone from the tobacco hornworm, *Manduca sexta*. Biochem. Biophys. Res. Commun. **146**: 746–750.
10. KONO, T., H. NAGASAWA, A. ISOGAI, H. FUGO & A. SUZUKI. 1991. Isolation and complete amino acid sequences of eclosion hormones of the silkworm, *Bombyx mori*. Insect Biochem. **21**: 185–195.
11. HORODYSKI, F. M., L. M. RIDDIFORD & J. W. TRUMAN. 1989. Isolation and expression of the eclosion hormone gene from the tobacco hornworm, *Manduca sexta*. Proc. Natl. Acad. Sci. USA **86**: 8123–8127.
12. KAMITO, T., H. TANAKA, B. SATO, H. NAGASAWA & A. SUZUKI. 1992. Nucleotide sequence of cDNA for the eclosion hormone of the silkworm *Bombyx mori*, and the expression in a brain. Biochem. Biophys. Res. Commun. **182**: 514–519.
13. HORODYSKI, F. M., J. EWER, L. M. RIDDIFORD & J. W. TRUMAN. 1993. Isolation, characterization and expression of the ecclosion gene of *Drosophilia melanogaster*. Eur. J. Biochem. **215**: 221–228.
14. KATAOKA, H., J. P. LI, A. S. T. LUI, S. J. KRAMER & D. A. SCHOOLEY. 1992. Complete structure of eclosion hormone of *Manduca sexta*. Assignment of disulfide bond location. Int. J. Peptide Res. **39**: 29–35.
15. COPENHAVER, P. F. & J. W. TRUMAN. 1986. Identification of the cerebral neurosecretory cells that produce eclosion hormone in the moth *Manduca sexta*. J. Neurosci. **6**: 1738–1747.
16. TRUMAN, J. W. & P. F. COPENHAVER. 1989. The larval eclosion hormone neurones in *Manduca sexta*: Identification of the brain-proctodeal neurosecretory system. J. Exp. Biol. **147**: 457–470.

17. RIDDIFORD, L. M., R. S. HEWES & J. W. TRUMAN. 1994. Dynamics and metamorphosis of an identifiable peptidergic neuron in an insect. J. Neurobiol. **25:** 819–830.
18. ICHIKAWA, T. 1992. Growth of axon collaterals of eclosion hormone neurons into a new release site during metamorphis of *Bombyx mori*. Neurosci. Lett. **138:** 14–18.
19. HEWES, R. S. & J. W. TRUMAN. 1991. The roles of central and peripheral eclosion hormone release in the control of ecdysis behavior in *Manduca sexta*. J. Comp. Physiol. **168A:** 697–707.
20. ZITNAN, D., T. G. KINGAN, J. L. HERMESAN & M. E. ADAMS. 1996. Identification of ecdysis-triggering hormone from an epitracheal endocrine system. Science **271:** 88–91.
21. TRUMAN, J. W., S. M. MUMBY & S. K. WELCH. 1979. Involvement of cyclic GMP in the release of stereotyped behavior patterns in moths by a peptide hormone. J. Exp. Biol. **84:** 201–212.
22. MORTON, D. B. & J. W. TRUMAN. 1985. Steroid regulation of the peptide-mediated increase in cyclic GMP in the nervous system of the hawkmoth, *Manduca sexta*. J. Comp. Physiol. **157A:** 423–432.
23. SCHWARTZ, L. M. & J. W. TRUMAN. 1984. Cyclic GMP may serve as a second messenger in peptide-induced muscle degeneration in an insect. Proc. Natl. Acad. Sci. USA **81:** 6718–6722.
24. MORTON, D. B. & J. W. TRUMAN. 1986. Substrate phosphoprotein availability regulates eclosion hormone sensitivity in an insect CNS. Nature **323:** 264–266.
25. MORTON, D. B. & J. W. TRUMAN. 1988. The EGPs—the eclosion hormone and cyclic GMP regulated phosphoproteins. I. Appearance and partial characterization in the CNS of *Manduca sexta*. J. Neurosci. **8:** 1326–1337.
26. TRUMAN, J. W., D. B. ROUNTREE, S. E. REISS & L. M. SCHWARTZ. 1983. Ecdysteroids regulate the release and action of eclosion hormone in the tobacco hornworm, *Manduca sexta* (L). J. Insect Physiol. **29:** 895–900.
27. MORTON, D. B. & J. W. TRUMAN. 1995. Effect of cycloheximide on eclosion hormone sensitivity and the developmental appearance of the eclosion hormone and cGMP regulated phosphoproteins in the CNS of the tobacco hornworm, *Manduca sexta*. J. Recept. Cell. Signal Transduction **15:** 773–786.
28. MORTON, D. B. & J. W. TRUMAN. 1988. The EGPs—the eclosion hormone and cyclic GMP regulated phosphoproteins. II. Regulation of appearance by the steroid 20-hydroxyecdysone in *Manduca sexta*. J. Neurosci. **8:** 1338–1345.
29. MORTON, D. B. & M. A. GIUNTA. 1992. Eclosion hormone stimulates cGMP levels in *Manduca sexta* nervous tissue via arachidonic acid metabolism with little or no contribution from the production of nitric oxide. J. Neurochem. **59:** 1522–1530.
30. SIEGEL, G. J., B. W. AGRANOFF, R. W. ALBERS & P. B. MOLINOFF. 1994. Basic Neurochemistry. 5th edit. Raven Press. New York.
31. TREMBLAY, J., R. GERZER & P. HAMET. 1988. Cyclic GMP in cell function. Adv. Second Messenger Phosphoprotein Res. **22:** 319–383.
32. SCHMIDT, H. H. H. W. & U. WALTER. 1994. NO at work. Cell **78:** 919–925.
33. MORTON, D. B. & P. J. SIMPSON. 1995. Eclosion hormone-stimulated cGMP levels in the central nervous system of *Manduca sexta*: Inhibition by lipid metabolism blockers, increase in inositol (1,4,5) trisphosphate and further evidence against the involvement of nitric oxide. J. Comp. Physiol. **165B:** 417–427.
34. SHIBANAKA, Y., H. HAYASHI, I. UMEMURA, Y. FUJISAWA, M. OKAMOTO, M. TAKAI & N. FUJITA. 1994. Eclosion hormone-mediated signal transduction in the silkworm abdominal ganglia: Involvement of a cascade inositol (1,4,5) trisphosphate to cyclic GMP. Biochem. Biophys. Res. Commun. **198:** 613–618.
35. TRUMAN, J. W., personal communication.
36. SHIBANAKA, Y., H. HAYASHI, M. TAKAI & N. FUJITA. 1993. Eclosion hormone activates phosphatidylinositol hydrolysis in silkworm abdominal ganglia during adult metamorphosis. Eur. J. Biochem. **211:** 427–430.
37. VERMA, A., D. J. HIRSCH, C. E. GLATT, G. V. RONNETT & S. H. SNYDER. 1993. Carbon monoxide: A putative neural messenger. Science **259:** 381–384.
38. SIMPSON, P. J. & D. B. MORTON, unpublished data.

39. LOUIS, J. C., M. O. REVEL & J. ZWILLER. 1993. Activation of soluble guanylate cyclase through phosphorylation by protein kinase C in intact PC12 cells. Biochim. Biophys. Acta **1177:** 299–306.

40. TRUMAN, J. W. 1978. Hormonal release of stereotyped motor programmmes from the isolated nervous system of the cecropia silkmoth. J. Exp. Biol. **74:** 151–173.

41. WEEKS, J. C. & J. W. TRUMAN. 1984. Neural organization of peptide-activated ecdysis behaviors during the metamorphosis of *Manduca sexta*. I. Conservation of the peristalsis motor pattern at the larval-pupal transformation. J. Comp. Physiol. **155A:** 407–422.

42. EWER, J., J. DE VENTE & J. W. TRUMAN. 1994. Neuropeptide induction of cyclic GMP increase in the insect CNS: Resolution at the level of single identifiable neurons. J. Neurosci. **14:** 7704–7712.

43. DAVIS, N. T., U. HOMBERG, H. DIRKSON, R. B. LEVINE & J. G. HILDEBRAND. 1993. Crustacean cardioactive peptide-immunoreactive neurons in the hawkmoth *Manduca sexta* and changes in their immunoreactivity during postembryonic development. J. Comp. Neurol. **338:** 612–627.

44. MORTON, D. B. 1996. Neuropeptide-stimulated cGMP immunoreactivity in the neurosecretory terminals of a neurohemal organ. J. Neurobiol. **29:** 341–353.

45. TAGHERT, P. H. & J. W. TRUMAN. 1982. Identification of the bursicon-containing neurones in abdominal ganglia of the tobacco hornworm, *Manducta sexta*. J. Exp. Biol. **98:** 385–401.

46. AGUILA, M. C. 1994. Growth-hormone releasing factor increases somatostatin release and mRNA levels in the rat periventricular nucleus via nitric oxide by activation of guanylate cyclase. Proc. Natl. Acad. Sci. USA **91:** 782–786.

47. DUVILANSKI, B. H., C. ZAMBRUNO, A. SEILICOVICH, D. PISERA, M. LASAGA, M. DEL C. DIAZ, N. BELOVA, V. RETTORI & S. M. MCCANN. 1995. Role of nitric oxide in control of prolactin release by the adenohypohysis. Proc. Natl. Acad. Sci. USA **92:** 170–174.

48. NIGHORN, A., D. RIVERS & D. B. MORTON. 1995. The cloning and localization of various forms of guanylyl cyclase in *Manduca sexta*. Soc. Neurosci. Abstr. **21:** 628.

Allatostatins: Diversity in Structure and Function of an Insect Neuropeptide Family

W. G. BENDENA,[a,c] C. S. GARSIDE,[b] C. G. YU,[b]
AND S. S. TOBE[b]

[a]Department of Biology
Queen's University
Kingston, Ontario, Canada K7L 3N6

[b]Department of Zoology
University of Toronto
Toronto, Ontario, Canada M5S 1A1

The juvenile hormones (JHs) are a unique group of sesquiterpenoids, identified definitively only in insects, that are responsible for the maintenance of juvenile characteristics. A reduction in JH titer is generally believed to be required for metamorphosis to the adult form. Reproductive functions in most adult female insect species are also regulated by JH, and oocyte growth and maturation, including vitellogenesis, show an absolute dependency on JH—in the absence of the hormone, oocyte growth is arrested.[1] JH titer is regulated in part by the rate of biosynthesis within the endocrine glands known as the corpora allata (CA), which can be considered analogous to the adenohypophysis of vertebrates. Based on the now classic experiments involving severance of nerve tracts originating in the brain and innervating the CA, Scharrer demonstrated that CA function was under close-range neural control. Scharrer suggested that signals in the hemolymph might also act to regulate the CA.[2] Consistent with Scharrer's original observations, the regulation of JH biosynthesis by CA is now known to be stimulated or inhibited by two groups of peptides, allatotropins and allatostatins (ASTs), respectively. This review will examine the ASTs which occur in multiple forms, are pleiotropic in function, and are widely distributed in both neural and nonneural tissue. The structural and functional features of the ASTs appear to parallel the vertebrate somatostatins and may provide one of the best examples of the parallel evolution of peptides.[2]

MOLECULAR ISOLATION AND CHARACTERIZATION OF THE COCKROACH ALLATOSTATIN PRECURSOR

The initial characterization of the first complete AST coding region was accomplished by polymerase chain reaction (PCR) amplifications[3] of specific sequence from cDNA derived from mRNA isolated from the brains of virgin females of the cockroach *Diploptera punctata*.[4] Initially, a pair of highly degenerate primers was used to produce an internal DNA consensus sequence representing the peptide sequence of Dip-AST2 (TABLE 1). This consensus sequence was then used in one-

[c] Corresponding author; e-mail: bendenaw@biology.queensu.ca

TABLE 1. Allatostatin Peptide Sequences [a]

Order/Species	Designation	−18	−17	−16	−15	−14	−13	−12	−11	−10	−9	−8	−7	−6	−5	−4	−3	−2	−1		Ref.
Dictyoptera																					
D. punctata	Dip-AST1													[b]Leu	Tyr	Asp	Phe	Gly	Leu	NH$_2$	4,6,13
D. punctata	Dip-AST2	Ala	Tyr	Ser	Tyr	Val	Ser	Glu	Tyr	Lys	Arg	Leu	Pro	Val	Tyr	Asn	Phe	Gly	Leu	NH$_2$	4,11
P. americana	Pea-AST2	Ala	Tyr	Ser	Tyr	Val	Ser	Glu	Tyr	Lys	Arg	Leu	Pro	Val	Tyr	Asn	Phe	Gly	Leu	NH$_2$	6,35
B. craniifer	Blc-AST2	Ala	Tyr	Ser	Tyr	Val	Ser	Glu	Tyr	Lys	Arg	Leu	Pro	Val	Tyr	Asn	Phe	Gly	Leu	NH$_2$	.
B. orientalis	Blo-AST2	Ala	Tyr	Ser	Tyr	Val	Ser	Glu	Tyr	Lys	Arg	Leu	Pro	Val	Tyr	Asn	Phe	Gly	Leu	NH$_2$	.
S. longepalpa	Sul-AST2	Ser	Tyr	Ser	Tyr	Val	Ser	Glu	Tyr	Lys	Arg	Leu	Pro	Val	Tyr	Asn	Phe	Gly	Leu	NH$_2$	.
D. punctata	Dip-AST3											[b]Ser	Lys	Met	Tyr	Gly	Phe	Gly	Leu	NH$_2$	4,6,35
D. punctata	Dip-AST4										Asp	Gly	Arg	Met	Tyr	Ser	Phe	Gly	Leu	NH$_2$	4,10,35
B. americana	Pea-AST4							Ser	Gly	Asn	Asp	Gly	Arg	Leu	Tyr	Ser	Phe	Gly	Leu	NH$_2$	6,35
B. craniifer	Blc-AST4							Ser	Gly	Asn	Asp	Gly	Arg	Leu	Tyr	Ser	Phe	Gly	Leu	NH$_2$	.
B. orientalis	Blo-AST4							Ser	Gly	Asn	Asp	Gly	Arg	Leu	Tyr	Ser	Phe	Gly	Leu	NH$_2$	.
S. longepalpa	Sul-AST4							Ala	Gly	Ser	Asp	Ser	Arg	Leu	Tyr	Ser	Phe	Gly	Leu	NH$_2$	.
B. germanica	Blg-AST4							Ala	Gly	Ser	Asp	Gly	Arg	Leu	Tyr	Ser	Phe	Gly	Leu	NH$_2$	13
D. punctata	Dip-AST5											Asp	Arg	Leu	Tyr	Ser	Phe	Gly	Leu	NH$_2$	4,5,35
P. americana	Pea-AST5											Asp	Arg	Met	Tyr	Ser	Phe	Gly	Leu	NH$_2$	6,35
B. craniifer	Blc-AST5											Asp	Arg	Leu	Tyr	Ser	Phe	Gly	Leu	MH$_2$	.
B. orientalis	Blo-AST5											Asp	Arg	Met	Tyr	Ser	Phe	Gly	Leu	NH$_2$	.
S. longepalpa	Sul-AST5											Glu	Arg	Leu	Tyr	Ser	Phe	Gly	Leu	NH$_2$	.
B. germanica	Blg-AST5											Asp	Arg	Leu	Tyr	Ser	Phe	Gly	Leu	NH$_2$	13
D. punctata	Dip-AST6											[b]Ala	Arg	Pro	Tyr	Ser	Phe	Gly	Leu	NH$_2$	4,6,35
D. punctata	Dip-AST7						Ala	Pro	Ser	Gly	Ala	Gln	Arg	Leu	Tyr	Gly	Phe	Gly	Leu	NH$_2$	4,5,35
P. americana	Pea-AST7						Ser	Pro	Ser	Gly	Met	Gln	Arg	Leu	Tyr	Gly	Phe	Gly	Leu	NH$_2$	6,12,35
B. craniifer	Blc-AST7						Ala	Pro	Ser	Gly	Thr	Gln	Arg	Leu	Tyr	Ala	Phe	Gly	Leu	NH$_2$	.
B. orientalis	Blo-AST7						Ser	Pro	Ser	Gly	Met	Gln	Arg	Leu	Tyr	Gly	Phe	Gly	Leu	NH$_2$	.
S. longepalpa	Sul-AST7					Ala	Pro	Ser	Ser	Gly	Val	Gln	Arg	Leu	Tyr	Gly	Phe	Gly	Leu	NH$_2$	.
B. germanica	Blg-AST7						Ala	Pro	Ser	Ser	Ala	Gln	Arg	Leu	Tyr	Ser	Phe	Gly	Leu	NH$_2$	13
D. punctata	Dip-AST8									Gly	Gly	Ser	Leu	Tyr	Ser	Phe	Gly	Leu	NH$_2$		4,5,35
P. americana	Pea-AST8									Gly	Gly	Ser	Met	Tyr	Ser	Phe	Gly	Leu	NH$_2$		6,35
B. craniifer	Blc-AST8								Ala	Gly	Ser	Ser	Leu	Tyr	Ser	Phe	Gly	Leu	NH$_2$		.
B. orientalis	Blo-AST8									Gly	Gly	Ser	Met	Tyr	Ser	Phe	Gly	Leu	NH$_2$		.
S. longepalpa	Sul-AST8									Gly	Gly	Ser	Leu	Tyr	Ser	Phe	Gly	Leu	NH$_2$		.
D. punctata	Dip-AST9								Gly	Asp	Gly	Arg	Leu	Tyr	Ala	Phe	Gly	Leu	NH$_2$		4,5,35
P. americana	Pea-AST9								Ala	Asp	Gly	Arg	Leu	Tyr	Ala	Phe	Gly	Leu	NH$_2$		6,12,35
B. craniifer	Blc-AST9								Gly	Glu	Gly	Arg	Leu	Tyr	Gly	Phe	Gly	Leu	NH$_2$		.
B. orientalis	Blo-AST9								Ala	Asp	Gly	Arg	Leu	Tyr	Ala	Phe	Gly	Leu	NH$_2$		.
S. longepalpa	Sul-AST9								Gly	Gly	Gly	Arg	Leu	Tyr	Ala	Phe	Gly	Leu	NH$_2$		.

Species	Peptide																	Ref.
D. punctata	Dip-AST10	Pro	Val	Asn	Ser	Gly	Arg	Ser	Ser	Gly	Ser	Arg	Phe	Asn	Phe	Gly	Leu	NH$_2$	4,35
P. americana	Pea-AST10	Pro	Val	Ser	Ser	Ala	Arg	Gln	Thr	Gly	Ser	Arg	Phe	Asn	Phe	Gly	Leu	NH$_2$	6,35
B. craniifer	Blc-AST10	Pro	Val	Asn	Ser	Gly	Arg	Ser	Ser	Gly	Ser	Arg	Phe	Asn	Phe	Gly	Leu	NH$_2$	·
B. orientalis	Blo-AST10	Pro	Val	Ser	Ser	Ala	Arg	Gln	Thr	Gly	Ser	Arg	Phe	Asn	Phe	Gly	Leu	NH$_2$	·
S. longepalpa	Sul-AST10	Pro	Val	Asn	Ser	Gly	Arg	Gln	Thr	Gly	Ser	Arg	Phe	Asn	Phe	Gly	Leu	NH$_2$	·
D. puntata	Dip-AST11						Tyr	Pro	Gln	Glu	His	Arg	Phe	Ser	Phe	Gly	Leu	NH$_2$	4,10,35
P. americana	Pea-AST11						Ser	Pro	Gln	Gly	His	Arg	Phe	Ser	Phe	Gly	Leu	NH$_2$	6,35
B. craniifer	Blc-AST11						Tyr	Pro	Gln	Glu	His	Arg	Phe	Ala	Phe	Gly	Leu	NH$_2$	·
B. orientalis	Blo-AST11						Ser	Pro	Gln	Ser	His	Arg	Phe	Ser	Phe	Gly	Leu	NH$_2$	·
B. germanica	Blg-AST11						Ser	Pro	Gln	Glu	His	Arg	Phe	Ser	Phe	Gly	Leu	NH$_2$	·
S. longepalpa	Sul-AST11						Phe	Pro	Gln	Asp	His	Arg	Phe	Ala	Phe	Gly	Leu	NH$_2$	·
P. americana	Pea-AST12									Ser	Leu	His	Tyr	Asn	Phe	Gly	Leu	NH$_2$	6,35
B. orientalis	Blo-AST12									Ser	Leu	His	Tyr	Asn	Phe	Gly	Leu	NH$_2$	·
D. punctata	Dip-AST12											Pro	Phe	Asn	Phe	Gly	Leu	NH$_2$	4,35
P. americana	Pea-AST13											Pro	Tyr	Asn	Phe	Gly	Leu	NH$_2$	6,35
B. orientalis	Blo-AST13											Pro	Tyr	Asn	Phe	Gly	Leu	NH$_2$	·
D. punctate	Dip-AST13									Ile	Pro	Met	Tyr	Asp	Phe	Gly	Ile	NH$_2$	4,35
P. americana	Pea-AST14									Ile	Pro	Met	Tyr	Asp	Phe	Gly	Ile	NH$_2$	6,35
B. orientalis	Blo-AST14									Ile	Pro	Met	Tyr	Asp	Phe	Gly	Ile	NH$_2$	·
Orthoptera																			
G. bimaculatus	Grb-ASTA1								Ala	Gln	His	Gln	Tyr	Ser	Phe	Gly	Leu	NH$_2$	14
	Grb-ASTA2							Ala	Gly	Gly	Arg	Gln	Tyr	Gly	Phe	Gly	Leu	NH$_2$	14
	Grb-ASTB1								Gly	Trp	Gln	Asp	Leu	Asn	Gly	Gly	Trp	NH$_2$	15
	Grb-ASTB2								Gly	Trp	Arg	Asp	Leu	Asn	Gly	Gly	Trp	NH$_2$	15
	Grb-ASTB3								Ala	Trp	Arg	Asp	Leu	Ser	Gly	Gly	Trp	NH$_2$	15
	Grb-ASTB4								Ala	Trp	Glu	Arg	Phe	His	Gly	Ser	Trp	NH$_2$	15
Diptera																			
C. vomitoria	Cav-AST1	Asp	Pro	Leu	Asn	Glu	Glu	Arg	Arg	Ala	Asn	Arg	Tyr	Gly	Phe	Gly	Leu	NH$_2$	20
	Cav-AST2			Leu	Asn	Glu	Glu	Arg	Arg	Ala	Asn	Arg	Tyr	Gly	Phe	Gly	Leu	NH$_2$	20
	Cav-AST3									Ala	Asn	Arg	Tyr	Gly	Phe	Gly	Leu	NH$_2$	20
	Cav-AST4									Asn	Arg	Pro	Tyr	Ser	Phe	Gly	Leu	NH$_2$	20
	Cav-AST5									Gly	Pro	Pro	Tyr	Asp	Phe	Gly	Met	MH$_2$	20
Lepidoptera																			
M. sexta	Mas-AST		pGlu	Val	Arg	Phe	Arg	Gln	Cys	Tyr	Phe	Asn	Pro	Ile	Ser	Cys	Phe		21,22

[a] Sequences derived from nucleotide sequence data (Bendena, Belles, Weaver, Edwards & Tobe, unpublished data).
[b] Identical peptide sequences were identified in all cockroach species examined.

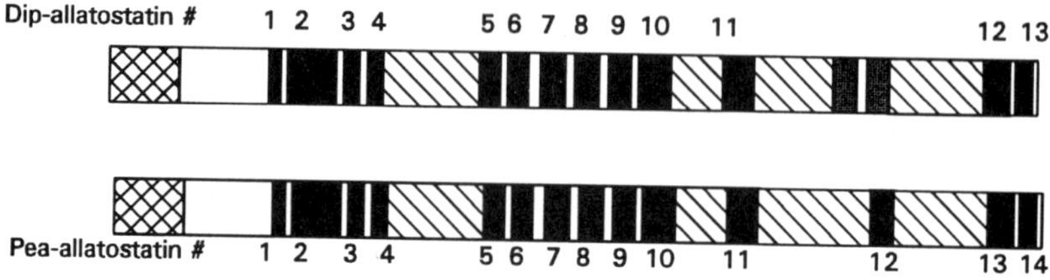

FIGURE 1. Schematic diagram of cockroach AST polypeptide precursors deduced from isolated DNA sequences. The polypeptide precursors of *D. punctata* **(upper panel)** and *P. americana* **(lower panel)** begin with a hydrophobic leader region (*cross-hatched*) that is presumably cleaved by signal endoproteases. Black boxes representing the individual AST peptides are numbered according to their position relative to the precursor N-terminus. Acidic regions are indicated by diagonal lines. Incorporated within the third acidic domain of the *D. punctata* AST precursor are sequences for two non-amidated peptides (*shaded boxes*).

sided 5′ end-specific PCR of a brain cDNA population that was ligated to a plasmid vector to provide a known reference point beyond the cDNA terminus where opposing primers could anneal. This plasmid cDNA library was never passed through a bacterial host after ligation. The amplification reaction produced the N-terminus of Dip-AST2 as well as Dip-AST1. This sequence was then used to design further specific oligonucleotide primers. These primers were used in combination with degenerate primers designed to amino acid sequences of four isolated ASTs.[5] With this approach, the entire coding region of the AST gene was obtained by successive amplification steps towards the 3′ end of the gene. Subsequently, the battery of primers that were generated in deriving the *D. punctata* AST gene were applied to generating the AST precursor structure of other species.[6] The open reading frame encodes a preproallatostatin polypeptide that begins with an amino-terminal signal peptide that serves to target secretory protein translocation into the lumen of the endoplasmic reticulum (see FIG. 1). The AST precursors of *D. punctata* and the American cockroach, *Periplaneta americana*, are remarkably similar in size (370 and 379 amino acids, respectively), sequence (71% amino acid identity), and structural organization (FIG. 1). The lowest degree of amino acid identity exists in the N-terminal region adjacent to the first putative dibasic KR endoproteolytic cleavage site (FIG. 1). The ASTs are found beyond this cleavage site, and following the C-terminal amino acid of each AST is the sequence GKR required for α-amidation and processing.[7] The predominant dibasic endoproteolytic cleavage site within the precursor is KR. A unique dibasic RR is conserved prior to DipAST12/PeaAST13. If endoproteolytic cleavage of this RR site occurs prior to cleavage of KR, then the final two peptides in each precursor may be released to follow a route separate from that of the body of the precursor. The basic endoproteolytic cleavage sites are balanced in the precursor by the presence of acidic spacer regions that in *D. punctata* have isoelectric points that range from 2.2 to 4.5 (FIG. 1). Between species, the acidic spacer domains have very limited sequence identity but the size and organization have been maintained. These acidic domains are similar to those found in the organization of the *Aplysia* FMRFamide precursor.[8] In contrast, the *Drosophila* FMRFamide precursor contains one or two D residues within the amino-terminal address sequence of the peptides themselves.[9]

ALLATOSTATIN PEPTIDES

The Dictyopteran AST peptides in their order on the precursor (TABLE 1) were derived from the complete DNA sequence of the *D. punctata*[4] and *P. americana*[6] AST genes and the available/parietal DNA sequence information for the AST precursors of the remaining species (unpublished data). The *P. americana* precursor is cleaved into 14 different AST peptides compared to 13 peptides that are derived from *D. punctata* precursor. The isolation and identification of seven AST peptides from *D. punctata*,[5,10,11] two from *P. americana*,[12] and four from *Blattella germanica*[13] indicate that the dibasic endoproteolytic cleavage sites are used *in vivo*. Five *P. americana* peptides, Pea-ASTs 1, 2, 3, 6, and 14 are identical to Dip-ASTs 1, 2, 3, 6, and 13 in both sequence and position within the precursor. Partial sequence information from other cockroach species has revealed that all share the same precursor organization and have AST peptides 1, 3, and 6 in common with *D. punctata* and *P. americana*. Peptide 2 is also invariant within all species tested with the exception of *S. longepalpa*, which has a single N-terminal amino acid substitution. The C-terminal region of the precursor has not been sequenced for all other cockroach species, and it is therefore uncertain whether the terminal peptide within the precursor is conserved. In all Dictyopteran ASTs thus far identified, there is a C-terminal consensus Y/FXFG/Ia. This C-terminal consensus sequence has also been found in two of the recently identified Orthopteran ASTs, Grb-AST-A1 and Grb-AST-A2.[14] Peptides with *in vitro* allatostatic activity have also been purified from *G. bimaculatus*,[15] Grb-AST B1–4 (TABLE 1), which are unique in sequence from the Dictyopteran ASTs. The existence of two structurally dissimilar peptides that show similar functions suggest that two separate receptors may have evolved in parallel or that a single receptor may have adapted to interact functionally with two distinct peptide sequence types. GrbASTs B1–B4 share sequence identity to locusta-myoinhibiting peptide (LomMIP) that functions in the inhibition of spontaneous contractions of hindgut and oviduct of *Locusta migratoria*.[16] Similarly, *Manduca sexta* allatotropin, which stimulates JH biosynthesis by the CA *in vitro*,[17] shows significant similarity to *L. migratoria* accessory gland myotropin I (Lom-AG-MT-I).[18] *M. sexta* allatotropin has also been shown to function as a cardioacceleratory peptide in *Manduca*.[19] Although GrbASTs have not as yet been shown to affect muscle contraction, it is likely that they will join a growing list of bi- or multifunctional peptides. Peptides that share the C-terminal consensus sequence of the Dictyopteran ASTs have also been isolated from the fly *Calliphora vomitoria* (Cav-AST1–5). Although Cav-ASTs have allatostatic activity on CA of *D. punctata*, none of these peptides inhibited the production of JH by CA of *C. vomitoria*.[20] Thus, peptides with similar structures appear to have been adapted to distinct functions.

A peptide of unique sequence with allatostatic activity has also been isolated from *M. sexta*, Mas-AST (TABLE 1). Mas-AST inhibits JH biosynthesis *in vitro* with CA from adult *M. sexta* and *Helicoverpa zea* by 100% (0.1 μM) and 77% (0.5 μM), respectively.[21] In contrast, inhibition of JH biosynthesis by CA from the true armyworm, *Pseudaletia unipuncta*, at 1 μM Mas-AST was only 60%.[2] Further studies should confirm whether this within-order interspecies difference reflects a functional shift in peptide activity or whether environmental conditions and/or developmental times of maximal AST sensitivity vary. Immunoreactive material has also been detected in brains of *P. unipuncta* with antibodies raised to Dip-ASTs, which suggests that other types of ASTs may occur in Lepidoptera (unpublished data).

Several lines of evidence have confirmed that the C-terminal pentapeptide of the *D. punctata* ASTs is required for functional AST activity *in vitro*. Extension of the C-terminus, deletion of C-terminal amino acids or removal of the C-terminal amide all abolish bioactivity.[11] Similarly, for Dip-AST5, the most dramatic effects on signal transmission to target tissues (potency) were observed following amino acid substitutions at positions L^8, F^6, and Y^4, respectively.[23] The active core sequence YSFGLa shows binding only at high concentrations and its binding efficiency is greatly reduced in brain receptor binding assays. Furthermore, removal of L^8-NH_2 from Dip-AST5 results in complete loss of binding activity (unpublished data). In contrast to the Dipteran extended-FMRFamide peptide family,[8,9] sequence data suggest that tandem copies of the peptides do not occur within the precursors of Dictyopteran ASTs. Each intraspecies N-terminal address sequence is unique. Removal of the N-terminal amino acids D^1R^2 from Dip-AST5 results in a 30% loss in bioactivity[24] and reduces receptor binding *in vitro* by 50% (unpublished data). The N-terminus appears to modify the specificity and affinity between the C-terminal region of the AST peptides and their corresponding receptors. Terminal amino acid substitutions may also alter the degradation rate of select peptides. In the majority of cases, interspecies amino acid substitutions arise by single point mutations and are conservative, maintaining either the hydrophobic or aromatic character of the position. Substitutions of methionine residues with leucine or alanine are common in many proteins and are thought to decrease the likelihood of oxidation.[25] Amino acid replacements can alter the potency of the peptide with respect to inhibition of JH biosynthesis *in vitro*.[23] The two amino acid differences in the address sequences of Dip/Pea AST-7 alter the magnitude of JH inhibition by 2–3 orders in the same species as compared to the effect in the reciprocal species.[26] AST4 is the most variable peptide of the family, and is shortened by three N-terminal amino acids relative to AST4 in all other cockroach species. Another major variation is associated with Pea/Blo AST12 in which the appropriate GKR sequence required for amidation and processing has been altered in *D. punctata* through deletion of the C-terminal G residue required for peptide amidation and modification of the processing site to RK rather than KR. Because all species tested are probably more ancient than the viviparous *D. punctata*, the differences in amino terminal extension in AST4 and the absence of a functionally processed peptide counterpart to Pea/Blo AST12 suggest that this loss has occurred through the evolutionary process. As more sequence information has become available, it will be of interest to determine whether this latter alteration is related to the reproductive mode of this insect group which ranges from oviparity to ovoviviparity to viviparity.[27,28]

DISTRIBUTION AND EXPRESSION OF COCKROACH ALLATOSTATINS

The existence of inhibitory substances originating in the brain and traveling to the CA through nerve tracts was first demonstrated by nerve severance experiments which showed that final stage *Leucophaea maderae* larvae underwent supernumerary larval molts.[29] Retrograde filling of axons from the CA with metal ions in *D. punctata*[30] and *P. americana*[31] showed that the CA are innervated by medial cells of the pars intercerebralis extending through the nervi corporis cardiaci I (NCC I) and lateral cells of the pars lateralis through the NCC II. Both cell types appear to be involved in the inhibition of JH biosynthesis because destruction of

                                                      59

either medial or lateral cells by radiofrequency cauterization resulted in increased JH biosynthesis in virgin females.[32] By use of antibodies raised to Dip-AST7, allatostatin immunoreactive cells have been detected in numerous regions of the brain. Thirty or more immunoreactive lateral cells in the protocerebrum were found to project in NCC II and to arborize terminally in the corpus cardiacum (CC) and CA. Four distinct medial cells of the pars intercerebralis were also immunoreactive but these cells did not project in NCC I, but rather extended axons that arborized and terminated adjacent to the lateral neurosecretory cell bodies and axons.[33,34] This localization suggests that ASTs may have interneuronal functions, and, as such, AST receptors would be expected to be found on postsynaptic neurons of the brain. AST immunoreactivity is also found in somata in the protocerebrum between the optic lobes, between the protocerebrum and deutocerebrum, and in the tritocerebrum. Numerous axonal tracts and arborizations are also found in the optic, central, and antennal lobes.[33,34] Immunoreactivity has also been detected in branches of NCC III that extend through the tritocerebrum to enter and branch extensively within the pulsatile organ muscle.[35,36] AST mRNA expression has also been detected in similar cell bodies in brains of *D. punctata* and *P. americana* by *in situ* hybridization.[4,6] In both species, strong hybridization was found in the two pairs of medial cells of the pars intercerebralis. Hybridization to lateral neurosecretory cells was also detected in both species, but signal strength was consistently lower than in the medial cells of *D. punctata*. In *P. americana*, AST mRNA expression was also detected in numerous cells near the optic lobes and within the tritocerebrum.[6] AST-immunoreactive neurons and fine arborizations that project to the antennal heart have also been found in the subesophageal ganglia (SOG). AST mRNA expression occurs in an unpaired medial neuron and in up to four lateral neurons of the SOG (unpublished data). Immunocytochemistry also revealed neurons in the ventral ganglia and terminal abdominal ganglia and terminal abdominal ganglia that project to the proctodeal nerve. The proctodeal nerve innervates the rectal dilator muscle, muscles of the rectum, anterior hindgut, and midgut.[36,37] In *D. punctata* midgut, in addition to the proctodeal innervation, immunoreactivity is also found in open-type endocrine cells which extend from the basal lamina to the lumen.[37,38] In contrast, antisera raised to Cav-AST (TABLE 1) only detects midgut endocrine cells in the Dipteran, *C. vomitoria*.[39] Dip-AST mRNA expression as visualized by *in situ* hybridization is similarly confined to endocrine cells in *D. punctata* midgut, which suggests that the immunoreactivity observed in nerve tracts is a result of peptide transport from cell bodies of the terminal abdominal ganglia.

Immediately after mating in *D. punctata*, the rate of JH biosynthesis increases to a maximum on day 4–5, then declines and remains at a low level after day 5.[40–42] Both ELISA with a Dip-AST7 antibody and bioassays have demonstrated that the changes in Dip-AST content of the brain correlate with changes in JH biosynthesis associated with the reproductive cycle of mated females. Similar changes were not observed in virgin females of similar age.[43] Similarly, with use of antibodies to Dip-ASTs 2, 7, 10, and 11, the Dip-AST content of mated female midgut and hindgut was lowest four days post-mating, then increased during the second half of the reproductive cycle.[44] The four peptides were found to be present in approximately equal quantities in each tissue. Because these peptides originate from three separate regions of the Dip-AST precursor, processing and release of the peptides (and possibly all 13 peptides) may occur simultaneously in these tissues.[44] The maximal Dip-AST content in brain (7.5 pmol/brain) is much higher than that found in midgut (340 fmol/midgut) and hindgut (250 fmol/hindgut). However, the weight of the combined gut tissue is much greater than the brain and thus the absolute amounts of immunoreactive Dip-ASTs in the two tissue types may be similar. Preliminary quantification of the steady-state levels of Dip-AST mRNA in day 6 mated female midgut also suggests that Dip-AST-specific mRNA levels are 20 times lower than those found in brains of similar stage (unpublished data).

Dip-AST immunoreactive material has also been found in the hemolymph,[43,45] and in mated females, the developmental profile of hemolymph Dip-AST content differs from that in brain and gut tissues. During the reproductive cycle, hemolymph Dip-AST content is low during vitellogenesis, sharply increases at oviposition, reaches a maximum (2.25 nM) early during pregnancy (day 13), and declines thereafter.[43] During the last half of the last larval stadium, hemolymph Dip-AST content is also elevated, reaching approximately 1.27 nM. The concentrations of Dip-ASTs in hemolymph during the last larval stadium and late in the first gonadotrophic cycle are sufficient to inhibit significantly JH biosynthesis *in vitro*. Thus, these peptides may function as humoral inhibitors *in vivo*, and this may explain the low rates of JH production exhibited by denervated larval glands (i.e., removed from neural inhibition).[46,47] ASTs may be released into the hemolymph, in part, from the NCC II, which branches extensively within the CC.[43,45,48] The midgut may also represent another site from which cockroach ASTs are released into hemolymph, and this release may be regulated, in part, by nutritional status.[37] However, the humoral Dip-ASTs are unlikely to contribute to the regulation of JH biosynthesis in adult CA because starvation does not affect the cycle of JH biosynthesis associated with the gonadotrophic cycle in *D. punctata* adult females.[49] In starved and dehydrated animals, a significant initial increase is found in the Dip-AST content of midgut.[37] This elevation could result from increased synthesis, inhibition of release or an enhanced transport of the peptides by nerves extending from hindgut to midgut or from changes in AST content or release from midgut endocrine cells. With increasing duration of nutritional stress, the Dip-AST midgut levels decline, presumably as a consequence of their release into the hemolymph.[37] Dip-ASTs have also been established as inhibitors of proctolin-induced and spontaneous myogenic contractions[36,50] of hindgut muscle. Because the minimum ED_{50} for Dip-ASTs in the hindgut *in vitro* myotropic assay is approximately 100 nM, hemolymph ASTs are unlikely to be primary regulators of hindgut contraction. However, the K_d of Dip-AST7 interaction with its corresponding midgut receptor is 8.4 nM, which is in the same range as the hemolymph concentration, suggesting that a humoral route for Dip-AST inhibition of proctolin-induced contraction cannot be excluded (unpublished data). Thus, during periods of food and water deprivation, Dip-ASTs may inhibit proctolin-induced activity, resulting in decreased metabolic activity and slower movement of material through the gut. It will be of interest to examine the functional significance of humoral ASTs. Humoral Dip-ASTs could, for example, enter the antennae and modify antennal sensory function[48] or could cross the perineurium and act on brain receptors, permitting communication between humoral and neural routes of expression.

FUNCTIONAL ACTIVITY OF ALLATOSTATIN PEPTIDES

The rank order of activity of Dip-ASTs for inhibition of JH biosynthesis *in vitro* differs from that for inhibition of proctolin-induced muscle contraction (TABLE 2). Dip-AST2 and Dip-AST5, which are the two most potent ASTs in inhibition of JH biosynthesis, are relatively ineffective as inhibitors of proctolin-induced muscle

TABLE 2. Rank Order of Effectiveness of *D. punctata* Allatostatins

| | | Effectiveness as an Inhibitor in | | | |
| | | Muscle Contraction[a] | | Juvenile Hormone Biosynthesis[b] | |
Peptide Designation	Residues (aa)	ED_{50}	Rank Order	Rank Order	ED_{50}
Dip-AST1	6	5.3×10^{-7}	7	nd	nd
Dip-AST2	18	1.8×10^{-6}	9	1	1.0×10^{-11}
Dip-AST3	8	1.9×10^{-6}	10	nd	nd
Dip-AST4	9	3.7×10^{-7}	5	8	2.0×10^{-8}
Dip-AST5	8	1.1×10^{-6}	8	2	1.6×10^{-10}
Dip-AST6	8	2.0×10^{-6}	11	4	2.3×10^{-9}
Dip-AST7	13	2.8×10^{-7}	3	3	4.1×10^{-10}
Dip-AST8	9	2.1×10^{-7}	2	7	9.4×10^{-9}
Dip-AST9	10	1.7×10^{-7}	1	6	7.2×10^{-9}
Dip-AST10	16	4.5×10^{-7}	6	nd	nd
Dip-AST11	11	3.5×10^{-7}	4	5	3.2×10^{-9}
Dip-AST12	6	$>1.3 \times 10^{-5}$	12	nd	nd
Dip-AST13	8	2.1×10^{-7}	2	nd	nd

[a] ED_{50}: Molar concentration of Dip-AST required to inhibit a 5×10^{-6} M proctolin-induced contraction by 50%.[50]

[b] ED_{50}: Molar concentration of Dip-AST required for 50% inhibition of JH release in a 3-h *in vitro* radiochemical assay with pairs of 2-day virgin CA compared to groups of controls ($n = 7$).[5,34,44]

nd = not determined.

contraction, ranking ninth and eighth, respectively. Conversely, Dip-AST9 and Dip-AST8 are the two most potent inhibitors in the hindgut assay but are relatively ineffective as inhibitors of JH biosynthesis (TABLE 2). Thus, the functional differences in each AST may reside at the level of receptors in different tissues. This notion is supported and extended by recent *in vitro* binding assays.[51] Dip-AST7 exhibits different binding affinities with membrane preparations from different tissues: the K_d of receptor binding for CA and hindgut is 7.2×10^{-10} M and 8.4×10^{-9} M, respectively, whereas brain membranes appear to have either two receptors or a single receptor with at least two binding sites because two K_d values of 1.5×10^{-9} and 3.8×10^{-9} were obtained. Dip-AST11 can bind to hindgut membranes with a $K_d = 1.1 \times 10^{-8}$ M and does not displace Dip-AST7 in competition assays (unpublished data), indicating that Dip-AST7 does in fact bind with greater affinity and corresponds to the order of inhibition of proctolin-induced muscle contraction. Similarly, Dip-AST5 binds with higher affinity than Dip-AST7 to brain membrane preparations which corresponds to the order of potency in terms of inhibition of JH biosynthesis. Surprisingly, a Dip-AST5 photoaffinity analogue has demonstrated a putative brain receptor protein of 37 kDa,[51] whereas a different photoaffinity analogue of Dip-AST7 detected two putative corpora allata receptors of 59 and 39 kDa.[52] Further study may reveal a family of receptor proteins of similar size. The availability of high-affinity ligands should permit the isolation of these membrane receptors through molecular expression studies.[53] CavASTs are also potent inhibitors of hindgut contractions in both *C. vomitoria*[39] and *D. punctata*.[50] Since Dipterans appeared more recently in evolutionary time, the loss of peptide action with respect to inhibition of JH biosynthesis in *C. vomitoria* may reflect a selection pressure against a responsive CA receptor.

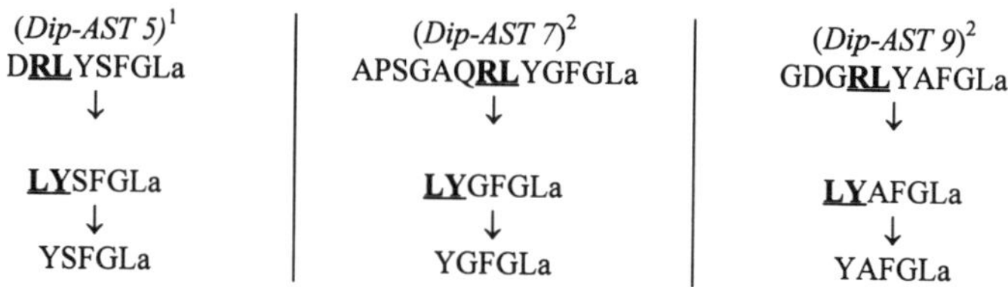

FIGURE 2. Metabolism of allatostatins by hemolymph. Dip-allatostatins were incubated from 0 to 180 min with hemolymph (diluted 100×). [1]Following phenyl RP-HPLC and liquid scintillation spectrometry, metabolites of 4.1 nM [^{3}H] Dip-AST5 incubations were identified by comparison to retention times of synthetic standards. [2]Following extraction and separation by C_{18} RP-HPLC, metabolites of 4–6 μM allatostatin incubations, comigrating with standards, were determined by amino acid analysis (PTC method). a = NH_2.

ALLATOSTATIN METABOLISM

ASTs exert rapid physiological effects on both JH biosynthesis and modulation of gut myotropic activity.[5,36,50] Equally important, the effects of the peptides are rapidly reversible, suggesting that mechanisms for inactivation of the peptides must exist, at least at the level of CA and of gut tissue. Hence, metabolism of peptides is likely to be a major pathway for the termination of biological effect, as has been demonstrated in the case of vertebrate peptide degradation.[54] The principal route of inactivation in the latter case appears to be hydrolysis of peptides by membrane-bound or soluble peptidases.[54,55]

The presence of multiple forms of the ASTs and the wide range of biological activities with respect to inhibition of JH biosynthesis (TABLE 2) suggest that this peptide system shows a high degree of complexity in terms of both mode of action and degradation. Hence, different tissues show different levels of sensitivity to the various ASTs; these differences also appear to be age- or stage-dependent.[35] In addition, the ability of different tissues to degrade the ASTs differs, (although at present we do not know if this ability changes as a function of stage), possibly as a consequence of differences in the complement of degradative enzymes. In addition, certain AST metabolites retain significant biological activity/potency, albeit reduced efficacy, as a consequence of the presence of the remaining core C-terminal pentapeptide responsible for biological activity.[23]

Metabolism of peptides differs dramatically between hemolymph and tissues, and this suggests that the ASTs may play a different role in the hemolymph than in the target tissues. In fact, the hemolymph may represent a transport vehicle for the peptides to target tissues, because metabolism of three representative ASTs (Dip-AST5, 7, and 9) always occurs near the N-terminus, rather than at the C-terminus, as occurs in the four target tissues examined (see FIGS. 2 and 3). It is significant that at least in the radiochemical assay for JH biosynthesis and in the myotropic assay, nonamidated ASTs, metabolites produced by the tissues (FIG. 3), lack biological activity,[11,35,36,50] whereas the C-terminal pentapeptide, the metabolite produced in hemolymph from Dip-ASTs (FIG. 2), shows full efficacy (but reduced potency); it appears to represent the active core sequence— Y/FXFGLa.[34] Why ASTs are released into the hemolymph rather than delivered directly to target tissues by neurosecretory axons remains uncertain but several functions can be envisaged. For example, the humoral ASTs may "prime" the

Hindgut	Midgut	Brain	CA
DRLYSF**GL**a	DRLYSF**GL**a	DRLYSF**GL**a	DRLYSF**GL**a
↓	↓	↓	↓
DRLYS**FG**-OH*	DRLYSFG-OH*	DRLYS**FG**-OH*	DRLYS**FG**-OH
↓		↓	↓
DRL**YS**F-OH		D**RL**YSF-OH	D**RL**YSF-OH
↓		↓	↓
DRLY-OH		LYSF-OH	LYSF-OH

FIGURE 3. Metabolism of Dip-allatostatin 5 by membrane preparations. [^{3}H] Dip-AST5 (4.1 nM) was incubated from 0 to 180 min with membrane preparations (25 ng protein/μL). Following phenyl RP-HPLC and liquid scintillation spectrometry, metabolites were identified by comparison to retention times of synthetic standards. *Metabolites confirmed by amino acid analysis of incubations of Dip-AST5 (6 μM).

receptors of the target tissues to increase their responsiveness to locally released peptides, either as a result of modulating metabolism or by directly interacting with receptors. The ASTs may also exert as yet unknown effects on other target tissues, such as modulation of neural activity.

Substrate concentration also appears to influence significantly the rates of metabolism (and half-life) of the peptides. The importance of using the peptides for metabolic studies at physiologically relevant concentrations cannot be overemphasized, but unless the peptides are available in radiolabeled form, such studies are laborious in insect systems in which large amounts of tissues are difficult to generate. This applies particularly to nervous and endocrine tissue such as brain, SOG, and CA. In addition, without a radiolabeled substrate, it is difficult to detect metabolites at low concentrations by high-performance liquid chromatography separation and amino acid analysis. For this reason, most previous studies have employed peptides at micromolar concentrations, a level considerably greater than the endogenous concentrations.[43] Estimates of AST half-life, using 4 and 6000 nM Dip-AST5, suggest that at the high concentration of the peptide, the rate of metabolism is slowed in hemolymph but accelerated in CA homogenates (unpublished data). However, at more physiological concentrations of Dip-AST5, the half-life is reduced in hemolymph approximately fourfold, whereas in CA homogenates the half-life appears to remain relatively constant.

The range of degradative enzymes that are able to use AST peptides as a substrate also appears to differ between the tissues. Not only does the hemolymph contain peptidases which hydrolyze at the amino (RL) terminus (aminopeptidases) whereas the tissues contain putative carboxypeptidases that initially appear to cleave at GL residues, but the enzymes in different tissues show differing sensitivities to known peptidase inhibitors. TABLE 3 shows, for example, that metallopeptidase inhibitors such as phosphoramidon, 1,10-phenanthroline, and EDTA are effective against peptidases in both hindgut and brain, but only marginally active in midgut. Serine peptidase inhibitors are also effective against hindgut peptidases but show variable effects on midgut and brain and are ineffective against the peptidases in the hemolymph. With the exception of midgut, aminopeptidase inhibitors are more effective in inhibiting AST degradation in the tissues than in the hemolymph. In general, peptidase inhibitors show greatest activity against enzymes in the hindgut and brain and reduced activity against peptidases in the midgut. This is not surprising because it is likely that the midgut, as the site of

TABLE 3. Inhibition of Dip-Allatostatin Metabolism by Peptidase Inhibitors

	Hemolymph			HG	MG	BR	CA
				(Membrane preparations) Dip-AST5			
	Dip-AST5	Dip-AST7	Dip-AST9				
Serine-PI							
TPCK	−	−	−	+++	+++	++	ND
PMSF	−	−	−	+++	+	+	ND
Metallo-PI							
PRAM	ND	ND	ND	+++	++	+++	ND
1,10-Phe	+++	++	++	+++	+	+++	+++
EDTA	ND	ND	ND	+++	−	+++	++
Amino-PI							
Amastatin	++	+	+	+++	−	+++	+++
Bestatin	+	+	−	ND	ND	ND	ND

NOTE: Hemolymph was diluted 100× with saline. Membrane preparations were prepared at 0.025 μg protein/μL incubation medium, except corpora allata (CA) which were prepared with 50 pairs. Inhibitors were pre-incubated with incubation medium for 20 min prior to assay. Allatostatins were incubated at 4–6 μM concentrations except Dip-AST5 which was incubated in hemolymph at 4 nM. PI = Peptidase inhibitors; HG = hindgut; MG = midgut; BR = brain; TPCK = L-1-chloro-3-(4-tosylamido)-4-phenyl-2-butanone; PMSF = phenylmethylsulfonyl fluoride; EDTA = (ethylenedinitrilo)tetraacetic acid; Pram = phosphoramidon; 1,10-Phe = 1,10-phenanthroline.

− symbol = no effect; + = <25% inhibition; ++ = 25–50% inhibition; +++ = >50% inhibition; ND = not determined.

synthesis and release of digestive enzymes, contains a wide range of peptidases, including chymotrypsin, only some of which are subject to inhibition. It is noteworthy that the serine peptidase inhibitor TPCK is quite effective against chymotrypsin, because it functions as an affinity label, alkylating the histidine residue 57 in the active site of the vertebrate enzyme. The cockroach enzyme also appears to be susceptible to inhibition by TPCK but is relatively resistant to PMSF inhibition (TABLE 3). Chymotrypsin cleaves at aromatic residues on the carboxyl side of the peptide; such cleavage in Dip-AST5 would be expected to yield SGLa or DRLYSF-OH or a smaller peptide, depending on cleavage at the C-terminus, and although this peptide was not detected from midgut preparations, a potential substrate DRLYSF-OH for chymotrypsin—with two potential cleavages sites **YS** and **FG**—was identified. However, DRLYSF-OH was detected as a product of hindgut, brain and CA incubations, suggesting the presence of chymotrypsin-like activity in these tissues.

Another cleavage site, observed in both hemolymph and tissue homogenate preparations, is the **RL** site; this site is found only in Dip-AST5, 7, and 9 at the 6,7 C-terminal position (although it does occur in Dip-AST2, as part of the dibasic cleavage site $K^9R^{10}L^{11}$). Such cleavage is known to be affected by enkephalinase (EC 3.4.24.11) in vertebrates and although this enzyme is generally regarded as a tissue enzyme,[56] enkephalinase-like activity clearly occurs in hemolymph of *D. punctata*. This represents an unusual cleavage for insect peptide metabolism which, although largely unreported, is clearly important in *D. punctata* peptide modification because such cleavage in the address sequence reduces potency but not efficacy of the cleavage products in terms of inhibition of JH biosynthesis.[23,24]

CONCLUSIONS

The occurrence of the large allatostatin family in the Dictyoptera and the high degree of conservation of the peptides suggest that they regulate a range of important physiological functions. There is a clear conservation of function with respect to both inhibition of JH production and modulation of myotropic activity within this family, but a divergence of structure in higher insect orders. The myomodularity function appears to be conserved in these orders, but with a loss of JH inhibitory activity. A new family of peptides has appeared to subsume this function. At present, the role for ASTs in neuromodulation (e.g., through interneurons) remains uncertain. It will be important to isolate both types of ASTs from a range of insects and other invertebrates and to identify the genes coding for these peptides, to permit an understanding of the diverse functions which they regulate.

REFERENCES

1. TOBE, S. S. & B. STAY. 1985. Adv. Insect. Physiol. **18:** 305–432.
2. SCHARRER, B. 1987. Annu. Rev. Entomol. **32:** 1–16.
3. SAIKI, R. K., S. GELFAND, S. STOFFEL, S. J. SCHARF, R. HIGUCHI, G. T. HORN, K. B. MULLIS & H. R. ERLICH. 1988. Science **239:** 487–491.
4. DONLY, B. C., Q. DING, S. S. TOBE & W. G. BENDENA. 1993. Proc. Natl. Acad. Sci. USA **90:** 8807–8811.
5. WOODHEAD, A. P., B. STAY, S. L. SEIDEL, M. A. KHAN & S. S. TOBE. 1989. Proc. Natl. Acad. Sci. USA **86:** 5997–6001.
6. DING, Q., B. C. DONLY, S. S. TOBE & W. G. BENDENA. 1995. Eur. J. Biochem. **234:** 737–746.
7. EIPPER, B. A., D. A. STOFFERS & R. E. MAINS. 1992. Annu. Rev. Neurosci. **15:** 57–85.
8. SCHAEFER, M., M. R. PICCIOTTO, T. KREINER, R. R. KALDANY, R. TAUSSIG & R. H. SCHELLER. 1985. Cell **41:** 457–467.
9. SCHNEIDER, L. E. & P. H. TAGHERT. 1988. Proc. Natl. Acad. Sci. USA **85:** 1993–1997.
10. WOODHEAD, A. P., M. A. KHAN, B. STAY & S. S. TOBE. 1994. Insect Biochem. Mol. Biol. **24:** 257–263.
11. PRATT, G. E., D. E. FARNSWORTH, K. F. FOK, N. R. SIEGEL, A. L. McCORMACK, J. SHAMBOWITZ, D. F. HUNT & R. FEYEREISEN. 1991. Proc. Natl. Acad. Sci. USA **88:** 2412–2416.
12. WEAVER, R. J., Z. A. FREEMAN, M. G. PICKERING & J. P. EDWARDS. 1994. Comp. Biochem. Physiol. C. Comp. Pharmacol. **107:** 119–127.
13. BELLES, X., J. L. MAESTRO, M. D. PIULACHS, A. H. JOHNSEN, H. DUVE & A. THORPE. 1994. Regul. Pept. **53:** 237–248.
14. LORENZ, M. W., R. KELLNER & K. H. HOFFMANN. 1995. Regul. Pept. **57:** 227–236.
15. LORENZ, M. W., R. KELLNER & K. H. HOFFMANN. 1995. J. Biol. Chem. **270:** 21103–21108.
16. SCHOOFS, L., J. VANDEN BROECK & A. DE LOOF. 1993. Insect Biochem. Mol. Biol. **23:** 859–891.
17. KATAOKA, H., A. TOSCHI, J. P. LI, R. L. CARNEY, D. A. SCHOOLEY & S. J. KRAMER. 1989. Science **243:** 1481–1483.
18. PAEMEN, L., A. TIPS, L. SCHOOFS, P. PROOST, J. VAN DAMME & A. DE LOOF. 1991. Peptides **12:** 7–10.
19. VEENSTRA, J. A., H. LEHMAN & N. T. DAVIS. 1994. J. Exp. Biol. **188:** 347–354.
20. DUVE, H., A. H. JOHNSEN, A. G. SCOTT, C. G. YU, K. J. YAGI, S. S. TOBE & A. THORPE. 1993. Proc. Natl. Acad. Sci. USA **90:** 2456–2460.
21. KRAMER, S. J., A. TOSCHI, C. A. MILLER, H. KATAOKA, G. B. QUISTAD, J. P. LI, R. L. CARNEY & D. A. SCHOOLEY. 1991. Proc. Natl. Acad. Sci. USA **88:** 9458–9462.

22. JANSONS, I. S., M. CUSSON, J. N. MCNEIL, S. S. TOBE & W. G. BENDENA. 1996. Insect Biochem. Mol. Biol. In press.
23. HAYES, T. K., X-C. GUAN, V. JOHNSON, A. STREY & S. S. TOBE. 1994. Peptides **15:** 1165–1171.
24. STAY, B., A. P. WOODHEAD, S. JOSHI & S. S. TOBE. 1991. *In* Insect Neuropeptides: ACS Symposia Series. J. J. Menn, T. J. Kelly & E. P. Masler, Eds. Vol. 453: 164–176. American Chemical Society. Washington, D.C.
25. PRICE, D. A., 1986. Am. Zool. **26:** 1007–1015.26.
26. WEAVER, R. J. 1991. J. Insect Physiol. **37:** 111–118.
27. ROTH, L. M. & B. STAY. 1961. J. Insect Physiol. **7:** 186–202.
28. TOBE, S. S. 1980. *In* Insect Biology in the Future. VBW 80. M. Locke & D. S. Smith, Eds.: 345–367. Academic Press. New York.
29. SCHARRER, B. 1952. Biol. Bull. **102:** 261–272.
30. LOCOCO, D. J. & S. S. TOBE. 1984. Int. J. Insect Morphol. Embryol. **13:** 65–76.
31. PIPA, R. L. 1978. Cell Tissue Res. **193:** 443–455.
32. RUEGG, R. P., D. J. LOCOCO & S. S. TOBE. 1983. Experientia **39:** 1329–1334.
33. STAY, B., K. K. CHAN & A. P. WOODHEAD. 1992. Cell Tissue Res. **270:** 15–23.
34. STAY, B., A. P. WOODHEAD & K. K. CHAN. 1992. *In* Insect Juvenile Hormone Research: Fundamental and Applied Approaches. B. Mauchamp, F. Couillaud & J. C. Baehr, Eds.: 257–263. INRA. Paris.
35. STAY, B., S. S. TOBE & W. G. BENDENA. 1994. Adv. Insect Physiol. **25:** 267–337.
36. LANGE, A., K. K. CHAN & B. STAY. 1993. Arch. Insect Biochem. Physiol. **24:** 70–92.
37. YU, C. G., B. STAY, Q. DING, W. G. BENDENA & S. S. TOBE. 1995. J. Insect. Physiol. **41:** 1035–1043.
38. REICHWALD, K., G. C. UNNITHAN, N. T. DAVIS, H. AGRICOLA & R. FEYEREISEN. 1994. Proc. Natl. Acad. Sci. USA **91:** 11894–11898.
39. DUVE, H. & A. THORPE. 1994. Cell Tissue Res. **276:** 367–379.
40. TOBE, S. S. & B. STAY. 1977. Gen. Comp. Endocrinol. **31:** 138–147.
41. TOBE, S. S., R. P. RUEGG, B. STAY, F. C. BAKER, C. A. MILLER & D. A. SCHOOLEY. 1985. Experientia **41:** 1028–1034.
42. STAY, B., S. JOSHI & A. P. WOODHEAD. 1992. J. Insect Physiol. **37:** 63–70.
43. YU, C. G., B. STAY, S. JOSHI & S. S. TOBE. 1993. J. Insect. Physiol. **39:** 111–122.
44. TOBE, S. S., C. G. YU & W. G. BENDENA. 1994. *In* Perspectives in Comparative Endocrinology. K. G. Davey, R. E. Peter & S. S. Tobe, Eds.: 12–19. National Research Council of Canada. Ottawa.
45. WOODHEAD, A. P., W. Y. ASANO & B. STAY. 1993. J. Insect. Physiol. **39:** 1001–1005.
46. PAULSON, C. R. & B. STAY. 1987. J. Insect Physiol. **33:** 613–622.
47. PAULSON, C. R., B. STAY, S. KIKUKAWA & S. S. TOBE. 1987. Insect Biochem. **17:** 961–964.
48. WOODHEAD, A. P., C. A. STOLZMAN & B. STAY. 1992. Arch. Insect Biochem. Physiol. **20:** 253–263.
49. ROTH, L. M. & B. STAY. 1961. J. Insect. Physiol. **7:** 186–202.
50. LANGE, A. B., W. G. BENDENA & S. S. TOBE. 1995. J. Insect Physiol. **41:** 581–588.
51. YU, C. G., T. K. HAYES, A. STREY, W. G. BENDENA & S. S. TOBE. 1995. Regul. Pept. **57:** 347–358.
52. CUSSON, M., G. D. PRESTWICH, B. STAY & S. S. TOBE. 1991. Biochem. Biophys. Res. Commun. **181:** 736–742.
53. REAGAN, J. D. 1994. J. Biol. Chem. **269:** 9–12.
54. MCKELVY, J. P. & S. BLUMBERG. 1986. Annu. Rev. Neurosci. **9:** 415–434.
55. TURNER, A. J., R. MATSAS & A. J. KENNY. 1985. Biochem. Pharmacol. **34:** 1347–1356.
56. KENNY, A. J. & A. J. TURNER. 1987. *In* Mammalian Ectoenzymes. A. J. Kenny & A. J. Turner, Eds.: 1–7. Elsevier. Amsterdam.

Insect Neuropeptides of the Pyrokinin/PBAN Family Accelerate Pupariation in the Fleshfly (*Sarcophaga bullata*) Larvae[a]

JAN ZDAREK,[b] RONALD J. NACHMAN,[c]
AND TIMOTHY K. HAYES[d]

[b]Institute of Organic Chemistry and Biochemistry
Academy of Sciences of the Czech Republic
Flemingovo nam. 2
166 10 Prague 6, Czech Republic

[c]Veterinary Entomology Research Unit
Food Animal Protection Research Laboratory, USDA-ARS
2881 F&B Road
College Station, Texas 77845

[d]Department of Entomology
Biotechnology Support Laboratory
Texas A&M University
College Station, Texas 77843

INTRODUCTION

Neuropeptides that regulate various aspects of insect homeostasis have been isolated from several orders of insects.[1] Comparison of the chemical sequences revealed some common features among many of these peptides. Therefore it is not surprising that the same class of neuropeptides may serve different physiological and behavioral functions. For example, the pyrokinin/PBAN family of neuropeptides, originally identified as myotropins from orthopteroid insects,[2,3] appears to be widespread in several insect orders. Its members were found to serve such diverse functions as the induction of sex pheromone production in various female moths,[4] contraction of the locust oviduct,[3] melanization and reddish coloration in caterpillars,[5] and egg diapause in the silkworm.[6] Although these pyrokinin peptides have different numbers of amino acids,[7] they share the common C-terminal sequence FXPRLamide (X = G, S, T or V). The pentapeptide fragment proved to be the active core required not only for myotropic activity in the cockroach hindgut assay,[8] but also for pheromonotropic[9,10] and melanotropic activity,[5] and egg diapause inducing activity in the silkworm.[6,11]

The existence of cross activity of pyrokinin neuropeptides prompted us to screen various synthetic arthropod neuropeptides, including those of the pyrokinin/PBAN family, for the ability to accelerate formation of the puparium (pupariation) in larvae of the fleshfly, *Sarcophaga bullata*, a developmental process known to be under neuroendocrine control.[12]

[a] This study was supported by a Fulbright Fellowship (J.Z.) and the U.S. Department of Agriculture (R.J.N.).

67

In cyclorrhaphous Diptera, pupariation is a complex process involving both cuticular changes known as sclerotization (deplasticization and phenolic tanning) and behavioral activities (immobilization, anterior segment retraction, longitudinal body contraction).[13,14] This complex morphogenetic process is orchestrated by compounds of neurosecretory origin. The active agent was found in the hemolymph of pupariating larvae and in extracts of various neural or neurohemal organs as well.[15] Both behavioral and cuticular events were greatly accelerated when the active component was injected into fleshfly larvae 2–3 hours before the onset of pupariation.[16] Two separate factors were recognized, and following partial purification were named the puparium tanning factor (PTF) and the anterior (segment) retraction factor (ARF), according to the effects that the assayed materials produced in the tests used at that time.[17] The chemical structures of these pupariation factors have never been determined, but their proteinaceous nature and neurosecretory origin have been established.[12,18,19]

In the present study, we examined representative members of eight classes of biologically active insect neuropeptides for their effects on acceleration of pupariation behavior and puparial tanning in fleshfly larvae; we found members of one peptide family highly active.

MATERIALS AND METHODS

Fleshfly Pupariation Bioassay

The insect neuropeptides used in this study were synthesized via FMOC solid-phase methodology on Milligen/BioSearch 9600 and Vega Coupler 250 peptide synthesizers, cleaved from the resin anchor, and purified by high-performance liquid chromatography, as previously described.[9,11,20—24] The peptide samples were dissolved in a buffered Ringer solution[25] for injection into fleshfly larvae. Larvae of the fleshfly, *Sarcophaga bullata* (Parker), that had been bred on beef liver were used for all experiments. The test peptides were injected 2–3 hours before pupariation, with a calibrated glass capillary with a finely drawn tip, into red-spiracle larvae, that had previously been immobilized by chilling on ice. Control larvae were injected with Ringer solution only. After removal from the ice, the injected larvae were kept at 25 °C, and the time of anterior retraction (R), longitudinal contraction (C), and the onset of tanning (T) was recorded. Normally retraction precedes contraction, and tanning starts after the white puparium has been formed. The effects of a tested peptide were expressed as the difference between the mean time after which the developmental change (R, C, or T) occurred in the control and experimental groups of larvae. For the evaluation of direct myotropic action, the tested peptides were injected into the posterior part of a ligated larva. Peripheral effects of the tested peptides on the neuromuscular system are manifested by spontaneous contractions of the abdominal somatic muscles. Details of the pupariation assay can be obtained in an earlier publication.[18] A ligature placed in about one-third of the body (i.e., behind the fused central nervous system) causes denervation of all somatic musculature and consequently paralysis of the hind part, but does not prevent tanning if applied after the critical period for molting hormone release. Tanning of the hind part is, however, delayed in relation to tanning of the anterior part, which contains the central nervous system (CNS). In our bioassay a positive effect is manifested by acceleration of posterior tanning in relation to the anterior portion.

TABLE 1. Sequences of Representative Members of Eight Insect Neuropeptide Families

Peptide Family	Code	Peptide Sequence
Insect kinin	CDP-1	NPFHSWGamide
Pyrokinin/PBAN	LPK [4–8]LPK	pQTSFTPRLamide FTPRLamide
FMRFamide-related	LMS	pQVDHVFLRFamide
Insectatachykinin	CTK-1	APSGFMGMRamide
Sulfakinin	[F^8, Nle^9]LSK [S(SO$_3$H)^2]LSK-II	DY(SO$_3$H)GF[Nle]RFamide pQS(SO$_3$H)DDY(SO$_3$H)GHMRFamide
Diuretic peptides	*Musca* CRF-DP	NKPSLSIVNPLDRVLQRLLLEIAR- RQMKENTR-QVELNRAILKNVamide
AKH/RPCH	Bld-HrTH	pQVNFSPGWGTamide
Eclosion hormone	Mas-EH	HNPAIATGYDRXEICEINCANCKKXL- GAZFEGPLCAESCIKFKGKLIPECED- FASIAPPFLNKL-OH (X = Nle, Z = 4-Naphyl-Ala)

RESULTS

A high pupariation accelerating activity at a level of one pmol or less was recorded only in the peptides of the pyrokinin/PBAN family (TABLES 1 and 2). Marginal activity was detected in a diuretic peptide isolated from the housefly (tanning only)[21] and in the adipokinetic hormone from the cockroach *Blaberus discoidalis*.[25] Peptides of other families did not exhibit any activity at the highest dose tested (TABLES 1 and 2).

The activity patterns of leucopyrokinin (LPK), originally isolated from the cockroach *Leucophaea maderae*, and the C-terminal pentapeptide fragment injected at high doses differed from those produced by marginal doses or by hemo-

TABLE 2. Pupariation Acceleration Activity for Representative Members of Eight Insect Neuropeptide Families

Peptide Code	Threshold Concentration for Pupariation Acceleration (pmoles)	
	Pupariation Behavior (R, C)	Puparial Tanning
CDP-1	NA	NA
LPK	0.25	0.25
[4–8]LPK	0.25	2.5
LMS	NA	NA
CTK-1	NA	NA
[F^8, Nle^9]LSK	NA	NA
[S(SO$_3$H)^2]LSK-II	NA	NA
Musca-CRF-DP	NA	25
Bld-HrTH	250	250
Mas-EH	NA	NA

NA, not active at 250-pmole level.

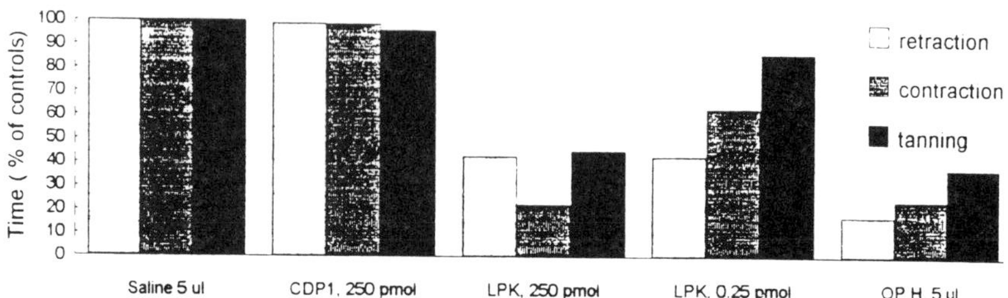

FIGURE 1. Pupariation accelerating activity of two synthetic neuropeptides, CDP (culekinin depolarizing peptide) and LPK (leucopyrokinin), compared with the activity of a hemolymph sample from fresh puparia (OPH). Mean times of anterior segment retraction, longitudinal body contraction, and onset of cuticular tanning were related to the mean times of saline-injected controls that were assigned a value of 100%.

lymph samples of pupariating larvae. Although in the latter cases the highest acceleration activity was recorded for anterior retraction, excessive doses of the synthetic LPK analogues accelerated longitudinal contraction to a greater degree than retraction of the anterior segments (FIG. 1). The analogue [4–8]LPK produced no appreciable muscular contractions when injected into isolated larval abdomens at a level of 25 pmoles. However, the analogue greatly accelerated tanning of larvae with ligated abdomens ($T_p/T_a = 0.46$, $n = 14$) in comparison with Ringer-injected controls ($T_p/T_a = 1.36$, $n = 15$; T_p and T_a represent mean times of tanning of the posterior and anterior parts, respectively).

DISCUSSION

Members of only one of the eight peptide families screened in this study accelerated puparium formation in fleshfly larvae at physiological concentrations. These active peptides belong to the pyrokinin/PBAN family, which is defined by the common pentapeptide fragment FXPRLamide (X = G, S, T or V) at the C-terminus. Four different physiological effects, namely, hindgut/oviduct myotropic, pheromonotropic, egg diapause induction, and melanization activities have been elicited experimentally by members of this family.[9] Here we report yet two more physiological effects of the pyrokinin peptides: acceleration of pupariation behavior and subsequent puparial tanning. By accelerating both aspects of puparium formation, LPK mimics the effects of the pupariation factors—the ARF that stimulates neuromuscular activity (behavior) and the PTF that is responsible for the timing of cuticular tanning and thus the synchronization of sclerotization of the cuticle with morphological changes during transformation of the larval body into an ovoid puparium. Because the ARF is an ethotropic factor (stimulates the release of stereotypic patterns of pupariation behavior), LPK is shown for the first time to exert an effect on the central motor neurons. Their possible direct peripheral action on the neuromuscular system has been ruled out by our finding that a ligated hind part (i.e., larval segments deprived of neural connection with the CNS) does not respond to exogenous LPK by muscular contractions. In contrast, tanning of the ligated hind parts was significantly accelerated, indicating that LPK elicits cuticular changes via a peripheral mechanism.

Our observation that the synthetic C-terminal pentapeptide fragment FTPRL-amide can elicit the activity of both pupariation factors suggests that this fragment, common to the pyrokinin/PBAN family, is sufficient for the induction of these effects, as it has been demonstrated for the other four physiological effects associated with the family.

The pupariation acceleration activity of the pyrokinin C-terminal sequence can explain earlier reports on the ARF/PTF activity of extracts of a large variety of neural and neurohemal organs of different arthropods (see ref. 19 for review), as well as semipurified neurohormones such as neurohormone D (Zdarek and Gersch, unpublished data). Despite considerable effort,[19] the chemical nature of pupariation factors remains elusive. They are believed to be rather large protein molecules with subunits of 26 kDa and 90 kDa for PTF and ARF, respectively, which are presumably released into the hemolymph from the peripheral nerve endings at the time of puparium formation.[17] The results of our present study suggest that the pupariation factors may also contain the C-terminal pentapeptide sequence FXPRLamide, or some portion thereof, shared by all members of the pyrokinin/PBAN family. This has not been confirmed to date, because isolation and chemical identification of the pupariation factors have been hampered by their large molecular size. However, the study of stage-specific immunoreactivity of larval and puparial hemolymph with LPK antibodies, currently under way, may yield indirect evidence for this hypothesis.

REFERENCES

1. HOLMAN, G. M., R. J. NACHMAN & M. S. WRIGHT. 1990. Insect neuropeptides. Annu. Rev. Entomol. **35:** 201–217.
2. HOLMAN, G. M., B. J. COOK & R. J. NACHMAN. 1986. Isolation, primary structure and synthesis of a blocked neuropeptide isolated from the cockroach, *Leucophaea maderae*. Comp. Biochem. Physiol. C **85:** 219–224.
3. SCHOOFS, L., G. M. HOLMAN, R. J. NACHMAN, T. K. HAYES & A. DELOOF. 1991. Isolation, primary structure, and synthesis of locustapyrokinin: A myotropic peptide of *Locusta migratoria*. Gen. Comp. Endocrinol. **81:** 97–104.
4. KITAMURA, A., H. NAGASAWA, H. KATAOKA, T. INOUE, S. MATSUMOTO, T. ANDO & A. SUZUKI. 1989. Amino acid sequence of pheromone-biosynthesis-activating-neuropeptide (PBAN) in the silkworm, *Bombyx mori*. Biochem. Biophys. Res. Commun. **163:** 520–526.
5. MATSUMOTO, S., A. KITAMURA, H. NAGASAWA, *et al.* 1990. Functional diversity of a neurohormone produced by the subeosophageal ganglian: Molecular identity of melanization and reddish coloration hormone and pheromone biosynthesis activating neuropeptide. J. Insect. Physiol. **36:** 427–432.
6. IMAI, K., T. KONNO, Y. NAKAZAWA, *et al.* 1991. Isolation and structure of diapause hormone of the silkworm moth, *Bombyx mori*. Proc. Jpn. Acad. **67:** 98–101.
7. NACHMAN, R. J., H. KUNIYOSHI, V. A. ROBERTS, G. M. HOLMAN & A. SUZUKI. 1993. Active conformation of the pyrokinin/PBAN neuropeptide family for pheromone biosynthesis in the silkworm. Biochem. Biophys. Res. Commun. **193:** 661–666.
8. NACHMAN, R. J., G. M. HOLMAN & B. J. COOK. 1986. Active fragments and analogs of the insect neuropeptide leucopyrokinin: Structure-function studies. Biochem. Biophys. Res. Commun. **137:** 936–942.
9. ABERNATHY, R. L., R. J. NACHMAN, P. E. A. TEAL, O. YAMASHITA & J. H. TUMLINSON. 1995. Pheromonotropic activity of naturally occurring pyrokinin insect neuropeptides (FXPRLamide) in *Helicoverpa zea*. Peptides **16:** 215–219.
10. KUNIYOSHI, H., H. NAGASAWA, T. ANDO, A. SUZUKI, R. J. NACHMAN & G. M. HOLMAN. 1992. Cross-activity between pheromone biosynthesis activating neuro-

peptide (PBAN) and myotropic pyrokinin insect peptides. Biosci. Biotech. Biochem. **56:** 167–168.

11. NACHMAN, R. J., G. M. HOLMAN, L. SCHOOFS & O. YAMASHITA. 1993. Silkworm diapause induction activity of myotropic pyrokinin (FXPRLamide) insect neuropeptides. Peptides **14:** 1043–1048.

12. ZDAREK, J. & P. SIVASUBRAMANIAN. 1991. Hormonal control of early metamorphosis in flies. *In* Morphogenetic Hormones of Arthropods. A. P. Gupta, Ed: 593–614. Rutgers University Press. New Brunswick, NJ.

13. ZDAREK, J. & G. FRAENKEL. 1972. The mechanism of puparium formation in flies. J. Exp. Zool. **179:** 315–324.

14. ZDAREK, J., K. SLAMA & G. FRAENKEL. 1979. Changes in internal pressure during puparium formation in flies. J. Exp. Zool. **207:** 187–196.

15. ZDAREK, J. & G. FRAENKEL. 1969. Correlated effects of ecdysone and neurosecretion in puparium formation in flies. Proc. Natl. Acad. Sci. USA **64:** 565–572.

16. FRAENKEL, G., J. ZDAREK & P. SIVASUBRAMANIAN. 1972. Hormonal factors in the CNS and haemolymph of pupariating fly larvae which accelerate puparium formation in flies. Biol. Bull. Woods Hole **143:** 127–139.

17. SIVASUBRAMANIAN, P., S. FRIEDMAN & G. FRAENKEL. 1974. Nature and role of proteinaceous hormonal factors acting during puparium formation in flies. Biol. Bull. Woods Hole **147:** 163–185.

18. ZDAREK, J. 1980. Neurohormonal factors involved in the control of pupariation. *In* Neurohumoral Techniques in Insects. T. A. Miller, Ed.: 154–178. Springer-Verlag. New York.

19. ZDAREK, J. 1985. Regulation of pupariation in flies. *In* Comprehensive Insect Physiology, Biochemistry and Pharmacology. G. A. Kerkut & L. I. Gilbert, Eds.: Vol. 8: 301–333. Pergamon Press. Oxford.

20. NACHMAN, R. J., G. M. HOLMAN, B. J. COOK, W. F. HADDON & N. LING. 1986. Leucosulfakinin-II, a blocked sulfated insect neuropeptide with homology to cholecystokinin and gastrin. Biochem. Biophys. Res. Commun. **140:** 357–364.

21. CLOTTENS, F. L., G. M. HOLMAN, G. M. COAST, N. F. TOTTY, T. K. HAYES, I. KAY, A. I. MALLET, M. S. WRIGHT, J. S. CHUNG, O. TRUONG & D. L. BULL. 1994. Isolation and characterization of a diuretic peptide common to the house fly and stable fly. Peptides **15:** 971–979.

22. NACHMAN, R. J., G. M. HOLMAN, T. K. HAYES & R. C. BEIER. 1993. Structure-activity relationships for inhibitory insect myosuppressins: Contrast with the stimulatory sulfakinins. Peptides **14:** 665–670.

23. HAYES, T. K., G. M. HOLMAN, T. L. PANNABECKER, M. S. WRIGHT, A. A. STREY, R. J. NACHMAN, D. F. HOEL, J. K. OLSEN & K. W. BEYENBACH. 1994. Culekinin depolarizing peptide: A mosquito leucokinin-like peptide that influences insect Malpighian tubule ion transport. Regul. Pept. **52:** 235–248.

24. HAYES, T. K., L. L. KEELEY & D. W. KNIGHT. 1986. Insect hypertrehalosemic hormone: Isolation and primary structure from *Blaberus discoidalis* cockroaches. Biochem. Biophys. Res. Commun. **140:** 674–678.

25. HAYES, T. K. & L. L. KEELEY. 1985. Properties of an in vitro bioassay for hypertrehalosemic hormone of *Blaberus discoidalis* cockroaches. Gen. Comp. Endocrinol. **57:** 246–256.

Pupariation Acceleration in Fleshfly (*Sarcophaga bullata*) Larvae by the Pyrokinin/PBAN Neuropeptide Family

Structure-Activity Relationships[a]

RONALD J. NACHMAN,[b,e] JAN ZDAREK,[c]
G. MARK HOLMAN,[b] AND TIMOTHY K. HAYES[d]

[b]Veterinary Entomology Research Unit
Food Animal Protection Research Laboratory
USDA-ARS
2881 F&B Road
College Station, Texas 77845

[c]Institute of Organic Chemistry and Biochemistry
Academy of Sciences of the Czech Republic
Prague 6, Czech Republic

[d]Department of Entomology
Texas A&M University
College Station, Texas 77843

INTRODUCTION

Pupariation in cyclorrhaphous Diptera has been recognized as a complex process involving both behavioral activities (immobilization, anterior segment retraction, longitudinal body contraction) and cuticular changes known as sclerotization (deplasticization and phenolic tanning).[1,2] This complex process has been shown to be controlled and regulated by proteinaceous products of the neuroendocrine system. The active principle(s) has been found in extracts of various neural or neurohemal organs, as well as in the hemolymph of pupariating larvae.[3] When injected into fleshfly (*Sarcophaga bullata*) larvae 2–3 hours before the onset of pupariation, both behavioral and cuticular events were greatly accelerated by these factors.[4] The chemical identity of these pupariation factors has not as yet been characterized, but their protein-like nature and neurosecretory origin have been firmly established.[5-7]

A recent study[8] has established that leucopyrokinin (LPK), a member of the pyrokinin/PBAN insect neuropeptide family, accelerates the onset of both aspects of pupariation following injection in larvae of the fleshfly. A representative of this family was tested in the pupariation assay not only because it is associated with such physiological activities as hindgut/oviduct contraction in the cockroach and locust[9,10] and pheromone biosynthesis in the corn earworm, tobacco budworm, and silkworm moths,[11-13] but specifically because it is associated with the develop-

[a] This study was supported in part by a Fulbright Fellowship (J.Z.) and the United States Department of Agriculture (R.J.N. and G.M.H.).

[e] Corresponding author.

mental processes of egg diapause induction and melanization in the silkworm.[14,15] Representatives of seven other insect neuropeptide families failed to elicit pupariation acceleration at physiological concentrations.[8] The pyrokinin family comprises peptides with 8 to 33 amino acid residues and shares the common C-terminal pentapeptide FXPRLa (X = G, S, T or V),[13] a fragment of sufficient length to elicit significant pupariation acceleration activity in the fleshfly.[8] In this study, we explore the relationship between the chemical structure and conformation of analogues of the pyrokinin/PBAN insect neuropeptide family and pupariation acceleration activity in the fleshfly *Sarcophaga bullata*.

MATERIALS AND METHODS

The insect peptide and pseudopeptide analogues used in this study were synthesized via FMOC solid-phase methodology on Milligen/BioSearch 9600 and Vega Coupler 250 peptide synthesizers, cleaved from the resin, and purified by high-performance liquid chromatography according to previously described procedures.[13,16,17] The peptide samples were dissolved in a buffered Ringer solution[18] for injection.

Pupariation Bioassay

Larvae of the fleshfly, *Sarcophaga bullata* (Parker), that had been bred on beef liver were used for all experiments. The peptide analogues were injected using a calibrated glass capillary with a finely drawn tip, into red-spiracle larvae 2–3 hours before pupariation; previously the larvae had been immobilized by chilling on ice. Control larvae were injected with Ringer solution only. After removal from the ice, the injected larvae were kept at 25 °C, and the time of the onset of tanning (T) and the behavioral aspects (B) of anterior retraction and longitudinal contraction was recorded. Details of the pupariation assay can be obtained from an earlier description.[7] The effects of a test peptide analogue were expressed as a difference between the mean time after which the developmental change occurred in the control versus the experimental groups of larvae. Peptide analogues showing strong puparium acceleration activity typically shortened the latency period to less than 50% of the control time period. Lowering the dose of such analogues usually gave rise to a dose-dependent increase of the latency period. To facilitate a quantitative comparison of various tested peptide analogues, we adopted the concept of threshold dose. The threshold dose was defined as the lowest dose capable of shortening the latency period to shorter or equal to 80% of that in control groups.

RESULTS AND DISCUSSION

The results of the evaluation of the analogues of the pyrokinin/PBAN insect neuropeptide family in a *Sarcophaga bullata* fleshfly pupariation assay are summa-

TABLE 1. Pupariation Acceleration Activity of Analogues of the Pyrokinin/ PBAN Neuropeptide Family in the Fleshfly *Sarcophaga bullata*

	Threshold Concentrations for Pupariation Acceleration (pmole)	
Peptide Sequence	Behavioral Aspect	Tanning Aspect
pGlu-Thr- Ser- Phe-Thr-Pro-Arg-Leu-NH$_2$[a]	0.25	0.25
Ser- Phe-Thr-Pro-Arg-Leu-NH$_2$	2.5	2.5
Phe-Thr-Pro-Arg-Leu-NH$_2$	0.25	2.5
Thr-Pro-Arg-Leu-NH$_2$	0.25	0.25
Pro-Arg-Leu-NH$_2$	2.5	2.5
Phe-Thr-Pro-Arg-NH$_2$	NA[b]	NA[b]
Tyr- Phe-Thr-Pro-Arg-Ala- NH$_2$	NA[b]	NA[b]
Tyr- Phe-Thr-Pro-Ala- Leu-NH$_2$	NA[b]	NA[b]
Tyr- Phe-Thr-Ala-Arg-Leu-NH$_2$	2.5	2.5
Tyr- Phe-Ala-Pro-Arg-Leu-NH$_2$	0.25	0.25
Tyr- Ala-Thr-Pro-Arg-Leu-NH$_2$	2.5	2.5
Tyr- Phe-Thr-Pro-Arg-Aib- NH$_2$	NA[b]	NA[b]
pGlu-Thr- Ser- Phe-Thr-Pro-Arg-Leu-OH	250	250
pGlu-Thr- Ser- Phe-Thr-Pro-Lys-Leu-NH$_2$	2.5	2.5
pGlu-Thr- Ser- Phe-Thr-Pro-Asn-Leu-NH$_2$	NA[b]	NA[b]
pGlu-Thr- Ser- Tyr-Thr-Pro-Arg-Leu-NH$_2$	2.5	2.5
pGlu-Thr- Ser- Phe-Thr-Gly-Arg-Leu-NH$_2$	(250)[c]	(250)[c]
pGlu-Thr- Ser- Phe-Thr-Pro-Arg-Ala- NH$_2$	25	250
pGlu-Thr- Ser- Phe-Thr-Pro-Arg-Trp- NH$_2$	NA[b]	NA[b]
pGlu-Thr- Ser- Phe-Ser-Pro-Arg-Leu-NH$_2$	2.5	250
pGlu-Thr- Thr- Phe-Thr-Pro-Arg-Leu-NH$_2$	0.25	0.25
Glu-Gly- Asp-Phe-Thr-Pro-Arg-Leu-NH$_2$[d]	0.25	0.25
Arg-Gln-Gln- Pro- Phe-Val-Pro-Arg-Leu-NH$_2$[e]	2.5	25
Arg-Leu-His-Gln-Asn-Gly-Met-Pro- Phe-Ser- Pro-Arg-Leu-NH$_2$[f]	2.5	250
cyclo[Asn-Thr- Ser- Phe-Thr-Pro-Arg-Leu][g]	2.5	25
Aib- Aib- Phe-Thr-Pro-Arg-Leu-NH$_2$[h]	<0.25	0.25
Hca-Thr-Pro-Arg-Leu-NH$_2$[i]	<0.25	0.25
4Pbm-tCpd-Arg-Leu-NH$_2$[j]	250	250

[a] Leucopyrokinin (LPK).[9]
[b] NA, Inactive at 250 pmoles.
[c] Inhibited pupariation.
[d] Lom-MT-II.[10]
[e] Lom-MT-III.[10]
[f] Lom-MT-IV.[10]
[g] *cyclo*LPK.[16]
[h] Aib = aminoisobutyryl-.
[i] Hca = hydrocinnamyl-.
[j] 4Pbm = 4-phenylbutyrylamino-; tCpd = *trans*-1,2-cylopentanedicarboxyl.

rized in TABLE 1. As with other pyrokinin bioassay systems, the C-terminal penta-peptide FXPRLa (X = T in this case), common to all members of the pyrokinin/ PBAN family, is sufficient to elicit acceleration of pupariation in fleshfly larvae. In the other pyrokinin assays (hindgut,[9] oviduct,[19] pheromone biosynthesis,[20,21] egg diapause,[22] and melanization[15]), the pentapeptide represents the "active core," although it elicits only a fraction of the activity of the parent and/or endogenous peptide factors. In contrast, the pentapeptide demonstrates equipotency in the behavioral aspect and about 10% of the potency of the tanning aspect of the pupariation response. Furthermore, the C-terminal tetrapeptide TPRLa is equipotent with the full LPK sequence for both aspects of the response. The pupariation acceleration activity of the pyrokinin C-terminal tripeptide PRLa demonstrates markedly reduced but significant activity at 25 pmoles, or 1% of the response of the parent LPK and the tetrapeptide fragment. Thus, the tripeptide would appear to be the "active core" for pupariation acceleration, although evaluation of the dipeptide remains to be carried out. The core sequence for a *full* response would be the tetrapeptide TPRLa. This represents a *major* difference in the activity profile observed in the other pyrokinin assays. As in other pyrokinin assays, truncation of the pentapeptide at the C-terminus (i.e., FTPRa) results in complete loss of activity.

The evaluation of an Ala replacement series of analogues demonstrates that the two C-terminal residues are the most critical for pupariation acceleration. The replacement of Arg with either Ala or Asn leads to inactive sequences. When Arg is replaced with another basic residue, Lys, activity is retained, though at a reduced level (10%). Therefore, a basic, positively charged group is required at this position for activity. The analogues YFTPRAa and pQTSFTPRAa, in which Ala replaces Leu, demonstrate no activity and very low activity (1% B, 0.1% T), respectively. This delineates the importance of the branched side chain character of Leu at this position. The activity of the pyrokinins is additionally sensitive to changes in steric bulk at the C-terminal position occupied by Leu. For instance, the replacement of Leu with larger Trp or sterically hindered Aib residues leads to inactivity. The C-terminal amide group is also of great importance to pupariation acceleration, because LPK acid (LPK-OH) demonstrates a large drop in threshold activity (0.1%). The residues Phe, Thr, and Pro in the N-terminal portion of the pyrokinin pentapeptide can be replaced with Ala without complete loss of activity. The Thr, occupying the variable X position of the pyrokinin pentapeptide, demonstrates the most tolerance to Ala substitution because the resulting analogue is equipotent with the parent LPK. Replacement of either Phe or Pro with Ala leads to a reduced but significant level of activity (10%). The conservative replacement of Phe with Tyr led to a similar reduction in activity (10%). Outside of the C-terminal pentapeptide region, the conservative replacement of Ser with Thr, both bearing hydroxyl groups on the side chain, led to an analogue equipotent to LPK (TABLE 1).

The pupariation acceleration activity of several naturally occurring, myotropic locust pyrokinins include Lom-MT-II (equipotent), as well as Lom-MT III (10% B, 1% T) and Lom-MT-IV (10% B, 0.1% T). Notably, the variable X position in the C-terminal pentapeptide region of the highly active Lom-MT-II is the same as in LPK (Thr). The X positions of Lom-MT-III and Lom-MT-IV are Val and Ser, respectively. Replacement of the Thr in the X position of the LPK sequence with Ser also led to a reduction in pupariation acceleration activity (10% B, 10% T) (TABLE 1).

Significant activity was elicited by the conformationally constrained LPK analogue *cyclo*[Asn-Thr-Ser-Phe-Thr-Pro-Arg-Leu] (10% B, 1% T). Spectroscopic studies coupled with computer molecular dynamics calculations indicate that this highly rigid analogue contains a type I β-turn over the four residues Thr-Pro-Arg-Leu within the C-terminal pentapeptide region (FIG. 1). The Pro was determined to be exclusively in the *trans* orientation.[16,23] The rigidity of the well-defined backbone structure suggests that a turn conformation is adopted by the pyrokinins at the pupariation acceleration receptor site. It is notable that the core sequence for full activity, TRPLa, coincides with the four residues that compromise the β-

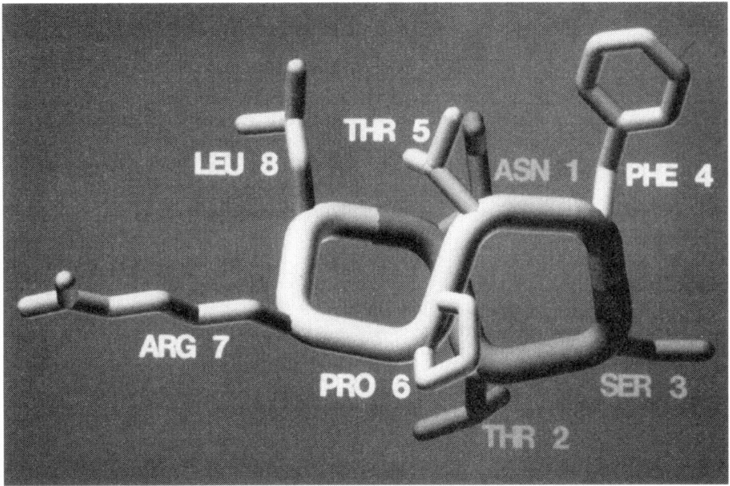

FIGURE 1. The backbone of the pyrokinin/PBAN analogue *cyclo*[Asn-Thr-Ser-Phe-Thr-Pro-Arg-Leu] is a conformationally constrained "saddle-shaped" fold as indicated by molecular dynamics and NMR data.[16,23] The residues Thr-Pro-Arg-Leu adopt a type I β-turn with an exclusively *trans* peptide bond between Thr and Pro. (From Nachman *et al.*[23] Reprinted with permission, from *Biochemical and Biophysical Research Communications*.)

turn. The analogue Aib-Aib-FTPRLa is extremely active (>100% B, 100% T). The aminoisobutyryl (Aib) residues at the N-terminus of the analogue each contain two methyl groups at the alpha carbon, whose steric interactions with neighboring residues promote turn formation in the rest of the sequence.[24,25] Indeed, this Aib analogue (in aqueous solution) demonstrates an α-helical type-C circular dichroism (CD) spectrum (Nachman, R. J., unpublished data) very similar to the one obtained from the rigid cycle LPK analogue[16] and indicative of the presence of a helical type I β-turn.[26] The high activity elicited by this analogue is probably due to resistance to aminopeptidases as well as its propensity to form a turn. The analogue Hca-TPRLa (Hca = hydrocinnamyl-) lacks the N-terminal amino group and is resistant to aminopeptidase M.[27,28] This pseudotetrapeptide also elicits an extremely potent pupariation acceleration response (>100% B, 10% T). The pseudo-dipeptide 4Pbm-tCpd-RLa (4Pbm = 4-phenylbutylamino-; tCpd = *trans*-1,2-cyclopentane-dicarboxyl-) contains a carbocyclic Pro-mimetic analogue that retains measurable activity (0.1% B, 0.1% T) despite the major structural modifications incorporated into its sequence. It retains an ability to form a hydrogen bond, involving the Leu NH, that stabilizes the β-turn in the rigid cyclic LPK analogue[17] (FIG. 1).

Replacement of the Pro residue in LPK with Gly led to an analogue with unusual biological properties. It did not demonstrate pupariation acceleration activity, but at a dose of 250 pmoles it delays the onset of both the behavioral and tanning aspects of pupariation in the fleshfly. The data suggests that the Gly analogue of LPK may operate as an antagonist of the pupariation response of LPK. This possibility will be the subject of future investigations. The inactivity of the Gly analogue also provides some evidence that more than the dipeptide RLa is required for pupariation acceleration activity.

The above structure-activity studies provide a possible explanation for the measurable tanning activity observed for housefly diuretic peptide *Musca*-CRF-DP (N.A. B, 1% T), which contains 44 amino acids.[29] The tripeptide Q^{16}-R^{17}-L^{18}, which resembles the pyrokinin tripeptide PRL, can be found in the middle of the *Musca*-CRF-DP sequence. It should be noted that the Pro residue can be replaced with an Ala without complete loss of activity. Therefore, the Pro can probably be replaced by other residues (e.g., Gln) as well.

The structure-activity data suggest that uncharacterized natural pupariation factors in extracts of insect neural and neurohemal organs as well as hemolymph[3–5] may contain the pyrokinin C-terminal tetrapeptide XPRLa (X = G, S, T, V or another residue) within its sequence. The study of stage-specific immunoreactivity of larval and puparial hemolymph with antibodies to LPK may yield indirect evidence for this hypothesis. These and future structure-activity and conformation-activity studies can aid in the design and synthesis of pseudopeptide and nonpeptide mimetic agonist and antagonist analogues of the pyrokinins with the potential to disrupt the pupariation of cyclorrhaphous Diptera. Such mimetic analogues, in isolation or in combination with other factors, could have potential use in future pest insect management strategies.

REFERENCES

1. ZDAREK, J. & G. FRAENKEL. 1972. The mechanism of puparium formation in flies. J. Exp. Zool. **179:** 315–324.
2. ZDAREK, J., K. SLAMA & G. FRAENKEL. 1979. Changes in internal pressure during puparium formation in flies. J. Exp. Zool. **207:** 187–196.
3. ZDAREK, J. & G. FRAENKEL. 1969. Correlated effects of ecdysone and neurosecretion in haemolymph of pupariating fly larvae which accelerate puparium formation and tanning. Proc. Natl. Acad. Sci. USA **64:** 565–572.
4. FRAENKEL, G., J. ZDAREK & P. SIVASUBRAMANIAN. 1972. Hormonal factors in the CNS and haemolymph of pupariating fly larvae which accelerate puparium formation and tanning. Biol. Bull. Woods Hole **143:** 127–139.
5. ZDAREK, J. & P. SIVASUBRAMANIAN. 1991. Hormonal control of early metamorphosis in flies. *In* Morphogenetic Hormones of Arthropods. A. P. Gupta, Eds.: 593–614. Rutgers University Press. New Brunswick, NJ.
6. ZDAREK, J. 1985. Regulation of pupariation in flies. *In* Comprehensive Insect Physiology, Biochemistry and Pharmacology. G. A. Kerkut & L. I. Gilbert, Eds. Vol. 8: 301–333. Pergamon Press. New York.
7. ZDAREK, J. 1980. Neurohormonal factors involved in the control of pupariation. *In* Neurohormonal Techniques in Insects. T. A. Miller, Ed.: 154–178. Springer Verlag. New York.
8. ZDAREK, J., R. J. NACHMAN & T. K. HAYES. 1996. Insect neuropeptides of the pyrokinin/PBAN family accelerate pupariation in the fleshfly (*Sarcophaga bullata*) larvae. Ann. N.Y. Acad. Sci. This volume.
9. NACHMAN, R. J., G. M. HOLMAN & B. J. COOK. 1986. Active fragments and analogs of the insect neuropeptide leucopyrokinin: Structure-function studies. Biochem. Biophys. Res. Commun. **137:** 936–942.
10. SCHOOFS, L., G. M. HOLMAN, T. K. HAYES, R. J. NACHMAN, J. P. KOCHANSKY & A. DELOOF. 1992. Isolation, identification and synthesis of locustamyotropin III and IV, two additional neuropeptides of *Locusta migratoria*: Members of the locustamyotropin family. Insect Biochem. Mol. Biol. **22:** 447–452.
11. RAINA, A. K., H. JAFFE, T. G. KEMPE, P. KEIM, R. W. BLACHER, H. M. FALES, C. T. RILEY, J. A. KLUN, R. L. RIDGEWAY & D. K. HAYES. 1989. Identification of a neuropeptide hormone that regulates sex pheromone production in female moths. Science **244:** 796–798.

12. KITAMURA, A., H. NAGASAWA, H. KATAOKA, T. INOUE, S. MATSUMOTO, T. ANDO & A. SUZUKI. 1989. Amino acid sequence of pheromone-biosynthesis-activating neuropeptide (PBAN) of the silkworm, *Bombyx mori*. Biochem. Biophys. Res. Commun. **163:** 520–526.

13. ABERNATHY, R. L., R. J. NACHMAN, P. E. A. TEAL, O. YAMASHITA & J. H. TUMLINSON. 1995. Pheromonotropic activity of naturally occurring pyrokininsect neuropeptides (FXPTLamide) in *Helicoverpa zea*. Peptides **16:** 215–219.

14. IMAI, K., T. KONNO, Y. NAKAZAWA, T. KOMIYA, M. ISOBE, K. KOGA, T. GOTO, T. YAGINUMA, K. SAKAKIBARA, K. HASEGAWA & O. YAMASHITA. 1991. Isolation and structure of diapause hormone of the silkworm, *Bombyx mori*. Proc. Jpn. Acad. B **67:** 98–101.

15. MATSUMOTO, S., A. FONAGY, L. SCHOOFS, *et al.* 1993. Induction of cuticular melanization in the armyworm larvae, *Pseudoletia separata*, by insect myotropic peptides possessing FXPRLamide at the C-terminus. J. Pestic. Sci. **18:** 127–129.

16. NACHMAN, R. J., V. A. ROBERTS, H. J. DYSON, G. M. HOLMAN & J.A. TAINER. 1991. Active conformation of an insect neuropeptide family. Proc. Natl. Acad. Sci. USA **88:** 4518–4522.

17. NACHMAN, R. J., V. A. ROBERTS, G. M. HOLMAN & R. C. BEIER. 1995. Pseudodipeptide analogs of the pyrokinin/PBAN (FXPRLa) insect neuropeptide family containing carbocyclic Pro-mimetic conformational components. Regul. Pept. **57:** 359–370.

18. HAYES, T. K. & L. KEELEY. 1985. Properties of an in vitro bioassay for hyperthalassemic hormone of *Blaberus discoidalis* cockroaches. Gen. Comp. Endocrinol. **57:** 246–256.

19. NACHMAN, R. J. & G. M. HOLMAN. 1991. Myotropic insect neuropeptide families from the cockroach *Leucophaea maderae*: Structure-activity relationships. *In* Insect Neuropeptides: Chemistry, Biology and Action. J. J. Menn, T. J. Kelly & E. P. Masler, Eds.: 194–214. American Chemical Society. Washington, DC.

20. RAINA, A. K. & T. G. KEMPE. 1990. A pentapeptide of the C-terminal sequence of PBAN with pheromonotropic activity. Insect Biochem. **18:** 849–851.

21. KUNIYOSHI, H., R. A. NAGASAWA, A. SUZUKI, R. J. NACHMAN & G. M. HOLMAN. 1992. Cross-activity between pheromone biosynthesis activating neuropeptide (PBAN) and myotropic pyrokinin insect peptides. Biosci. Biotechnol. Biochem. **56:** 167–168.

22. NACHMAN, R. J., G. M. HOLMAN, L. SCHOOFS & O. YAMASHITA. 1993. Silkworm diapause induction activity of myotropic pyrokinin (FXPRLamide) insect peptides. Peptides **14:** 1043–1048.

23. NACHMAN, R. J., H. KUNIYOSHI, V. A. ROBERTS, G. M. HOLMAN & S. SUZUKI. 1993. Active conformation of the pyrokinin/PBAN neuropeptide family for pheromone biosynthesis in the silkworm. Biochem. Biophys. Res. Commun. **193:** 661–666.

24. TALLON, M., D. RON, D. HALLE, P. AMODEO, G. SAVIANO, P. A. TEMUSSI, Z. SELINGER, F. NAIDER & M. CHOREV. 1993. Synthesis, biological activity, and conformational analysis of [pGlu[6], NMePhe[8], Aib[9]]substance P (6–11): A selective agonist for the NK-3 receptor. Biopolymers **33:** 915–926.

25. TONIOLO, C., G. M. BONORA, A. BAVOSO, E. BENEDETTI, B. DI BLASIO, V. PAVONE & C. PEDONE. 1983. Preferred conformations of peptide analogs containing α,α-disubstituted α-amino acids. Biopolymers **22:** 205–214.

26. GIERASCH, L. M., C. M. DIEBER, V. MADISON, C. H. NIU & E. R. BLOUT. 1981. Conformations of (X-L-Pro-Y)$_2$ cyclic hexapeptides. Preferred β-turn conformers and implications for β turns in proteins. Biochemistry **20:** 4730–4738.

27. NACHMAN, R. J., G. M. HOLMAN, T. K. HAYES & R. C. BEIER. 1993. Acyl, pseudotetra-, tri- and dipeptide active-core analogs of insect neuropeptides. Int. J. Peptide Protein Res. **42:** 372–377.

28. NACHMAN, R. J., G. M. HOLMAN, W. F. HADDON & W. H. VENSEL. 1991. An active pseudopeptide analog of the leucokinin insect neuropeptide family. Int. J. Peptide Protein Res. **37:** 220–223.

29. CLOTTENS, F. L., G. M. HOLMAN, G. M. COAST, N. F. TOTTY, T. K. HAYES, I. KAY, A. I. MALLET, M. S. WRIGHT, J. S. CHUNG, O. TRUONG & D. L. BULL. 1994. Isolation and characterization of a diuretic peptide common to the house fly and stable fly. Peptides **15:** 971–979.

Neurally Derived Factors That Affect *in Vitro* Ecdysteroid Production by Prothoracic Glands of the Gypsy Moth, *Lymantria dispar*

Possible Developmental Regulators

EDWARD P. MASLER[a]

Insect Neurobiology and Hormone Laboratory
U.S. Department of Agriculture
Agricultural Research Service
Beltsville, Maryland 20705-2350

INTRODUCTION

Early in this century, the Polish biologist Stephan Kopec conducted a series of experiments on the development of the gypsy moth, *Lymantria dispar*. He was particularly interested in the phenomenon of metamorphosis where *L. dispar*, like other holometabolous insects, undergoes dramatic morphological changes as it proceeds through larval and pupal instars and into the adult stage. Kopec isolated portions of the larval body from the influence of the head by various measures. Using ligatures to limit the humoral contact with the head of specific body sections, he observed that those sections that had the ligature between themselves and the head remained larval. In the same animals, those sections in humoral contact with the head proceeded to develop pupal features. In other experiments, removal of the brain resulted in larvae that could live for weeks and did not develop into pupae, but instead remained as larvae. In larvae where the brain was left in place, but posterior nervous connections were severed, metamorphosis into pupae occurred. These observations led Kopec to suggest that ". . . the brain should be considered as an organ of internal secretion."[1] This prophetic statement foreshadowed the advent of studies on neurosecretion, and Kopec's report was the first on any animal system to suggest that the nervous and endocrine systems were anatomically and functionally linked. Kopec's basic conclusion was supported experimentally by numerous laboratories,[2-8] and contributions from insect studies have influenced the field of neurobiology and elucidated the involvement of neuropeptides in development.[9]

Neuropeptides, involved in essentially all physiological processes in insects, have been and continue as the subjects of intensive research.[10-13] The cerebral neuropeptide responsible for much of the development that Kopec observed is now known to be an ecdysiotropic neuropeptide, prothoracicotropic hormone (PTTH),[14] which stimulates the larval prothoracic glands to produce ecdysteroids

NOTE: Mention of a proprietary product does not constitute endorsement by the USDA.

[a] Address correspondence to Dr. E. P. Masler, USDA, ARS, INHL, 10300 Baltimore Avenue, B-306, R-322, BARC-East, Beltsville, MD 20705-2350. E-mail: emasler@asrr.arsusda.gov

necessary for larval growth and development. This neuropeptide has been widely studied in a number of lepidopterans and fulfills a similar ecdysiotropic role in each of them.[12,14–16] Cerebral ecdysiotropins are also known to be involved in ovarian maturation in dipterans.[17] In addition, noncerebral ecdysiotropins have been described in dipterans,[18] and a noncerebral site has been implicated in the control of molting and ecdysteroid production *in vivo* in *L. dispar*.[19] A factor that inhibits ecdysteroid production ("ecdysiostatin") and follicle development has been described in mature ovaries from the dipteran *Neobellieria bullata*.[20] Presented here, for the first time, is evidence which suggests that, in addition to cerebral PTTH, a lepidopteran, *L. dispar*, contains both noncerebral ecdysiotropic activity and ecdysiostatic factors.

MATERIALS AND METHODS

Insect Rearing

The culture of *L. dispar* used in this study originated at the USDA/APHIS Methods Development Laboratory, Otis Air Force Base, Massachusetts, and is now maintained in the laboratory according to Bell.[21] A diapausing strain is used and eggs are chilled (5–7 °C, 180 days) to break diapause. Eggs are transferred to a rearing chamber and held at 26 °C, 50–60% relative humidity, and a 16 hL: 8 hD photoperiod. Larvae hatch 2–3 days later and are reared on a high wheat-germ diet[21] under the above conditions. Female larvae undergo four molts during the period 14–16 days after hatching. Late fourth instar larvae are examined each day between 12 noon and 4 P.M. (lights on at 6 A.M.) for evidence of an imminent molt (e.g., head capsule slippage[19]). Newly molted fifth instars are designated as day-1 larvae (L_5D_1). Larvae used in this study were day 5 (L_5D_5), and adults were 1–2 days old (post-eclosion).

Tissue Collection

Tissues used for the extraction of ecdysiotropic and ecdysiostatic activities were dissected from CO_2-anesthetized animals. Larvae were pinned ventral side up in the dissecting dish flooded with saline,[22] and a posterior-to-anterior incision made along the ventral midline. For collection of the brain-subesophageal ganglion (brain-SOG) complex, an additional incision was made in the head. Eleven ganglia were identified along the ventral nerve cord (VNC) and collected in separate pools. Adult brain-SOG complexes were collected by removing the head from anesthetized animals, cleaning off cuticular scales, and making an incision in the centerline of the head capsule. All tissues were collected in polypropylene tubes on dry ice and stored at −80 °C until use.

Bioassay and Radioimmunoassay

Prothoracic glands were removed from L_5D_5 animals, anesthetized and dissected as above, and collected in saline. Glands were then transferred to 25 μL drops of Grace's medium (Gibco, Grand Island, NY) at 1 gland/drop and incubated for 1 h at 26 °C. Medium was removed and replaced with fresh medium for the second hour. This second-hour medium was collected and saved. Fresh medium or medium containing tissue extracts or test samples was added for the third hour and collected. Media from the second- and third-hour incubations were analyzed by RIA for ecdysteroid content. The difference in ecdysteroid levels between second-hour (pre-treatment) and third-hour (post-treatment) media (net synthesis) was used to indicate levels of stimulation. RIA was conducted as previously described.[23,24]

Tissue Extraction and Fractionation

Tissue extracts for incubation were prepared by adding Grace's medium to the polypropylene storage tubes and homogenizing by hand using a tight-fitting polypropylene pestle. For each incubation experiment, typically 5–10 tissues (ganglia or brain-SOG) were extracted in 25 μL medium/tissue. Homogenates were centrifuged at 16,000 × g for 3 min at room temperature, and the supernatants used for incubation. Dilutions were prepared using Grace's medium. Tissues used for fractionation were homogenized in saline using a Polytron homogenizer (Brinkman Instruments, Long Island, NY). Adult brains (day 1–2) were homogenized (100 brain-SOGs in 2.5 mL) and the homogenate centrifuged at 20,000 × g, 5 min, 4 °C. The supernatant was collected, heated 5 min in a boiling water bath, chilled on ice, then centrifuged as above. The supernatant was concentrated to 100 μL under vacuum centrifugation (Speed-Vac, Savant Instruments, Farmingdale, NY), diluted with 400 μL of 0.1% TFA (Aldrich Chemical Co., Milwaukee, WI), and centrifuged as above. The supernatant was fractionated using a TSK 2000SW$_{XL}$ HP-SEC column (7.8 × 300 mm; TosoHaas, Montgomeryville, PA) fitted with a guard column (3.9 × 10 mm). Elution conditions were 0.5 mL/min, 40% CH_3CN (Fisher Scientific, Springfield, VA) in 0.1% aqueous trifluoroacetic acid (TFA). Standards used for column calibration were carbonic anhydrase (29,000), cytochrome C (12,300), aprotinin (6,500), and insulin chain B (3,196), all from Sigma Chemical Co., St. Louis, Missouri, and gastrin (2,126) and locust adipokinetic hormone-I (AKH, 1,159), both from Peninsula Laboratories, Belmont, California. Fractions were collected each minute from 12–24 min (6–12 mL elution volume) into 1.5 mL polypropylene tubes, then dried in the vacuum centrifuge. For bioassay of the ecdysiostatic activity of the column fractions, dried fractions were dissolved in brain-SOG extract (prepared as above at 0.02 brain-SOG equivalent/μL), and dilutions of the column fractions prepared in the same extract.

RESULTS

In vitro *Prothoracic Gland Stimulation*

Larval (L_5D_5) prothoracic glands respond to exposure to adult brain-SOG extract by producing ecdysteroid in a dose-dependent manner (FIG. 1). Stimulation is initially observed at a dose near 0.0015 brain-SOG equiv/μL (mean ± SE; 28 ± 11 pg 20-OH-ecdysone equiv/gland/h), and is maximal at a dose near 0.015

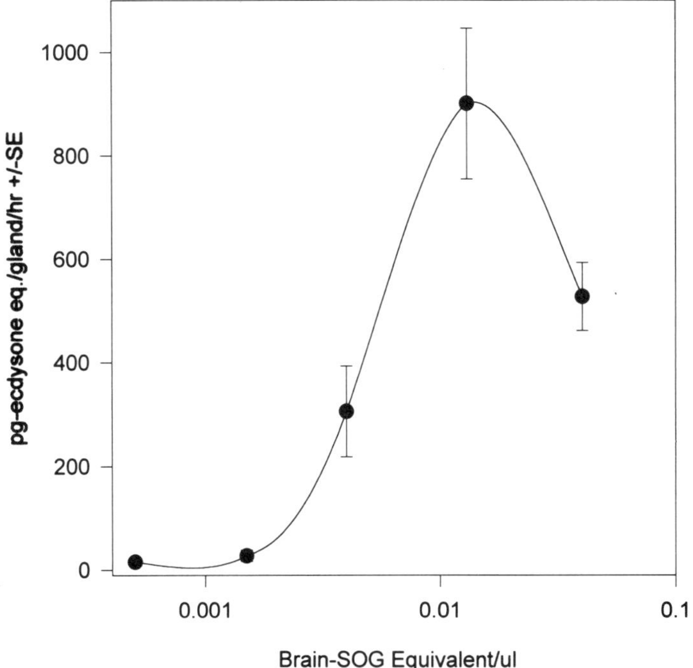

FIGURE 1. *In vitro* dose-response characteristics of larval prothoracic glands challenged with adult brain-SOG extract. Prothoracic glands were removed from L_5D_5 *L. dispar* larvae and incubated as described in MATERIALS AND METHODS. Adult brain-SOG complexes were collected and extracted in Grace's medium (see MATERIALS AND METHODS), and dilutions were prepared with Grace's medium. Post-incubation ecdysteroid was quantified by RIA and activity is expresseed as picograms of 20-OH-ecdysone equivalents produced per hour by a single prothoracic gland. Each data point represents the mean ± SE of 10 separate replicate incubations. Errors for the lowest extract doses were too small to illustrate.

brain-SOG equiv/μL (900 ± 146 pg 20-OH-ecdysone equiv/gland/h). Significant reduction of ecdysteroid output is observed at 0.04 brain-SOG equiv/μL (528 ± 66 pg 20-OH-ecdysone equiv/gland/h).

Ecdysiotropic Activity

Eleven ganglia were identified along the ventral nerve cord and were referred to as ganglia 1–11 from anterior to posterior. Ganglia 1–3 were found in the thorax and ganglia 4–11 in the abdomen. Ganglion 11 is in the terminal abdominal segment and is considered to be the terminal abdominal ganglion (TAG). Each of the ganglia examined had ecdysiotropic activity (FIG. 2), with highest levels of activity observed in ganglion 6 (390 pg 20-OH-ecdysone equiv) and the TAG (620 pg). A pattern of decrease and increase in ecdysiotropic activity was observed along the length of the VNC. Relative levels of ecdysiotropic activity were high in the brain-SOG (420 pg), declined through ganglia 1 and 2 (200 pg), increased through ganglion 6, declined again through ganglia 7, 8, and 9 (180 pg), then increased through the TAG.

Ecdysiostatic Activity

Column fractions 7–9 (elution volume 10–12 mL) each inhibited nearly all prothoracic gland ecdysteroid production when tested at a dose of 0.8 fractionated brain equivalent per microliter (FIG. 3). The molecular weight range of the inhibitory factor(s) was ~1200 to less than 1000 (FIG. 3, inset). Essentially no inhibition was observed in any of the other fractions tested. At the dose of brain-SOG extract used (0.02 equiv/μL) prothoracic glands are typically stimulated to produce between at least 900 (FIG. 1) and as much as 2000 pg 20-OH-ecdysone equiv/

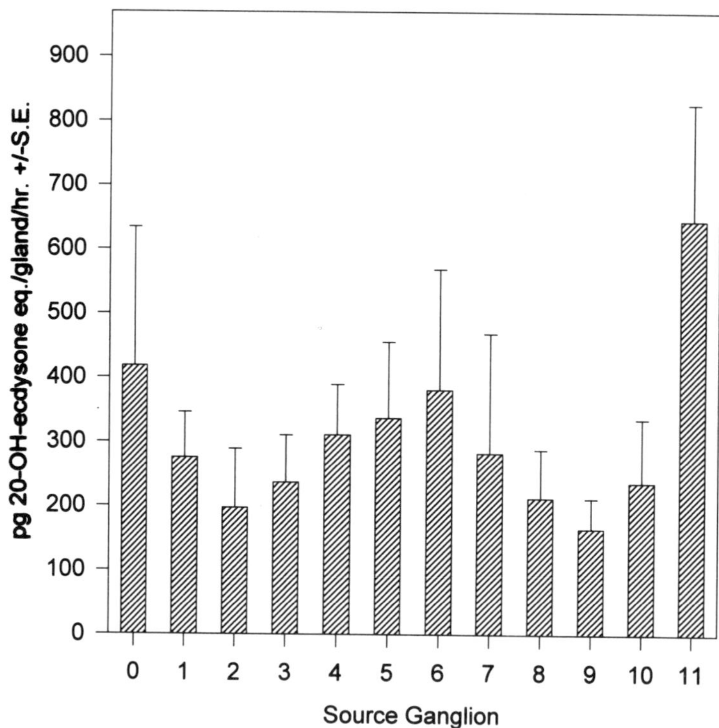

FIGURE 2. Ecdysiotropic activity in the central nervous system of larval *L. dispar*. Extracts of segmental ganglia (1–11) or L_5D_5 brain-SOG (0) were incubated (0.02 tissue equiv/μL) with prothoracic glands from L_5D_5 larvae of *L. dispar* as described in MATERIALS AND METHODS. Post-incubation ecdysteroid was quantified by RIA and activity is expressed as picograms of 20-OH-ecdysone equivalents produced per hour by a single prothoracic gland. Each bar represents the mean ± SE of six to seven separate replicate incubations. Ganglia 1–3 are thoracic; ganglia 4–11 are abdominal. Ganglion 11 is in the terminal abdominal segment and is considered as the terminal abdominal ganglion.

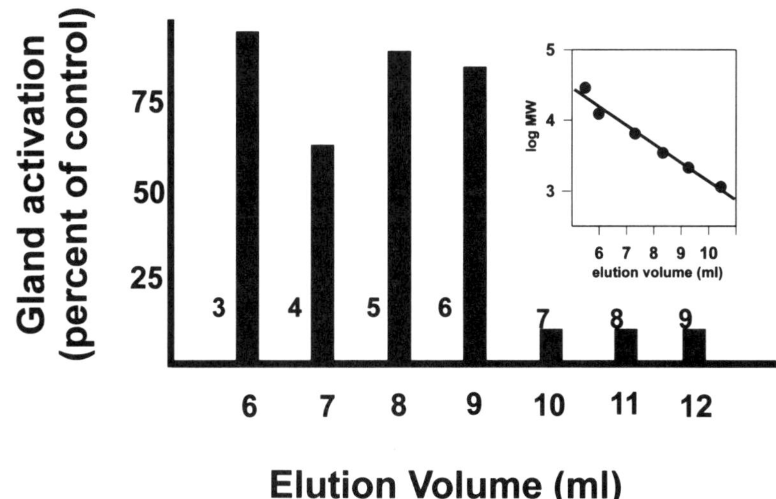

FIGURE 3. Ecdysiostatic activity in adult *L. dispar* brain-SOG. Brain-SOG extracts for incubation were prepared as described in MATERIALS AND METHODS at 0.02 brain-SOG equiv/μL. Brain-SOG extract fractions were prepared by fractionating by HP-SEC (TSK G2000SW). Aliquots of fractions were dried, then dissolved in brain-SOG extract and incubated with prothoracic glands (see MATERIALS AND METHODS). Fractionated brain-SOG extract was tested at 0.8 equiv/μL. Numbers next to the vertical bars are fraction numbers. Activation data are expressed as the percent of activation by mixtures of whole brain-SOG extract (0.02 equiv/μL) plus column fraction aliquot (0.8 equiv/μL) relative to whole brain-SOG extract alone. **Inset:** HP-SEC column calibration with carbonic anhydrase (29,000), cytochrome C (12,300), aprotinin (6,500), insulin chain B (3,196), gastrin (2,126), and adipokinetic hormone (1,159).

gland/h (FIG. 4). When the contents of fractions 7, 8, or 9 were present along with the brain-SOG extract, ecdysteroid production in each case was less than 50 pg 20-OH-ecdysone equiv/gland/h (FIG. 3). Each of the inhibitory fractions also displayed dose-response characteristics. From the dose-response curves (FIG. 4), fractions 7 and 9 show less potency than fraction 8. Fraction 7 inhibited completely only at the highest dose tested (0.7 fraction equiv/μL), and fraction 9 lost most of its inhibition near 0.02 fraction equiv/μL (FIG. 4). At 0.004 fraction equiv/μL, fractions 7 and 9 showed essentially no ecdysiostatic activity whereas fraction 8 inhibited approximately 75%. From the dose responses obtained (FIG. 4), it is not possible to estimate an IC_{50} for fraction 8. However, by using 2000 pg/h/gland as the maximum level of prothoracic gland stimulation in this experiment, a 50% inhibition (= 1000 pg/h/gland) yields IC_{50}'s of 0.50 and 0.04 fraction equiv/μL for fractions 7 and 9, respectively. An IC_{50} for fraction 8 would clearly be less than 0.004 fraction equiv/μL and establishes it as the most potent of the ecdysiostatic fractions tested. This activity elutes at a molecular weight near 900 (FIG. 3 and inset).

DISCUSSION

Cerebral control of insect development is a long-accepted tenet of insect physiology. The humoral factor generally regarded as centrally responsible for molting and metamorphosis, PTTH, exerts its control through stimulation of ecdysteroid production. In a parallel system, cerebral ecdysiotropin in adult dipterans stimulates ecdysteroid production by the gonads.[17,25] In each system, however, evidence exists for noncerebral factors that influence ecdysteroid production and development.[18,19,26] The larval proctodea (hindguts) of two lepidopterans, *Ostrinia nubilalis* and *L. dispar*, contain ecdysiotropic peptides[26] capable of stimulating ecdysteroid production by prothoracic glands *in vitro*. In addition, proctodeal extracts stimulate pupal development in *L. dispar* larvae, which are neck-ligated to prevent pupation.[26] In related experiments,[19] the terminal abdominal segment (TAS) of fifth instar *L. dispar* larvae was found to be necessary for normal development. Ligation of the TAS prevented pupation as did ligature of the VNC anterior to the TAS. A specific area of the TAS, the anal papilla,[19] contains tissue which stimulates pupation in neck-ligated larvae and showed evidence of ecdysiotropic activity. Our current data (FIG. 2) clearly show the presence of an ecdysiotropin in the TAG (ganglion 11). The question of whether this ecdysiotropin is the same as or different from the cerebral PTTH, as has been suggested,[19] remains unanswered.

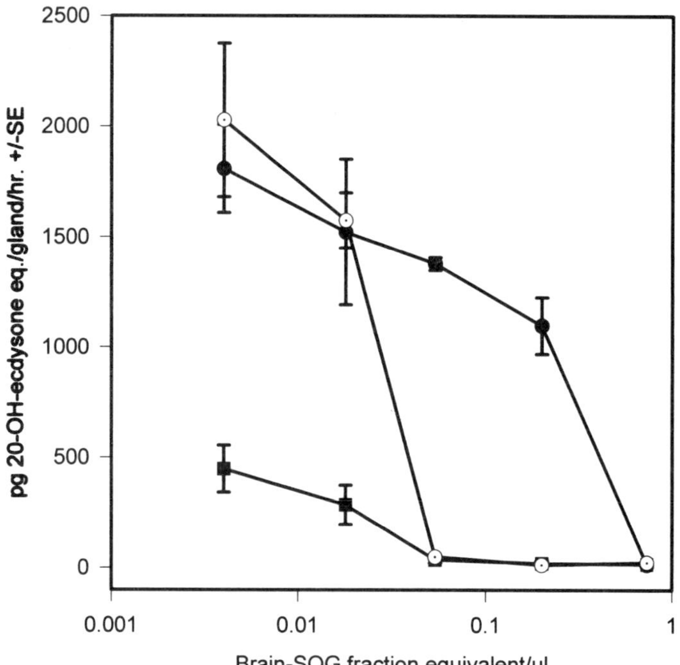

FIGURE 4. Dose response of ecdysiostatic activity. Dried column fraction aliquots from HP-SEC of adult day 1–2 brain-SOG extracts were prepared at various doses in adult day 1–2 brain-SOG extract (0.02 equiv/μL) and incubated with prothoracic glands as indicated in MATERIALS AND METHODS. Fractions 7 (*filled circles*), 8 (*filled squares*), and 9 (*open circles*) were tested. Data are presented as picograms of 20-OH-ecdysone equivalents produced in one hour by a single prothoracic gland. Each data point represents the mean ± SE of three separate incubations.

The level of ecdysiotropic activity in the TAG is similar to that in the brain-SOG, however. The presence of PTTH-like activity in each ganglion along the VNC elicits the question: Why does neck ligation prevent ecdysteroid-dependent development?[1,19,27] Ecdysiotropins present in VNC ganglia can stimulate ecdysteroid production by prothoracic glands *in vitro* (FIG. 2), but neck ligation inhibits ecdysteroid production and pupation.[27] The prothoracic glands of ligated larvae are capable of responding to PTTH *in vitro*,[27] suggesting that ligation does not affect the glands directly.

Further complicating the issue is the observation[19] that ligature at a number of different locations along the *L. dispar* VNC axis inhibits pupation. Finally, ligature of the VNC itself[19] prevents pupation and results in low hemolymph ecdysteroid titer. Thus, there appears to be a nervous as well as humoral component to the production of ecdysteroid, presumably by the prothoracic glands, *in vivo*. To explain the contradiction that PTTH-like activity is present along the length of the VNC but does not compensate for cerebral PTTH in ligated larvae, one can argue that the ganglionic ecdysiotropins are not released after neck ligation or may be ineffective in the physiological milieu of a ligated larva and may merely be supplemented to cerebral PTTH. Thus, our *in vitro* stimulation of prothoracic glands may occur at doses of ecdysiotropin that are higher than those normally occurring *in vivo*. These questions can be addressed with *in vivo* testing of ganglion extracts and the purification of cerebral and ganglionic PTTH.

The presence of an ecdysiostatin has been suspected for some time in flies and mosquitos, and its recent isolation and characterization[20] provide evidence for ecdysteroid promoters and inhibitors in the Diptera. Although lepidopteran PTTH has been extensively investigated, a lepidopteran ecdysiostatin has not been described. From observations of experiments with ligated larvae and the effect of implanted brains and prothoracic glands on pupation, the presence of a developmental inhibitor in the brain was proposed.[19] In dose-response tests using *L. dispar* brain extracts and prothoracic glands *in vitro*, we noticed some inhibition of gland response at high doses (≥ 0.04 brain-SOG equiv/μL) of extract (FIG. 1). Further investigation confirmed the presence of an ecdysiostatic factor(s) in *L. dispar* brain-SOG extracts (FIGS. 3 and 4). The factor is fairly potent. Although we have not determined an IC_{50}, it must be well below 0.004 fractionated brain equiv/μL (FIG. 4). In a typical prothoracic gland dose response, maximum stimulation with brain extract is obtained near 0.015 brain-SOG equiv/μL (FIG. 1). One-fourth of that dose in terms of fractionated brain-SOG equivalents can inhibit prothoracic gland response by 75% (FIG. 4). The molecular weight estimated for *Lymantria* ecdysiostatin ($\sim$900) is similar to that reported for *Neobellieria* folliculostatin,[20] and both are adult-derived factors which inhibit ecdysteroid production by larval tissues. Of current interest is the characterization of *Lymantria* ecdysiostatin, the question of whether ecdysiostatins are present in ganglia other than the brain-SOG (i.e., noncerebral ecdysiostatins) and in immature stages, and how these inhibitors are integrated with the ecdysiotropin in the regulation of prothoracic gland response and *L. dispar* development.

REFERENCES

1. KOPEC, S. 1922. Studies on the necessity of the brain for the inception of insect metamorphosis. Biol. Bull. Woods Hole **42:** 323–342.
2. FUKUDA, S. 1940a. Induction of pupation in silkworm by transplanting the prothoracic gland. Proc. Imp. Acad. Japan **16:** 414–416.

3. FUKUDA, S. 1940b. Hormonal control of molting and pupation in the silkworm. Proc. Imp. Acad. Japan **16:** 417–420.
4. GERSCH, M. 1962. The activation hormone of the metamorphosis of insects. Gen. Comp. Endocrinol. **1:** 322–329.
5. WIGGLESWORTH, V. B. 1934. The physiology of ecdysis in *Rhodnius prolixus* (Hemiptera). II. Factor controlling moulting and 'metamorphosis.' Q. J. Microsc. Sci. **77:** 191–222.
6. WIGGLESWORTH, V. B. 1940. The determination of characters at metamorphosis in *Rhodnius prolixus* (Hemiptera). J. Exp. Biol. **17:** 201–222.
7. WILLIAMS, C. M. 1947. Physiology of insect diapause. II. Interaction between the pupal brain and prothoracic glands in the metamorphosis of the giant silkworm, *Platysamia cecropia*. Biol. Bull. Mar. Biol. Lab. Woods Hole **93:** 89–98.
8. WILLIAMS, C. M. 1952. Physiology of insect diapause. IV. The brain and prothoracic glands as an endocrine system in the *Cecropia* silkworm. Biol. Bull. Mar. Biol. Lab. Woods Hole **103:** 120–138.
9. SCHARRER, B. 1987. Neurosecretion: Beginnings and new directions in peptide research. Annu. Rev. Neurosci. **10:** 1–16.
10. MASLER, E. P., T. J. KELLY & J. J. MENN. 1993. Insect neuropeptides: Discovery and application in insect management. Arch. Insect Biochem. Physiol. **22:** 87–111.
11. ISHIZAKI, H. 1993. Recent advances in insect neuropeptides. Comp. Biochem. Physiol. **106C:** 295–300.
12. ISHIZAKI, H. & A. SUZUKI. 1994. The brain secretory peptides that control moulting and metamorphosis of the silkmoth, *Bombyx mori*. Int. J. Dev. Biol. **38:** 301–310.
13. KELLY, T. J., E. P. MASLER & J. J. MENN. 1990. Insect neuropeptides: Current status and avenues for pest control. *In* Natural and Engineered Pest Management Agents. P. A. Hedin, J. J. Menn & R. M. Hollingsworth, Eds. ACS Symp. Ser. Vol. 551: 292–318. American Chemical Society. Washington, DC.
14. BOLLENBACHER, W. E. & N. A. GRANGER. 1985. Endocrinology of the prothoracicotropic hormone. *In* Comprehensive Insect Physiology, Biochemistry and Pharmacology. G. A. Kerkut & L. A. Gilbert, Eds. Vol. 7: 109–151. Pergamon Press. Oxford.
15. KELLY, T. J., E. P. MASLER, R. A. BELL, B. S. THYAGARAJA, R. E. DAVIS, H. W. FESCEMYER & A. B. BORKOVEC. 1991. Gypsy moth prothoracicotropic hormones: Progress toward identification. *In* Insect Neuropeptides: Chemistry, Biology and Action. J. J. Menn, T. J. Kelly & E. P. Masler, Eds. ACS Symp. Ser. Vol. 453: 27–37. American Chemical Society. Washington, DC.
16. ISHIZAKI, H. & A. SUZUKI. 1984. Prothoracicotropic hormone of *Bombyx mori*. *In* Biosynthesis, Metabolism and Mode of Action of Invertebrate Hormones. J. Hoffmann & M. Porchet, Eds.: 63–77. Springer-Verlag. Berlin-Heidelberg.
17. HAGEDORN, H. H. 1985. The role of ecdysteroids in reproduction. *In* Comprehensive Insect Physiology, Biochemistry and Pharmacology. G. A. Kerkut & L. A. Gilbert, Eds. Vol. 8: 205–262. Pergamon Press. Oxford.
18. MASLER, E. P. & T. J. KELLY. 1995. Non-cerebral ecdysiotropic and gonadotropic activities from the mosquito *Aedes aegypti* (Diptera: Culicidae). Eur. J. Entomol. **92:** 113–122.
19. THYAGARAJA, B. S., D. B. GELMAN, E. P. MASLER, T. J. KELLY, R. A. BELL & R. B. IMBERSKI. 1993. A role for the terminal abdominal segment in the larval and pupal molts of the gypsy moth *Lymantria dispar* L: Discovery of a developmental control center in the anal papilla pouch. J. Insect Physiol. **39:** 969–980.
20. BYLEMANS, D., Y-J. HUA, S-J. CHIOU, J. KOOLMAN, D. BOROVSKY & A. DELOOF. 1995. Pleiotropic effects of trypsin modulating oostatic factor (Neb-TMOF) of the fleshfly *Neobellieria bullata* (Diptera: Calliphoridae). Eur. J. Entomol. **92:** 143–149.
21. BELL, R. A., C. D. OWENS, M. SHAPIRO & J. R. TARDIF. 1981. Development of mass-rearing technology. *In* The Gypsy Moth: Research Toward Integrated Pest Management. C. C. Doane & M. L. McManus, Eds.: 599–633. U.S. Dept. Agr. Tech. Bull. 1584. Washington, DC.

22. OKUDA, M., S. SAKURAI & T. OHTAKI. 1985. Activity of the prothoracic gland and its sensitivity to prothoracicotropic hormone in the penultimate and last-larval instar of *Bombyx mori.* J. Insect Physiol. **31:** 455–461.
23. KELLY, T. J., M. J. BIRNBAUM, C. W. WOODS & A. B. BORKOVEC. 1984. Effects of house fly oostatic hormone on egg development neurosecretory hormone action in *Aedes atropalpus.* J. Exp. Zool. **229:** 491–496.
24. MASLER, E. P. & T. S. ADAMS. 1986. An ecdysiotropin from *Musca domestica. In* Host-Regulated Developmental Mechanisms in Vector Arthropods. D. Borovsky & A. Spielman, Eds.: 60–65. University of Florida Press. Vero Beach, FL.
25. MATSUMOTO, S., M. R. BROWN, A. SUZUKI & A. O. LEA. 1989. Isolation and characterization of ovarian ecdysteroidogenic hormones from the mosquito, *Aedes aegypti.* Insect Biochem. **19:** 651–656.
26. GELMAN, D. B., B. S. THYAGARAJA, T. J. KELLY, E. P. MASLER, R. A. BELL & A. B. BORKOVEC. 1991. The insect gut: A new source of ecdysiotropic peptides. Experientia **47:** 77–80.
27. THYAGARAJA, B. S., T. J. KELLY, E. P. MASLER, D. B. GELMAN, R. A. BELL & R. B. IMBERSKI. 1992. Kopec revisited—Neuroendocrine regulation of metamorphosis in the gypsy moth (*Lymantria dispar*): Development of a larval *in vivo* assay for prothoracicotropic hormone. J. Insect Physiol. **38:** 925–938.

Distribution and Processing of Secretoneurin in the Developing Rat Brain[a]

ALOIS SARIA,[b,e] WALTER A. KAUFMANN,[b]
JOSEF MARKSTEINER,[b] BERND LEITNER,[c]
HAROLD TRAURIG,[b,d] REINER FISCHER-COLBRIE,[c]
AND HANS WINKLER[c]

[b]Neurochemical Laboratory
Department of Psychiatry, University Hospital
Innsbruck, Austria

[c]Department of Pharmacology
University of Innsbruck Medical School
Innsbruck, Austria

[d]Department of Anatomy and Neurobiology
University of Kentucky Medical School
Lexington, Kentucky

INTRODUCTION

Secretoneurin (SN) is an endoproteolytic product (amino acids 156–186) of secretogranin II (SgII, formerly named chromogranin C).[1,2] SgII is one of the proteins of the secretogranin-chromogranin family, which includes chromogranin A and chromogranin B. These proteins have been detected in large, dense core vesicles of neurons and endocrine cells and have been analyzed using immunocytochemistry, immunoblot, reversed-phase high-performance liquid chromatography, high-performance gel filtration, and radioimmunoassay.[3] Previous data revealed that most SgII in brain undergoes posttranslational processing suggesting that SgII is a precursor for SN and possibly for other brain proteins. Furthermore, *in vitro* experiments have shown that SN is released from spinal cord and hypothalamus by potassium stimulation and evokes dopamine release from striatum.[4–7] SN is present in relatively higher concentrations in brain compared with chromogranin A and chromogranin B. Its distribution does not closely overlap the distribution patterns associated with other transmitters or brain peptides; however, SN is prevalent in phylogenetically older forebrain structures. Overlapping distribution of SN-immunoreactivity (IR) with substance P immunoreactivity in certain areas of the brain stem have been observed.[2,8,9] In the adult rat forebrain, high densities of markers for SgII mRNA and SN-IR have been detected in a continuum of neurons and fibers referred to as the "extended amygdala," which includes the central and medial amygdala (CA and MA), the bed nucleus stria terminalis, the

[a] This work was supported by the Austrian Science Foundation (Grant No. F00206).

[e] Address correspondence to Prof. Dr. Alois Saria, Neurochemical Laboratory, Department of Psychiatry, University Hospital, Anichstr. 35, A-6020 Innsbruck, Austria. E-mail: alois.saria@uibk.ac.at

ventromedial (shell) portion of the nucleus accumbens, and interconnecting fibers.[10]

The goal of the present study was to examine the distributions of SN and SgII expressions in the fetal and newborn rat brain with special attention to forebrain structures and components of the extended amygdala.

METHODS

Pregnant Sprague-Dawley rats were used. The first day of pregnancy was determined by the observation of sperm in vaginal lavages. Pregnant rats were anesthetized, perfused through the arterial system with saline containing heparin followed by 4% paraformaldehyde. Fetuses were collected, postfixed in fixative overnight, and immersed in Tris-buffered saline containing 20% sucrose and frozen at -70 °C. Rat pups at 1, 5, 10, and 15 days (PD) of age were anesthetized and their brains collected as follows. PD1 and PD5 brains were removed and fixed by immersion overnight and stored as above. PD10 and PD15 rats were first perfused with heparinized saline followed by fixative and cryoprotection. Horizontal and sagittal brain cryostat sections (60 μm) were obtained and immunostained using a free-floating section procedure. Other sections (14 μm) were mounted on glass slides and processed for immunostaining and *in situ* hybridization histochemistry. SN antiserum was generated in rabbits using a synthetized peptide fragment (amino acids 156–186) of rat SgII. This antiserum has been previously characterized and reported.[1] Sections were processed for SN-IR using the PAP technique (1 : 250 dilution of SN primary antiserum at 4 °C for 24 h). Controls included omission of SN primary antiserum and antiserum preabsorbed with synthetic SN peptide (10 μM).[8,9] *In situ* hybridization was performed using a modified technique.[11] An oligonucleotide complementary to the nucleotide sequence 193–240 of rat SgII mRNA was labeled with terminal deoxynucleotidyl transferase and [^{35}S]α-thio dATP. Control sections were pretreated with ribonuclease. Sections were dipped in Kodak NTB-2 emulsion, exposed for 4 weeks, developed, lightly stained, and mounted.[2] High-performance gel filtration and radioimmunoassay were carried out as previously described.[1]

RESULTS

SN-IR could clearly be identified in various structures of the developing brain as early as fetal day 14 (FD14). An example of SN-IR developing neurons of different intensities in the ventral pallidum is shown in FIGURE 1. At FD20, development of axons with varicosities appeared in the caudate-putamen (FIG. 1). At this time point, massive expression of SgII mRNA could be observed in the central and medial amygdaloid nuclei (FIG. 2). At PD15, full development of the extended amygdala marked with SN-IR could be observed. Thus, FIGURE 3 illustrates a horizontal section of rat ventral forebrain. The staining pattern of SN-IR is similar to that of the adult rat brain. Here, SN-IR is apparent in the medial or shell portion of nucleus accumbens and the ventral extension of the striatopallidum. The SN-IR continues into the bed nucleus stria terminalis (FIG. 3). In addition, SN-IR extending laterally from the anterior hypothalamus, overlapping in the nucleus accumbens and ventral pallidum, could be seen (FIG. 3). From

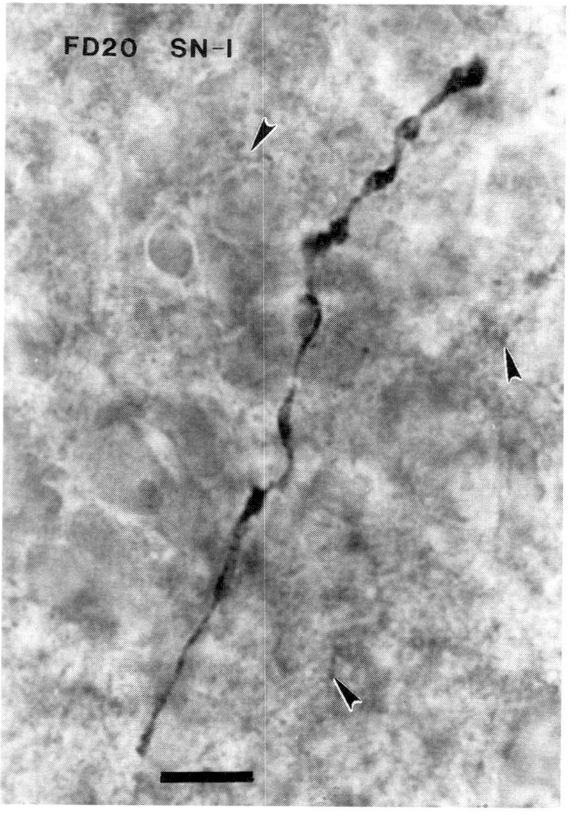

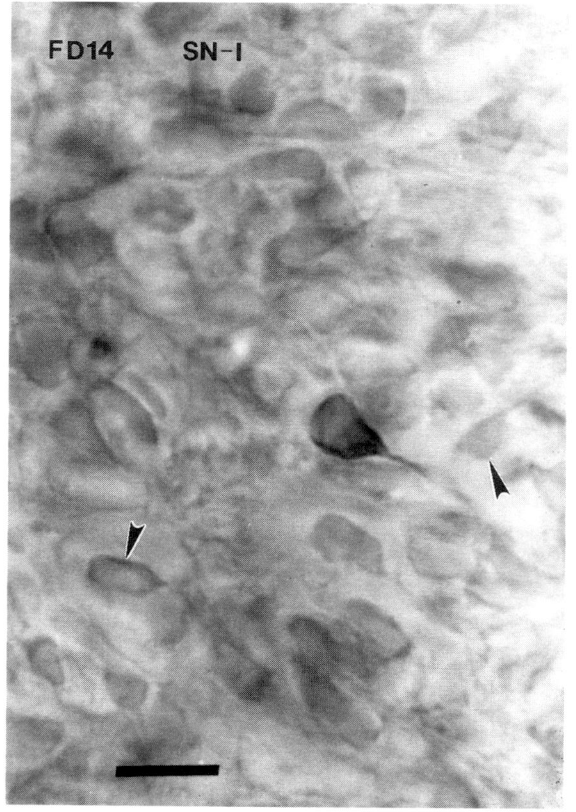

FIGURE 1. (*Left*) Intensely stained secretoneurin-immunoreactive (SN-IR) nerve fiber in a horizontal section of caudate-putamen at FD20. A light, punctate SN-IR is evident among developing neurons (*arrowheads*). Bar = 10 μm. (*Right*) Intensely stained SN-IR developing neuron in the ventral pallidum at FD14. Other more lightly SN-IR developing neurons are evident (*arrowheads*). Bar = 10 μm.

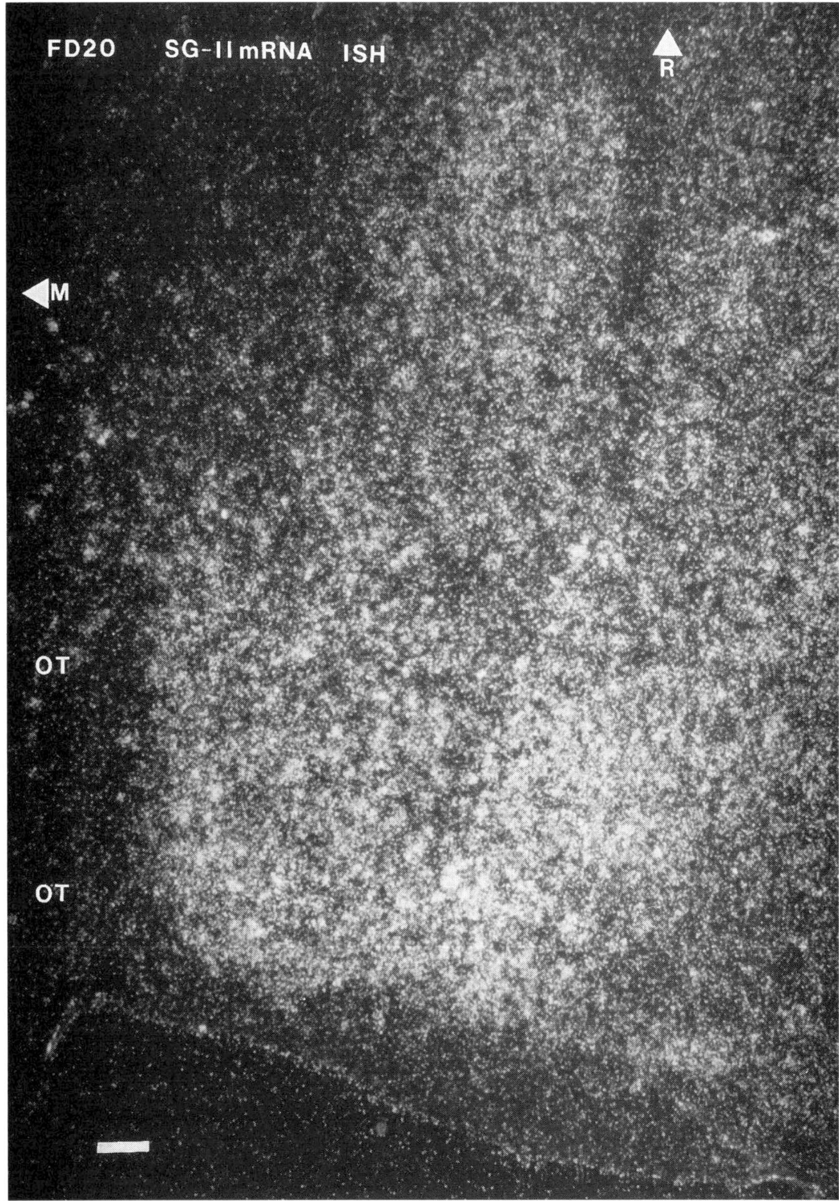

FIGURE 2. Secretogranin II (SgII) mRNA expression by FD20 developing neurons in central and medial amygdaloid nuclei which fill the field. The illustration is a horizontal section processed for *in situ* hybridization with [^{35}S]-labeled oligonucleotide probe complementary to a nucleotide sequence (193–240) of rat SgII mRNA. Following autoradiography the section was photographed in darkfield illumination. The white dots represent a reaction signal associated with developing neurons. OT, optic tract; R, rostral; M, medial. Bar = 50 μm.

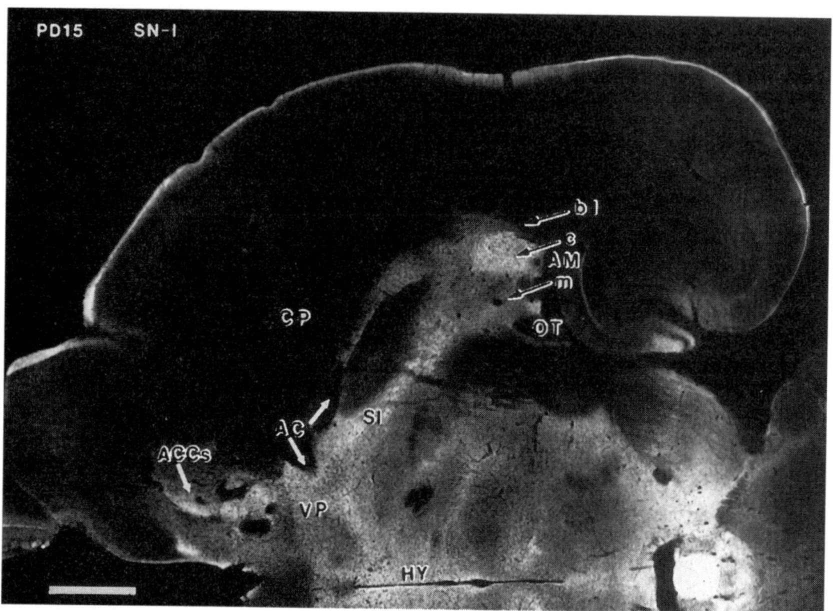

FIGURE 3. Illustration of a horizontal section of PD15 rat ventral forebrain obtained using secretoneurin (SN)-immunostained slides as negatives; therefore, SN-immunoreactive (IR) structures appear white. By PD15 the pattern of SN-IR is similar to that of the adult rat brain. SN-IR is apparent in the medial or shell portion of nucleus accumbens (ACCs) and the ventral extension of the striatopallidum (VP). The SN-IR continues into the bed nucleus stria terminalis, which lies in a more dorsal plane (not shown here). Also, note SN-IR extending laterally from the anterior hypothalamus (HY) and overlapping the ACCs and VP. From these forebrain regions SN-IR extends caudally through substantia innominata (SI) to the amygdaloid nuclei (AM). Note the intense SN-IR in the medial and central AM (Amm, Amc), while relatively less SN-IR is apparent in lateral and basal lateral AM (Ambl). Little SN-IR is evident in the caudate-putamen (CP) at PD15. OT, optic tract. Bar = 1 mm.

the latter forebrain regions, SN-IR extends caudally through substantia innominata to the amygdaloid nuclei. SN-IR staining in the medial and central amygdaloid nuclei was more intense than in the lateral and basal amygdaloid nuclei (FIG. 3). FIGURE 4 represents a schematic representation of the results of FIGURE 3 to illustrate that SN labels a region suggested to constitute a functional unit.

To investigate the processing of SgII, high-performance gel filtration of extracts of developing rat brains was performed. Interestingly, authentic SN was the major peak found in brains removed at any fetal or postnatal day, including fetal day 14, the earliest day of investigation. Only minute amounts of SgII and immunoreactive products of intermediate size could be detected. None of the chromatograms obtained at FD14, 16, 18, and 20 or PD1, 5, 10, and 15 differed from the chromatogram of extracts of adult rat brains (not shown).

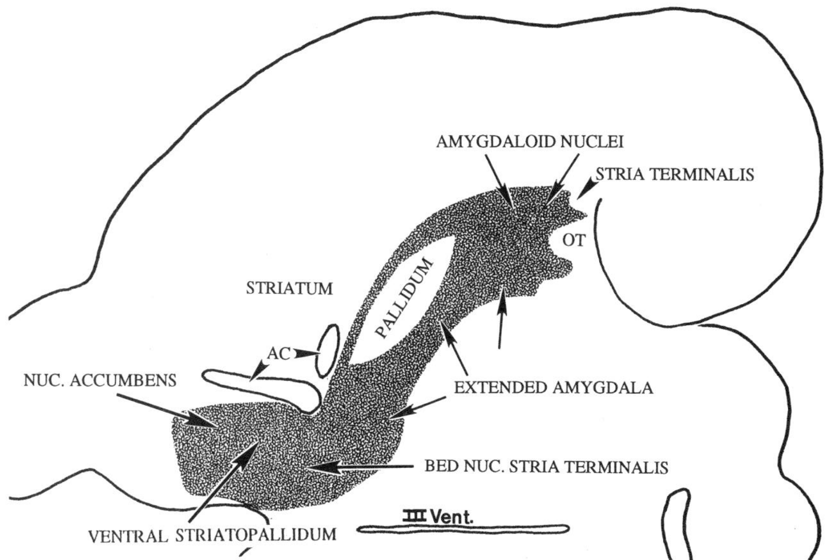

FIGURE 4. Schematic representation of intense SN-IR labeling (derived from FIG. 3) of the extended amygdala and the ventral striatopallidum (*shaded area*).

CONCLUSIONS

Our data provide evidence that SgII mRNA and SN-IR are expressed as early as FD14 in components of the extended amygdala in rat ventral forebrain. The chromatographic analysis shows that the processing is already complete at FD14 and not different from adult rat brain. This indicates that all necessary processing enzymes should be developed by that time point. In fact, it has been recently demonstrated that various prohormone convertases, suggested to be responsible for the processing of a variety of neuropeptides, can already be detected at FD14.[12,13] The expressions of SgII and its proteolytic product SN probably reflect developments of components of the extended amygdala which contain circuits that link hypothalamic, limbic, and striatopallidal functions.

REFERENCES

1. KIRCHMAIR, R., R. HOGUE-ANGELETTI, J. GUTIERREZ, R. FISCHER-COLBRIE & H. WINKLER. 1993. Secretoneurin—A neuropeptide generated in brain, adrenal medulla and other endocrine tissues by proteolytic processing of secretogranin II (chromogranin C). Neuroscience **53:** 359–366.
2. MARKSTEINER, J., R. KIRCHMAIR, S. K. MAHATA, M. MAHATA, R. FISCHER-COLBRIE, R. HOGUE-ANGELETTI, A. SARIA & H. WINKLER. 1993. Distribution of secretoneurin, a peptide derived from secretogranin II, in rat brain. Neuroscience **54(4):** 923–944.
3. WINKLER, H. & R. FISCHER-COLBRIE. 1992. The chromogranins A and B: The first 25 years and future perspectives. Neuroscience **49:** 497–528.

4. KIRCHMAIR, R., J. MARKSTEINER, J. TROGER, S. K. MAHATA, M. MAHATA, J. DONNERER, R. AMANN, R. FISCHER-COLBRIE, H. WINKLER & A. SARIA. 1994. Human and rat primary C-fibre afferents store and release secretoneurin, a novel neuropeptide. Eur. J. Neurosci. **6:** 861–868.

5. TROGER, J., R. KIRCHMAIR, J. MARKSTEINER, C. V. SEIDL, R. FISCHER-COLBRIE, A. SARIA & H. WINKLER. 1994. Release of secretoneurin and noradrenaline from hypothalamic slices and its differential inhibition by calcium channel blockers. Naunyn-Schmiedebergs Arch. Pharmacol. **349:** 565–569.

6. SARIA, A., J. TROGER, R. KIRCHMAIR, R. FISCHER-COLBRIE, R. HOGUE-ANGELETTI & H. WINKLER. 1993. Secretoneurin releases dopamine from rat striatal slices: A biological effect of a peptide derived from secretogranin II (chromogranin C). Neuroscience **54(1):** 1–4.

7. AGNETER, E., H. H. SITTE, S. STÖCKL-HIESLEITNER, R. FISCHER-COLBRIE, H. WINKLER & E. A. SINGER. 1995. Sustained dopamine release induced by secretoneurin in the striatum of the rat. J. Neurochem. **65:** 622–625.

8. MARKSTEINER, J., S. K. MAHATA, R. PYCHA, M. MAHATA, A. SARIA, R. FISCHER-COLBRIE & H. WINKLER. 1994. Distribution of secretoneurin immunoreactivity in the spinal cord and lower brainstem in comparison with that of substance P and calcitonin gene-related peptide. J. Comp. Neurol. **340:** 243–254.

9. MARKSTEINER, J., A. SARIA & H. HINTERHUBER. 1994. Distribution of secretoneurin-like immunoreactivity in comparison with that of substance P in the human brainstem. J. Chem. Neuroanat. **7:** 253–270.

10. MARKSTEINER, J., E. LASSNIG, S. TELSER, S. KROESEN, R. KIRCHMAIR, R. FISCHER-COLBRIE, C. MILLER & A. SARIA. 1995. Evidence for a high density of secretoneurin-like immunoreactivity in the extended amygdala of the rat. J. Comp. Neurol. **353:** 275–290.

11. MAHATA, S. K., M. MAHATA, J. MARKSTEINER, G. SPERK, R. FISCHER-COLBRIE & H. WINKLER. 1991. Distribution of mRNAs for chromogranins A and B and secretogranin II in rat brain. Eur. J. Neurosci. **3:** 895–904.

12. ZHENG, M., R. D. STRECK, R. E. M. SCOTT, N. G. SEIDAH & J. E. PINTAR. 1994. The developmental expression in rat of proteases furin, PC 1, PC 2, and carboxypeptidase E: Implications for early maturation of proteolytic processing capacity. J. Neurosci. **14(8):** 4656–4673.

13. MARCINKIEWICZ, M., R. DAY, N. G. SEIDAH & M. CHRETIEN. 1993. Ontogeny of the prohormone convertases PC1 and PC2 in the mouse hypophysis and their colocalization with corticotropin and α-melanotropin. Proc. Natl. Acad. Sci. USA **90:** 4922–4926.

Expression of Cytokine Genes during Ontogenesis of the Central Nervous System

F. POUSSET,[a] J. FOURNIER, AND P. E. KEANE

[b]*Department of Neuropsychiatry Research*
Sanofi Recherche
195 Route d'Espagne
31036 Toulouse Cedex, France

The cytokines are soluble proteins characterized by their autocrine or paracrine activities, by the diversity of their biological action, and by a certain redundancy in some of their effects.[1] The cytokines have been divided into a number of different families, as a function of their biological activities, the most representative of these being the interleukins (IL), the chemokines, the tumor necrosis factors (TNF), the interferons, and the colony-stimulating factors (CSF). For many years, most immunologists considered the cytokines as factors that were specific to the immune system. In a similar way, neurobiologists considered the central nervous system (CNS) to be a part of the body devoid of any immune reactivity, in which the neurotransmitters were the only mediators of cerebral function. We now know that these concepts are no longer tenable, mainly because of the discovery that many cytokines are produced with numerous nonimmune cells, and particularly in brain cells.

The occurrence of cytokines within the CNS was first proposed by Fontana *et al.*,[2] who demonstrated an IL1-like activity in the supernatant of primary cultures of mouse astrocytes that had been stimulated by bacterial lipopolysaccharides. The study of cytokines has been gradually facilitated by the cloning of their genes and the production of recombinant proteins and corresponding antibodies. This has led to an increasing number of cytokines being identified in cerebral cells in primary cultures, where they have been found to occur either constitutively or after stimulation by mitogenic agents.[3] The astrocytes and microglia have been the most extensively studied cell types until now, and they are known to synthesize a large number of cytokines from a number of different families. For example, neuronal cells have been found to produce IL3 and IFNγ.[4,5] In addition, both isolated glial and neuronal cells express mRNA for the receptors of some of these cytokines, including IL1 type I and type II receptors, and IL6 receptors.[6,7]

Studies *in vivo* have found that cytokines are produced within the brain, although the levels of expression are often quite low, and difficult to quantify. Most *in vivo* studies have shown a preponderance of neuronal expression of cytokines and their receptors. For example, IL1β mRNA is associated with mossy fibers in the hippocampus, and IL6 mRNA is principally localized in the pyramidal and granular layers of the hippocampus.[8,9] In contrast, most *in vitro* studies on primary cultures of brain cells have localized cytokines and their receptor to glial cells. Most studies have been performed on adult animals, and most frequently the

---

[a] Present address: INSERM U. 394, Rue Camille St. Saens, 33077 Bordeaux, France.
[b] Address for correspondence.

expression of cytokine mRNA has been evaluated, using a variety of methodological approaches (Northern blot, *in situ* hybridization, S1 nuclease protection, reverse transcription-polymerase chain reaction [RT-PCR]). These studies have led to the characterization of the occurrence of cytokines in different brain areas with, it seems, different degrees of expression, although this could possibly be attributed to the different methodological approaches applied by the different research groups. Studies on the cytokine proteins have been performed for a small number of cytokines, including IL1 and IL6. Similarly, a limited number of cytokine receptors have been studied in the CNS, including IL1α/β, IL2, and IL6 receptors.[10-12] Overall, these studies have examined cytokines in a range of species (mouse, rat, and human) and ages, and have generally evaluated the expression of the mRNA independently of that of the corresponding protein. For these reasons, it is difficult at present to establish from the existing data a precise distribution of cytokines within the CNS, even for the most frequently studied of them, and the localization of many members of this class of compounds still remains to be studied.

CYTOKINES AND THEIR RECEPTORS DURING DEVELOPMENT OF THE CENTRAL NERVOUS SYSTEM

The study of cytokines during the ontogenesis of the CNS has been undertaken only recently and has concentrated on the expression of their mRNAs. The corresponding proteins have only been studied for a small number of cytokines. Most of the studies have been of a qualitative nature, with the exception of those on TGFα, which have been quantified using the nuclease S1 protection method. The mRNA levels of this cytokine attained 3 pg/μg total RNA between day 14 of gestation (G14) and the first postnatal day (P1).[13]

TGFα and the members of the TGFβ superfamily (genes and proteins) are among the earliest cytokines expressed in mammalian embryos.[14-16] TGFα/β immunoreactivity was high in the embryonic CNS, choroid plexus, and meninges of the rat, mouse, and chicken.[17,18] TFGα showed marked differential expression as a function of time[19,20]: its mRNA levels rose between G14 and P1, and then fell by day P10.[13] TGFβ appears to be least intense in the ventricular zone, a region of active proliferation, and most intense in the marginal zone, where proliferation has ceased and neuronal differentiation is occurring.[21] The mRNA isoforms of TGFβ (β1, β3) have a continuous high expression from postnatal day P8 to P18 and during adult life.[21,22] TGFβ2 mRNAs (4.0 and 4.5 kb) were detected by Northern blot analysis in mouse embryos: levels were high at days G15–16, then fell somewhat, before increasing at postnatal day P6.[23] Other cytokines, including TNF, IL6, and MCSF, have been studied in whole brain extracts from embryonic and newborn mice. Their transcripts were expressed in the developing nervous system as early as the second week of embryonic life, and the expression continued into adult life.[23] Some cytokines—for example, IL6—seem to have a constant expression during embryonic life and the first postnatal weeks, whereas other transcripts (TNF, MCSF) demonstrated specific patterns of expression at various stages of development.[23] For example, TNFα mRNA was detected by Northern blot in high amounts in the mouse brain from G15 to P9, its intensity increasing somewhat after G16. TNFα immunoreactivity was observed in neuroepithelial cells and Purkinje cells of the developing brain, although no positive immunocytochemical staining for this protein could be observed in the adult brain.[24]

The presence and the expression of several cytokines have been studied during

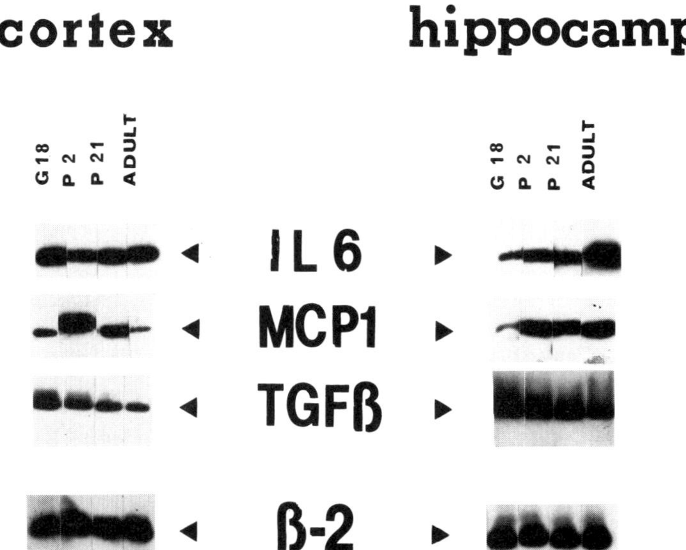

FIGURE 1. Autoradiographic analyses of the expression of IL6, MCP1, TGFβ, and β2-microglobulin mRNAs in cortex (*left*) and hippocampus (*right*) of embryonic (day G18), postnatal (day 2 and day 21), and adult rat. RNA samples (50 ng) were purified, reverse transcribed, and amplified with primers specific for each cytokine.[25] The β2-microglobulin was used as a control. Sizes were determined by comparison with a 123 DNA ladder; negative results were obtained for PCR without RT (not shown). Similar results were obtained in two independent experiments.

ontogenesis in certain brain regions. IL1α, IL2, IL6, TGFβ, TNFα, and MCP1 were studied in rat cortex and hippocampus, using RT-PCR.[11,25] The results of PCR amplification of IL6, TGFβ, and MCP1 are shown in FIGURE 1. These mRNAs were detected in the developing brain regions examined, together with the β2-microglobulin. They were detectable at the earliest embryonic day tested (G18). At this stage, the expression of IL6 was high in the embryonic cortex, dropped off during the postnatal period, and then increased slightly in the adult. This level of expression was lower in the embryonic hippocampus but increased during postnatal development, with a maximal expression in the adult. Gadient and Otten[11] also noted the presence of this transcript in other brain areas, including the striatum, pons-medulla, and cerebellum. The pattern of MCP1 mRNA expression was different from that of IL6. The cortical expression of MCP1 mRNA was clearly detectable at G18, increased to a maximal level at P2, and subsequently fell at later ages. In the hippocampus, MCP1 mRNA appeared to be expressed at relatively equal levels during the early postnatal period and in the adult. TGFβ mRNA was very marked in the cortex at day G18, and then gradually fell to reach its lowest level in the adult. In the hippocampus, the TGFβ mRNA is expressed at a constant level in the whole period from G18 to adult. In cerebral cortex of mice, MCSF transcript is clearly detectable from the beginning of the second week of gestation (G14) until two weeks after birth (P13) by Northern blot and the S1 nuclease protection assay.[26,27] IL1α biological activity, determined using the murine thymocyte proliferation assay, has been identified in cerebral cortical extracts from rats of G20 and P14. At the embryonic stage G14 and in normal adult brain, there was no measurable IL1 activity. The highest concentrations of IL1α in the normally developing cortex of rat appeared during the perinatal period from embryonic stage G18 to the time of birth.[28] These observations show that the expression of each of the cytokines varies during the different stages of development and that these changes are different for each brain region.

A limited number of receptors for cytokines have also been studied in the CNS. The receptor for MCSF, c-fms, has been found from the embryonic stage G14 to the postnatal P13 day, using the S1 nuclease protection assay.[27] The mRNA coding for the IL6 receptor has been detected using RT-PCR in different rat brain areas (cerebral hemispheres, hippocampus, striatum, pons-medulla) from the second postnatal day (P2), and also at days P20 and P70.[11] These receptors are concomitantly expressed with their corresponding cytokines, at the same periods of time and in the same regions. At present, however, no studies of cytokine receptors during CNS development have been performed using binding techniques.

Possibly as a result of the diversity of the techniques employed and their different sensitivities, some observations in the literature are contradictory in nature. For example, studies with the S1 nuclease assay have not detected IL1α and IL6 transcripts between the embryonic stage G13 and the adult, whereas Northern blot and RT-PCR have enabled the detection of the transcript in the whole brain.[23,27] Similarly, the observations are not always identical from one species to another: IL1α, IL2, and MCSF mRNAs have been detected in mouse brain and failed to be identified in rat brain, suggesting a possible species-specific distribution.[25,26]

It is not at present possible to establish any correlations between mRNA expression and protein synthesis, or to establish any reliable maps of cytokine distribution during brain development. This would require a detailed study of each cytokine in all the main brain areas, in order to define for each one their earliest occurrence in each area and the evolution of their expression during the various stages of development. Ideally, for each cytokine, it would be necessary to study their mRNA, the cytokine protein, and the receptor as a function of the brain area, developmental period, cell type, and species. Nevertheless, we already have sufficient information to suggest that at least some cytokines play distinct roles at different stages of development, and the temporal characteristics of their expression suggest an involvement in the processes of neuro- or gliogenesis.

POSSIBLE ROLES OF CYTOKINES IN DEVELOPMENT OF THE CENTRAL NERVOUS SYSTEM

The question of the functional importance of cytokines has been recently approached through the study of mutant mice in which certain genes for cytokines had been deleted or disrupted. Homozygous TGFα null mice, generated by gene targeting in embryonic stem cell, were viable and fertile, but were characterized by abnormalities of the eyes (open eyelids at birth, reduced eyeball size, retinal defects, corneal inflammation), skin, and hair (pronounced waviness of the hair and whiskers).[29,30] The MCSF deficiency in op/op mice results in a depletion in the number of monocytes and macrophages but does not affect the morphology or number of microglia which proliferate in the presence of this cytokine.[31,32]

However, these mice presented functional deficits: a significant increase in neuron vulnerability to ischemia and considerably reduced microglial response to neuron injury. Architectural abnormalities in the CNS, however, were not observed, suggesting that the deleted or non-produced cytokine might be partially substituted by other CNS factors.[27]

At present, the exact roles of the cytokines in brain development *in vivo* remain unknown. Their functions have been deduced from studies performed *in vitro* in several culture systems, including immortalized cell lines, as well as progenitor cell and primary cerebral cell cultures derived from fetal or early postnatal brain. Such experiments with isolated cell populations exposed to cytokines *in vitro* have led to the description of diverse activities.

The cytokines have multiple biological activities on different cell types. Functions such as proliferation, differentiation, or cytotoxicity have been demonstrated on microglia, astrocytes, oligodendrocytes, and neurons. IL1α/β, IL6, and TGFβ mainly induce the proliferation of astrocytes and microglia.[33] TNFα is toxic for oligodendrocytes.[34] The effects of cytokines on neurons tend to be more specific. For example, IL6 can increase the survival of neurons with a cholinergic phenotype, and neurite outgrowth in rat hippocampal neurons can be induced by TGFβ.[35,36] Furthermore, cytokines can modulate neuronal excitability by acting on phenomena such as long-term potentiation and neurotransmitter release. Depending on the type of cell and their level of differentiation, the effects of cytokines can vary considerably.[37] For example, TGFβ can stimulate the proliferation of astrocytes, but suppresses the proliferation of microglia *in vitro*.[38,39] As a general rule, such apparently contradictory effects of cytokines depend on their concentration and on possible interactions with serum proteins present in the culture medium.[40]

The cytokines could constitute an autocrine or paracrine signal in the course of development, possibly during the stages of proliferation, migration, differentiation, and maturation of the different cell types. The process of cell proliferation is observed at different stages of development: neuroblasts and glioblasts proliferate early, and astrocytes continue to proliferate up to the P9 postnatal stage. Microglial cells present an intense phase of proliferation from E18 until P1 *in vivo* and then fall markedly in number by the second postnatal week. The effects of IL1α, IL6, MCSF, and TGF on the proliferation of astrocytes and microglia *in vitro*, together with the detection of their transcripts during development, suggest that production of these cytokines in the brain might stimulate the proliferation of cerebral macrophages and astrocytes observed *in situ* until postnatal day 9. IL6, MCP1, and TGFβ mRNA expressions appear in the cortex and hippocampus at the embryonic stage, which corresponds to the proliferation of cerebral cells, suggesting that they may contribute to this process. This proliferative activity is not restricted to the embryonic stage in the two structures studied, where a high degree of postnatal cytogenesis exists.

Cell migration is also an intrinsic part of CNS development, but the factors involved in the guidance of cell migration have not yet been identified. The formation of extracellular matrix components (fibronectin, laminin, collagen) is an important process in embryogenesis.[41] *In vitro*, TGFβ participates in this process, and its very early presence in the embryonic brain might suggest its involvement in the matrix formation *in vivo*. Some cytokines, including TGFα, MCSF, and MCP1, have been shown to possess chemotactic activity in peripheral tissues. TGFα and MCP1 are potent chemoattractants for rat heart endothelial cells[42] and monocytes,[43,44] respectively. However, their chemotactic effects on CNS cells remain to be demonstrated. The early detection of TGFα in embryos could be

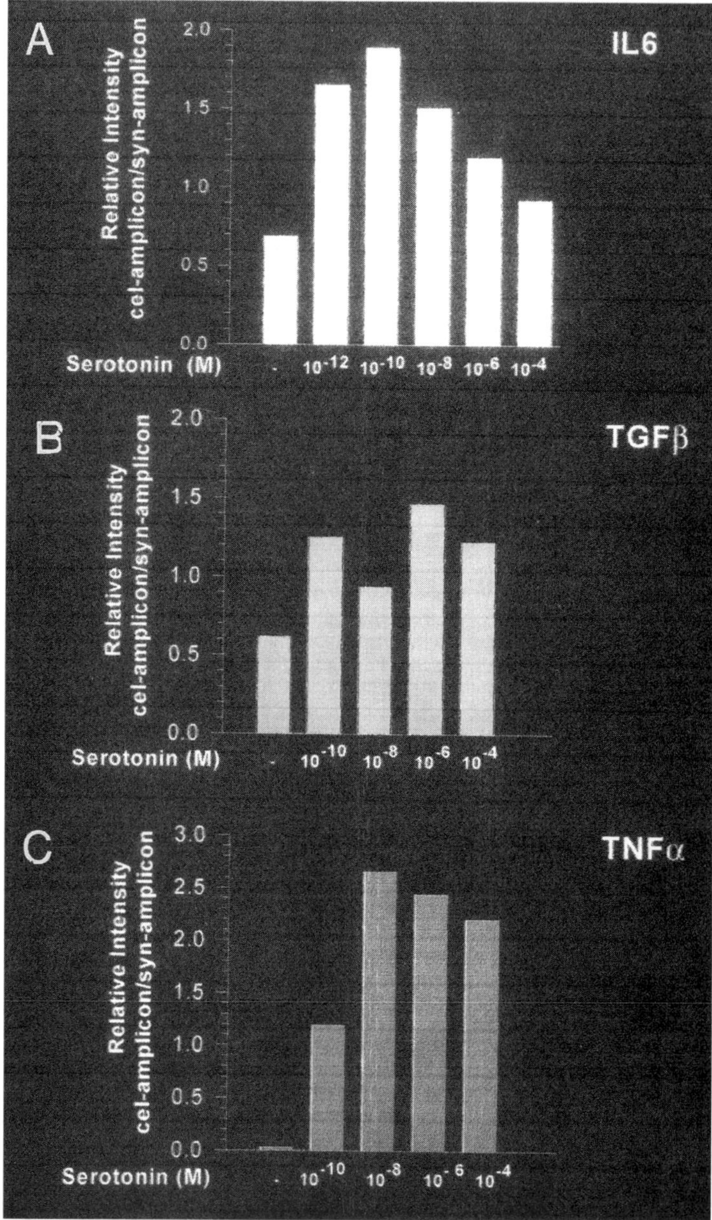

FIGURE 2. Semiquantitative analysis of cytokine mRNAs in rat hippocampal astrocytes incubated in the presence or absence of serotonin for 6 h. Cellular (cel)-RNA [900, 400, and 100 ng for IL6 **(A),** TGFβ **(B),** and TNFα **(C),** respectively] was mixed with synthetic multispecific RNA standard (syn-RNA) (20 fg), reverse transcribed, and subjected to 30 cycles of PCR.[64] Reaction mixtures were electrophoresed, the specific bands were excised, and the radioactivity determined. Background values were subtracted. The ratio of cellular-amplimer/synthetic-amplimer, corrected for their respective deoxycytidine contents, was plotted against the concentration of serotonin.

related to the migration of glial precursor cells and radial astrocytes from ventricular and subventricular layers throughout the CNS to reach their final destination. TGFβ immunoreactivities have been observed in processes and end feet of astrocytes and might contribute to radial glial cell functions in organizing the migration of neuroblasts during the development of the CNS.[45] MCP1 could also promote the migration of neuroblasts to their specific adult location and might also be implicated in initial axon elongation.

The final phase of development is the differentiation of the various populations of CNS cells. IL6 could be involved in the lineage commitment of cells *in vivo*, in view of its known effects on progenitor cells.[46,47] However, the most documented effect of IL6 is its neurotrophic activity on both catecholaminergic and cholinergic neurons.[48] Other cytokines such as TGFβ exert trophic actions on neurons *in vitro*, and might have a similar effect *in vivo* on their terminal differentiation.[49,50] In addition, IL2 seems to be implicated in oligodendrocyte maturation.[51]

A key event in neuronal ontogenesis is programmed cell death (PCD).[52] Some neuropoietic cytokines of the IL6 family, such as ciliary neurotrophic factor and leukemia inhibitory factor, can initiate PCD in primary cultures of sympathetic neurons.[53] In addition, TNFβ is known to induce PCD of oligodendrocytes.[54] These effects are mediated by cellular factors such as heat shock protein (hsp) or bcl-2 proto-oncogene. A relationship between neuronal cell death and expression of hsp has been well documented. Induction of hsp has been reported to prevent PCD in cultured neurons.[55] Kessler *et al.*[53] have suggested that the number of neurons can be regulated both by factors that initiate PCD and by factors that prevent it. This could explain the presence of some cytokines with a possible anti-apoptotic role at the end of the embryonic stage (G18), such as IL1α and TNFα, which have been shown to induce hsp 70 in mixed glial cell cultures.[56] In addition, at this step of development, cytokines may represent the signal that triggers microglial activation to phagocytose the dying cells.

Cytokines have also been identified in the adult brain when all the events related to the organization of the cerebral structures are completed. This might suggest a physiological role of cytokines in the functions of the adult brain, possibly in situations such as learning and memorization, where synaptic connections may be perpetually remodeled.

INTERACTIONS BETWEEN CYTOKINES AND OTHER SOLUBLE FACTORS OF THE CENTRAL NERVOUS SYSTEM

The development of the CNS is regulated by a number of factors including neurotrophic factors, growth factors, and neurotransmitters. The recent studies on cytokines increase the number of soluble factors participating in CNS development. The variety of these factors lends credence to the idea that they might cooperate, positively or negatively, in order to control ontogenetic processes. Some examples are already known of sequential effects of multiple factors on cell development. For example, the differentiation of primary cultures of oligodendroblasts and their survival are at first assured by platelet-derived growth factor (PDGF); then this role is taken over by a neuropoietic cytokine, leukemia inhibitory factor.[57]

At present, the most studied interactions between cytokines and other soluble factors have concerned the control of their synthesis. For example, IL1 and IL6 can induce the secretion of nerve growth factor (NGF) by astrocytes, which would suggest an indirect neurotrophic role for these cytokines. IL1β can modulate the secretion of neurotransmitters such as acetylcholine, noradrenaline, and dopamine in the anterior hypothalamus of the rat.[58] Conversely, noncytokines may well modify the production of cytokines, as glial cells express receptors for a number of neurotransmitters and neurotrophic factors. In fact, some neuropeptides (substance P, vasoactive intestinal peptide), neurotrophins (NGF), and neurotransmitters (histamine) can increase the production of proinflammatory cytokines (IL1, IL6, and TNFα) in astrocytes.[59–62]

Interactions between cytokines and neurotransmitters such as noradrenaline[63] or serotonin[64] have recently been studied. The application of a method of semiquantitative RT-PCR has shown that serotonin at sub-nanomolar concentrations can increase the expression of IL6 and TGFβ mRNA, and induce the expression of TNFα mRNA in rat hippocampal astrocytes in primary culture (FIG. 2). It is likely that many other similar interactions remain to be discovered.

CONCLUSION

Cytokines and their receptors are synthesized in the CNS from early embryogenesis to the adult stage, and their presence in the CNS during ontogenesis extends the list of factors that may be implicated in neuro- or gliogenesis. The cytokines present spatiotemporal profiles of expression which suggest that they mediate specific functional roles at each stage of brain development. Studying the exact functions of the cytokines during embryogenesis has proved difficult *in vivo*, and the roles that have been attributed to them are based on their effects *in vitro*. These studies have underlined the diversity of effects of cytokines on different brain cells, as well as the differences in regulation occurring between cytokines and other soluble factors, notably, in terms of the control of their synthesis and in the occurrence of synergistic effects on brain cells. Examination of the expression of cytokines and their receptors in the brain and in cerebral cultures has confirmed that these signals are present at the appropriate times to mediate their different actions (neurotrophism, proliferation or differentiation). This suggests that cytokines may be involved in many aspects of CNS physiology, including intercellular communication between cerebral cells and involvement in cellular ontogenetic events and in the structural organization of the CNS. All these effects occur in a context of interactions with numerous other soluble factors, including neurotrophic factors, growth factors, and neurotransmitters, within a highly complex system of cellular regulation.

REFERENCES

1. METCALF, D. 1993. Hematopoietic regulators: Redundancy or subtlety? J. Am. Soc. Hematol. **82:** 3515–3523.
2. FONTANA, A., F. KRISTENSEN, R. DUBS, D. GEMSA & E. WEBER. 1982. Production of prostaglandin E and an interleukin-1 like factor by cultured astrocytes and C6 glioma cells. J. Immunol. **129:** 2413–2419.
3. LIEBERMAN, A. P., P. PITHA, H. S. SHIN & M. L. SHIN. 1989. Production of tumor necrosis factor and other cytokines by astrocytes stimulated with lipopolysaccharide or a neurotropic virus. Proc. Natl. Acad. Sci. USA **86:** 6348–6352.
4. KONISHI, Y., M. KAMEGAI, K. TAKANASHI, T. KUNISHITA & T. TABIRA. 1994. Produc-

tion of interleukin-3 by murine central nervous system neurons. Neurosci. Lett. **182:** 271–274.

5. OLSSON, T., S. KELIC, C. EDLUND, M. BAKHIET, B. HOJBERG, P. VAN DER MEIDE, J. LJUNGDAHL & K. KRISTENSSON. 1994. Neuronal interferon gamma immunoreactive molecule: Bioactivities and purification. Eur. J. Immunol. **24:** 308–314.

6. SCHÖBITZ, B., D. A. VOORHUIS & E. R. DE KLOET. 1992. Localization of interleukin-6 mRNA and interleukin-6 receptor mRNA in rat brain. Neurosci. Lett. **136:** 189–192.

7. SAWADA, M., Y. ITOH, A. SUZUMURA & T. MARUNOUCHI. 1993. Expression of cytokine receptors in cultured neuronal and glial cells. Neurosci. Lett. **60:** 131–134.

8. BREDER, C. D., C. A. DINARELLO & C. B. SAPER. 1988. Interleukin-1 immunoreactive innervation of the human hypothalamus. Science **240:** 321–324.

9. GADIENT, R. A. & U. OTTEN. 1994. Identification of interleukin-6 expressing neurons in the cerebellum and hippocampus of normal adult rats. Neurosci. Lett. **182:** 243–246.

10. BAN, E., G. MILON, N. PRUDHOMME, G. FILLION & F. HAOUR. 1991. Receptor for ILlα and β in mouse brain: Mapping and neuronal localization in hippocampus. Neuroscience **43:** 21–30.

11. GADIENT, R. A. & U. OTTEN. 1994. Expression of interleukin-6 and interleukin-6 receptor mRNAs in rat brain postnatal development. Brain Res. **637:** 10–14.

12. EISENBERG, O., A. FABER-ELMAN, M. LOTAN & M. SCHWARTZ. 1995. Interleukin-2 transcripts in human and rodent brains: Possible expression by astrocytes. J. Neurochem. **64:** 1928–1936.

13. LAZAR, L. M. & M. BLUM. 1992. Regional distribution and developmental expression of epidermal growth factor transforming factor a-mRNA in mouse brain by a quantitative nuclease protection assay. J. Neurosci. **12:** 1688–1697.

14. WILCOX, J. N. & R. DERYNCK. 1988. Developmental expression of transforming growth factors alpha and beta in mouse fetus. Mol. Cell. Biol. **8:** 3415–3422.

15. GATHERER, D., P. TEN DIJKE, D. BAIRD & J. AKHURST. 1990. Expression of TGFβ isoforms during first embryogenesis. Development **110:** 445–460.

16. ROELEN, B. A., H. Y. LIN, V. KNEZEVIC, E. FREUND & C. L. MUMMERY. 1994. Expression of TGFβs and their receptors during implantation and organogenesis of the mouse embryo. Dev. Biol. **166:** 716–728.

17. PELTON, R., B. SAXENA, M. JONES, H. MOSES & L. GOLD. 1991. Immunohistochemical localization of TGFβ1, TGFβ2 and TGFβ3 in the mouse embryo: Expression patterns suggest multiple roles during embryonic development. J. Cell Biol. **115:** 1091–1105.

18. DIAZ-RUIZ, C., R. PÉREZ-TOMAS, J. DOMINGO & I. FERRER. 1993. Immunohistochemical localization of transforming growth factor-α in choroid plexus of the rat and chicken. Neurosci. Lett. **164:** 44–46.

19. LEE, D. C., R. ROCHFORD, G. TODARO & L. VILLAREAL. 1985. Developmental expression of rat transforming growth factor-alpha mRNA. Mol. Cell. Biol. **5:** 3644–3646.

20. SCHMID, P., D. COX, G. BILBE, R. MAIER & G. MASTER. 1991. Differential expression of TGFβ1, β2 and β3 genes during mouse embryogenesis. Development **111:** 117–130.

21. FLANDERS K., G. LÜDECKE, S. ENGELS, D. CISSEL, A. ROBERTS, P. KONDAIAH, R. LAFAYATIS, M. SPORN & K. UNSICKER. 1991. Localization and actions of transforming growth factor-βs in the embryonic nervous system. Development **113:** 183–191.

22. HEINE, U., E. F. MUNOZ, K. C. FLANDERS, L. R. ELLINGSWORTH, H. Y. PETER-LAM, N. L. THOMPSON, A. ROBERTS & M. B. SPORN. 1987. Role of transforming growth factor-β in the development of the mouse embryo. J. Cell Biol. **105:** 2861–2876.

23. BURNS, T., J. CLOUGH, R. KLEIN, G. WOOD & N. BERMAN. 1993. Developmental regulation of cytokine expression in the mouse brain. Growth Factors **9:** 252–258.

24. GENDRON, R. L., F. P. NESTEL & W. S. LAPP. 1991. Expression of TNFα in the developing nervous system. Int. J. Neurosci. **60:** 129–136.

25. POUSSET, F. 1994. Developmental expression of cytokine genes in the cortex and hippocampus of the rat central nervous system. Dev. Brain Res. **81:** 143–146.

26. THÉRY, C., E. HETIER, C. EVRARD & M. MALLAT. 1990. Expression of macrophage colony stimulating factor gene in the mouse brain during development. J. Neurosci. Res. **26:** 129–133.

27. CHANG, Y., S. ALBRIGHT & F. LEE. 1994. Cytokines in the central nervous system: Expression of macrophage colony stimulating factor and its receptor during development. J. Neuroimmunol. **52:** 9–17.

28. GIULIAN, D., D. G. YOUNG, J. WOODWARD, D. C. BROWN & L. B. LACHMAN. 1988. Interleukin-1 as an astroglial growth factor in the developing brain. J. Neurosci. **8:** 709–714.

29. LUETTECKE, N. C., T. H. QIU, R. L. PEIFFER, P. OLIVER, O. SMITHIES & D. C. LEE. 1993. TGFα deficiency results in follicle and eye abnormalities in targeted and waved-1 mice. Cell **73:** 263–278.

30. MANN, G. G., K. J. FOWLER, A. GABRIEL, E. C. NICE, R. L. WILLIAMS & A. R. DUNN. 1993. Mice with a null mutation of the TGFα gene have abnormal skin architecture, wavy hair and curly whiskers and often develop inflammation. Cell **73:** 249–261.

31. SAWADA, M., A. SUZUMURA, H. YAMAMOTO & T. MARUNOUCHI. 1990. Activation and proliferation of the isolated microglia by colony stimulating factor-1 and possible involvement of protein kinase C. Brain Res. **509:** 119–124.

32. BEREZOVSKAYA, O., D. MAYSINGER & S. FEDEROFF. 1995. The hematopoietic cytokine, colony stimulating factor 1, is also a growth factor in the CNS—Congenital absence of CSF-1 in mice results in abnormal microglial response and increased neuron vulnerability to injury. Int. J. Dev. Neurosci. **13:** 285–299.

33. GIULIAN, D. & L. B. LACHMAN. 1985. Interleukin-1 stimulation of astroglial proliferation after brain injury. Science **228:** 497–499.

34. ROBBINS, D., Y. SHIRAZI, B. DRYSDALE, A. LIEBERMAN, H. SHIN & M. SHIN. 1987. Production of cytotoxic factor for oligodendrocytes by stimulated astrocytes. J. Immunol. **139:** 2593–2597.

35. HAMA, T., M. MIYAMOTO, H. TSUKUI, C. NISHIO & H. HATANAKA. 1990. IL6 as a neurotrophic factor for promoting the survival of cultural basal forebrain cholinergic neurons from postnatal rats. Neurosci. Lett. **104:** 340–344.

36. ISHIHARA, A., H. SAITO & K. ABE. 1994. Transforming growth factor-β1 and -β2 promote neurite sprouting and elongation of cultured rat hippocampal neurons. Brain Res. **639:** 21–25.

37. MEHLER, M. F., R. MARMUR, R. GROSS, P. C. MABIE, Z. Y. ZHANG, A. PAPAVASILIOU & J. A. KESSLER. 1995. Cytokines regulate the cellular phenotype of developing neural lineage species. Int. J. Dev. Neurosci. **13:** 213–240.

38. DA CUNHA, A. & L. VITKOVIC. 1992. TGFβ1 expression and regulation in rat cortical astrocytes. J. Neuroimmunol. **36:** 157–169.

39. SUZUMARA, A., M. SAWADA, J. YAMAMOTO & T. MARUNOUCHI. 1993. Transforming growth factor-β suppresses activation and proliferation of microglia in vitro. J. Immunol. **151:** 2150–2158.

40. BOGLER, O., D. WREN, S. C. BARNETT, H. LAND & M. NOBLE. 1990. Co-operation between two growth factors promotes extended self-renewal and inhibits differentiation of oligodendrocyte-type-2-astrocyte (O-2A) progenitor cells. Proc. Natl. Acad. Sci. USA **87:** 6368–6372.

41. ROVASIO, R., A. DELOUVEE, K. YAMADA, R. TIMPL & J. P. THIERY. 1983. Neural crest cell migration requirements for exogenous fibronectin and high cell density. J. Cell Biol. **96:** 462–473.

42. GROTENDORST, G. R., Y. SOMA, K. TAKEHARA & M. CHARLETTE. 1989. EGF and TGF alpha are potent chemoattractants for endothelial cells and EGF-like peptides are present at sites of tissue regeneration. J. Cell. Physiol. **139:** 617–623.

43. LEONARD, E. J. & T. YOSHIMURA. 1990. Human monocyte chemoattractant protein-1 (MCP1). Immunol. Today **11:** 97–101.

44. WANG, J. M., J. D. GRIFFIN, A. RAMBALDI, Z. G. CHEN & A. MANTOVANI. 1988. Induction of monocyte migration by recombinant macrophage colony stimulating factor. J. Immunol. **141:** 57–79.

45. LEVITT, P. & P. RAKIC. 1980. Immunoperoxidase localization of glial fibrillary acidic protein in radial glial cells and astrocytes of the developing rhesus monkey brain. J. Comp. Neurol. **193:** 815–840.

46. MEHLHER, M. F., R. ROZENTAL, M. DOUGHERTY, D. SPRAY & J. KESSLER. 1993. Cytokine regulation of neuronal differentiation of hippocampal progenitor cells. Nature **362:** 62–65.

47. KAHN, M. A. & J. DEVELLIS. 1994. Regulation of an oligodendrocyte progenitor cell line by the interleukin-6 family of cytokines. Glia **12:** 87–98.

48. KUSHIMA, Y., T. HAMA & H. HATANAKA. 1992. Interleukin-6 as a neurotrophic factor for promoting the survival of cultured catecholaminergic neurons in a chemically defined medium from fetal and postnatal rat midbrains. Neurosci. Res. **13:** 267–280.

49. ALEXI, T. & F. HEFTI. 1993. Trophic actions of transforming growth factor α on mesencephalic dopaminergic neurons developing in culture. Neuroscience **55:** 903–918.

50. ZHANG, M., D. D. WOO & B. D. HOWARD. 1990. Transforming growth factor α and PC12 derived growth factor induce neurites in PC12 cells and enhance survival of embryonic brain neurons. Cell Regul. **1:** 511–521.

51. BENVENISTE, E. N. & J. E. MERILL. 1986. Stimulation of oligodendrocyte proliferation and maturation by interleukin-2. Nature **321:** 610–613.

52. FERRER, I., T. SERRANO & E. SORIANO. 1990. Naturally occurring cell death in the subicular complex and hippocampus in the rat during development. Neurosci. Res. **8:** 60–66.

53. KESSLER, J. A., W. H. LUDLAM, M. M. FREIDIN, D. H. HALL, M. D. MICHAELSON, D. C. SPRAY, M. DOUGHERTY & D. K. BATTER. 1993. Cytokine induced programmed death of cultured sympathetic neurons. Neuron **11:** 1123–1132.

54. SELMAJ, K., C. RAINE, M. FAROOQ, W. NORTON & C. BROSNAN. 1990. Cytokine cytotoxicity against oligodendrocytes: Apoptosis induced by lymphotoxin. J. Immunol. **147:** 1522–1529.

55. RORDORF, G., W. KOROSHETZ & J. BONVENTRE. 1991. Heat shock protects cultured neurons from glutamate toxicity. Neuron **7:** 1043–1051.

56. D'SOUZA, S. D., J. P. ANTEL & M. FREEDMANN. 1994. Cytokine induction of heat shock protein expression in human oligodendrocytes: An interleukin-1 mediated mechanism. J. Neuroimmunol. **50:** 17–24.

57. GARD, A. L., M. R. BURELL, S. E. PFEIFFER, J. S. RUDGE & W. C. WILLIAMS. 1995. Astroglial control of oligodendrocyte survival mediated by PDGF and leukemia inhibitory factor-like protein. Development **121:** 2187–2197.

58. SHINTANI, F., S. KANBA, T. NAKAKI, M. NIBUYA, N. KINOSHITA, E. SUZUKI, G. YAGI, R. KATO & N. ASAI. 1993. Il1β augments release of norepinephrine, dopamine and serotonin in the rat anterior hypothalamus. J. Neurosci. **13:** 3574–3581.

59. ALHEIM, K., C. ANDERSSON, S. TINGSBORG, M. ZIOLKOWSKA, M. SCHULTZBERG & T. BARTFAI. 1991. Interleukin-1 expression is inducible by nerve growth factor in PC 12 pheochromocytoma cells. Proc. Natl. Acad. Sci. USA **88:** 9302–9306.

60. MARTIN, F. C., A. C. CHARLES, M. J. SANDERSON & J. E. MERILL. 1992. Substance P stimulates IL1 production by astrocytes via intracellular calcium. Brain Res. **599:** 13–18.

61. LÜBER-NAROD, J., R. KAGE & S. E. LEEMAN. 1994. Substance P enhances the secretion of tumor necrosis factor alpha from neuroglial cells stimulated with lipopolysaccharides. J. Immunol. **152:** 819–823.

62. CADMAN, E., D. WHITTE & L. CHI-MING. 1994. Regulation of the release of interleukin-6 from human astrocytoma cells. J. Neurochem. **63:** 980–987.

63. MAIMONE, D., C. CIONI, S. ROSA, G. MACCHIA, F. ALOISI & F. ANNUNZIATA. 1993. Norepinephrine and vasoactive intestinal peptide induce IL6 secretion by astrocytes—Synergism with IL1β and TNFα. J. Neuroimmunol. **47:** 73–82.

64. POUSSET, F., J. FOURNIER, P. LEGOUX, P. E. KEANE, D. SHIRE & P. SOUBRIÉ. 1996. Effect of serotonin on cytokine mRNA expression in rat hippocampal astrocytes. Mol. Brain Res. **38:** 54–62.

Ontogeny of Prohormone Convertases in Rat Prenatal Development[a]

MIN ZHENG[b] AND JOHN E. PINTAR[c]

Department of Neuroscience and Cell Biology
UMDNJ—Robert Wood Johnson Medical School
Piscataway, New Jersey 08854

INTRODUCTION

Many hormones, neurotransmitters, and growth factors are peptides that act as chemical messengers regulating essentially all aspects of physiological and developmental processes. Nearly all known peptides are synthesized as larger precursors that require limited proteolytic cleavages and other posttranslational modifications to liberate bioactive peptides. Among the different steps in proprotein conversion, the endoproteolytic cleavage is obviously a critical event because it dictates susceptibility of peptide intermediates to subsequent modifications and to a large extent shapes the final peptide products. The nature of the enzymes responsible for the endoproteolytic cleavages of proproteins had remained elusive until recently. Progress was first made from studies of yeast *Saccharomyces cerevisiae* because it offered an experimental system more amenable to genetic analysis. In yeast, the endoproteolytic cleavages directed at Lys-Arg sites are required to liberate multiple copies of peptide pheromone α-mating factors from the two precursors. The kex2 gene was successfully cloned by genomic complementation of kex2 mutant strains.[1] Analyzing its sequence revealed that kex2 contains near its N-terminus a region homologous (60% when conservative amino acids are considered) to the catalytic domain of bacterial protease subtilisins[2] (FIG. 1). The relevance of kex2 to mammalian proprotein processing was first raised when incubation of membrane fractions enriched with kex2 together with proalbumin resulted in the cleavage of the latter into the mature form.[3] Furthermore, kex2 was shown to be able to cleave POMC in gene transfer experiments.[4] This indicated that kex2 is a functional homologue of mammalian proprotein processing enzymes and, therefore, would be structurally related to the mammalian proprotein convertases. Furin was subsequently identified as the first potential mammalian proprotein convertase by virtue of its protein sequence alignment with that of kex2[5] (FIG. 1). Thereafter, several additional endoproteases, including PC1/PC3, PC2, PACE4, PC4, and PC5/PC6, were isolated using a polymerase chain reaction (PCR)-based technique with degenerative primers corresponding to the sequence surrounding to the subtilisn-like catalytic domains of kex2 and furin[6–13] (FIG. 1).

[a] This work was supported by research grants HD-18592 and DA-08622 from the National Institutes of Health to J.E.P.

[b] Present address: Department of Pathology, Columbia University College of Physicians and Surgeons, 630 West 168th Street, New York, NY 10032.

[c] Address correspondence to John Pintar, 675 Hoes Lane, CABM 326, Piscataway, New Jersey 08854. E-mail: pintar@mbcl.rutgers.edu

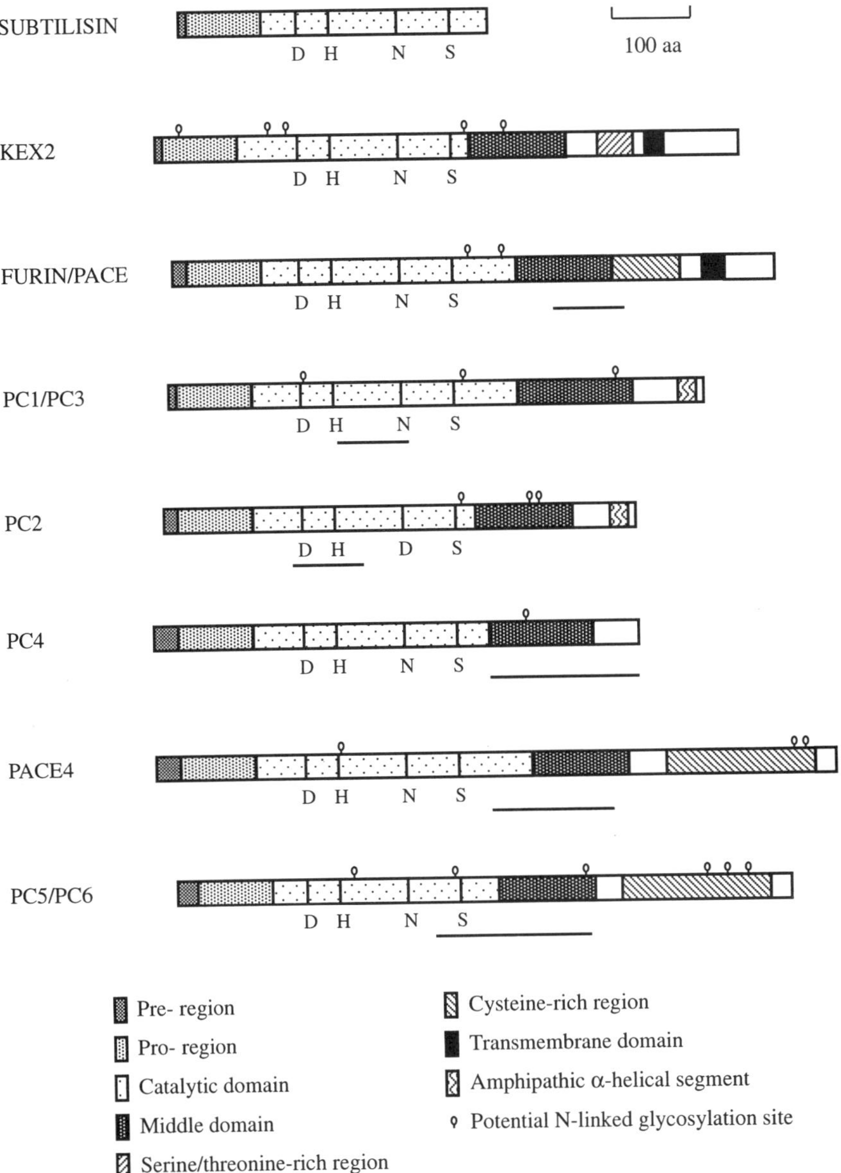

FIGURE 1. Schematic representation of protein structures of mammalian subtilisin-like serine endoproteases in comparison with bacterial subtilisin and yeast kex2. The positions of the catalytically important amino acid residues are indicated with one-letter abbreviations. The bars represent the corresponding regions of cRNA probes used in the study.

Given the past difficulties in isolating these enzymes and the repeated false enzymes encountered, it remains useful to summarize the evidence that these prohormone convertases (PCs) really represent bona fida proprotein convertases *in vivo*.

1. PC1 and PC2 expressions are restricted to the neuroendocrine system as would be expected for neuropeptide-specific prohormone convertases. For instance, both PC1 and PC2 are expressed at particularly high levels in certain brain regions active in peptide biosynthesis, such as the hypothalamus and hippocampus.[6,9,14] In contrast, furin is almost ubiquitously expressed,[15,16] indicating that it may be an enzyme involved in the precursor processing in the constitutive secretory pathway.
2. PC1 and PC2 exhibit properties of endoproteases residing in the regulated secretory pathway, catalyzing endoproteolytic conversion of substrates in a Ca^{2+}-dependent manner, with a pH optimum at 5.5, and exhibit the expected inhibitor profiles.[17,18] In agreement with its role in the constitutive processing of proproteins, furin activity is characterized as Ca^{2+}-dependent with a pH optimum at 7.0,[19] and is predominantly localized in a Golgi-related compartment.[20,21]
3. In gene transfer experiments, PC1 and PC2 are capable of processing prohormones and neuropeptide precursors at dibasic sites *in vitro*, whereas furin is effective in the processing of growth factors, growth factor and hormonal receptors, serum proteins, and viral surface glycoproteins (for reviews, see refs. 22 and 23).
4. PC expression levels are coregulated with their candidate substrates. Parallel changes in PC1, PC2, and POMC mRNA levels in the pituitary occur in response to secretagogues, which are consistent with their role as prohormone processing enzymes.[24–26]
5. Disruption of PC expression blocks proprotein processing. Antisense inhibition of PC1 expression in AtT-20 cells substantially reduces POMC processing.[25]

Multiple growth factor ligand and receptor gene products have been shown to play pivotal roles in embryonic and fetal development, such as those for insulin-like growth factors (IGFs), transforming growth factors (TGFs), and platelet-derived growth factor (PDGF).[27–32] Many neuropeptides, such as vasoactive intestinal peptide (VIP) and gonadotropin-releasing hormone (GnRH), in addition to serving as hormones and neurotransmitters, exhibit trophic functions in stimulating mitosis, promoting cellular growth, and permitting neural cell survival.[33–35] Given the known role of PCs in mediating proteolytic cleavage and the plethora of peptide precursors requiring such a modification, it is interesting to evaluate the prenatal expression profile of individual PCs. Specific questions that can be addressed are whether one or more members of PCs are present during embryogenesis, whether they are distinctively expressed, and whether their expression patterns correlate to any specific peptide substrate. In this report, we summarize our previous studies[36] and recent studies[37] detailing the prenatal expression profiles of rat PCs.

MATERIALS AND METHODS

For *in situ* hybridization, embryos, dated embryonic days 6–21 (e6–e21), were obtained from 46 timed pregnant Sprague-Dawley female rats in accordance with the principles and procedures outlined in the National Institutes of Health Guide for the Care and Use of Laboratory Animals. Uteri (for embryos e6–e9.5), whole embryos (e10–e15), or whole fetuses or fetal head and trunk (e16–e21) were collected. Embryos younger than e12 were fixed by immersion in 4% paraformaldehyde for 6–12 h and equilibrated in 20% sucrose solution before embedding. Older embryos or fetuses (e12.5–e21) were freshly frozen and embedded directly in OCT compound without prior fixation. Both sagittal and transverse cryostat sections (e6–e21) as well as frontal cranial sections (e14–e21) were prepared. Both sense and antisense [^{35}S]UTP-labeled cRNA were prepared using an *in vitro* transcription system (Promega, Madison, WI). Corresponding positions of cRNA probes for PCs used in this study are shown in FIGURE 1. *In situ* hybridization experiments were performed as previously described.[38]

RESULTS

Cellular distribution of PCs and related genes in the early stages of rat gestation (e7–e12) are summarized in TABLE 1; those of middle to late stages of gestation (e13–e21) in TABLE 2. Furin, PC5, PACE4, and CPE are differentially expressed in uterine tissues following implantation, with the expression of PACE4 in the decidua being the most prominent (TABLE 1; FIG. 2). Furin is the first PC expressed in the embryo proper, detectable at an early postimplantation stage (e7; TABLE 1). The widespread distribution of furin transcripts in the mesoderm and endoderm, as well as their derivatives, is maintained throughout prenatal development, with the highest level observed in the heart and liver (TABLE 2; a representative section is shown in FIG. 3C). PC1 expression is initiated at a midgestation (e13) and is restricted to hypothalamus, brain stem, peripheral ganglia, pituitary, and pancreas (TABLE 2; FIG. 3A). Expression of PC2 is more widespread in the developing nervous system than that of PC1, being detected in the telencephalon (cerebral cortex, striatum, hippocampus), diencephalon (thalamus, hypothalamus), mesencephalon, metencephalon, myelencephalon, peripheral ganglia, pituitary, and pancreas (TABLE 2; FIG. 3B). A highly restricted expression of PC5 in several embryonic primordia in the early gestational stages (TABLE 1) expands to multiple organ systems in the middle to late gestational stages (TABLE 2; FIG. 3D). Similar to PC5, PACE4 is also expressed both in several brain regions and in multiple peripheral tissues, with its expression in the lung and gut complementary to that of PC5 (TABLE 2; FIG. 3E; compare FIG. 4E and F). CPE is expressed at a high level in nearly all regions of the brain and several peripheral organs, with a notable absence of expression in liver (TABLE 2; FIG. 3F). During middle and late gestations, 7B2 is expressed at a high level in the peripheral ganglia, low to undetectable in the thalamus and hippocampus, and at a moderate level in the rest of the brain (TABLE 2). Although multiple PCs are expressed in the developing nervous system, their expression domain is usually region-specific. PC2 expression in both thalamus and pons, for instance, is restricted to the intermediate zones of neuroepithelium, whereas that of PACE4 is restricted to the ventricular zones (FIG. 4A–D).

DISCUSSION

Many peptides modulating cellular growth and differentiation in development, such as growth factors, growth factor receptors, and neuropeptides, are first

TABLE 1. PC Gene Expression in the Early Gestational Stages of Rat Development

	Uterus, Placenta	e7	e9	e10	e12
Furin	Undecidualized endometrium, chorion, amnion, placenta	Intra-/extraembryonic mesoderm and endoderm	Myocardium and somatic mesoderm surrounding pericardial coelem	Cardiac primordium, hepatic primordium	All tissues outside of nervous system, particularly high in heart and liver
PC1	—	—	—	—	—
PC2	—	—	—	—	—
PC5	Implantation site	—	Restricted regions of the neural tube, caudal myotomes, fetal-maternal junction	Optic vesicles, otic vesicles, roof of midbrain, trunk myotomes	n/a
PACE4	Metametrial side of decidua	—	—	—	—
CPE	Primary decidua	—	—	Neural tube	Neural tube, ganglia, optic vesicle, branchial archer, endocardium, mesenchyme
7B2	—	—	—	—	—

TABLE 2. PC Gene Expression in the Middle and Late Gestational Stages of Rat Development[a]

	Furin	PC1	PC2	PC5	PACE4	CPE	7B2
Nervous system							
Caudate-putamen	−	−	++	−	+	++	+
Cerebral cortex	−	−	++	+	−	++	+
Hippocampus	−	−	++	+	++	++	−
Thalamus	−	−	++	+	+	++	−
Hypothalamus	−	++	++	++	+	++	+
Pons	−	+	++	+	+	++	+
Medulla	−	+	++	+	+	++	+
Cerebellar cortex	−	−	+	+	+	++	+
Spinal cord	−	−	++	+	+	++	+
Peripheral ganglia		++	++	n/a	n/a	++	++
Peripheral tissues							
Pituitary	−	+	+	n/a	n/a	++	+
Pancreas	−	+	+	n/a	n/a	+	n/a
Heart	++	−	−	−	+	+	−
Liver	++	−	−	−	+	−	−
Lung	+	−	−	+[b]	+[c]	+	−
Gut	+	−	−	+[d]	+[e]	+	−
Adrenal	+	−	−	+	−	+	−
Kidney	+	−	−	+	−	+	−
Cartilage	+	−	−	+	+	+	−

[a] ++ = high level; + = low level; − = undetectable.
[b] Splanchnic mesenchyme.
[c] Respiratory epithelium.
[d] Mucosa.
[e] Submucosa.

synthesized as inactive precursors that require proteolytic cleavages.[39,40] Proteolytic processing is performed by endoproteases usually at sites marked by dibasic or single basic residues, followed by exoproteolytic removal of exposed basic residues by a carboxypeptidase. These processing events, together with other posttranslational modifications, such as acetylation and amidation, lead to the final release of bioactive peptides. Because of their extremely low abundance *in vivo*, the identities of endoproteases had remained elusive for decades despite extensive biochemical and enzymatic studies. Only in recent years has significant progress been made in elucidating the molecular identities of the endoproteases involved in proprotein processing. Thus far, by virtue of PCR-based techniques, six members of the "prohormone convertase" family of endoproteases have been isolated, including PC1, PC2, furin, PC4, PC5, and PACE4.[6–13]

Because of the universal requirement of proteolytic processing activity in development for proprotein maturation, endoproteases and other processing enzymes are inevitably employed in development. Since multiple members of subtilisin-like convertases have been isolated, we attempted to address the questions of whether they are all employed during development and whether they are functionally redundant or distinct in processing different sets of substrates by assaying the comparative ontogenies of their gene expression. Using *in situ* hybridization, we established that all PCs (except PC4) are present in various regions of the embryo during the period of active morphogenesis and establishment of cellular

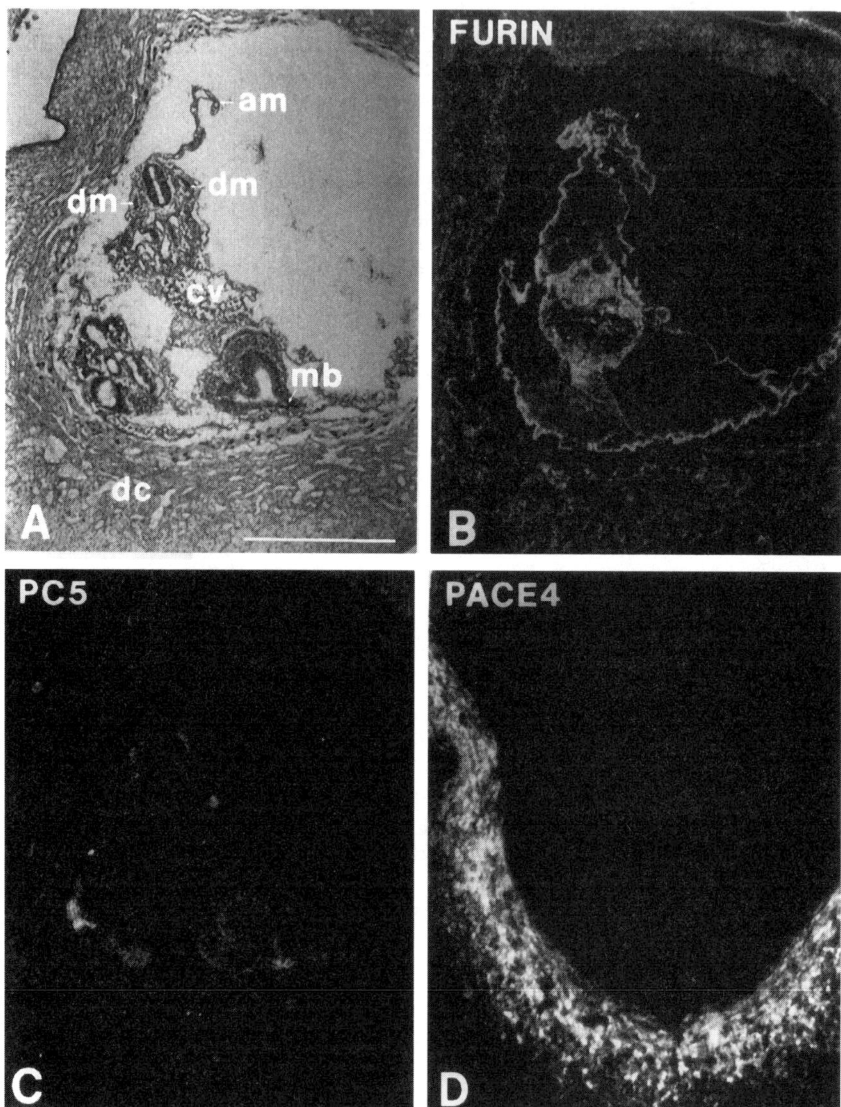

FIGURE 2. PC5, PACE4, and furin are expressed in distinct patterns in extraembryonic and uterine cells at e10. Note the prominent expressions of furin in the embryo proper **(B)** and PACE4 in the decidua **(D)**. In contrast, PC5 is expressed in a highly restricted manner in several primordial structures **(C)**. am, amnion; cv, common ventricle (heart); dc, decidua; dm, dermomyotome; mb, midbrain (roof). Scale bar: 1 mm.

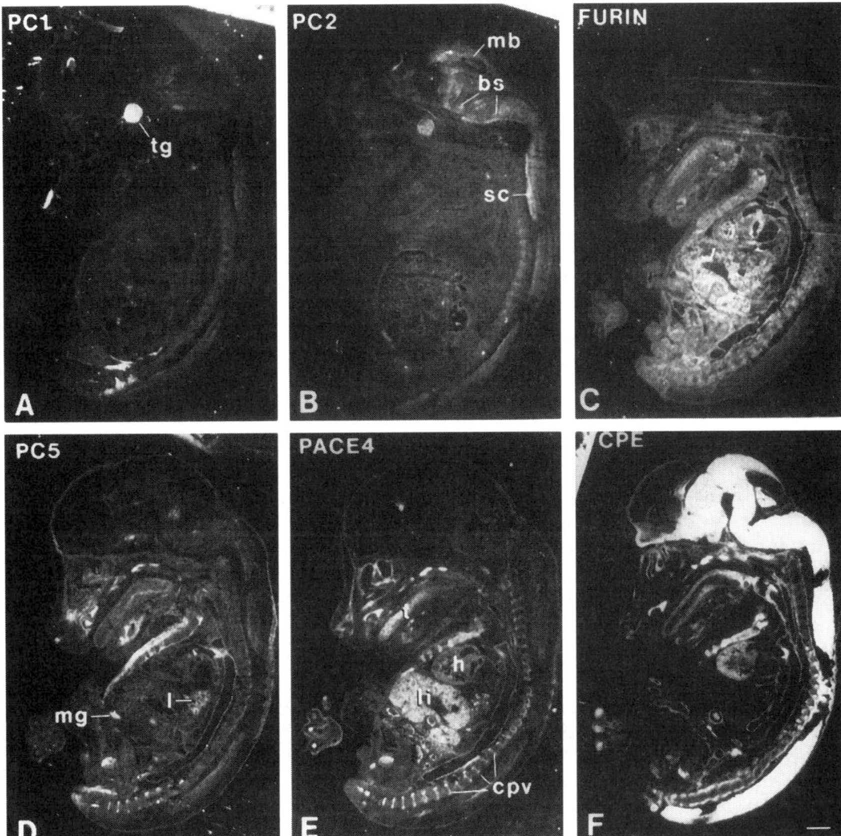

FIGURE 3. Comparative expressions of PCs at e15.5. PC1 **(A)** gene expression is mostly detectable in the peripheral nervous system, such as trigeminal ganglion (tg). PC2 **(B)** is expressed in multiple regions of the developing nervous system, such as the midbrain (mb), brain stem (bs), and spinal cord (sc). Furin **(C)** is expressed in nearly all regions of the embryo outside of the nervous system. Restricted expression of PC5 **(D)** is observed in several embryonic structures, such as the lung (l) and midgut (mg). PACE4 **(E)** expression in the peripheral tissues is more widespread than that of PC5, and is detected in the intrinsic muscle of the tongue (t), heart (h), liver (li), and cartilage primordia of vertebra (cpv). A high level of CPE **(F)** is observed in nearly all regions of the developing nervous system and selected regions of the peripheral tissues. Scale bar: 1 mm.

functions. In early postimplantation stages, for instance, furin expression is detected, indicating that this enzyme is preferentially employed in early embryogenesis. By the middle and late gestational stages, a general pattern of expression emerges such that PC1 and PC2 are expressed in the neuroendocrine system, PC5 and PACE4 transcripts are present in both the nervous system and peripheral tissues, and those of furin predominate in tissues outside of the nervous system. Restricted expressions of PC1 and PC2 in the neuroendocrine system suggest that they may be specifically involved in the processing of prohormones and neuropeptide precursors. By comparison, widespread expression of furin, PC5, and PACE4 in peripheral tissues suggests that they may be involved in the processing of fetal growth factors and their receptors. The expressions of PC5 and PACE4 are largely distinct and often complementary in several organ systems, suggesting that they might be involved in the processing of distinctive sets of proproteins in the peripheral tissues.

Comparison of the expression patterns of PCs in embryogenesis with their distribution in adult suggests that individual PCs may undergo significant developmental changes before assuming the adult pattern of expression. Prenatal expression of PC1, once initiated, is largely restricted to the hypothalamus, whereas only a low level of PC1 is detected in several extrahypothalamic regions, such as the brain stem. In comparison, in the adult, a moderate to high level of PC1 expression is reported in multiple brain regions, such as the caudate-putamen, hippocampus, thalamus, and spinal cord.[14] This suggests that the majority of brain regions expressing PC1 in the adult are established at a later stage, presumably during postnatal development. A similar delay is also observed for furin. During embryogenesis, expression of furin is widespread in the peripheral tissues, whereas the developing central nervous system (CNS) is devoid of detectable furin expression. In comparison, a region-specific expression of furin is observed in the adult brain.[14,41] In contrast to PC1 and furin, relatively high-level and widespread expressions for PC2 and PC5 are observed in both adult and in the embryo.[14,36,42] In fact, comparison of PC2 gene expression with those of other PCs makes it clear that PC2 represents the major endoprotease presence in both fetal and adult brain. In development, the expression of PC2 also characterizes the intermediate zone, instead of ventricular zone, suggesting that PC2 expression is switched on after the neural cells withdraw from the cell cycle and begin differentiation. The ventricular zone, which contains proliferating progenitor neural cell populations, expresses PACE4 in many brain regions, including the thalamus and pons. Therefore, the prenatal expressions of PC2 and PACE4 in many brain regions are complementary. The preference for PACE4 to be present in the proliferating zone of the neuroepithelium is similar to the peptidylglycine α-amidating monooxygenase (PAM; Zhang, Eipper, and Pintar, in preparation), suggesting that this cell population may express specific substrates participating in developmental decisions that require this specific set of processing enzymes. Strikingly, a particularly high level of PACE4 gene expression is observed in the area containing proliferating

---

FIGURE 4. Distinct and complementary expressions of PCs in embryogenesis. Adjacent transverse sections of e13 head (**A, B**) and coronal sections of e17 head (**C, D**) and lung (**E, F**) were hybridized with PC2 (**A, C**), PACE4 (**B, D, F**), and PC5 (**E**). Note the nonoverlapping expressions of PC2 and PACE4 in the brain. Specifically, PC2 is expressed in the intermediate zones of the thalamus (t) and pons (p). In contrast, PACE4 is expressed in the ventricular zones of the thalamus and pons, next to the third (3v) and fourth (4v) ventricle, respectively. PACE4 is also expressed at a high level in the hippocampus (hi) at e13. At e17 thalamic region, PC2 is detected in the lateral habenular nucleus and ventrolateral nuclear complex, whereas PACE4 is observed in the epithalamic and intermediate thalamic neuroepithelium. PC2 is also expressed at a moderate level in the amygdala (am) and hypothalamus (hy). PC5 is expressed in the splanchnic mesenchyme of the lung (l), whereas PACE4 is expressed in the epithelial components. b, bronchus; cpr, cartilage primordium of rib; lv, lateral ventricle. Scale bar: 1 mm.

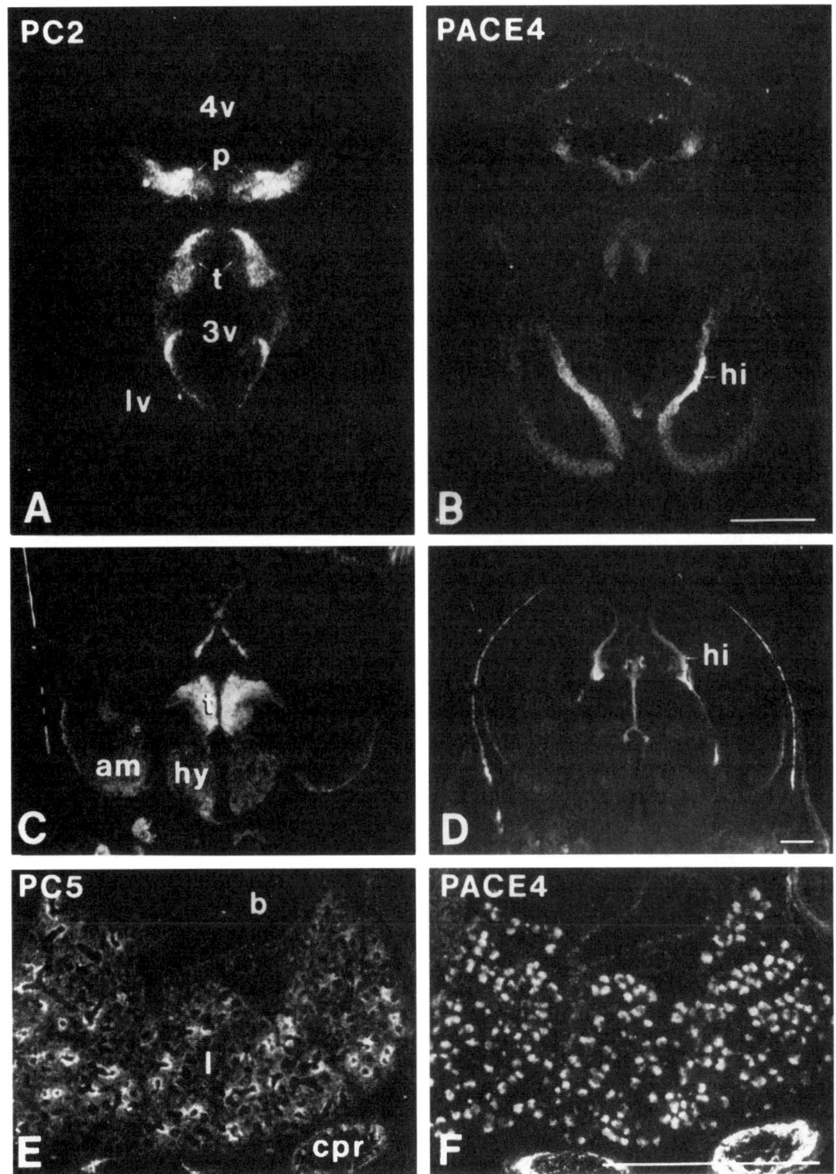

FIGURE 4.

progenitors of the hippocampal formation in early and middle gestational stages. In contrast, PACE4 is not detected in the hippocampus in the adult.[42] This suggests that, for this brain region, PACE4 expression peaks during the proliferating phase of its development and withdraws as neural cells in this region mature. It should be noted that PACE4 is also expressed in the multiple brain regions outside of the ventricular zone in development. In these regions, PACE4 gene expression, once established, is either similar to the adult level (e.g., caudate-putamen, hypothalamus, and brain stem) or lower (cerebellar cortex). Taken together, these data suggest that the transcriptional regulation of PACE4 gene expression may be differentially controlled in a tissue-specific manner by at least two alternative mechanisms, one of which is cell cycle-related.

Assaying the ontogeny of convertase gene expression provides a unique opportunity to advance our general understanding of the cellular mechanism of proprotein processing. Analysis of PC gene expression in an embryo section can be efficiently correlated to expression patterns of proproteins. Correlated expression of a particular PC with a proprotein is suggestive of an enzyme-substrate relationship. The comparably high levels of furin and PACE4 expressed in the liver, for instance, suggest that these two enzymes, but not other known PCs, may be involved in the proteolytic maturation of proalbumin, IGF-II, TGF-β1, and complement factor precursors synthesized in this organ. Embryogenesis also offers a temporal dimension which, when gene expression is carefully traced throughout, can provide additional clues about functional substrates. For example, for furin, its early initiation of gene expression, its surge in its level in regions outside of the developing nervous system at midgestational stages, and the eventual decline in its level at late gestation in many structures closely match the temporal changes in the levels of expression of several fetal proproteins, such as IGF-II and TGF-β1.[43,44] These results should in turn encourage more gene transfer experiments to test the *in vitro* potentiality of furin in the proteolytic cleavage of IGF-II and TGF-βs. Many tentative spatial correlations of individual PCs in the developing nervous system with neuropeptide precursors can also be made. In the middle and late gestational stages, for instance, the expressions of PC1, PC2, and PC5 in the hypothalamus overlap with those of CRH,[45] CGRP,[46] neurotensin,[47] neuropeptide Y,[48] somatostatin,[49] substance P,[50] and vasopressin.[51] The expression of PC2 in the basal ganglia overlaps with those of neuropeptide Y,[48] somatostatin,[49] and substance P.[50] The expressions of PC1, PC2, and PC5 in the brain stem overlap with those of CGRP,[46] neurotensin,[47] neuropeptide Y,[48] and somatostatin.[49] The expressions of PC2, PC5, and PACE4 in the spinal cord overlap with those of neuropeptide Y,[48] somatostatin,[49] substance P,[50] and VIP.[52] Since embryonic brain displays immense functional diversity, further inference of enzyme-substrate relationship should preferentially come from comparative gene expression analysis at the cellular or subcellular levels. This calls for extensive colocalization studies, such as using [^{35}S]UTP-labeled convertase probe and digoxigenin-UTP-labeled neuropeptide probe in the *in situ* hybridization experiments. This type of experiment has so far been successful both in adult[26] and neonatal[36] tissues. Different PCs, such as PC1 and PC2, display distinct cleavage selectivity when multiple dibasic or monobasic sites are presented in the same precursor.[53,54] When these results are coupled with their overlapping but distinct expression patterns in the pituitary, it suggests that manipulating the stoichiometry of PC1 and PC2 may be responsible for the tissue-specific and developmentally regulated POMC processing in this organ.[26,36,55] Therefore, careful analysis of not only the distinct distribution pattern of PCs, but also the comparable level of their expression in the regions of overlap, should be emphasized in future studies. Different ratios of PCs may well be the underlying mechanism in mediating numerous other examples of tissue specificity and developmental plasticity in proprotein processing.[56]

Regulated biosynthesis, processing, and release of peptide hormone is essential in maintaining the homeostasis of an organism. Regulation of pituitary POMC derivatives, either using the AtT-20 model system or by *in vivo* studies, has been a focus of intensive research. In pituitary, POMC is subject to a complex regulatory network which exerts its effect both at the transcriptional and posttranslational levels.[57–59] Certain conditions, such as chronic stress treatments of whole animal or long-term CRH exposure of pituitary primary culture, qualitatively alter the pattern of pituitary POMC processing and release.[60–62] It would be interesting to find out whether quantitative changes in the ratio of different processing enzymes such as PC1 and PC2 may mediate alterations in POMC processing under these conditions. Studies of responsiveness of POMC to various regulatory mechanisms have also been extended to the neonatal and prenatal period.[63–66] Parallel evaluation of POMC and PC expression under various conditions can be accomplished by a convenient *in vitro* primary culture system, in which neonatal pituitaries are dissected, lobes separated, and individually cultured.[38,66] The effects of physiologically relevant stimuli applied to the culture, such as that of CRH, on the transcriptional levels of POMC and PCs, can be assayed by a quantitative measure, such as RNase protection assay.

Because the biological activity of PCs can be modulated at the posttranslational level, functional evaluation of PCs should also consider potential regulators. We have begun to examine the expression of PC2 along with its regulator 7B2[67–69] in development. At late gestational stages, for instance, the expression of 7B2 is dramatically reduced in multiple brain regions where PC2 expression is high, such as the thalamus, hippocampus, and caudate-putamen (Zheng, Day, Seidah, and Pintar, unpublished observations). These differences could result in significant differential stoichiometry of 7B2 and PC2 proteins in these regions, which in turn could lead to modulation of PC2 bioactivity in these brain regions.

Studies of ontogeny of CPE gene expression have shown that it is expressed early on throughout the neural tube, at a time when no known PCs are expressed in the nervous system except for PC5 in the floor plate. In middle and late gestational stages, CPE is expressed by nearly all neural tissues in the developing nervous system. In contrast, those PCs that are expressed in CNS, such as PC1, PC2, PC5, and PACE4, display a more restricted pattern of expression. Certain regions of the brain, such as part of cerebral cortex and medial region of ventral thalamus, lack apparent expression of any known PCs. Because the expression of CPE should reflect the functional potentiality of neural cells to process proproteins, a low level of PCs under the threshold of detection by *in situ* hybridization, or additional PC members not yet identified, may be present in these regions. It is conceivable that using the current PCR-based techniques, new PCs may be isolated in the future. Conversely, several peripheral organ systems express PCs but lack detectable CPE expression. Liver, for instance, expresses high levels of furin and PACE4 with no apparent accompanying CPE expression. This suggests that in these regions other exopeptidase(s) may be employed. It would be interesting to find out whether several newly identified carboxypeptidases, such as carboxypeptidase D (CPD)[70] and AEBP1,[71] are expressed in these embryonic regions. Indeed, preliminary analysis suggests a discrete and highly restricted expression of CPD in some previously unnoticed areas devoid of CPE expression (Zheng, Arai, Fricker, and Pintar, unpublished observations).

With the advent of gene targeting techniques, individual PCs can be ablated from the genome. This, coupled with characterization of the processing status of proteins, will provide the definitive evidence of the involvement of PCs in the processing of an individual proprotein. Since each PC can be involved in the proteolysis of a set of substrates, initial candidates for gene targeting are likely those PCs with a restricted pattern of gene expression, which therefore are potentially involved in the processing of a smaller pool of substrates. PC4, for instance, is only expressed by sperm cells at a particular stage of spermatogenesis and can be chosen as the target of gene disruption. Processing of proenkephalin is likely to be one of the first proproteins to be assayed in PC4-deficient mice because it has been colocalized with PC4 in the spermatocyte.[11] Targeting of more broadly expressed PCs, such as PC2, may result in the failure of generation of numerous species of bioactive peptides, leading to a complex array of phenotypes. In addition, blockade of prohormone processing by depleting prohormone convertase would result in the absence of circulating hormones and distal effect in their target organs. As a prerequisite for the phenotypical analysis of such a PC-deficient mouse, detailed knowledge of sites of PC gene expression, especially in development, and its colocalization with specific substrate should be obtained. Studies presented here have furthered our understanding in proprotein processing, which in turn should help make PC-deficient mice amenable for analysis.

SUMMARY

It has been well established that peptide precursors usually undergo limited proteolysis at pairs or single basic amino acids during their biosynthetic process. This posttranslational modification paradigm is common for numerous membrane-spanning and secreted proteins, neuropeptides, and peptide hormones of physiological significance, in which endoproteolytic cleavage is invariably essential for the accurate biosynthesis and full activity of the mature products. Establishment of an effective peptide profile is dependent on not only the presence of peptide precursor, but also the presence and the enzymatic specificities of cleavage enzymes. We have, therefore, characterized the spatial and temporal patterns of six subtilisin-like serine endoproteases known to be involved in proprotein processing, including furin, PC1, PC2, PC4, PC5, and PACE4, in rat prenatal development and related the results to the expression patterns of several peptide precursors. We have observed largely distinct and sometimes complementary expression patterns of individual PCs in various embryonic structures, suggesting PCs may be functionally distinct in processing different sets of proprotein substrates in development. From these studies, numerous tentative enzyme-substrate relationships in various embryonic structures have been proposed and should encourage more studies to test the *in vitro* cleavage potentialities of individual PCs toward these precursors. In the future, knowledge gained from these studies, when combined with insights gained from *in vivo* perturbation and genetic ablation studies, should lead to the final comprehensive understanding of specific precursors cleaved by specific enzymes at specific cleavage sites in known spatial and temporal expression patterns during development.

ACKNOWLEDGMENTS

The authors thank Drs. Nabil Seidah, Robert Day, and Lloyd Fricker for providing probes used in these studies. We also thank Ming-Sing Hsu for technical advice and assistance during these studies.

REFERENCES

1. JULIUS, D., A. BRAKE, L. BLAIR, R. KUNISAWA & J. THOMER. 1984. Isolation of the putative structural gene for the lysine-arginine-cleaving endopeptidase required for processing of yeast prepro-α-factor. Cell **37:** 1075–1089.

2. MIZUNO, K., T. NAKAMURA, T. OHSHIMA, S. TANAKA & H. MATSUO. 1988. Yeast Kex2 gene encodes an endopeptidase homologous to subtilisin-like serine proteases. Biochem. Biophys. Res. Commun. **156:** 246–254.

3. BATHURST, I. C., S. O. BRENNAN, R. W. CARRELL, L. S. COUSENS, A. J. BRAKE & P. J. BARR. 1987. Yeast KEX2 protease has the properties of a human proalbumin converting enzyme. Science **235:** 348–349.

4. THOMAS, G., B. A. THORNE, L. THOMAS, R. G. ALLEN, D. E. HRUBY, R. FULLER & J. THORNER. 1988. Yeast KEX2 endopeptidase correctly cleaves a neuroendocrine prohormone in mammalian cells. Science **241:** 226–230.

5. FULLER, R. S., A. J. BRAKE & J. THORNER. 1989. Intracellular targeting and structural conservation of a prohormone-processing endoprotease. Science **246:** 482–486.

6. SEIDAH, N. G., L. GASPAR, P. MION, M. MARCINKIEWICZ, M. MBIKAY & M. CHRÉTIEN. 1990. cDNA sequence of two distinct pituitary proteins homologous to kex2 and furin gene products: Tissue-specific mRNAs encoding candidates for pro-hormone processing proteinases. DNA Cell Biol. **9:** 415–424.

7. SMEEKENS, S. & D. F. STEINER. 1990. Identification of a human insulinoma cDNA encoding a novel mammalian protein structurally related to the yeast dibasic processing protease kex2. J. Biol. Chem. **265:** 2997–3000.

8. KIEFER, M. C., J. E. TUCKER, R. JOH, K. E. LANDSBERG, D. SALTMAN & P. J. BARR. 1991. Identification of a second human subtilisin-like protease gene in the fes/fps region of chromosome 15. DNA Cell Biol. **10:** 757–769.

9. SEIDAH, N. G., M. MARCINKIEWICZ, S. BANJANNET, L. GASPAR, G. BEAUBIEN, M. G. MATTEL, C. LAZURE, M. MBIKAY & M. CHRÉTIEN. 1991. Cloning and primary sequence of a mouse candidate prohormone convertase PC1 homologous to PC2, furin, and kex2: Distinct chromosomal localization and messenger RNA distribution in brain and pituitary compared to PC2. Mol. Endocrinol. **5:** 111–122.

10. NAKAYAMA, K., W.-S. KIM, S. TORII, M. HOSAKA, T. NAKAGAWA, J. IKEMIZU, T. BABA & K. MURAKAMI. 1992. Identification of the fourth member of the mammalian endoprotease family homologous to the yeast kex2 protease. J. Biol. Chem. **267:** 5897–5990.

11. SEIDAH, N. G., R. DAY, J. HAMELIN, A. GASPAR, M. W. COLLARD & M. CHRÉTIEN. 1992. Testicular expression of PC4 in the rat: Molecular diversity of a novel germ cell-specific kex2/subtilisin-like proprotein convertase. Mol. Endocrinol. **6:** 1559–1570.

12. LUSSON, J., D. VIEAU, J. HAMELIN, R. DAY, M. CHRÉTIEN & N. G. SEIDAH. 1993. cDNA structure of the mouse and rat subtilisin/kexin-like PC5: A candidate proprotein convertase expressed in endocrine and nonendocrine cells. Proc. Natl. Acad. Sci. USA **90:** 6691–6695.

13. NAKAGAWA, T., M. HOSAKA, S. TORII, T. WATANABE, K. MURAKAMI & K. NAKAYAMA. 1993. Identification and functional expression of a new member of the mammalian kex2-like processing endoprotease family: Its striking structural similarity to PACE4. J. Biochem. **113:** 132–135.

14. SCHÄFER, M. K., R. DAY, W. E. CULLINAN, M. CHRÉTIEN, N. G. SEIDAH & S. J. WATSON. 1993. Gene expression of prohormone and proprotein convertases in the rat CNS: A comparative in situ hybridization analysis. J. Neurosci. **13:** 1258–1279.

15. HATSUZAWA, K., M. HOSAKA, T. NAKAGAWA, M. NAGASE, A. SHODA, K. MURAKAMI & K. NAKAYAMA. 1990. Structure and expression of mouse furin, a yeast kex2-related protease. J. Biol. Chem. **25:** 22075–22078.

16. BARR, P. J., O. B. MASON, K. E. LANDBERG, P. A. WONG, M. C. KIEFER & A. J. BRAKE. 1991. cDNA and gene structure for a human subtilisin-like protease with cleavage specificity for paired basic amino acid residues. DNA Cell Biol. **10:** 319–328.

17. SHENNAN, K. I. J., S. P. SMEEKEN, D. F. STEINER & K. DOCHERTY. 1991. Characterization of PC2, a mammalian Kex2 homologue, following expression of the cDNA in microinjected *Xenopus* oocytes. FEBS Lett. **284:** 277–280.

18. ZHOU, Y. & I. LINDBERG. 1993. Purification and characterization of the prohormone convertase PC1 (PC3). J. Biol. Chem. **268:** 5615–5623.
19. HATSUZAWA, K., M. NAGAHAMA, S. TAKAHASHI, K. TAKADA, L. MURAKAMI & K. NAKAYAMA. 1992. Purification and characterization of furin, a kex2-like processing endoprotease, produced in Chinese hamster ovary cells. J. Biol. Chem. **267:** 16094–16099.
20. MISUMI, Y., K. ODA, T. FUJIWARA, N. TAKAMI, K. TASHIRO, & Y. IKEHARA. 1991. Functional expression of furin demonstrating its intracellular localization and endoprotease activity for processing of proalbumin and complement pro-C3. J. Biol. Chem. **266:** 16954–16959.
21. MOLLOY, S. S., L. THOMAS, J. K. VANSLYKE, P. E. STENBERG & G. THOMAS. 1994. Intracellular trafficking and activation of the furin proprotein convertase: Localization to the TGN and recycling from the cell surface. EMBO J. **13:** 18–33.
22. SEIDAH, N. G. & M. CHRÉTIEN. 1992. Proprotein and prohormone convertases of the subtilisin family. Trends Endocrinol. Metab. **3:** 133–140.
23. VAN DE VEN, W. J. M., J. VOORBERG, R. FONTIJN, H. PANNEKOEK, A. M. W. VAN DE OUWELAND, H. L. P. VAN DUIJNHOVEN, J. M. ROEBROEK & R. J. SIEZEN. 1990. Furin is a subtilisin-like proprotein processing enzyme in higher eukaryotes. Mol. Biol. Rep. **14:** 265–275.
24. BIRCH, N., H. L. TRACER, D. J. HAYKES & Y. P. LOH. 1991. Coordinate regulation of mRNA levels of pro-opiomelanocortin and the candidate processing enzymes PC2 and PC3, but not furin, in rat pituitary intermediate lobe. Biochem. Biophys. Res. Commun. **179:** 1311–1319.
25. BLOOMQUIST, B. T., B. A. EIPPER & R. E. MAINS. 1991. Prohormone-converting enzymes: Regulation and evaluation of function using antisense RNA. Mol. Endocrinol. **5:** 2014–2024.
26. DAY, R., M. K.-H. SCHÄFER, S. J. WATSON, M. CHRÉTIEN & N. G. SEIDAH. 1992. Distribution and regulation of the prohormone convertases PC1 and PC2 in the rat pituitary. Mol. Endocrinol. **6:** 485–497.
27. MURPHY, L. J. & D. J. BARRON. 1993. The IGFs and their binding proteins in murine development. Mol. Reprod. Dev. **35:** 376–381.
28. HEYNER, S., C. Z. SHI, W. T. GARSIDE & R. M. SMITH. 1993. Functions of the IGFs in early mammalian development. Mol. Reprod. Dev. **35:** 421–425.
29. ADAMSON, E. D. 1993. Activities of growth factors in preimplantation embryos. J. Cell. Biochem. **53:** 280–287.
30. SHULL, M. M. & T. DOETSCHMAN. 1994. Transforming growth factor-beta 1 in reproduction and development. Mol. Reprod. Dev. **39:** 239–246.
31. WALL, N. A. & B. L. HOGAN. 1994. TGF-beta related genes in development. Curr. Opin. Genet. Dev. **4:** 517–522.
32. HARVEY, M. B., K. J. LECO, M. Y. ARCELLANA-PANLILIO, X. ZHANG, D. R. EDWARDS & G. A. SCHULTZ. 1995. Roles of growth factors during peri-implantation development. Hum. Reprod. **10:** 712–718.
33. SCHWARTZ, J. P. 1992. Neurotransmitters as neurotrophic factors: A new set of functions. Int. Rev. Neurobiol. **34:** 1–23.
34. MUSKE, L. E. 1993. Evolution of gonadotropin-releasing hormone (GnRH) neuronal systems. Brain Behav. Evol. **42:** 215–230.
35. WASCHEK, J. A. 1995. Vasoactive intestinal peptide: An important trophic factor and development regulator? Dev. Neurosci. **17:** 1–7.
36. ZHENG, M., R. D. STRECK, R. E. M. SCOTT, N. G. SEIDAH & J. E. PINTAR. 1994. The developmental expression in rat of proteases furin, PC1, PC2, and carboxypeptidase E: Implications for early maturation of proteolytic processing capacity. J. Neurosci. **14:** 4656–4673.
37. ZHENG, M., N. G. SEIDAH & J. E. PINTAR. 1997. The developmental expressions in rat of proteases PC5 and PACE4: Their comparison with other proprotein processing enzymes. Dev. Biol. In press.
38. ZHENG, M. & J. E. PINTAR. 1995. Analysis of ontogeny of processing enzyme gene expression and regulation. *In* Methods in Neurosciences. I. A. Smith, Ed. Vol. 23: 45–64. Academic Press. London.

39. DOUGLASS, J., O. CIVELLI & E. HERBERT. 1984. Polyprotein gene expression: Generation of diversity of neuroendocrine peptides. Annu. Rev. Biochem. **53:** 665–715.

40. MAINS, R. E., I. M. DICKERSON, V. MAY, D. A. STOFFERS, S. N. PERKINS, L. H. OUAFIK, E. J. JUSTEN & B. A. EIPPER. 1990. Cellular and molecular aspects of peptide hormone biosynthesis. Front. Neuroendocrinol. **11:** 52–89.

41. DAY, R., M. K.-H. SCHÄFER, W. E. CULLINAN, S. J. WATSON, M. CHRÉTIEN & N. G. SEIDAH. 1993. Region specific expression of furin mRNA in the rat brain. Neurosci. Lett. **149:** 27–30.

42. DONG, W., M. MARCINKIEWICZ, D. VIEAU, M. CHRÉTIEN, N. G. SEIDAH & R. DAY. 1995. Distinct mRNA expression of the highly homologous convertases PC5 and PACE4 in the rat brain and pituitary. J. Neurosci. **15:** 1778–1796.

43. STYLIANOPOULOU, F., A. EFSTRATIADIS, J. HERBERT & J. E. PINTAR. 1988. Pattern of the insulin-like growth factor II gene expression during rat embryogenesis. Development **103:** 497–506.

44. WILCOX, J. N. & R. DERYNCK. 1988. Developmental expression of transforming growth factors alpha and beta in mouse fetus. Mol. Cell. Biol. **8:** 3415–3422.

45. DAIKOKU, S. & S. HISANO. 1992. Development of corticotropin-releasing factor in rat brain. *In* Handbook of Chemical Neuroanatomy. A. Bjorklund, T. Jokfelt & M. Tohayama, Eds. Vol. 10: 477–520. Elsevier. Amsterdam.

46. INAGAKI, S. 1992. Ontogeny of calcitonin gene-related peptide (CGRP). *In* Handbook of Chemical Neuroanatomy. A. Bjorklund, T. Jokfelt & M. Tohayama, Eds. Vol. 10: 432–456. Elsevier. Amsterdam.

47. KIYAMA, H., M. SATO & P. C. EMSON. 1992. Ontogeny of neurotensin immunoreactivity and mRNA in the rat central nervous system. *In* Handbook of Chemical Neuroanatomy. A. Bjorklund, T. Jokfelt & M. Tohayama, Eds. Vol. 10: 399–431. Elsevier, Amsterdam.

48. FOSTER, G. A. & P. L. WOODHAMS. 1992. Neuropeptide Y. *In* Handbook of Chemical Neuroanatomy. A. Bjorklund, T. Jokfelt & M. Tohayama, Eds. Vol. 10: 521–546. Elsevier. Amsterdam.

49. SHIOSAKA, S. 1992. Ontogeny of the central somatostatinergic system. *In* Handbook of Chemical Neuroanatomy. A. Bjorklund, T. Jokfelt & M. Tohayama, Eds. Vol. 10: 369–398. Elsevier. Amsterdam.

50. SAKANAKA, M. 1992. Development of neuronal elements with substance P-like immunoreactivity in the central nervous system. *In* Handbook of Chemical Neuroanatomy. A. Bjorklund, T. Jokfelt & M. Tohayama, Eds. Vol. 10: 197–255. Elsevier. Amsterdam.

51. BUIJS, R. M. 1992. The development of vasopressin and oxytocin systems in the brain. *In* Handbook of Chemical Neuroanatomy. A. Bjorklund, T. Jokfelt & M. Tohayama, Eds. Vol. 10: 547–571. Elsevier. Amsterdam.

52. HARES, K. & G. A. FOSTER. 1992. Vasoactive intestinal polypeptide and peptide histidine isoleucine. *In* Handbook of Chemical Neuroanatomy. A. Bjorklund, T. Jokfelt & M. Tohayama, Eds. Vol. 10: 457–476. Elsevier. Amsterdam.

53. BENJANNET, S., N. RONDEAU, R. DAY & N. G. SEIDAH. 1991. PC1 and PC2 are proprotein convertases capable of cleaving proopiomelanocortin at distinct pairs of basic residues. Proc. Natl. Acad. Sci. USA **88:** 3564–3568.

54. THOMAS, L., R. LEDUC, B. A. THORNE, S. P. SMEEKENS, D. F. STEINER & G. THOMAS. 1991. Kex2-like endoproteases PC2 and PC3 accurately cleave a model prohormone in mammlian cells: Evidence for a common core of neuroendocrine processing enzymes. Proc. Natl. Acad. Sci. USA **88:** 5297–5301.

55. MARCINKIEWICZ, M., R. DAY, N. G. SEIDAH & M. CHRÉTIEN. 1993. Ontogeny of the prohormone convertases PC1 and PC2 in the mouse hypophysis and their colocalization with corticotropin and α-melanotropin. Proc. Natl. Acad. Sci. USA **90:** 4922–4926.

56. DICKERSON, I. M. & G. NOEL. 1991. Tissue-specific peptide processing. *In* Peptide biosynthesis and processing. L. D. Fricker, Ed.: 71–109. CRC Press. Boca Raton, FL.

57. LUNDBLAD, J. R. & J. L. ROBERTS. 1988. Regulation of proopiomelanocortin gene expression in pituitary. Endocrin. Rev. **9:** 135–158.

58. AUTELITANO, D. J., J. R. LUNDBLAD, M. BLUM, & J. L. ROBERTS. 1989. Hormonal regulation of POMC gene expression. Annu. Rev. Physiol. **51:** 715–726.
59. ROBERTS, J. L., N. LEVIN, D. LORANG, J. R. LUNDBLAD, S. DERMER & M. BLUM. 1993. Regulation of pituitary proopiomelanocortin gene expression. *In* Handbook Experimental Pharmacology. H. Akil & E. J. Simon, Eds. Vol. 104: 1–39. Springer-Verlag. Berlin.
60. HAM, J. & D. G. SMYTH. 1986. Chronic stimulation of anterior pituitary cell cultures with CRF leads to the secretion of lipotropin. Neuroendocrinology **44:** 533–538.
61. SHIOMI, H., S. J. WATSON, J. E. KELSEY & H. AKIL. 1986. Pretranslational and posttranslational mechanisms for regulating β-endorphin-adrenocorticotropin of the anterior pituitary lobe. Endocrinology **119:** 1793–1799.
62. WAND, G. S., V. MAY & B. A. EIPPER. 1988. Comparison of acute and chronic secretagogue regulation of proadrenocorticotropin/endorphin synthesis, secretion and messenger ribonucleic acid production in primary cultures of rat anterior pituitary. Endocrinology **123:** 1153–1161.
63. SATO, S. M. & R. E. MAINS. 1986. Regulation of adrenocorticotropin/endorphin-related peptides secretion in neonatal rat pituitary cultures. Endocrinology **119:** 793–801.
64. HOTTA, M., T. SHIBASAKI, A. MASUDA, T. IMAKI, H. DEMURA, H. OHNO, S. DAIKOKU, R. BENOIT, N. LING & K. SHIZUME. 1988. Ontogeny of pituitary responsiveness to corticotropin-releasing hormone in rat. Regul. Pept. **21:** 245–252.
65. GRINO, M., J. BURGUNDER, R. L. ESKAY & L. E. EIDEN. 1989. Onset of glucocorticoid responsiveness of anterior pituitary corticotrophs during development is scheduled by corticotropin-releasing factor. Endocrinology **124:** 2686–2692.
66. SCOTT, R. E. M. & J. E. PINTAR. 1993. Developmental regulation of proopiomelanocortin gene expression in the fetal and and neonatal rat pituitary. Mol. Endocrinol. **7:** 585–596.
67. BRAKS, J. A. & G. J. MARTENS. 1994. 7B2 is a neuroendocrine chaperone that transiently interacts with prohormone convertase PC2 in the secretory pathway. Cell **78:** 263–273.
68. MARTENS, G. J., J. A. BRAKS, D. W. EIB, Y. ZHOU & I. LINDBERG. 1994. The neuroendocrine polypeptide 7B2 is an endogenous inhibitor of prohormone convertase PC2. Proc. Natl. Acad. Sci. USA **91:** 5784–5787.
69. BENJANNET, S., D. SAVARIA, M. CHRÉTIEN & N. G. SEIDAH. 1995. 7B2 is a specific intracellular binding protein of the prohormone convertase PC2. J. Neurochem. **64:** 2303–2311.
70. SONG, L. & L. D. FRICKER. 1995. Purification and characterization of carboxypeptidase D, a novel carboxypeptidase E-like enzyme, from bovine pituitary. J. Biol. Chem. **270:** 25007–25013.
71. HE, G.-P., A. MUISE, A. W. LI & H.-S. RO. 1995. A eukaryotic transcriptional repressor with carboxypeptidase activity. Nature **378:** 92–96.

Use of Nonpeptide Antagonists to Explore the Physiological Roles of Neurotensin

Focus on Brain Neurotensin/Dopamine Interactions[a]

WILLIAM ROSTENE,[b,d] MOUNIA AZZI,[b]
HELENE BOUDIN,[b] ISABELLE LEPEE,[b]
FREDERIQUE SOUAZE,[b] MILAGROS MENDEZ-UBACH,[b]
CATALINA BETANCUR,[b] AND DANIELLE GULLY[c]

[b]INSERM U.339, Hôpital St. Antoine
184 Rue du Fg St. Antoine
75012 Paris, France

[c]Sanofi Recherche
195 Route d'Espagne
31036 Toulouse Cedex, France

INTRODUCTION

Neurotensin (NT), a 13 amino acid peptide (pGlu-Leu-Tyr-Glu-Asn-Lys-Pro-Arg-Arg-Pro-Tyr-Ile-Leu) first isolated from bovine hypothalamus,[1] has been demonstrated to be widely distributed in the brain and peripheral tissues of several mammalian species including man.[2,3]

In the central nervous system, the focus of this review, NT acts as a neuromodulator specifically in the dopaminergic (DA) transmission of the nigrostriatal and mesolimbic pathways.[4,5] Thus, when administered carefully, NT can antagonize the behavioral action of DA in a manner similar, but not strictly identical, to antipsychotic drugs (see below). NT has even been considered an endogenous neuroleptic-like compound.[6]

Although NT has been shown to produce various effects depending on its site of injection in the brain, it seems to be implicated in the physiopathology of several brain diseases such as schizophrenia, Huntington's, and Parkinson's diseases.[5,7,8] Additionally, NT plays a role in nociception, hypothermia, control of anterior pituitary hormone secretion, and muscle relaxation.[4] In the periphery where 90% of the NT concentration is found, NT was shown to be mainly involved in the regulation of the gastrointestinal and cardiovascular systems.[9]

In this review, the ontogenic development of brain NT systems will be considered as well as the interaction between central NT and DA neurons. This interaction, mediated by an action of NT on NT receptors (NTR), and the effect of

[a] This work was supported by INSERM (grant to U.339, postdoctoral fellowships to M.M.-U. and C.B.), the Ministry of Education (fellowships to I.L., H.B. and F.S.), the Algerian government (fellowship to M.A.), and an INSERM-Sanofi joint program.

[d] E-mail: rostene@idf.ext.jussieu.fr

blockade of the NT neurotransmission by means of potent, recently developed, nonpeptide receptor antagonists will be presented.

ONTOGENIC DEVELOPMENT OF NEUROTENSIN IN THE BRAIN

It has been shown that NT and its related active peptide neuromedin N (N) are encoded by the same precursor molecule consisting of a 169-residue polypeptide.[10] In the rat, expression of NT/N mRNA and NT immunoreactivity varies during brain ontogeny in a region-specific manner.[11–13] In most brain regions, NT/ N mRNA is first detected in the perinatal period, and levels gradually increase to reach a plateau by adulthood. In some regions, including the subiculum and CA1 subfield of the hippocampus, NT/N mRNA is first detected prenatally, reaches a maximum level of expression during the first postnatal week, and then decreases gradually to reach levels in the adult that are a small fraction of postnatal levels.[14] Finally, in regions like the olfactory bulb and hypoglossal nucleus, NT/N mRNA is found relatively early in embryonic development (embryonic day 14), reaches a maximum level perinatally, and decreases to very low or undetectable levels within several weeks of birth.[12,13] Transient expression of NT/N mRNA in certain brain cells with a peak around the time of maximal synaptogenesis (first postnatal week) suggests that NT and/or N released from these cells might subserve a trophic function, although there is as yet no direct evidence for this in the central nervous system. In the human brain, NT immunoreactive cells and fibers have been detected in the fetus and in the infant brain in concentrations much higher than those observed in the adult.[15,16]

EVIDENCE FOR NEUROTENSIN/DOPAMINE INTERACTION IN THE BRAIN

Effects of Centrally Administered Neurotensin on Dopamine-related Behaviors

When administered intraventricularly in rats or in mice, NT shares many central properties with atypical neuroleptics.[17] It can induce hypolocomotion[18] and affect behavior induced by systemic administration of DA agonists, for example, yawning and penile erection produced by low doses of agonist, or climbing evoked by higher doses, without modifying orofacial movements.[19,20] It also reverses hyperlocomotion and rearing induced by direct injection of DA agonists in the nucleus accumbens (mesolimbic projection area).[21] In contrast, bilateral injections of NT into the ventral tegmental area (mesolimbic cell bodies) increase locomotion.[22] Experiments performed with NT protected from enkephalinase degradation by thiorphan co-administration revealed a dual effect on rat locomotion: hypolocomotion at the lower dose (30 ng) and hyperlocomotion at the higher dose (750 ng). This hyperlocomotion can be reversed by DA uptake inhibitor (GBR 12783) or DA receptor antagonist (haloperidol) treatment that unmasked the hypokinetic response.[23] As with antipsychotic agents, NT potentiates barbitone- and ethanol-induced sedation, induces hypothermia and muscular relaxation, and inhibits the frequency of intracranial self-stimulation and locomotor activity.[6]

Neurotensin Distribution and Regulation in the Brain by Dopamine Innervation

Radioimmunoassay and immunohistochemical analyses concerning the brain distribution of NT reveal that the peptide is found mainly in areas where DA systems are present, such as the hypothalamus (A12 tuberoinfundibular neurons) and the midbrain [mesolimbic (A 10) and nigrostriatal (A 9) DA systems]. Since 1980, numerous studies have illustrated the increase of immunoreactive-like NT concentrations in DA-rich brain regions after chronic neuroleptic treatment in rats, implying that the neuropeptide is a selective intermediate for the response to prolonged DA receptor blockade.[24] It has been widely proposed that the blockade of DA D_2 receptors in the extrapyramidal regions (caudate nucleus) may be responsible for the motor side effects associated with the use of these drugs, whereas DA D_2 receptors in the limbic regions (nucleus accumbens) may be involved in the therapeutic action of neuroleptics. In fact, classical neuroleptics well known to produce tardive dyskinesia induce an increase of NT mRNA in both the nucleus accumbens and in the caudate nucleus, and, conversely, the so-called atypical neuroleptics, devoid of this side effect, only induce an increase of NT mRNA in the former region.[25,26] This regional specificity determined in both NT-like immunoreactivity and NT gene expression argue for the involvement of NT in the therapeutic effect of antipsychotic drugs. Although the involvement of NT has not been directly demonstrated in clinical studies, it is supported by the observed normalization of cerebrospinal fluid concentrations of NT by treatment with a course of antipsychotic drugs.[7,8] However, NT is not an endogenous DA receptor antagonist; as expected, it is unable to interact directly with D_1 (adenylate cyclase-linked) or D_2 (spiperone-labeled) receptors. In addition, its selective properties such as analgesia, vasopressive, and neuroendocrine effects, as well as numerous gastrointestinal actions, suggest a specific role for this peptide different from that of neuroleptic compounds.

In summary, central injection of NT elicited a confusing mixture of DA-mimetic and neuroleptic-like actions. In fact, these actions have been often studied by observing the effects of local central administration of the peptide. Such an approach is open to criticism when administered concentrations largely exceed those naturally involved. A useful complementary approach consists of using NT receptor antagonists (see below).

NEUROTENSIN RECEPTORS

It is now well established that the physiological effects of NT are mediated through specific high-affinity NTR, which have been revealed by radioligand binding experiments in rat, mouse, and human brains.[9,27,28] NTR are particularly abundant in brain regions rich in DA cell bodies, such as the substantia nigra and the ventral tegmental area, and in projection areas of both nigrostriatal and mesolimbic DA pathways, such as the striatum, the nucleus accumbens, and the frontal cortex.[28–30]

In the adult rat and mouse brains—in contrast to those of guinea-pig, rabbit, and man—NT can bind to two different binding sites, which can be distinguished by their sensitivity to levocabastine, an histamine H1 receptor antagonist.[31,32] It is believed that the high-affinity sites are responsible for the physiological effects of NT.[5,33] Some evidence suggests that the low- and high-affinity sites do not correspond to different affinity states of the same receptor, but constitute distinct molecular entities. Indeed, their distribution and their ontogenic pattern are quite different. In contrast to the high-affinity NTR, which appears in the brain as early as 18 days in rat embryo,[34] the levocabastine-sensitive site first appears after birth (tenth day in the murine brain), increasing progressively to adult levels.[35,36]

Further understanding of the molecular structure of NTR has resulted from the recent cloning, sequencing, and expression of the functional high-affinity rat NTR.[37] These studies have shown that the high-affinity NTR correspond to a 424 amino acid protein belonging to the family of G-protein–coupled receptors with seven membrane-spanning domains inducing the activation of several transduction pathways.[38] More recently, a human colon adenocarcinoma cell line, HT29 cells, was used to sequence the human NTR, which possesses 84% sequence similarity with that of the rat NTR.[39] The human cDNA NTR was reported to encode a 418 amino acid protein.[39] Recently, a cDNA clone encoding the human NTR was also isolated from a human substantia nigra cDNA library that differs by three bases from that obtained in HT29 cells.[40] Both rat and human receptors, however, when separately, stably transfected into cell lines, express only one type of NTR that has similar affinities for NT, NT fragments, and some antagonists.[39,41]

The cloning of NTR also allowed initial localization of NTR mRNA-expressing cells by *in situ* hybridization, with both RNA and oligonucleotide probes demonstrating a much higher expression in the juvenile brain[42] than in the adult brain.[43] Developmental changes in the density of NT binding sites may be partly determined by concurrent changes in the abundance of NT. A dramatic increase in NT expression coincides with a similarly dramatic increase in NTR expression during the first two weeks after birth in the rat[34,42] and human brain.[44]

Investigations of the various molecular forms of the NTR and their occurrence in the different functional states of this receptor can be greatly facilitated by the development of antibodies toward different domains of the NTR. Moreover, the development of such antibodies is a necessary step to measure and visualize NTR proteins and to study their possible regulation. We thus recently developed such antibodies against the cloned rat NTR.[45,46] When immunoblot analysis was carried out on membrane extracts from the cerebral cortex of 7-day-old rats, two immunoreactive bands were detected with a specific antibody raised against the third intracellular loop of the rat NTR[45]: one at 52 kDa and the other at 54 kDa. This two-band pattern was reminiscent of what was previously obtained in photoaffinity-labeling experiments on adult rat brain membranes, where two proteins with a molecular mass of 49 kDa and 51 kDa, respectively, were identified.[47]

NEUROTENSIN RECEPTOR ANTAGONISTS

The discovery of new therapeutic agents and a better understanding of the complex effects of NT may result from the use of highly potent, selective, orally active antagonists of NTR. Several pharmaceutical companies such as Pfizer with the UK-73,093, Merck with the L-734836, and Parke-Davis with the PD-156425 compounds have recently tried to discover such antagonists.[48] However, the first reported potent and selective nonpeptide NTR antagonist, SR48692,[49] was developed by Sanofi and has provided researchers with a tool for investigating the action of NT in numerous models, both *in vitro* and *in vivo*.

This small molecule (2-{[1-(7-chloroquinolin-4-yl)-5-(2,6-dimethoxyphenyl)-1H-pyrazole-3-carbonyl]-amino}-adamantane-2-carboxylic acid), obtained by opti-

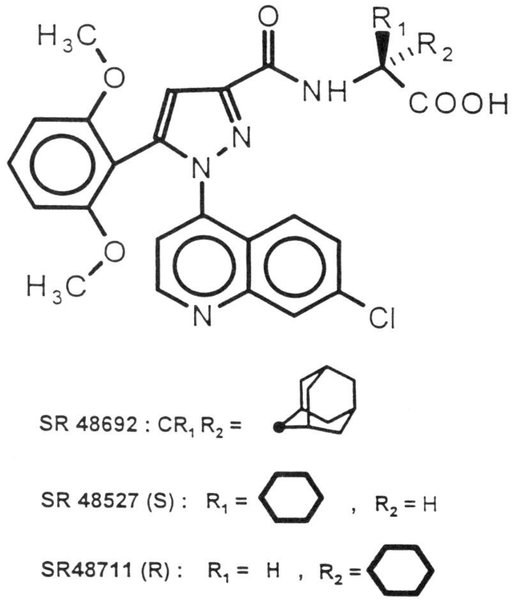

FIGURE 1. Chemical structures of different NTR antagonists: SR48692 and its two chiral analogues, SR48527 and SR49711.

mization of a lead compound discovered by random screening[49,50] (FIG. 1), has revealed a surprising neuropharmacological profile.

In Vitro *Studies*

SR48692 competitively inhibits, with K_i values in the nanomolar range, the binding of [^{125}I]NT in several models such as homogenates of guinea-pig, rat, mouse, and human brain, primary cultures of rat mesencephalic neurons, and various cell lines transfected with cloned human and rat NTRs.[49,51,52] In transfected cell lines as in the case of the NTR expressed in insect Sf9 cells, NT can induce inositol phosphate production, an effect blocked by SR48692[41] (FIG. 2). A slight reduction in basal inositol phosphate levels by SR48692 can be noticed, which may suggest that the antagonist could inhibit under certain conditions the activation of endogenous NTR.[41] Such an inverse agonist effect of an antagonist has been already described for some aminergic receptors.[53] As shown in FIGURE 3, SR48692 can dose-dependently inhibit the binding of [^{125}I]NT in the human midbrain (substantia nigra and ventral tegmentum) with complete inhibition observed at a similar concentration as with unlabeled NT. The antagonist inhibits NT facilitation of K^+-induced DA release in rat mesencephalic neurons in culture as well as in striatal slices from guinea pig[51,52] (TABLE 1).

The high degree of selectivity of SR48692 for NTR was demonstrated by its lack of activity in several binding assays with receptors of nonpeptide and peptide ligands at concentrations of up to 1 μM.[51] In order to study the stereoselectivity

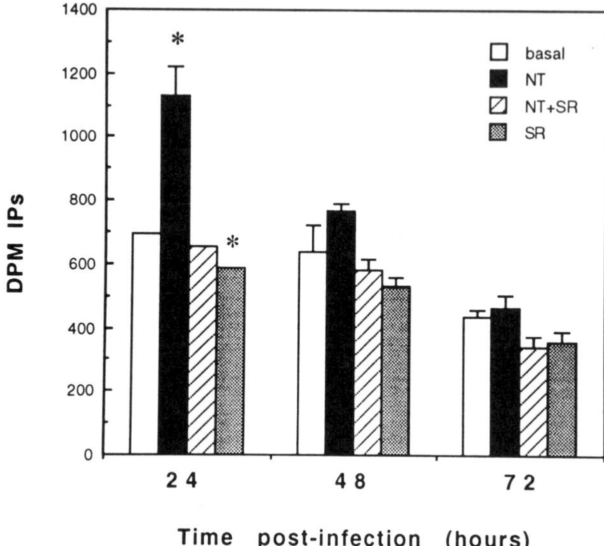

FIGURE 2. Functional coupling of the rat NTR to phospholipase C in NTR-transfected insect Sf9 cells with a recombinant baculovirus procedure. Total [³H]-labeled inositol phosphates (IPs) formation was measured following a 24-h preincubation of the cells with [³H]myo-inositol in the presence of 10^{-7} M NT and/or 10^{-5} M SR48692. Number of disintegrations/min (DPM) was normalized for 2.5×10^5 cells. Each value represents the mean ± SEM of triplicate determinations. *p <0.05 vs. basal values without drug treatment. (Adapted from Boudin *et al.*[41])

of the ligand, chiral analogues of SR48692 (FIG. 1) were selected: SR48527 (S) was about 100-fold more potent than SR49711 (R) in binding experiments carried out on guinea-pig brain homogenates (TABLE 1). This stereospecificity was confirmed in the model of stimulation by NT of K^+-evoked release of tritiated DA from guinea-pig striatal slices[52] (TABLE 1).

In addition, using the NTR antagonist as radioligand ([³H]SR48692), recent autoradiographic data demonstrated that the distribution of [³H]SR48692- and [¹²⁵I]NT-labeled binding sites in the guinea-pig brain closely resembled that of the DA neuronal systems.[54] Similar to what has been previously shown for [¹²⁵I]NT sites in rat and human brains,[27,29] both radioligands showed a high labeling density in the ventral tegmental area and the substantia nigra pars compacta, suggesting that in the guinea pig, as in other species, [³H]SR48692 and [¹²⁵I]NT sites are present in brain regions associated with the mesolimbic and nigrostriatal DA pathways. Furthermore, [³H]SR48692 and [¹²⁵I]NT sites were also observed in the amygdaloid and septal complexes, caudate putamen, nucleus accumbens, as well as entorhinal, piriform, cingulate and frontal cortices, which constitute projection sites of nigrostriatal, mesolimbic, and mesocortical DA pathways. Indeed, *in vitro* studies have demonstrated that in the rat, as well as in the human brain, NT binding sites are intimately associated with DA cell bodies and dendrites.[29,30,55–58] A similar presence of NT receptors on DA neurons in the guinea-pig brain is in agreement with the stimulatory effect of NT on DA release in this species, an action antagonized by SR48692, as shown in TABLE 1.

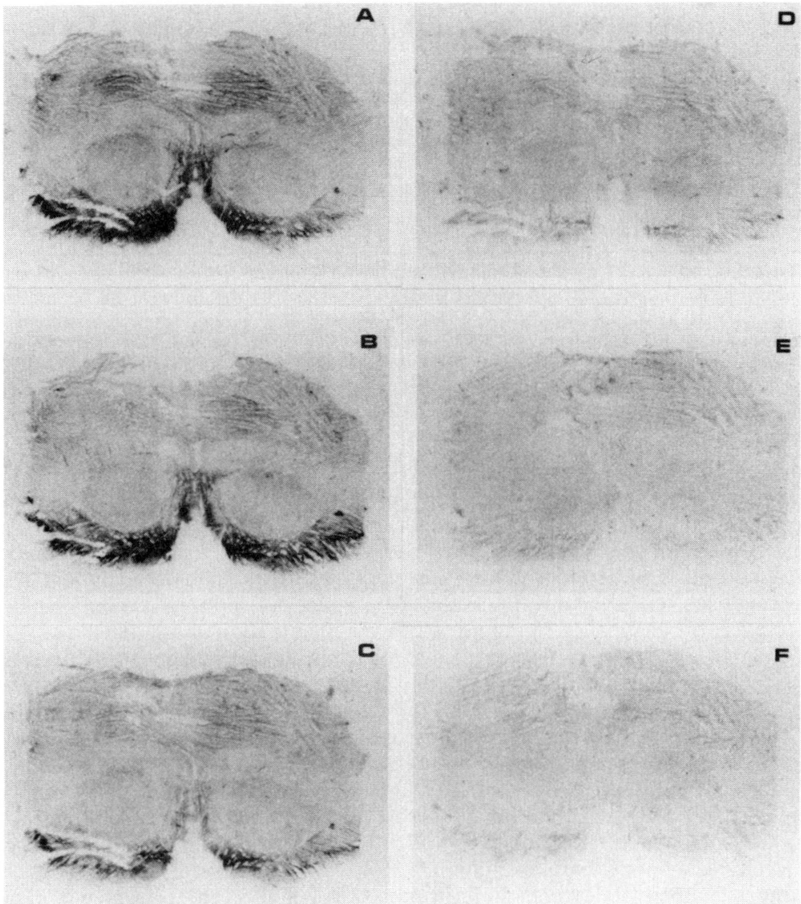

FIGURE 3. Effect of increasing concentrations of SR48692 and of an excess (10^{-6} M) of unlabeled NT on the binding of [^{125}I]NT to coronal sections of the human brain at the level of the mesencephalon. The dark areas represent structures where a high density of [^{125}I]NT binding sites is found (substantia nigra, nucleus paranigralis, and ventral tegmentum). Binding procedure and autoradiograms were carried out according to Sadoul *et al.*[27] **(A)** Total binding with 0.1 nM [^{125}I]NT; **(B)** [^{125}I]NT with SR48692 10^{-9} M; **(C)** [^{125}I]NT with SR48692 10^{-8} M; **(D)** [^{125}I]NT with SR48692 10^{-7} M; **(E)** [^{125}I]NT with SR48692 10^{-6} M; **(F)** [^{125}I]NT with NT 10^{-6} M.

In Vivo *Studies*

Antagonism by SR48692 of the Central Effects of Exogenous Neurotensin

Related to close relationship between NT and DA systems, pretreatment with SR48692 (80 mg/kg, i.p. or p.o.) blocked the turning behavior induced in mice by

                    ANNALS NEW YORK ACADEMY OF SCIENCES

TABLE 1. K_i Values and Hill Coefficients (nH) for the Inhibition of [^{125}I]NT Specific Binding to Adult Guinea-Pig Brain Membranes by Unlabeled NT, SR48692, SR48527, and SR49711

| | [^{125}I]NT Binding | | K$^+$-Evoked [^{3}H]DA Release |
Drug	K_i (nM)	nH	IC$_{50}$ (nM)
NT	0.37 ± 0.01	0.98 ± 0.20	—
SR48692	3.92 ± 0.20	0.96 ± 0.03	0.46 ± 0.02
SR48527	6.95 ± 0.66	1.06 ± 0.20	3.75 ± 0.27
SR49711	322.79 ± 92.06	0.94 ± 0.10	>1000

NOTE: K_i values can be directly compared to the inhibition (IC$_{50}$) of NT stimulation of potassium (K$^+$)-evoked [^{3}H]DA release from striatal slices by the different SR compounds. IC$_{50}$ is defined as the concentration decreasing by 50% the effect of NT. Each value is the mean ± SEM of three determinations.
(Adapted from Azzi *et al.*[51])

unilateral intrastriatal injection of NT (10 pg)[59] in a stereospecificity similar to that shown in TABLE 1. At higher doses (0.16–0.32 mg/kg), a reinstatement of rotations occurred that was abolished by spiroperidol or 6-hydroxydopamine lesions, suggesting a DA regulatory effect.[59] The injection of NT into the rat ventral tegmental area induced contralateral rotations that were partially counteracted by SR48692[60] at oral doses very similar to those found to reduce NT-induced turning in mice.[59] However, SR48692 (0.1–10 mg/kg, p.o.) was unable to inhibit either DA release or changes in DOPAC/DA ratios elicited in the nucleus accumbens by NT injection into the ventral tegmental area.[60] In rats, SR48692 can counteract the hypomotility induced by i.c.v. injection of NT[18] and the vacuous chewing induced by intrastriatal administration of NT, which has been proposed as a model of tardive dyskinesia.[61] Finally, SR48692, taken orally, also reversed in a dose-dependent manner (0.1–1 mg/kg) the inhibitory effect of intra-accumbens NT administration on i.p. amphetamine-induced hyperlocomotion.[60] These data are in accordance with those reporting that immunoneutralization of central NT facilitates both amphetamine-induced DA release and behavioral excitation.[62] However, as reported for the DA release in the accumbens following administration of NT in the ventral tegmental area,[60] not all the effects of exogenous NT are antagonized by SR48692. Indeed, SR48692 does not block the hypothermic effect obtained after i.c.v. administration of a high dose of NT in rats and mice.[18,63] SR48692 is also unable to counteract the antinociception induced by i.c.v. injection of NT in rats.[18] Recent data demonstrate that NT injection into the nucleus raphe magnus in rat elicits a biphasic effect, hyperalgesia and analgesia at respectively low and high doses,[64] and that SR48692 can block the former effect.[65]

Antagonism by SR48692 of Dopaminergic-related Effects

In addition to the blockade of NT actions, the effects of SR48692 on well-characterized behavioral responses to DA receptor stimulation have been investigated extensively. SR48692 antagonizes turning behavior induced by intrastriatal administration of a D_1 receptor agonist [(+) SKF 38393], a D_2 receptor agonist (bromocriptine), a mixed D_1/D_2 receptor antagonist (apomorphine), and amphetamine.[66] Administered orally 60 min before DA agonists, SR48692 significantly reduces yawning induced by both apomorphine (0.07 mg/kg, s.c.) and bromocriptine (2 mg/kg, i.p.). Other apomorphine-induced effects in mice and rats such as climbing, hypothermia, hypo- and hyperlocomotion, penile erections, and stereotypies are not, however, significantly modified by SR48692.[66] Interestingly, SR48692 potentiates the facilitatory effect of haloperidol on evoked DA release in the rat nucleus accumbens, thereby suggesting that this antagonist may act as an antipsychotic drug under certain conditions.[67] Taken together, these data suggest that endogenous NT may play a role in the expression of some but not all behavioral responses to DA receptor modification.

NEUROTENSIN RECEPTOR SUBTYPES

The observation that not all *in vivo* behavioral effects of NT can be counteracted by SR48692 has prompted several investigators to postulate the existence of multiple subtypes of NTRs, by analogy with other neuropeptide receptors.[18,68,69] Of interest is the recent observation obtained on guinea-pig brain membranes that, by means of [^{3}H]SR48692, NT agonists interact with [^{3}H]SR48692 binding with a much lower affinity than that observed with [^{125}I]NT.[54] In contrast, NT antagonists exhibit similar abilities to inhibit the binding of both [^{3}H]SR48692 and [^{125}I]NT. These experiments suggest the existence of different affinity states of the NTR or of multiple subtypes of NTRs, one being selectively sensitive to SR48692.[54] A short recent comprehensive review recalls the various evidence for possible NTR subtypes.[70] Recent mutagenesis studies carried out on transfected cell lines overexpressing the NTR have revealed the importance of the extracellular amino-terminus of the NTR in the binding of NT agonists. Indeed, deletion of amino acids 45–60 induced a loss of [^{125}I]NT binding with no change in the binding of antagonist compounds, suggesting different binding sites for the agonist and the antagonist.[71] However, the fact that NT can competitively inhibit the binding of the antagonist suggests that in brain tissue, the binding sites for NT agonists and antagonists may also share common domains of the NTR.[54]

EFFECT OF CHRONIC TREATMENT WITH SR48692 ON BRAIN NEUROTENSIN SYSTEMS

While developing NTR antagonists as possible new therapeutic drugs, an essential question relates to their potential effects following chronic administration. Such approaches may also bring new information on the role of endogenous NT in various cerebral functions.

Thus, we first examined the regulation of NTR following a chronic pharmacological blockade of the NT transmission with the antagonist SR48692. Our results showed that treatment of the rats for five days with SR48692, at doses from 0.25 to 1 mg/kg, i.p., induced an increase of both the number of binding sites for [^{125}I]NT (with no change in the dissociation constant) to whole brain membrane homogenates and NTR mRNA levels in the ventral mesencephalon.[72] This study was the first evidence for an *in vivo* up-regulation of a neuropeptide receptor following its pharmacological blockade, as previously reported for classical neurotransmitter substances such as DA or cholinergic receptors.[73] It suggests furthermore that endogenous NT exerts a tonic inhibitory control on mesencephalic NTR

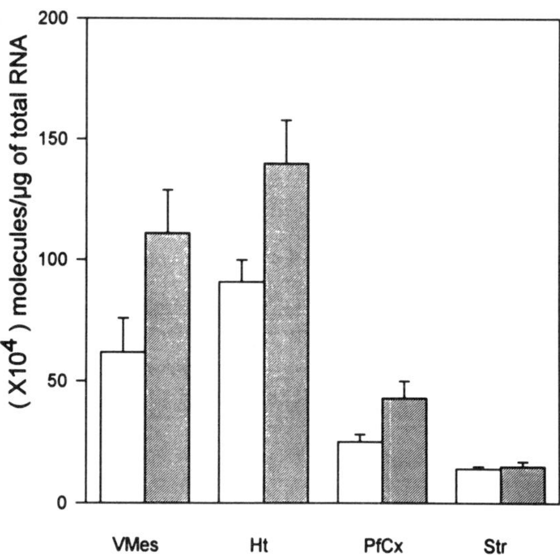

FIGURE 4. Effect of a two-week treatment with SR48692 (1 mg/kg, i.p.) in the adult rat on the amount of NTR mRNA measured by quantitative RT-PCR.[76] Results are expressed as number of NTR molecules/μg of total RNA in control (*white bars*) or SR48692-treated animals (*grey bars*) in the ventral mesencephalon (VMes), hypothalamus (Ht), prefrontal cortex (PfCx), and striatum (Str).

mRNA levels through either an inhibition of NTR gene expression or a decrease of NTR mRNA stability. The tonic inhibitory effect of NT may be related to the presence of a dense NT innervation at the level of the rat midbrain tegmentum.[74] The possibility cannot be excluded, however, that long-term SR48692 may directly regulate NTR mRNA levels.

Interestingly, when SR48692 was administered for 15 days at a dose (1 mg/kg, i.p.) that was shown to increase both [125I]NT binding and NTR mRNA after five days,[72] a much higher expression of NTR mRNA was observed as shown by quantitative *in situ* hybridization. The increase of the signal was seen in several areas of the brain, such as the anterior cingulate, perirhinal and retrosplenial cortices, the suprachiasmatic nucleus, the ventral tegmental area, the substantia nigra, and the posterior cortical nucleus of the amygdaloid complex.[75] Moreover, treatment with SR48692 reveals the presence of NTR mRNA expression in many nuclei of the diencephalon where it could not be detected under basal conditions. This is the case, for example, in the lateral hypothalamic area, the paraventricular and anterodorsal thalamic nuclei, and in the bed nucleus of the stria terminalis.[75] It is thus possible that in these different areas, neurons express low levels of NTR mRNA which cannot be normally detected, and that sustained treatment with SR48692, by stimulating their expression, facilitates their detection. Such important increases in NTR mRNA were confirmed using quantitative RT-PCR[76] (FIG. 4). It is interesting to note, however, that the effect of SR48692 treatment is not a general phenomenon and remains restricted to certain brain regions (FIG. 4).

Immunoblot analysis with specific antibodies directed against the cloned NTR[45] revealed a parallel increase in NTR protein following long-term treatment with

SR48692, which corresponds to one immunoreactive band at 55 kDa in both control and SR48692-treated rats, consistent with the molecular mass previously reported in the rat NTR in the adult brain.[47]

Unexpectedly and in contrast to the effects of SR48692 on NTR mRNA and NTR protein, the number and the affinity constant of NT binding sites determined on brain membrane homogenates remained unchanged after 15 days of treatment with SR48692, even after membrane permeabilization with low concentrations of digitonin, a detergent which allows the detection of possible vesicular receptors not readily accessible to hydrophilic NT.[71]

The regulation of NTR by its own ligand has essentially been studied *in vitro*. As for other neurotransmitter receptors, it has been demonstrated that exposure to NT induces a decrease of NTR from the cell surface and internalization of NT into the cytoplasm.[77–79] After a short exposure to the agonist, the receptor may be recycled to the membrane by a process not requiring *de novo* protein synthesis. However, for longer periods, as a second step in the process, the recovery of NT binding sites requires new receptor synthesis, suggesting a down-regulation of NT receptors.[80]

The lack of change in [^{125}I]NT binding following a two-week treatment with SR48692[75] is at variance with the increase in the number of [^{125}I]NT binding sites we observed after a short-term treatment (5 days) with the antagonist.[72] These data suggest that functional NT receptors present an adaptation to the absence of endogenous ligand and stabilize their expression to the basal levels after a long treatment with the NTR antagonist. They suggest that part of the newly synthesized NT receptor could represent a pool of spare receptors ready to respond to further stimulation by NT, as previously shown for other transmitter systems.[81] To our knowledge no similar study has been performed for other neuropeptides, even if some new potent neuropeptide antagonists capable of crossing the blood–brain barrier have been recently developed.[82] It will be of great interest to establish whether such regulation in response to long-term transmission blockade can be extended to other neuropeptides.

Related to the above data, we have observed recently that treatment of newborn rats with SR48692 for either 5 or 15 days, periods corresponding to the important increases in both NT and NTR levels,[11,29] did not modify the specific binding of [^{125}I]NT in the brain (FIG. 5).

CONCLUSION

With its neuropharmacological profile, SR48692 confirms the neuromodulatory role of NT in the central nervous system, in particular in the regulation of DA neurons. Although many questions still remain unanswered, recent progress has been made in understanding the central actions of NT. Even if a lot of data indicates that the DA system is involved in the etiology and/or symptomatology of a variety of neuropsychiatric disorders, the involvement of other neurotransmitters such as neuropeptides in such pathologies is strongly suggested by the fact that therapeutic effects of neuroleptics take days, weeks, and even months to develop. NT is a possible candidate with contrasting site-specific effects in the brain and with the ability for modulating behavioral states in a biphasic manner. The recent discovery of potent and specific nonpeptide NTR antagonists such as SR48692 represents not only a tool to explore the physiology of NT and to reveal yet unknown NTR subtypes, but also opens the way for a new therapeutic approach

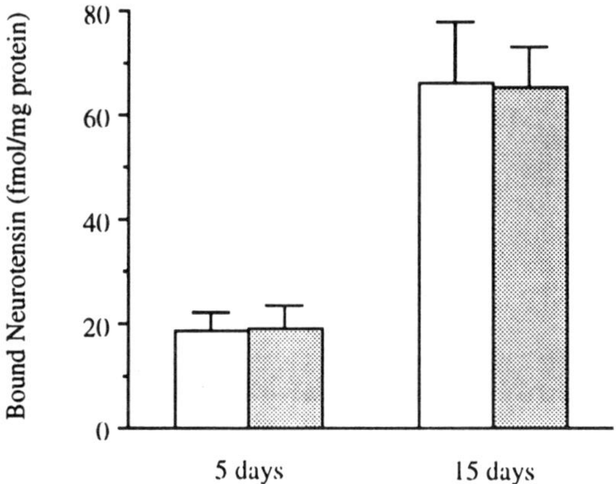

FIGURE 5. Effect of treatment from birth for 5 days and 15 days with SR48692 (1 mg/kg, s.c.) on specific binding of [^{125}I]NT in brain homogenates. Control rats (*white bars*) and SR48692-treated juvenile rats (*grey bars*).

in the functional disturbances of DA transmission associated with several neuro-psychiatric disorders.

ACKNOWLEDGMENTS

The authors are grateful to A. Berod, P. Forgez, J. P. Maffrand, D. Pelaprat, and P. Soubrie for their important contribution in the experimental data presented in this chapter.

REFERENCES

1. CARRRAWAY, R. & S. E. LEEMAN. 1973. The isolation of a new hypotensive peptide, neurotensin, from bovine hypothalami. J. Biol. Chem. **248:** 6854–6861.
2. EMSON, P. C., M. GOEDERT & P. W. MANTYH. 1985. Neurotensin-containing neurons. *In* Handbook of Chemical Neuroanatomy. GABA and Neuropeptides in the CNS. Vol. 4. A. Björklund & T. Hökfelt, Eds.: 355–405. Elsevier. Amsterdam.
3. MAI, J., K. J. TRIEPEL & J. METZ. 1987. Neurotensin in the human brain. Neuroscience **22:** 499–524.
4. KITABGI, P. & C. B. NEMEROFF, Eds. 1992. The Neurobiology of Neurotensin. Ann. N. Y. Acad. Sci. **668:** 1–374.
5. ROSTENE, W., A. BROUARD, C. DANA, Y. MASUO, F. AGID, M. VIAL, A. M. LHIAUBET & D. PELAPRAT. 1992. Interaction between neurotensin and dopamine in the brain. Morphological and clinical evidence. Ann. N. Y. Acad. Sci. **668:** 217–231.
6. ERVIN, G. N. & C. B. NEMEROFF. 1988. Interactions of neurotensin with dopamine-containing neurons in the central nervous system. Neuropsychopharmacol. Biol. Psychiatry **12:** S53–S69.
7. GARVER, D. L., G. BISSETTE, J. K. YAO & C. B. NEMEROFF. 1992. Relation of CSF neurotensin concentrations to symptoms and drug response of psychotic patients. Am. J. Psychiatry **148:** 484–488.
8. WIDERLÖV, E., L. H. LINDSTRÖM, G. BESEV, P. J. MANBERG, C. B. NEMEROFF, G. R. BREESE, J. S. KISER & A. J. PRANGE, JR. 1982. Subnormal CSF levels of

neurotensin in a subgroup of schizophrenic patients: Normalization after neuroleptic treatment. Am. J. Psychiatry **139:** 1122–1126.

9. KITABGI, P., F. CHECLER, J. MAZELLA & J. P. VINCENT. 1985. Pharmacology and biochemistry of neurotensin receptors. Rev. Clin. Basic Pharmacol. **5:** 397–486.

10. DOBNER, P. R., D. L. BARBER, L. VILLA-KOMAROFF & C. MCKIERNAN. 1987. Cloning and sequence analysis of cDNA for the canine neurotensin/neuromedin N precursor. Proc. Natl. Acad. Sci. USA **84:** 3516–3520.

11. HARA, Y., S. SHIOSAKA, E. SENBA, M. SAKANAKA, S. INAGAKI, H. TAKAGI, Y. KAWAI, K. TAKATSUKI, T. MATSUZAKI & M. TOHYAMA. 1982. Ontogeny of the neurotensin-containing neuron system of the rat: Immunohistochemical analysis. I. Forebrain and diencephalon. J. Comp. Neurol. **208:** 177–195.

12. KIYAMA, H., M. SATO & P. C. EMSON. 1992. Ontogeny of neurotensin immunoreactivity and mRNA in the rat central nervous system. *In* Handbook of Chemical Neuroanatomy. Ontogeny of Transmitters and Peptides in the CNS. Vol. 10. A. Björklund, T. Hökfelt & M. Tohyama, Eds.: 399–430. Elsevier. Amsterdam.

13. SATO, M., H. KIYAMA, S. YOSHIDA, T. SAIKA & M. TOHYAMA. 1991. Postnatal ontogeny of cells expressing prepro-neurotensin/neuromedin N mRNA in the rat forebrain and midbrain: A hybridization histochemical study involving isotope-labeled and enzyme-labeled probes. J. Comp. Neurol. **310:** 300–315.

14. NICOT, A., A. BEROD & W. ROSTENE. 1992. Distribution of prepro-neurotensin/neuromedin N mRNA in the young and adult rat forebrain. Ann. N. Y. Acad. Sci. **668:** 361–364.

15. MAILLEUX, P. & J. J. VANDERHAEGEN. 1988. Transient neurotensin in the human inferior olive during development. Brain Res. **456:** 199–203.

16. SAKAMOTO, N., J. P. MICHEL, N. KOPP & J. PEARSON. 1987. Neurotensin like immunoreactive neurons in the human infant diencephalon. Brain Res. **403:** 31–42.

17. NEMEROFF, C. B. 1980. Neurotensin: Perhaps an endogenous neuroleptic. Biol. Psychiatry **15:** 283–302.

18. DUBUC, I., J. COSTENTIN, J. P. TERRANOVA, M. C. BARNOUIN, P. SOUBRIE, G. LE FUR, W. ROSTENE & P. KITABGI. 1994. The nonpeptide neurotensin antagonist SR 48692 as a tool reveal putative neurotensin receptor subtypes. Br. J. Pharmacol. **112:** 352–354.

19. JOLICOEUR, F. B., S. ST-PIERRE, C. AUBE, R. RIVEST & M. A. GAGNE. 1984. Relationships between structure and duration of neurotensin's central action: Emergence of long acting analogs. Neuropeptides **4:** 467–476.

20. JOLICOEUR, F. B., M. A. GAGNE, R. RIVEST, A. DRUMHELLER & S. ST-PIERRE. 1993. Atypical neuroleptic-like behavioral effects of neurotensin. Brain Res. Bull. **32:** 487–491.

21. ERVIN, G. N., C. B. NEMEROFF & A. J. PRANGE. 1981. Neurotensin blocks certain amphetamine-induced behaviours. Nature **291:** 73–76.

22. KALIVAS, P. W., S. K. BURGESS, C. B. NEMEROFF & A. J. PRANGE. 1993. Behavioural and neurochemical effects of neurotensin microinjection into the ventral tegmental area of the rat. Neuroscience **8:** 495–505.

23. NOUEL, D. & J. COSTENTIN. 1991. Inhibition of apomorphine-induced yawning and penile erection by neurotensin. Peptides **12:** 755–759.

24. KILTS, C. D., C. M. ANDERSON, G. BISSETTE, T. D. ELY & C. B. NEMEROFF. 1988. Differential effects of antipsychotic drugs on the neurotensin concentration of discrete rat brain nuclei. Biochem. Pharmacol. **37:** 1547–1554.

25. MERCHANT, K. M., D. J. DOBIE, F. M. FILLOUX, M. TOTZKE, M. ARAVAGIRI & D. M. DORSA. 1994. Effects of chronic haloperidol and clozapine treatment on neurotensin and c-fos mRNA in rat neostriatal subregions. J. Pharmacol. Exp. Ther. **27:** 460–471.

26. MERCHANT, K. M. 1994. c-fos Antisense oligonucleotide specifically attenuates haloperidol—induces increases in neurotensin/neuromedin N mRNA expression in rat dorsal striatum. Mol. Cell. Neurosci. **5:** 336–344.

27. SADOUL, J. L., P. KITABGI, W. ROSTENE, F. JAVOY-AGID & J. P. VINCENT. 1984. Characterization and visualization of neurotensin binding to receptor sites in human brain. Biochem. Biophys. Res. Commun. **120:** 206–213.

28. MOYSE, E., W. ROSTENE, M. VIAL, K. LEONARD, J. MAZELLA, P. KITABGI, J. P. VINCENT & A. BEAUDET. 1987. Distribution of neurotensin binding sites in rat brain: A light microscopic radioautographic study using monoiodo [^{125}I]Tyr$_3$-neurotensin. Neuroscience 22: 525–536.

29. PALACIOS, J. M. & M. J. KUHAR. 1981. Neurotensin receptors are located on dopamine-containing neurons in the rat midbrain. Nature 294: 587–589. Brain Res. 1988. 457: 212–218.

30. QUIRION, R., C. CHIUEH, H. EVERIST & A. PERT. 1985. Comparative localization of neurotensin receptors on nigrostriatal and mesolimbic dopaminergic terminals. Brain Res. 327: 385–389.

31. SCHOTTE, A., W. ROSTENE & P.M. LADURON. 1988. Different subcellular localization of neurotensin receptor and neurotensin acceptor sites in the rat brain dopaminergic system. J. Neurochem. 50: 1026–1033.

32. KITABGI, P., W. ROSTENE, M. DUSSAILLANT, A. SCHOTTE, P. M. LADURON & J. P. VINCENT. 1987. Two populations of neurotensin binding sites in murine brain: Discrimination by the antihistamine levocabastine reveals markedly different radio-autographic distribution. Eur. J. Pharmacol. 140: 285–293.

33. ROSTENE, W. & M. ALEXANDER. 1996. Neurotensin involvement in neuroendocrine regulation. Front. Neuroendocrinol. In press.

34. PALACIOS, J. M., A. PAZOS, M. M. DIETL, M. SCHLUMPF & W. LICHTENSTEIGER. 1988. The ontogeny of brain neurotensin receptors studied by autoradiography. Neuroscience 25: 307–317.

35. HERMANS, E., A. JEANJEAN, P. LADURON, J. N. OCTAVE & J. M. MALOTEAUX. 1993. Postnatal ontogeny of the rat brain neurotensin receptor mRNA. Neurosci. Lett. 157: 45–48.

36. SCHOTTE, A., & P. M. LADURON. 1987. Different postnatal ontogeny of two [^{3}H] neurotensin binding sites in rat brain. Brain Res. 408: 326–328.

37. TANAKA, K., M. MASU & S. NAKANISHI. 1990. Structure and functional expression of the cloned rat neurotensin receptor. Neuron 4: 847–854.

38. OURY-DONAT, F., O. THURNEYSSEN, N. GONALONS, P. FORGEZ, D. GULLY, G. LE FUR & P. SOUBRIE. 1995. Characterization of the effect of SR48692 on inositol monophosphate, cyclic GMP and cyclic AMP responses linked to neurotensin receptor activation in neuronal and non-neuronal cells. Br. J. Pharmacol. 116: 1899–1905.

39. VITA, N., P. LAURENT, S. LEFORT, P. CHALON, X. DUMONT, M. KAGHAD, D. GULLY, G. LE FUR, P. FERRARA & D. CAPUT. 1993. Cloning and expression of a complementary DNA encoding a high affinity human neurotensin receptor. FEBS Lett. 317: 139–142.

40. WATSON, M., P. J. ISACKSON, M. MAKKER, M. S. YAMADA, M. YAMADA, B. CUSACK & E. RICHELSON. 1993. Identification of a polymorphism in the human neurotensin receptor gene. Mayo Clin. Proc. 68: 1043–1048.

41. BOUDIN, H., J. LABRECQUE, A. M. LHIAUBET. M. DENNIS, W. ROSTENE & D. PELAPRAT. 1996. Pharmacological and molecular characterization of the neurotensin receptor expressed in Sf9 cells. Biochem. Pharmacol. 51: 1243–1246.

42. SATO, M., H. KIYAMA & M. TOHYAMA. 1992. Different postnatal development of cells expressing mRNA encoding neurotensin receptor. Neuroscience 48: 137–149.

43. NICOT, A., W. ROSTENE & A. BEROD. 1994. Neurotensin receptor expression in the rat forebrain and midbrain: A combined analysis by in situ hybridization and receptor autoradiography. J. Comp. Neurol. 341: 407–419.

44. ZSURGER, N., J. CHABRY, A. COQUEREL & J. P. VINCENT. 1992. Ontogenesis and binding properties of high-affinity neurotensin receptors in human brain. Brain 586: 303–310.

45. BOUDIN, H., A. GRUAZ-GUYON, M. P. FAURE, P. FORGEZ, A. M. LHIAUBET, M. DENNIS, A. BEAUDET, W. ROSTENE & D. PELAPRAT. 1995. Immunological recognition of different forms of the neurotensin receptor in transfected cells and rat brain. Biochem. J. 305: 277–283.

46. BOUDIN, H., D. PELAPRAT, W. ROSTENE & A. BEAUDET. 1996. Cellular distribution of the neurotensin receptor in rat brain: Immunohistochemical study using an antipeptide antibody against the cloned high-affinity receptor. J. Comp. Neurol. 373: 76–89.

47. MAZELLA, J., P. KITABGI & J. P. VINCENT. 1985. Molecular properties of neurotensin receptors in rat brain. J. Biol. Chem. **260:** 508–514.

48. CHAKRAVARTY, P. K. & R. W. RANSOM. 1995. Neurotensin receptor antagonists. Curr. Pharm. Design **1:** 317–324.

49. GULLY, D., M. CANTON, R. BOIGEGRAIN, F. JEANJEAN, J. C. MOLIMARD, M. PONCELET, C. GUEUDET, M. HEAULME, R. LEYRIS, A. BROUARD, D. PELAPRAT, C. LABBE-JULLIE, J. MAZELLA, P. SOUBRIE, J. P. MAFFRAND, W. ROSTENE, P. KITABGI & G. LE FUR. 1993. Biochemical and pharmacological profile of a potent and selective non-peptide antagonist of neurotensin receptor. Proc. Natl. Acad. Sci. USA **90:** 65–69.

50. MAFFRAND, J. P., D. GULLY, R. BOIGEGRAIN, P. SOUBRIE, P. KITABGI, W. ROSTENE & G. LE FUR. 1993. Neurotensin receptor agonists and antagonists. Drugs of the Future **18**(12): 1137–1141.

51. AZZI, M., D. GULLY, M. HEAULME, A. BEROD, D. PELAPRAT, P. KITABGI, R. BOIGEGRAIN, J. P. MAFFRAND, G. LE FUR & W. ROSTENE. 1994. Neurotensin receptor interaction with dopaminergic systems in the guinea-pig brain shown by neurotensin receptor antagonists. Eur. J. Pharmacol. **255:** 167–174.

52. BROUARD, A., M. HEAULME, R. LEYRIS, D. PELAPRAT, D. GULLY, P. KITABGI, G. LE FUR & W. ROSTENE. 1994. SR 48692 inhibits neurotensin-induced [^{3}H]dopamine release in rat striatal slices and mesencephalic cultures. Eur. J. Pharmacol. **253:** 289–291.

53. BARKER, E. L., R. S. WESTPHAL, D. SMITH & E. SANDERS-BUSH. 1994. Constitutively active 5-hydroxytryptamine 2C receptors reveal novel inverse agonist activity of receptor ligands. J. Biol. Chem. **269:** 11687–11690.

54. BETANCUR, C., M. CANTON, D. GULLY, V. GEMA, D. PELAPRAT & W. ROSTENE. 1995. Characterization and distribution of binding sites for a new neurotensin receptor antagonist ligand, ^{3}H SR 48692, in the guinea pig brain. J. Pharmacol. Exp. Ther. **273:** 1450–1458.

55. BROUARD, A., D. PELAPRAT, C. DANA, M. VIAL, A. M. LHIAUBET & W. ROSTENE. 1992. Mesencephalic dopaminergic neurons in primary cultures express functional neurotensin receptors. J. Neurosci. **12:** 1409–1415.

56. HERVE, D., J. P. TASSIN, J. M. STUDLER, C. DANA, P. KITABGI, J. P. VINCENT, J. GLOWINSKI & W. ROSTENE. 1986. Dopaminergic control of ^{125}I-labeled neurotensin binding site density in corticolimbic structures of the rat brain. Proc. Natl. Acad. Sci. USA **83:** 6203–6207.

57. SADOUL, J. L., F. CHECLER, P. KITABGI, W. ROSTENE, F. JAVOY-AGID & J. P. VINCENT. 1984. Loss of high affinity neurotensin receptors in substantia nigra from parkinsonian subjects. Biochem. Biophys. Res. Commun. **125:** 395–404.

58. SZIGETHY, E. & A. BEAUDET. 1989. Correspondence between high affinity neurotensin binding site and dopaminergic neurons in the rat substantia and ventral tegmental area: A combined radioautographic and immunocytochemical light microscopic study. J. Comp. Neurol. **279:** 128–137.

59. PONCELET, M., C. GUEUDET, D. GULLY, P. SOUBRIE & G. LE FUR. 1994. Turning behavior induced by intrastriatal injection of neurotensin in mice: Sensitivity to non peptide neurotensin antagonists. Naunyn-Schmiedebergs Arch. Pharmacol. **349:** 57–60.

60. STEINBERG, R., P. BRUN, J. FOURNIER, J. SOUILHAC, D. RODIER, G. MONS, J. P. TERRANOVA, G. LE FUR & P. SOUBRIE. 1994. SR 48692, a non-peptide neurotensin receptor antagonist differentially affects neurotensin-induced behaviour and changes in dopaminergic transmission. Neuroscience **59:** 921–929.

61. STOESSL, A. J. 1995. Effects of neurotensin in a rodent model of tardive dyskinesia. Neuropharmacology **34:** 457–462.

62. WAGSTAFF, J. D., L. G. BUSH, J. W. GIBB & G. R. HANSON. 1994. Endogenous neurotensin antagonizes methamphetamine-enhanced dopaminergic activity. Brain Res. **665:** 237–244.

63. PUGSLEY, T. A., H. C. AKUNNE, S. Z. WHETZEL, S. DEMATTOS, A. E. CORBIN, J. N. WILEY, D. J. WUSTROW, L. D. WISE & T. G. HEFFNER. 1995. Differential

effects of the nonpeptide neurotensin antagonist, SR 48692, on the pharmacological effects of neurotensin agonists. Peptides **16:** 37–44.

64. URBAN, M. O. & D. J. SMITH. 1994. Localization of the antinociceptive and antianalgesic effect of neurotensin within the rostral ventromedial medulla. Neurosci. Lett. **174:** 21–25.

65. URBAN, M. O., D. J. SMITH & R.C. BYRD. 1994. The non-peptide antagonist, SR 48692, reveals the antianalgesic action of neurotensin neuronal projections from the PAG to the RMg. Can. J. Physiol. Pharmacol. **72**(Suppl. 1): 414.

66. PONCELET, M., J. SOUILHAC, C. GUEUDET, J. P. TERRANOVA, D. GULLY, G. LE FUR & P. SOUBRIE. 1994. Effects of SR 48692, a selective non peptide neurotensin receptor antagonist, on two dopamine-dependent behavioural responses in mice and rats. Psychopharmacology **116:** 237–241.

67. BRUN, P., R. STEINBERG, G. LE FUR & P. SOUBRIE. 1995. Blockade of neurotensin receptor by SR 48692 potentiates the facilitatory effect of haloperidol on the evoked in vivo dopamine release in the rat nucleus accumbens. J. Neurochem. **64:** 2073–2079.

68. LABBE-JULLIE, C., I. DUBUC, A. BROUARD, S. DOULUT, E. BOURDEL, D. PELAPRAT, J. MAZELLA, J. MARTINEZ, W. ROSTENE, J. COSTENTIN & P. KITABGI. 1994. In vivo and in vitro structure-activity studies with peptide and pseudopeptide neurotensin analogs suggest the existence of distinct central neurotensin receptor subtypes. J. Pharmacol. Exp. Ther. **268:** 328–336.

69. AL-RODHAN, N. R. F., E. RICHELSON, J. A. GILBERT, D. J. McCORMICK, K. S. KANBA, M. A. PFENNING, E. W. LARSON & T. L. YAKSH. 1991. Structure-antinociceptive activity of neurotensin and some novel analogues in the periaqueductal gray region of the brainstem. Brain Res. **557:** 227–235.

70. LE, F., B. CUSACK & E. RICHELSON. 1996. The neurotensin receptor: Is there more than one subtype? TIPS **17:** 1–3.

71. LABBE-JULLIE, C., J. M. BOTTO, M. V. MAS, J. CHABRY, J. MAZELLA, J. P. VINCENT, D. GULLY, J. P. MAFFRAND & P. KITABGI. 1995. ^{3}H SR 48692, the first nonpeptide neurotensin antagonist radioligand: Characterization of binding properties and evidence for distinct agonist and antagonist binding domains on the rat neurotensin receptor. Mol. Pharmacol. **47:** 1050–1056.

72. AZZI, M., A. NICOT, D. GULLY, P. KITABGI, A. BEROD & W. ROSTENE. 1994. Increase in neurotensin receptor expression in rat brain induced by chronic treatment with the nonpeptide neurotensin receptor antagonist SR 48692. Neurosci. Lett. **172:** 97–100.

73. CREESE, I. & D. P. SIBLEY. 1981. Receptor adaptations to centrally acting drugs. Annu. Rev. Pharmacol. Toxicol. **21:** 357–391.

74. WOULFE, J. & A. BEAUDET. 1989. Immunocytochemical evidence for direct connections between neurotensin-containing axons and dopaminergic neurons in the rat ventral midbrain tegmentum. Brain Res. **479:** 402–406.

75. AZZI, M., H. BOUDIN, N. MAHMUDI, D. PELAPRAT, W. ROSTENE & A. BEROD. 1996. In vivo regulation of neurotensin receptors following long-term pharmacological blockade with a specific receptor antagonist. Mol. Brain Res. **42:** 213–221.

76. MENDEZ, M., F. SOUAZE, M. NAGANO, P. A. KELLY, W. ROSTENE & P. FORGEZ. 1996. In vivo differential regulation of neurotensin receptor mRNA expression by SR48692 in rat brain and peripheral tissues. Soc. Neurosci. **22:** Abstr. 515.3.

77. FAURE, M. P., A. ALONSO, D. NOUEL, G. GAUDRIAULT, M. DENNIS, J. P. VINCENT & A. BEAUDET. 1995. Somatodendritic internalization and perinuclear targeting of neurotensin in the mammalian brain. J. Neurosci. **15:** 4140–4147.

78. MAZELLA, J., K. LEONARD, J. CHABRY, P. KITABGI, J. P. VINCENT & A. BEAUDET. 1991. Binding and internalization of iodinated neurotensin in neuronal cultures from embryonic mouse brain. Brain Res. **564:** 249–255.

79. VANISBERG, M. A., J. M. MALOTEAUX, J. N. OCTAVE & P. M. LADURON. 1991. Rapid agonist-induced decrease of neurotensin receptors from the cell surface in rat cultured neurons. Biochem. Pharmacol. **42:** 2265–2274.

80. DI PAOLA, D. E., B. CUSACK, M. YAMADA & E. RICHELSON. 1993. Desensitization

and down-regulation of neurotensin receptors in murine neuroblastoma N1E-115 by [D-Lys^8] neurotensin (8-13). J. Pharmacol. Exp. Ther. **264:** 1–5.

81. GOLDSTEIN, A., L. ARONOW & S. M. KALMAN. 1974. Consequences of drug-receptor interactions: Analysis of the graded dose-response relationship. *In* Principles of Drug Action: The Basis of Pharmacology, 2nd edit.: 82–111. Wiley. New York.

82. ROSTENE, W. & G. LE FUR. 1995. Peptide receptor antagonists. *In* Pharmacological Sciences: Perspectives for Research and Therapy in the Late 1990s. A. C. Cuello & B. Collier, Eds.: 29–36. Birkhäuser Verlag. Basel.

Metal-Ion Sites as Structural and Functional Probes of Helix–Helix Interactions in 7TM Receptors

CHRISTIAN E. ELLING,[a] KENNETH THIRSTRUP,[a]
SØREN M. NIELSEN,[a] SIV A. HJORTH,[a]
AND THUE W. SCHWARTZ[a,b]

[a]*Laboratory for Molecular Pharmacology*
The Laboratory Center, Rigshospitalet 6321
Copenhagen, Denmark

[b]*Department of Protein Chemistry*
Institute of Molecular Biology
University of Copenhagen
Copenhagen, Denmark

INTRODUCTION

An important group of integral membrane proteins is the G-protein–coupled receptor superfamily, which today constitutes more than 100 cloned receptors. They signal information across the membrane for a diverse collection of ligand molecules ranging from, for example, metal-ions, classical transmitters, and neuropeptides to larger glycoprotein hormones.[1,2] Despite the large variety of ligands binding to these receptors, the seven transmembrane segments of G-protein–coupled receptors (7TM) are believed to define a common fold of presumably seven α-helices traversing the membrane in an anti-parallel way. Even though the number of solved structures of soluble proteins has been steadily increasing, there has been no similar success in the solution of membrane protein structures. As a consequence, structural information is currently pursued mainly by noncrystallographic methods. Hence, for the 7TM receptor superfamily, detailed information on the receptor structure is scarce, and the only real structural information available is the electron projection density maps of bovine and frog rhodopsin[3,4] and the low-resolution three-dimensional data for rhodopsin.[5] With the limited resolution of these structures, the assignment of individual helices to the electron density remains tentative,[1,6] and the detailed understanding of the intramolecular signal transduction mechanism still remains obscure. Important progress does, however, appear to have been made in recent years in respect of some general overall molecular principles.[7–10]

METAL-IONS CAN SERVE AS STRUCTURAL PROBES OF HELIX–HELIX INTERACTIONS

Various molecular biology and spectroscopic methods have been applied to investigate the helical interactions in membrane receptors.[11–13] We have used an

---

[a] Address for correspondence: Laboratory for Molecular Pharmacology, Rigshospitalet 6321, Blegdamsvej 9, DK-2100 Copenhagen, Denmark.

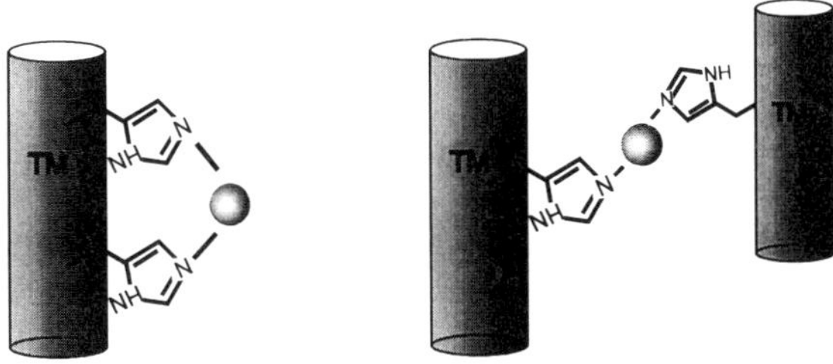

FIGURE 1. Schematic representation of minimal intra- and interhelical bis-histidine metal-ion binding sites.

approach in which metal-ion binding sites were introduced in the transmembrane segments of the tachykinin NK-1[14] and κ-opioid receptor[15] by protein engineering. The coordination of metals to metal-ion binding sites in proteins is well characterized in solved protein structures. Information regarding, for example, the mean distance from the chelating atoms of the side chain to the metal-ion is known.[16–18] Hence, by introducing metal-ion binding sites into the transmembrane segments of a 7TM receptor, the proximity of the involved residues may possibly be inferred and consequently used to direct the register of the involved helices in receptor models.

In the simplest case, a metal-ion binding site could be defined by two residues from the protein capable of donating electrons to the coordination sphere, a bidentate metal-ion site.[19] The remaining ligands for the metal-ion in the coordination sphere could presumably be solvent (water/hydroxide). In principle, a bidentate site in the transmembrane segment could be engineered either by designing an intrahelical or an interhelical metal-ion binding site, FIGURE 1. Intrahelical metal-ion sites can in some cases conveniently be introduced by substituting residues placed at an i and $i + 4$ position in the helical structure.[16] Interhelical metal-ion sites obviously require that the substituted residues are placed in the correct spatial proximity in the tertiary structure of the receptor. Thus, because the structural knowledge on 7TM receptors is very limited, an approach is required where residues are systematically substituted to locate residues in proximity. The spatial information about the residues obtained using this approach can then be used to subsequently increase the number of protein ligands in the coordination sphere, for example, from two to three in an attempt to further increase the affinity of the metal-ion site. Several amino acid side chains have the potential to chelate metal-ions. However, histidine, a known amino acid ligand in metal-ion coordination,[16] is expected to be better tolerated in the transmembrane helices than, for example, carboxylic acid side chains. Cysteine, another naturally occurring metal-ion ligand, could also be used; this residue, however, has the potential to form disulfide bridges which would complicate the interpretation of the data. Although cysteine has been used with success in cross-linking studies of receptors,[20–23] it does not appear to be an optimal choice for designing metal-ion binding sites in transmembrane helices (unpublished information).

By screening the receptor constructs in a competition binding assay, for example, with radiolabeled agonists or antagonists, only metal-ion sites that successfully compete for the receptor are selected. The advantage of this procedure is that the sites identified yield not only structural but also functional information about the receptor. However, the obvious drawback is that "silent" sites, which do not disturb the radioligand binding, are not detected in this screening procedure. In the tachykinin NK-1 receptor we initially introduced a high-affinity metal-ion site in the presumed binding site for the nonpeptide NK-1 receptor antagonist CP96,345.[14] By systematically substituting residues at the top of transmembranes V (TM-V) and VI (TM-VI) for histidines, the nonpeptide antagonist binding site was gradually converted into a high-affinity metal-ion site. This site was presumably of a tridentate nature composed of histidine residues located at positions V:01, V:05 (the naturally occurring His[197]), and VI:24 ([E193H;Y272H]NK-1 receptor) using the previously suggested generic nomenclature.[1] Importantly, metal-ions in the [E193H;Y272H]NK-1 receptor construct could compete with radiolabeled agonist substance P, TABLE 1, and nonpeptide antagonist, FIGURE 2, for the receptor with an apparent submicromolar affinity as determined in a competitive binding assay. The same phenotype of the construct was obtained regardless of whether the binding assay was performed in COS-7 cells transiently expressing the receptor[14] or in CHO cells stably expressing the receptor, TABLE 1. The apparent affinity of CP96,345 is decreased more than 200-fold with a concomitant increase in apparent affinity for Zn(II) ions of more than 500-fold, whereas the affinity of the natural ligand substance P is only slightly affected. Testing of a series of chemically distinct nonpeptide antagonists on this construct demonstrated that the binding site for CP96,345 was rather selectively exchanged, TABLE 2. These data support the notion that these compounds apparently exert their effect by binding to different subepitopes on the receptor.[24]

Combined with previously reported results,[14] it appears that local conformational effects or vertical positioning of potential metal-ion ligands in the helix may complicate the rational design of metal-ion sites. Thus, introducing histidine residues in an i and $i + 4$ manner in helical segments of 7TM receptors gave a high-affinity metal-ion site at the inner face of the top of TM-V,[14,15] but not one helical turn further down at the top of TM-VI (positions VI:17 and VI:21).[14] In the tachykinin NK-1 and the κ-opioid receptor the most optimal bis-histidine metal-ion sites gave an apparent affinity of approximately 5–15 μM for Zn(II), FIGURE 3, regardless of whether the site involved residues on the same or on two helices. It can be speculated that metal-ion sites located deeper in the transmembrane segments are more inaccessible to solvent and solutes. Consequently, the apparent affinity obtained in a competition binding assay as described above will be dependent on the relative position of the potential metal-ion site within the receptor structure.

HIGH-AFFINITY METAL-ION SITES CAN BE TRANSFERRED AMONG 7TM RECEPTORS

It is generally assumed that at least rhodopsin-like 7TM receptors have a common overall arrangement of the transmembrane segments. To verify that the spatial arrangement of residues as inferred from metal-ion sites is correct, we introduced histidines at the corresponding positions in the κ-opioid receptor[15] as those defining the high-affinity site in the NK-1 receptor. In this way the high-affinity metal-ion site was transferred to the κ-opioid receptor, as metal-ions competed with both radiolabeled agonist and antagonist with affinities comparable to those obtained on the NK-1 receptor. The metal-ion site in the κ-opioid receptor in fact functioned as a "zinc switch" of radioligand binding because the introduction of the histidine residues per se did not affect ligand binding.[15]

TABLE 1. Ligand Binding Data for the [E193H;Y272H]NK-1 Receptor in Chinese Hamster Ovary Cells

	Substance P			CP96,345				$ZnCl_2$				$CuCl_2$			
	K_d (nM)	±SEM		F_{mut}	K_i (nM)	±SEM		Fold Dec.	K_i (μM)	±SEM		Fold Inc.	K_i (μM)	±SEM		Fold Inc.
hNK-1	0.12	±0.03	(3)	1.0	0.13	±0.01	(3)	1	362	±74	(4)	1	189	±32	(4)	1
[E193H;Y272H]-2	0.54	±0.03	(3)	4.5	29.4	±3.1	(2)	226	0.66	± 0.04	(3)	548	19	± 4	(3)	10
[E193H;Y272H]-11	1.14	±0.09	(3)	9.5	32.3	±2.0	(3)	248	0.70	± 0.11	(3)	517	21	± 1	(3)	9

NOTE: Stable clones of the [E193H;Y272H]NK-1 receptor were established in Chinese hamster ovary cells. Two individual clones were tested. Cells were assayed in a competition binding assay using the radiolabeled agonist [125I]-Bolton-Hunter-substance P as previously described.[14,57] F_{mut} = K_d-mutant/K_d-wild type. Fold dec. (Fold decrease) = K_i-mutant/K_i-wild type. Fold inc. (Fold increase) = K_i-wild type/K_i-mutant. Numbers in parentheses indicate number of experiments.

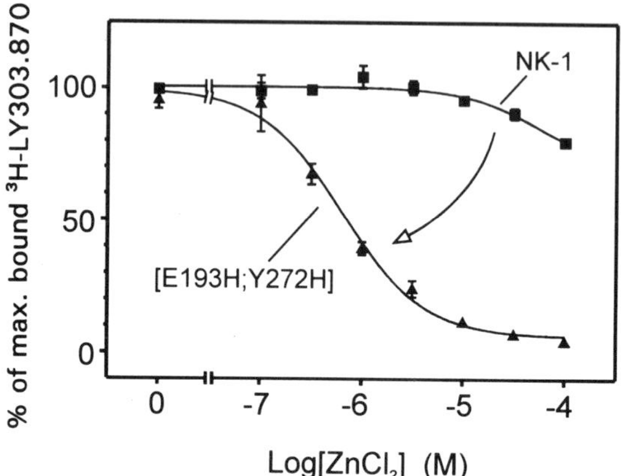

FIGURE 2. Gain of Zn(II) binding following introduction of His residues at positions 193 and 272 in the human NK-1 receptor. Wild-type and mutant receptors were transiently expressed in COS-7 cells and assayed using a competition binding assay[14,57] with the radiolabeled nonpeptide antagonist [^{3}H]LY303.870.

The direct transfer of the metal-ion site between two only distantly related receptors of the rhodopsin-like family (approximately 30% identity in the transmembrane segments) confirms the assumption that the transmembrane segments must have a common overall arrangement within the rhodopsin-like family. In the rhodopsin molecule, Oprian and co-workers[23] have used an elegant cysteine cross-linking approach in split-receptors to demonstrate the proximity of two of the corresponding positions that defined the high-affinity metal-ion binding site in the NK-1 receptor.[14] Thus, artificial metal-ion sites and disulfide bridges may be useful in general in the determination of interactions of amino acids in membrane proteins, and represent an alternative to other noncrystallographic methods.[12–14,20–23,25]

TABLE 2. Selective Conversion of the Binding Site for CP96,345 into a Metal-Ion Site

	NK-1			[E193H;Y272H]			Ratio
	K_i (nM)	±SEM		K_i (nM)	±SEM		Fold Decrease
CP96,345	0.25	±0.02	(10)	34.3	± 5.8	(8)	137
CP99,994	0.39	±0.13	(4)	2.48	± 0.65	(4)	6
CGP49,823	0.65	±0.13	(4)	0.69	± 0.04	(4)	1
FK888	1.59	±0.45	(3)	2.08	± 0.54	(3)	1
SR140,333	0.40	±0.03	(3)	1.18	± 0.37	(3)	3
LY303,870	0.27	±0.03	(4)	1.37	± 0.16	(4)	5
RP67,580	14.7	±1.9	(4)	228	±25	(4)	16

NOTE: Wild-type and mutant receptors were transiently expressed in COS-7 cells and assayed with a series of chemically distinct high-affinity NK-1 receptor antagonists in competition binding assay using radiolabeled agonist [^{125}I]-Bolton-Hunter-substance P. Numbers in parentheses indicate number of experiments.

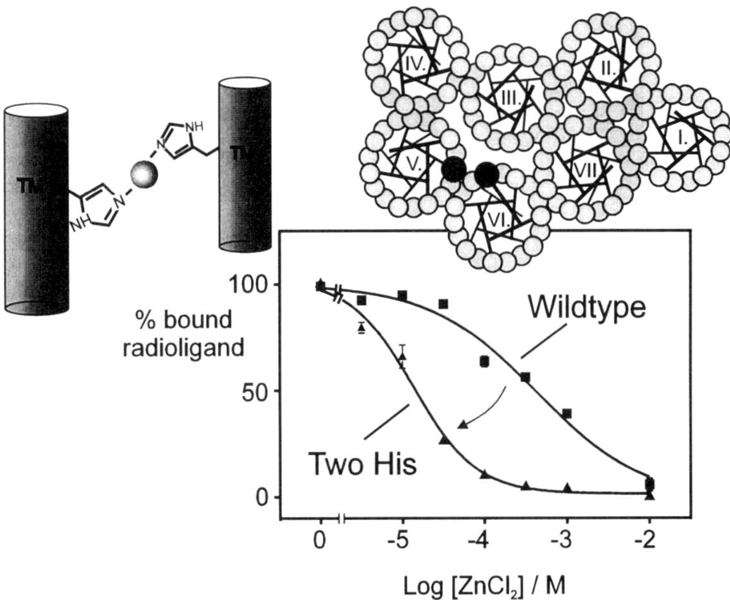

FIGURE 3. A prototype interhelical bis-His metal-ion site in a 7TM receptor. A helical wheel diagram of the transmembrane segments of a 7TM receptor of the rhodopsin-like family[1] is shown with a representative binding curve for the competition of Zn(II) with a radiolabeled ligand (substance P on the NK-1 receptor). Curves for wild type and a construct containing an interhelical bis-histidine metal-ion site are shown. The residues defining the bis-histidine site are indicated by *black circles* in the helical wheel diagram and are schematically depicted chelating a metal-ion in the *left panel*.

STOICHIOMETRY OF METAL-ION BINDING SITES IN 7TM RECEPTORS

Although the true affinity of Zn(II) is not known for the [E193H;Y272H]-NK-1 construct, the K_i value of approximately 0.6 μM fits well with previously reported affinities for tridentate binding sites,[26] 36 nM–3.7 μM (reviewed in ref. 19). Interestingly, affinities of 2 fM–25 nM have been reported for tetrahedral Zn(II) binding sites.[27,28] Thus, the metal-ion site in the [E193H;Y272H]NK-1 receptor more closely resembles a tridentate site. Zn(II) is most often found to coordinate with four ligands in a tetrahedral arrangement.[17] A possible fourth ligand in the coordination polyhedron in the [E193H;Y272H]NK-1 construct could be water/hydroxide in analogy with catalytic Zn(II) binding sites. The observation that the metal-ion site can be directly and stepwise transferred to the κ-opioid receptor by introducing the *three* histidines as two plus one at the corresponding positions supports the presumed tridentate nature of the site.

THE DESIGNED HIGH-AFFINITY METAL-ION BINDING SITE IN THE NK-1 RECEPTOR IS RELATIVELY SELECTIVE FOR ZN(II)

Although a qualitative analysis of metal-ion binding sites in proteins has been made,[18,29] the precise determinants of metal-ion specificity is unclear.[17] Some

TABLE 3. Metal-Ion Specificity of the High-Affinity Metal-Ion Binding Site in the [E193H;Y272H]NK-1 Construct

	NK-1			[E193H;Y272H]			Ratio
	K_i (μM)	$\pm$SEM		K_i (μM)	$\pm$SEM		Fold Increase
ZnCl$_2$	492	$\pm$42	(7)	0.62	$\pm$ 0.10	(7)	794
CuCl$_2$	346	$\pm$49	(6)	37	$\pm$10	(6)	9
NiCl$_2$	>1,000		(3)	52	$\pm$ 2	(3)	—
MgCl$_2$	>1,000		(3)	>1,000		(4)	—
MnCl$_2$	>1,000		(3)	>1,000		(3)	—
FeCl$_2$	527	$\pm$16	(4)	610	$\pm$ 0.05	(3)	1
FeCl$_3$	855	$\pm$28	(2)	713	$\pm$74	(3)	1
PbCl$_2$	>1,000		(4)	423	$\pm$94	(4)	—
CaCl$_2$	>1,000		(3)	>1,000		(4)	—
LiCl	>1,000		(3)	>1,000		(3)	—
KCl	>1,000		(3)	>1,000		(3)	—

NOTE: Wild-type and mutant receptors were transiently expressed in COS-7 cells and assayed in a competition binding assay using radiolabeled agonist [^{125}I]-Bolton-Hunter-substance P. Numbers in parentheses indicate number of experiments.

important factors appear to be the polarizability of the metal-ion and ligands, size and geometry of the binding site, and metal-ligand separation. A recent report demonstrated that metal-ligand distance and torsion angles of the chelating side chains are important determinations.[28] Originally we observed that the metal-ion site in the [E193H;Y272H]NK-1 receptor bound Zn(II) with 60-fold higher apparent affinity than Cu(II).[14] As shown in TABLE 3, among a series of other cations only Ni(II) showed a modest increase in apparent affinity in this construct as compared to the wild-type receptor as determined in competition binding experiments with radiolabeled substance P. However, it is very likely that ions such as Cd(II) and Hg(II), which we have not tested here, would also bind with a relatively high affinity similar to Zn(II) as has been demonstrated for other Zn(II) sites.[30]

METAL-ION SITES CAN BE USED TO DIRECT MOLECULAR MODELING ATTEMPTS

Most structural information on 7TM receptors has thus far mainly been inferred from mutagenesis studies identifying major contact points especially for monoamines. However, the potential flexibility of the ligands and side chains rarely allows a precise determination of distances between presumed contact points for the ligand. In the NK-1 receptor the presumed binding site for CP96,345, a nonpeptide antagonist of an approximate size of 6 × 9 × 12 Å, had previously been located to a binding pocket defined by residues on TM-III, TM-V, and TM-VI,[10,24,31–43] with possible additional contributing residues from transmembrane segments IV and VII. The spatial determination of residues using designed metal-ion sites simplifies the interpretation of the data. In contrast to a peptide or a nonpeptide ligand, a metal-ion, for example, Zn(II), is comparatively simpler with

Non-peptide NK-1 Antagonist

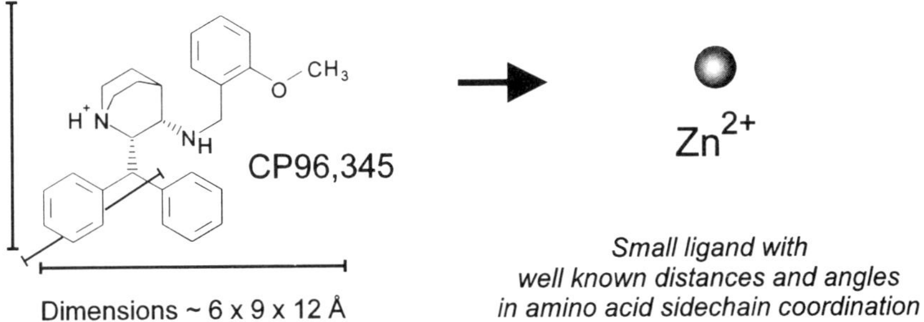

FIGURE 4. Comparison of the nonpeptide antagonist CP96,345 and Zn(II)—two structurally very different receptor ligands. The approximate size of CP96,345 in three dimensions is indicated. A single Zn(II) ion is shown out of scale for clarity.

well-characterized binding sites (see above) (FIG. 4). Thus, by assuming that the coordination of metals to the designed metal-ion site at the top of TM-V and TM-VI in the NK-1 and κ-opioid receptor (see above) is near optimal, this places the atoms (presumably N_ε[16]) of the chelating histidines within 2 Å[17,18] of the metal-ion. This information can subsequently be used to direct molecular modeling of the receptors.[15] Speculatively, unknown basic features of 7TM receptors like overall helical arrangement and connectivity could be assigned by systematically introducing metal-ion sites in the receptor structure.

THE MECHANISM OF ACTION OF LIGANDS MAY BE TESTED WITH DESIGNED METAL-ION SITES

In view of the hypothesis that ligands obtain their pharmacological properties—for example, agonism versus antagonism—by stabilizing appropriate receptor conformations,[44] it could be expected that the functional properties of the antagonist ligand could be transferred to the metal-ion. In the [E193H;Y272H]NK-1 construct, properties of the nonpeptide antagonist were in fact transferred to the metal-ion by redesigning the antagonist binding site to a metal-ion site using the positions corresponding to presumed contact points for the nonpeptide antagonist as guidance points. In this receptor system two different mechanisms of action had been proposed to explain the competitive antagonism of CP96,345. One model explained the competitive antagonism by CP96,345 by a volume exclusion effect;[34] thus, upon binding to the receptor, the antagonist would sterically exclude the binding of the agonist. According to this model the binding sites for the agonist and antagonist are thought to be overlapping in space, although sharing no major contact points on the receptor itself. A second model,[10,14,24,44] which we favor, argues that the binding of the agonist and antagonist are mutually exclusive. The two ligands compete for the whole receptor by binding to distinct sites presented by different receptor conformations. In the [E193H;Y272H]NK-1 receptor, the metal-ion is thought to bind and stabilize a receptor conformation that is inactive and cannot bind the agonist with high affinity. Conversely, the agonist is supposed to stabilize a receptor conformation that is active and cannot bind the antagonist with high affinity. The notion that receptors should be in an equilibrium between active and inactive conformations is supported by several observations.[7-9,45-56] Hence, instead of the ligand inducing the conformation, the ligand may rather select already present conformations of the receptor. By having different affinities for different conformations, the ligand biases an already established equilibrium between active and inactive conformations of the receptor. We believe that the small size and unrelated structure of the metal-ion compared to CP96,345 argue in favor of such an allosteric competitive model.

In conclusion, the engineering of metal-ion binding sites has previously been used extensively to study soluble globular proteins of known three-dimensional structure. It is proposed that metal-ion binding sites can be used to study structural and functional properties of 7TM receptors, whose structure currently is known only to a very limited resolution.

REFERENCES

1. SCHWARTZ, T. W. 1994. Curr. Opin. Biotechnol. **5:** 434–444.
2. PROBST, W. C., L. A. SNYDER, D. I. SCHUSTER, J. BROSIUS & S. C. SEALFON. 1992. DNA Cell Biol. **11:** 1–20.
3. SCHERTLER, G. F. X., C. VILLA & R. HENDERSON. 1993. Nature **362:** 770–772.
4. SCHERTLER, G. F. X. & P. A. HARGRAVE. 1995. Proc. Natl. Acad. Sci. USA **92:** 11578–11582.
5. UNGER, V. M. & G. F. X. SCHERTLER. 1995. Biophys. J. **68:** 1776–1786.
6. BALDWIN, J. M. 1993. EMBO J. **12:** 1693–1703.
7. SAMAMA, P., T. COTECCHIA, T. COSTA & R. J. LEFKOWITZ. 1993. J. Biol. Chem. **268:** 4625–4636.
8. LEFKOWITZ, R. J., S. COTECCHIA, P. SAMAMA & T. COSTA. 1993. TIPS **14:** 303–307.
9. SAMAMA, P. S., G. PEI, T. COSTA, S. COTECCHIA & R. J. LEFKOWITZ. 1993. Mol. Pharmacol. **45:** 390–394.
10. ROSENKILDE, M. M., M. CAHIR, U. GETHER, S. A. HJORTH & T. W. SCHWARTZ. 1994. J. Biol. Chem. **269:** 28160–28164.
11. ZHOU, W., C. A. FLANAGAN, J. A. BALLESTEROS, K. KONVICKA, J. S. DAVIDSON, H. WEINSTEIN, R. P. MILLAR & S. C. SEALFON. 1994. Mol. Pharmacol. **45:** 165–170.
12. HUBBEL, W. L. & C. ALTENBACH. 1994. Curr. Opin. Struct. Biol. **4:** 566–573.
13. VOSS, J., W. L. HUBBEL & H. R. KABACK. 1995. Proc. Natl. Acad. Sci. USA **92:** 12300–12303.
14. ELLING, C. E., S. M. NIELSEN & T. W. SCHWARTZ. 1995. Nature **374:** 74–77.
15. THIRSTRUP, K., C. E. ELLING, S. A. HJORTH & T. W. SCHWARTZ. 1996. J. Biol. Chem. **271:** 7875–7878.
16. CHAKRABARTI, P. 1990. Protein Eng. **4:** 57–63.
17. GLUSKER, J. P. 1991. Adv. Protein Chem. **42:** 1–76.
18. GREGORY, D. S., A. C. R. MARTIN, J. C. CHEETHAM & A. R. REES. 1993. Protein Eng. **6:** 29–35.
19. REGAN, L. 1995. TIBS **20:** 280–285.
20. LYNCH, B. A. & D. E. KOSHLAND, JR. 1991. Proc. Natl. Acad. Sci. USA **88:** 10402–10406.
21. FALKE, J. J. & D. E. KOSHLAND, JR. 1987. Science **237:** 1596–1600.
22. MILLIGAN, D. L. & D. E. KOSHLAND, JR. 1988. J. Biol. Chem. **263:** 6268–6275.
23. YU, H., M. KONO, T. D. MCKEE & D. D. OPRIAN. 1995. Biochemistry **46:** 14963–14969.
24. GETHER, U., X. EMONDS-ALT, J.-C. BRELIÉRE, T. FUJII, D. HAGIWARA, L. PRADIER, C. GARRET, T. E. JOHANSEN & T. W. SCHWARTZ. 1994. Mol. Pharmacol. **45:** 500–508.
25. JUNG, K., J. VOSS, M. HE, W. L. HUBBEL & H. R. KABACK. 1995. Biochemistry **34:** 6272–6277.
26. CROWDER, M. W., J. D. STEWART, V. A. ROBERTS, C. J. BENDER, E. TEVELRAKH,

J. PEISACH, E. D. GETZOFF, B. J. GAFFNEY & S. J. BENKOVIC. 1995. J. Am. Chem. Soc. **117:** 5627–5634.

27. REGAN, L. & N. D. CLARKE. 1990. Biochemistry **29:** 10878–10883.
28. IPPOLITO, J. A., T. T. BAIRD, JR., S. A. McGEE, D. W. CHRISTIANSON & C. A. FIERKE. 1995. Proc. Natl. Acad. Sci. USA **92:** 5017–5021.
29. YAMASHITA, M. M., L. WESSON, G. EISENMAN & D. EISENBERG. 1990. Proc. Natl. Acad. Sci. USA **87:** 5648–5652.
30. WELL, T. N. C., F. COULIN, M. A. PAYTON & A. E. I. PROUDFOOT. 1993. Biochemistry **32:** 1294–1301.
31. ZOFFMANN, S., U. GETHER & T. W. SCHWARTZ. 1993. FEBS Lett. **336:** 506–510.
32. HUANG, R. C., H. YU, C. D. STRADER & T. M. FONG. 1994. Mol. Pharmacol. **45:** 690–695.
33. FONG, T. M., M. A. CASCIERI, H. YU, A. BANSAL, C. SWAIN & C. D. STRADER. 1993. Nature **362:** 350–353.
34. HUANG, R. C., H. YU, C. D. STRADER & T. M. FONG. 1994. Biochemistry **33:**3007–3013.
35. FONG, T. M., R. C. HUANG & C. D. STRADER. 1992. J. Biol. Chem. **267:** 25664–25667.
36. GETHER, U., T. E. JOHANSEN, R. M. SNIDER, J. A. LOWE III, S. NAKANISHI & T. W. SCHWARTZ. 1993. Nature **362:** 345–348.
37. FONG, T. M., H. YU, M. A. CASCIERI, D. UNDERWOOD, C. J. SWAIN & C. D. STRADER. 1994. J. Biol. Chem. **269:** 2728–2732.
38. GETHER, U., Y. YOKOTA, X. EMONDS-ALT, J.-C. BRELIERE, J. A. LOWE III, R. M. SNIDER, S. NAKANISHI & T. W. SCHWARTZ. 1993. Proc. Natl. Acad. Sci. USA **90:** 6194–6198.
39. JENSEN, C. J., N. P. GERARD, T. W. SCHWARTZ & U. GETHER. 1994. Mol. Pharmacol. **45:** 294–299.
40. GETHER, U., L. NILSSON, J. A. LOWE & T. W. SCHWARTZ. 1994. J. Biol. Chem. **269:** 23959–23964.
41. FONG, T. M., H. YU & C. D. STRADER. 1992. J. Biol. Chem. **267:** 25668–25671.
42. FONG, T. M., H. YU, M. A. CASCIERI, D. UNDERWOOD, C. J. SWAIN & C. D. STRADER. 1994. J. Biol. Chem. **269:** 14957–14961.
43. SACHAIS, B. S., R. M. SNIDER, J. A. LOWE III & J. E. KRAUSE. 1993. J. Biol. Chem. **268:** 2319–2323.
44. SCHWARTZ, T. W., U. GETHER, H. T. SCHAMBYE & S. A. HJORTH. 1995.Curr. Pharm. Design **1:** 355–372.
45. PARMA, J., L. DUPREZ, J. SANDE VAN, P. COCHAUX, C. GERVY, J. MOCKEL, J. DUMONT & G. VASSART. 1993. Nature **365:** 649–651.
46. SHENKER, A., L. LAUE, S. KOSUGI, J. J. MERENDINO, T. MINEGISHI & G. B. CUTLER, JR. 1993. Nature **365:** 652–654.
47. ROBBINS, L. S., J. H. NADEAU, K. R. JOHNSON, M. A. KELLY, L. ROSELLI-REHFUSS, E. BAACK, K. G. MOUNTJOY & R. D. CONE. 1993. Cell **72:** 1–20.
48. LEFKOWITZ, R. J. 1993. Nature **365:** 603–604.
49. REN, Q., H. KUROSE, R. J. LEFKOWITZ & S. COTECCHIA. 1993. J. Biol. Chem. **268:** 16483–16487.
50. KJELSBERG, M. A., S. COTECCHIA, J. OSTROWSKI, M. G. CARON & R. J. LEFKOWITZ. 1992. J. Biol. Chem. **267:** 1430–1433.
51. ALLEN, L. F., R. J. LEFKOWITZ, M. G. CARON & S. COTECCHIA. 1991. Proc. Natl. Acad. Sci. USA **88:** 11354–11358.
52. BOONE, C., N. G. DAVIS & J. SPRAGUE. 1993. Proc. Natl. Acad. Sci. USA **90:** 9921–9925.
53. BARKER, E. L., R. S. WESTPHAL, D. SCHMIDT & E. SANDERS-BUSH. 1994. J. Biol. Chem. **269:** 11687–11690.
54. CHIDIAC, P., T. E. HEBERT, M. VALIQUETTE, M. DENNIS & M. BOUVIER. 1993. Mol. Pharmacol. **45:** 490–499.
55. BOND, R. A., P. LEFF, D. JOHNSON, C. A. MILANO, H. A. ROCKMAN, T. R. McMINN, S. APPARSUNDARAM, M. F. HYEK, T. P. KENAKIN, L. F. ALLEN & R. J. LEFKOWITZ. 1995. Nature **374:** 272–276.
56. BLACK, J. W. & N. P. SHANKLEY. 1995. Nature **374:** 214–215.
57. GETHER, U., T. E. JOHANSEN & T. W. SCHWARTZ. 1993. J. Biol. Chem. **268:**7893–7898.

Growth Factor Properties of VIP during Early Brain Development

Whole Embryo Culture and *in Vivo* Studies[a]

PIERRE GRESSENS,[b,d] BÉNÉDICTE PAINDAVEINE,[b]
JOANNA M. HILL,[c] DOUGLAS E. BRENNEMAN,[c] AND
PHILIPPE EVRARD[b]

[b]*Laboratoire de Neurologie du Développement
and Service de Neuropédiatrie
Hôpital Robert-Debré
48 Blvd. Sérurier
F-75019 Paris, France*

[c]*Section on Molecular and Developmental Pharmacology
National Institute of Child Health and Human Development
National Institutes of Health
Building 49, Room 5A-38
Bethesda, Maryland 20892*

INTRODUCTION

Cell production and cell death are the two ontogenic events determining the final neural cell number, which is a key factor in brain function. Our knowledge of the molecular mechanisms controlling the different steps of the mitotic cycle of cultured eukaryotic cells has greatly improved in the last decade (for a review see refs. 1 and 2). In contrast, little is known about the specific signaling molecules stimulating and/or inhibiting these recognized molecular regulatory mechanisms at the different steps of *in vivo* brain development.

The short developmental period between the isolation of the neural groove and the onset of neuronal migration is of crucial importance for the subsequent neural development: intense cell multiplication producing precursors for the whole brain occurs during this period in the primitive neuroepithelium along with the commitment of cells to the glial lineage.[3] In mouse, this critical ontogenic period (developmental phase II on FIG. 1) extends from the embryonic day (E) 8 to day E11.5, from 0 to 45 somites. Similar events take place in humans between 4 and 10 weeks of gestation (for a review see ref. 4).

Vasoactive intestinal peptide (VIP) has *in vitro* trophic and mitogenic properties[5,6] on embryonic neural tissues, but inhibits growth and mitosis in certain tumors.[7] VIP receptors are localized to the neuroepithelium of the early postimplantation embryo of both rat and mouse[8–10] and exhibit distribution patterns related to ontogenic events.[11] Furthermore, VIP is present in higher concentrations in the early postimplantation rat embryo than later during gestation. However,

[a] This work was supported in part by the Fonds National de la Recherche Scientifique (Belgium).
[d] E-mail: Gressens@msn.com

152 "

the mRNA for VIP was undetectable at this time of development.[10] The VIP concentration of pregnant rat serum exhibited a peak at days E10–E12 (comparable in development to days E9–E11 in the mouse[12,13]) reaching levels 6–10-fold higher than during the remainder of the pregnancy. In addition, undegraded VIP was detected in the E10 embryo after intravenous administration of radiolabeled VIP to pregnant mice.[10] Together, these data suggest that maternal VIP could act as a signaling molecule regulating early neurogenesis. This hypothesis has been addressed by culturing whole mouse embryos in the presence of various concentrations of VIP[8] and by blocking *in vivo* effects of VIP by administration of a specific VIP antagonist.[9]

EFFECTS OF VIP ON WHOLE CULTURED MOUSE EMBRYOS

The technique of whole embryo culture,[14,15] which permits the culture of E8 to E10 rodent embryos for a limited period, has been shown to be a precious tool to investigate the fine effects of exogenous substances on the developing neural tube and on its cellular components.[3] Controlled exposure of whole embryos of known somitic stage permits the circumvention of two major problems which are inherent to the *in vivo* administration of a drug to a pregnant animal: a lack of precise staging of embryos and variation in the amount of drug exposure.

In this model, 10^{-10} to 10^{-7} M VIP dramatically stimulates growth of E9.5 mouse embryos cultured for four hours.[8] Treatment with VIP produces a concentration-dependent increase in somitic number, a marker of growth and maturation (an average of 5.2 new somites are produced with 10^{-7} M VIP, whereas control embryos acquire an average of 2.2 new somites during the culture period). Similar VIP growth-stimulating effects (11–63% increase from control) (FIG. 2) are evident in embryonic volume. Both the macroscopic and microscopic examination of VIP-treated embryos reveal no apparent abnormalities. Furthermore, VIP does not interfere with the early commitment of cells to the glial lineage as assessed by the normal labeling with radial cell 2 (RC2) antibody (a specific glial marker[16]) of radial glial cells in the primitive neuroepithelium of VIP-treated embryos. In comparison to control animals in culture, VIP increases the DNA and protein content by 103% and 63%, respectively, implying that cell division is the major mechanism that accounts for VIP-induced enlargement of the embryos. When cells are labeled in S-phase with bromodeoxyuridine (BRDU), a thymidine analogue that is integrated in dividing cells at the S-phase, VIP produces a dramatic concentration-dependent increase in the number of cells in S-phase (up to a five- to six-fold increase) in both neural and nonneural tissues. These data indicate that VIP acts at the G1-S transition checkpoint. Experimental evidence from our laboratory suggests that G1- and S-phases are shortened by VIP, resulting in dramatic reduction of cell cycle length.[17]

Three indications strongly suggest that the observed growth-stimulating effects are mediated through VIP-specific mechanism and binding sites: (1) Similar to rat embryos,[11] most of the specific binding of [^{125}I]VIP in the mouse embryo explant is restricted to the central nervous system (CNS), including spinal cord and cephalic vesicles. Addition of VIP to the culture medium produces a concentration-dependent decrease in binding site density. (2) Secretin, another peptide of the VIP family, and pituitary adenylate cyclase activating peptide (PACAP 1-38), a VIP-related peptide which has been shown to act potently on some VIP receptors,[18] have no growth effect in this embryo culture paradigm (Gressens, Hill, and Brenneman, unpublished data). (3) A specific VIP antagonist[19,20] (VA) (FIG. 3) partially

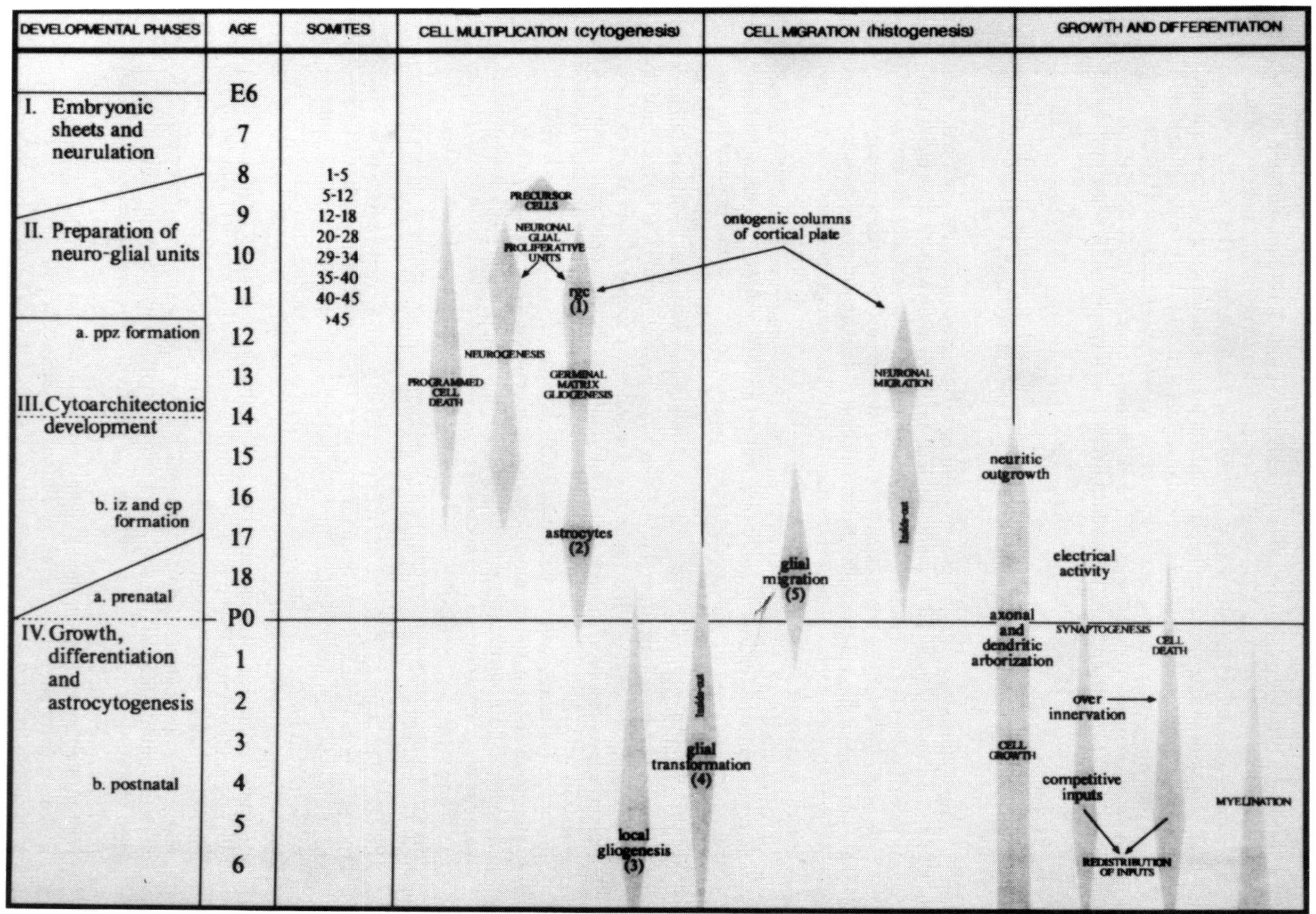
DEVELOPMENTAL PHASES
AGE
SOMITES
CELL MULTIPLICATION (cytogenesis)
CELL MIGRATION (histogenesis)
GROWTH AND DIFFERENTIATION
I. Embryonic sheets and neurulation
II. Preparation of neuro-glial units
a. ppz formation
III. Cytoarchitectonic development
b. iz and cp formation
a. prenatal
IV. Growth, differentiation and astrocytogenesis
b. postnatal
E6
7
8
9
10
11
12
13
14
15
16
17
18
P0
1
2
3
4
5
6
1-5
5-12
12-18
20-28
29-34
35-40
40-45
>45
PRECURSOR CELLS
NEURONAL GLIAL PROLIFERATIVE UNITS
rgc (1)
NEUROGENESIS
PROGRAMMED CELL DEATH
GERMINAL MATRIX GLIOGENESIS
astrocytes (2)
glial transformation (4)
local gliogenesis (3)
ontogenic columns of cortical plate
NEURONAL MIGRATION
glial migration (5)
neuritic outgrowth
electrical activity
axonal and dendritic arborization
SYNAPTOGENESIS
CELL DEATH
over innervation
CELL GROWTH
competitive inputs
REDISTRIBUTION OF INPUTS
MYELINATION

FIGURE 1. Schematic chronology of the main ontogenic events of the mouse neocortical development. E, embryonic day; P, postnatal day; ppz, primitive plexiform zone; iz, intermediate zone; cp, cortical plate; rgc, radial glial cell; numbers 1–5 refer to the successive steps of cortical gliogenesis. This figure is based on data in the literature, on our own studies, and on our interpretation of data obtained in other species.

blocks the VIP-stimulated increases in embryonic growth and the VIP-induced down-regulation of VIP binding sites.

However, when embryos are treated with a combination of VIP and VA, the VIP-induced increase in BRDU-positive cells in the CNS is completely prevented by the VIP antagonist, whereas in nonneural tissues the antagonist blocked a maximum of 38% of the increase. The tissue-specific effect of the VIP antagonist on BRDU-positive cells implies that regulation of mitosis occurs through pharmacologically distinct VIP receptors. This co-treatment with VIP and VA also induces embryonic malformation, including head deformation and irregularity of the neuroepithelium; this abnormal morphology may be related to disproportionate growth secondary to the differential selectivity of VA for receptors regulating the CNS versus non-CNS effects. Because most of the VIP receptors are localized in the CNS, the mechanism by which VIP stimulates nonneural tissue growth is unclear. Based on the observation that two pharmacologically distinct receptors mediate the VIP growth responses, two alternative mechanisms are feasible: (1) that tissue-specific VIP receptors mediate the regional responses—here a direct action of VIP on the few non-CNS receptors could result in growth because VIP has been shown to stimulate the mitosis of cultured nonneural cells,[21,22] or (2) that two pharmacologically distinct VIP receptors are present in the CNS, one that mediates effects on CNS cells and another that acts indirectly by influencing the non-CNS responses through secondary molecules. In the latter context, stimulation of cultured astrocytes with VIP results in the release of diffusible trophic substances.[23,24] An analogous mechanism could occur in cultured embryos, with glial CNS cells[3] releasing substances that influence non-CNS cell growth.

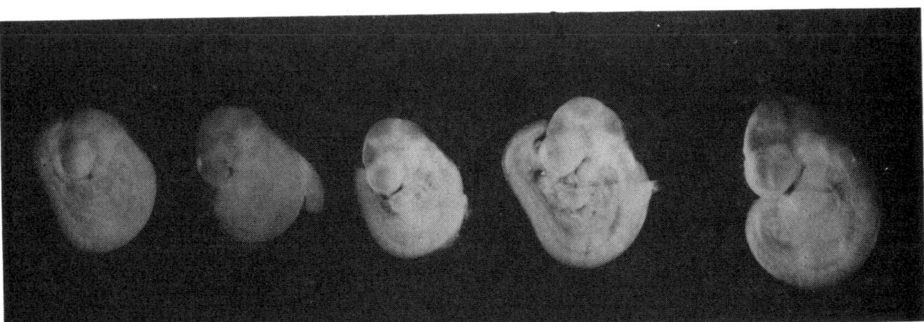

FIGGURE 2. VIP promotes embryonic growth without inducing macroscopic abnormalities. These 4-h cultured embryos were processed in the same experiment and represent the typical macroscopic aspect observed in repeated cultures with 10^{-10} M **(3)**, 10^{-9} M **(4)**, or 10^{-7} M **(5)** VIP and 10^{-7} M secretin **(2)** when compared with control cultured embryos **(1)**. Bar: 0.25 mm.

HYBRID VIP ANTAGONIST

VIP Deletion

His - Ser - Asp - Ala - Val - Phe -

Neurotensin Substitution   VIP

| Lys - Pro - Arg - Arg - Pro - Tyr | Thr - Asp - |
| Asn - Tyr - Thr - Arg - Leu - Arg - Lys - Gln - |
| Met - Ala - Val - Lys - Lys - Tyr - Leu - Ans - |
| Ser - Ile - Leu - Asn - NH2 |

FIGURE 3. Amino acid sequence of the specific VIP antagonist (VIP-neurotensin hybrid)[19,20] used in the studies described in the text.

IN VIVO STUDIES

To assess the role of VIP on early CNS growth, a specific VA[18,19] (FIG. 3) was administered twice daily to pregnant mice from E9.3 to E11.8.[9] VA induced a dose-dependent decrease in the DNA (52% of controls) and protein (43% of controls) content in the heads of E11.8 embryos. In contrast, the DNA and protein content of the body were less affected (89 and 72% of controls, respectively). Similar VA-induced reduction of CNS growth (59% of controls) was evident on E17.8 brain weight (FIG. 4); the weight loss was less marked in the body (82% of controls). Increased duration of treatment with VA from E9.3 to E17.8 did not

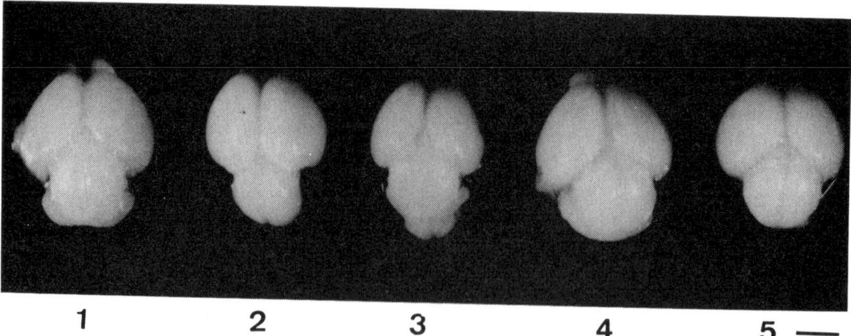

FIGURE 4. VA inhibits brain growth. Typical macroscopic appearance of E17.8 embryo brain treated between E9.3 and E11.8 with PBS alone **(1)**, 2 μg/g VA **(2–3)**, 0.2 μg/g VA plus 2 μg/g VIP **(4)**, and 0.2 μg/g VA plus 2 μg/g PACAP 1-38 **(5)**. Bar: 0.25 cm.

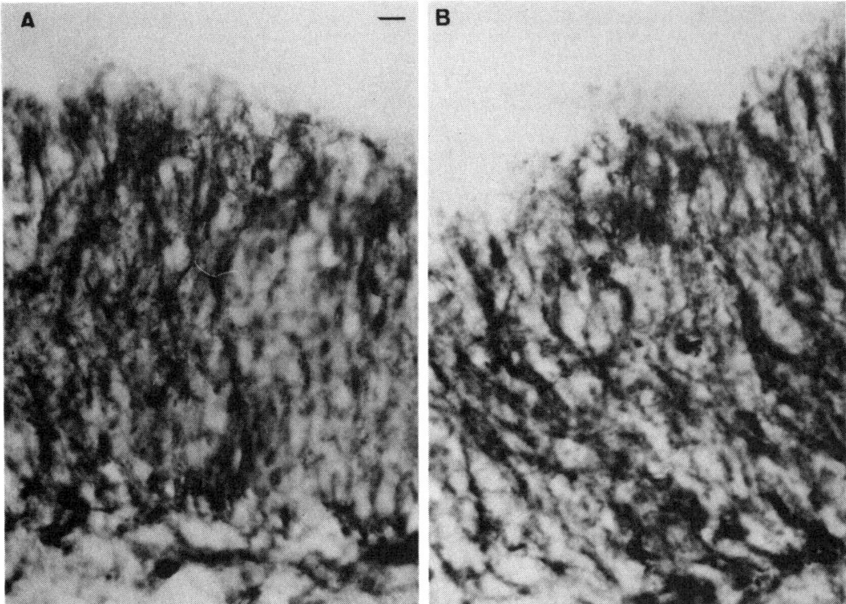

FIGURE 5. Microcephalic brains display normal gliogenesis. RC2 staining of radial glial fibers in the E11 neocortex of embryos treated between E9 and E11 with VA **(A)** or saline **(B)**. Bar: 5 μm.

significantly change the extent of microcephaly from that obtained from E9.3 to E11.8. Furthermore, no significant reduction in brain size was observed in the embryos of mice injected between E12.3 and E15.8 or E15.3 and E17.8 (see FIG. 1), demonstrating a stage dependency in the growth-regulating effects of VIP and identifying E9.3 to E11.8 as a critical VIP-sensitive period. In experiments where dams were allowed to deliver, VA-treated pups were born alive at term. Secretin had no effect on the parameters under study. Co-treatment with a 10-fold molar excess of VIP abolished the VA-induced impairment of CNS and body growth, demonstrating the specificity of the observed effects. However, co-treatment with PACAP 1-38 did not prevent the impairment of growth produced by VA. Furthermore, PACAP 6-27, a partial PACAP antagonist,[25] failed to inhibit embryonic growth, suggesting that VA inhibits a specific VIP function not shared by PACAP.

Histological examinations of the microcephalic brains revealed normal morphology including glial differentiation of radial fibers as demonstrated by immunohistochemistry using the RC2 antibody (FIG. 5). Parallel reduction in DNA and protein content of VA-treated embryos as well as the normal histology and cytoarchitecture observed after this treatment suggested a mechanism of mitotic inhibition. To test this hypothesis, BRDU was injected into pregnant dams 20 minutes before sacrifice. In E9.8 embryos, VA produced a decrease in the number of cells in S-phase. This mitotic inhibition affected predominantly neural (18% of control) rather than nonneural (4% of control) tissues.

Autoradiographic studies of [^{125}I]VIP binding were also conducted on the embryos from VA-treated dams to help establish the site of action and specificity of the drug-induced changes in growth. VA elicited an increase in the density of VIP binding sites in E9.8 embryos (136% of controls). The increased VIP binding after treatment with VA may be explained by a receptor supersensitivity response in the germinal neuroepithelium resulting from blockade of the endogenous agonist. VIP receptors in the germinal neuroepithelium were shown to be present at E9.8, but were reduced by E11.8. Similar measurements of VIP receptors in developing neuroepithelium in rat have indicated transient binding sites.[11] Together these observations provide a rationale to the stage-specific nature of the VIP-stimulated mitosis and of its inhibition by VA. Furthermore, these studies strongly imply that the growth inhibition was a receptor-mediated phenomenon acting directly on the embryo rather than a toxic or ischemic effect on maternal systems.

Based on the disappearance of detectable VIP binding sites in the germinative zone around E12, we can hypothesize that the VIP-mediated mitogenic effect on neural precursors happens during the ontogenic period between E8 and E11. In the prosencephalon, this period corresponds to the premigratory period during which neuroepithelial cells are thought to divide symmetrically (both daughter cells reenter the mitotic cycle). We have shown that VIP acts at the G1-S transition; this checkpoint is thought to be important in the decision of the cell to reenter the mitotic cycle or to leave G0. For neuronal cells, this last option means to migrate out of the germinative zone toward the cortical plate. Accordingly, VIP, by promoting cells to reenter S-phase, might prevent cells from leaving the mitotic cycle. Rakic[26] has proposed that the cortical surface is determined by the number of symmetrical divisions of neural precursors, whereas the cortical thickness depends on the number of asymmetrical divisions (one daughter cell reenters the cycle and the other one migrates to the cortical plate). In this context, we propose an ontogenic and phylogenic hypothesis for VIP function during cortical development. Within a given species, cortical surface is regulated by, among other yet-to-be identified factors, the concentration of VIP, which could be determined by maternal environment (see above); and by the duration of VIP receptor expression by the germinative cells, which represents the limiting factor since VIP is available throughout the whole pregnancy. The differences of brain surface observed within species might be related to differences in the duration of VIP receptor expression by germinative cells: since in early developmental stages, the mitotic cycle is relatively short (around 8–10 h), a variation of 24 hours in the presence of VIP receptors could largely increase cell number while preventing neural cells from migrating to the cortical plate.

CONCLUDING REMARKS

VIP appears to be a crucial determinant of embryonic growth and maturation as well as of early neurogenesis, bringing together the fields of neurobiology and mammalian development. Although the dramatic growth spurt after *in vitro* treatment with high (0.1 mM) concentrations of VIP is likely to be pharmacological in terms of magnitude and kinetics, *in vivo* studies suggest a physiological role for VIP in ontogenesis and provide a new model of microcephaly. The blockade of VIP functions in the primitive neuroepithelium during a limited development period results in inhibition of mitosis followed by severe growth retardation in the CNS. These data imply that VIP or a VIP-like molecule has a vital endocrine role in the *in vivo* regulation of brain ontogenesis. Furthermore, the observation that VIP (or VA) effects are not reproduced (or protected) by PACAP strongly suggests the existence of a specific embryonic VIP receptor not shared by PACAP.

SUMMARY

Vasoactive intestinal peptide (VIP), a 28 amino acid neuropeptide widely distributed in the mammalian nervous system, has potent growth-related actions that influence cell division, neuronal survival, and neurodifferentiation. To address the potential effects of VIP on embryonic growth, whole postimplantation embryo cultures were used. After a 4-hour incubation, VIP stimulated growth as assessed by the following increases from control: embryonic volume (63%), DNA (103%), and protein content (63%), as well as the number of cells in S-phase (490%). No apparent histological abnormalities are produced by VIP.

To assess the *in vivo* function of VIP in early CNS growth, a VIP antagonist (VA) was injected i.p. between E9 and E11. VA induced a dose reduction of the DNA (84% of controls) and protein (80% of controls) contents of the E11 head and a decrease of E17 brain weight (87% of controls). In contrast, body growth was less affected by the antagonist. Injections of VA for a longer period (E9 to E17) did not increase the severity of the microcephaly. By *ex vivo* autoradiography, GTP-insensitive VIP binding sites were detected in the germinative neuroepithelium between E9 and E11, but not between E13 and E15, during neuronal migration.

These data demonstrate that VIP regulates mitogenic activity in the premigratory neuroepithelium. Although this effect is limited to a short ontogenic period, blockade of VIP by a specific antagonist induces a severe microcephaly.

REFERENCES

1. LOHKA, M. J. 1989. J. Cell Sci. **92:** 131–135.
2. NURSE, P. 1990. Nature **344:** 503–508.
3. GRESSENS, P., F. GOFFLOT, G. VAN MAELE-FABRY. J. P. MISSON, J. J. PICARD & P. EVRARD. 1992. J. Neuropathol. Exp. Neurol. **51:** 206–219.
4. EVRARD, P., N. MILADI, C. BONNIER & P. GRESSENS. 1992. Normal and abnormal development of the brain. *In* Handbook of Neuropsychology. I. Rapin & S. Segalowitz, Eds. Vol. 6: 11–44. Elsevier Science Publishers. Amsterdam.
5. BRENNEMAN, D. E., L. E. EIDEN & R. E. SIEGEL. 1985. Peptides **6:** 35–39.
6. BRENNEMAN, D. E. & L. E. EIDEN. Proc. Natl. Acad. Sci. USA **83:** 1159–1162.
7. POSTON, G. J. 1988. Pancreas **3:** 439–443.
8. GRESSENS, P., J. M. HILL, B. PAINDAVEINE, I. GOZES, M. FRIDKIN & D. E. BRENNEMAN. 1993. Nature **362:** 155–158.
9. GRESSENS, P., J. M. HILL, B. PAINDAVEINE, I. GOZES, M. FRIDKIN & D. E. BRENNEMAN. 1994. J. Clin. Invest. **94:** 2020–2027.
10. HILL, J. M., S. K. MCCUNE, R. J. ALVERO, G. W. GLAZNER, K. A. HENINS, S. F. STANZIALE, J. KEIMOWITZ & D. E. BRENNEMAN. 1996. J. Clin. Invest. **97:** 1–7.
11. HILL, J. M., D. AGOSTON, P. GRESSENS & S. K. MCCUNE. 1994. J. Comp. Neurol. **342:** 186–205.
12. ATLAS, M. & V. P. BOND. 1963. J. Cell Biol. **26:** 19–24.
13. THEILER, K. 1989. The house mouse: Atlas of embryonic development. Springer-Verlag. New York.
14. NEW, D. A. T. 1971. Methods for the culture of postimplantation embryos of rodents. *In* Methods in Mammalian Embryology. J. C. Daniel, Jr., Ed.: 305–319.

15. VAN MAELE-FABRY, G., J. P. DEBRAS, I. FRANCOIS, M. BOUCAU & J. J. PICARD. 1988. Arch. Biol. **99:** 431–453.
16. MISSON, J. P., M. A. EDWARDS, M. YAMMAMOTO & V. S. CAVINESS. 1988. Dev. Brain Res. **44:** 427–431.
17. GRESSENS, P., B. PAINDAVEINE, J. M. HILL, P. EVRARD & D. E. BRENNEMAN. 1995. Soc. Neurosci. Abstr. **607:** 21.
18. MASUO, Y., T. OHTAKI, Y. MASUDA, M. TSUDA & M. FUJINO. 1992. Brain Res. **540:** 319–321.
19. GOZES, I., E. MELTZER, S. RUBINRAUT & D. E. BRENNEMAN. 1989. Endocrinology **125:** 2945–2949.
20. GOZES, I., S. K. MCCUNE, L. JACOBSON, D. WARREN, T. W. MOODY & D. E. BRENNEMAN. 1991. J. Pharmacol. Exp. Ther. **257:** 959–966.
21. HAEGERSTRAND, A., B. JONZON, C. J. DALSGAARD & J. NILSSON. 1989. Proc. Natl. Acad. Sci. USA **86:** 5993–5996.
22. SCHOLAR, E. M. & S. PAUL. 1991. Cancer **67:** 1561–1564.
23. BRENNEMAN, D. E., E. A. NEALE, G. A. FOSTER, S. D'AUTREMONT & G. L. WESTBROOK. 1987. J. Cell Biol. **104:** 1603–1610.
24. GLAZNER, G. W., J. Y. WU, P. GRESSENS, I. GOZES, D. E. BRENNEMAN & J. M. HILL. 1995. Soc. Neurosci. Abstr. **607:** 20.
25. ROBBERECHT, P., M. C. WOUSSEN-COLLE, P. DE NEEF, P. GOURLET, L. BUSCAIL, A. VANDERMEERS, M. C. VANDERLEERS-PIRET & J. CHRISTOPHE. 1991. FEBS Lett. **286:** 133–136.
26. RAKIC, P. 1988. Science **241:** 170–176.

Neuropeptides and Neuronal Survival: Neuroprotective Strategy for Alzheimer's Disease[a]

I. GOZES,[b,e] A. BARDEA,[b] M. BECHAR,[b] O. PEARL,[b]
A. RESHEF,[b] R. ZAMOSTIANO,[b] A. DAVIDSON,[b]
S. RUBINRAUT,[c] E. GILADI,[b] M. FRIDKIN,[c]
AND D. E. BRENNEMAN[d]

[b]*Department of Clinical Biochemistry*
Sackler School of Medicine
Tel Aviv University
Tel Aviv 69978, Israel

[c]*Department of Organic Chemistry*
Weizmann Institute of Science
Rehovot 76100, Israel

[d]*Section on Developmental and Molecular Pharmacology*
Laboratory of Developmental Neurobiology
National Institute of Child Health and Human Development
National Institutes of Health
Bethesda, Maryland 20892

INTRODUCTION

Neuropeptides generally exhibit multiple roles in the maintenance of homeostasis. Classically, neuropeptides were found to exert neurohormonal and neurotransmitter (neuromodulator) effects in the central and peripheral nervous system. It is becoming increasingly apparent that neuropeptides also act as regulators of cell division, differentiation, and survival. In the past we reviewed neuropeptides as growth factors.[1,2] Recent studies indicate that neuronal cells, in response to axotomy, increase expression of neuropeptides that are associated with the promotion of survival and regeneration (see ref. 3 for review). Examples include corticotropin-releasing factor, dynorphin, calcitonin gene-related peptide, cholecystokinin, galanin, vasoactive intestinal peptide (VIP), neuropeptide Y, and others. At the same time, down-regulation of neurotransmitters and substances related to the secretion of neurotransmitters may occur. Similarly, neurotrophins of the family of nerve growth factors (NGF) increase the expression of specific neuropeptides, for example, brain-derived neurotrophic factor enhanced neuropeptide Y and somatostatin levels in cortical neurons, whereas NGF did not.[4] In the periph-

[a] This research was supported in part by the Fujimoto Corporation and the U.S.-Israel Binational Science Foundation.

This paper was written while I.G. was a scholar-in-residence at the Fogarty International Center for Advanced Study in the Health Sciences, National Institutes of Health, Bethesda, Maryland.

[e] Corresponding author.

eral nervous system, a similar phenomenon is observed with down-regulation of excitatory peptides following axotomy and up-regulation of inhibitory peptides such as galanin and neuropeptide Y.[5] The neuropeptides exerting neuroprotection may be synthesized, in part, by astroglial cells.[6]

Studies aimed at understanding the mechanism of neuroprotection offered by neuropeptides suggest protection against free radicals via the maintenance of Ca^{2+} homeostasis.[7] Peptide fragments derived from the β-amyloid precursor protein may have trophic effects,[8] depending on the structure of the secreted peptide fragment. The β-amyloid fragment that arises from alternative processing of the β-amyloid precursor protein forms free radical peptides and aggregates that desta-bilize Ca^{2+} levels and make neurons vulnerable to metabolic insults,[7] which, in the case of Alzheimer's disease, lead to neurodegeneration. The neurotrophic factors and neuropeptides may protect against this neurodegeneration process. One pathway of second messengers that was associated with neuroprotection involves the activation of the Ras protein.[9] Other contributing factors probably include proteases and protease inhibitors,[10,11] as well as the complement system[12] and cytokines.[13]

VIP AS A NEURONAL PROTECTOR

VIP, a regulator of cell division, neurodifferentiation, neuronal survival,[14,15] and embryonic growth,[16] is widely distributed in brain areas associated with learning and memory.[17] Functional studies with animals treated either with VIP antagonists[18] or genetically manipulated (transgenic) to express less VIP,[19] revealed impairments in learning abilities. *In vitro*, VIP has been shown to exhibit neuroprotection against a variety of insults including electrical blockade[14] and neurodegeneration associated with the human immunodeficiency virus envelope protein.[20] We have now extended these studies to neuroprotection against β-amyloid neurotoxicity.[21]

VIP PROTECTS AGAINST ALZHEIMER'S-RELATED NEURONAL CELL DEATH *IN VITRO*

Our experimental paradigm included dissociated cerebral cortical cultures treated with the β-amyloid peptide (amino acids 25–35) shown previously to cause neuronal cell death *in vitro*.[22,23] Cultures treated (for 5 days) with the β-amyloid peptide exhibited a dose-dependent decrease in neuronal cell counts in comparison to controls [the maximal cell death (50–70%) was observed with 25 μM β-amyloid peptide]. In control cultures, a maximal reduction of 20% was found in neuronal cell counts between day 9 and day 14. Co-treatment with VIP plus β-amyloid peptide prevented the neuronal death associated with the amyloid peptide (control cultures had 385 $\pm$ 8; β-amyloid-treated cultures had 191 $\pm$ 5; and β-amyloid + VIP (10^{-11} M)-treated cells had 427 $\pm$ 7; $p < 0.05$, one-way analysis of variance, Student-Neuman-Keuls method). The VIP neuroprotective effect was exhibited over a very narrow range of concentrations.

A NOVEL VIP ANALOGUE ST-NLE-VIP WITH GREATER POTENCY AND EFFICACY THAN VIP

A lipophilic analogue of VIP was devised which has increased potency and bioavailability.[24] Containing a single amino acid substitution (methionine 17 to norleucine 17) and the addition of a fatty acyl moiety (attachment of stearic acid to the N-terminus), the new VIP analogue (St-Nle-VIP) exhibited both a greater potency than VIP and specificity for a VIP receptor associated with neuronal survival in dissociated spinal cord cultures.[25] St-Nle-VIP protected cortical neurons from the toxic effect of the β-amyloid peptide fragment with severalfold greater potency than that observed for VIP exhibiting maximal potency at 10^{-14} M ($p < 0.001$, one-way analysis of variance, Student-Neuman-Keuls method). The neuroprotective action of St-Nle-VIP was efficacious over a broader range of concentrations than that of VIP. In cultures not treated with the β-amyloid peptide, a dose-response curve to St-Nle-VIP indicated a 20% protection against naturally occurring cell death at concentrations of 10^{-11} M of the lipophilic peptide and at 10^{-9} M for VIP ($p < 0.05$, one-way analysis of variance, Student-Neuman-Keuls method).

ST-NLE-VIP PROTECTS AGAINST ALZHEIMER'S-RELATED RETARDATION OF LEARNING AND MEMORY *IN VIVO*

Because a major class of neurons known to be lost in Alzheimer's disease are cholinergic neurons, cholinergic blockade, resulting in impairment of learning and memory, has been employed as a model of this disease. Ethylcholine aziridium (AF64A) is a blocker of choline uptake; it is well established that intracerebroventricular (i.c.v.) administration of this drug can induce loss in cholinergic neurons at the basal forebrain.[26]

For the evaluation of learning and memory abilities, rats were tested in a Morris water maze as previously described.[18,19] The latency (in seconds) of reaching the submerged platform of a circular water pool (the water maze) was recorded for each rat, and the changes over days of training reflected learning and memory. Results indicated that rats treated with the cholinergic blocker exhibited a nine-day delay in their spatial learning capacity in comparison to controls. The ability of St-Nle-VIP to improve learning and memory capacities was thereafter tested in AF64A-treated animals. Daily i.c.v. injections of St-Nle-VIP completely prevented the learning impairment in animals treated with the cholinergic blocker (after seven days of training the AF64A-treated animals were significantly different from both control animals and those treated with AF64A + St-Nle-VIP, $p < 0.001$); furthermore, results from AF64A + St-Nle-VIP–treated rats were similar to those of the control animals, i.e., the animals not exposed to the cholinotoxin. Thus, St-Nle-VIP offers protection against cholinergic deficiencies.

INTRANASAL ADMINISTRATION OF ST-NLE-VIP PROTECTS AGAINST ALZHEIMER'S-RELATED RETARDATION OF LEARNING AND MEMORY

Since St-Nle-VIP was originally designed to cross lipophilic barriers,[24] we also tested the possibility of intranasal administration of the potent VIP analogue as a novel route of lipophilic peptide administration. Initially, the penetration of [^{125}I]St-Nle-VIP into the brain following intranasal administration was evaluated. Results have indicated that 15 minutes after intranasal application intact radioactive St-Nle-VIP was still detected in the brain (although less than 1% of the applied material was actually incorporated).

Administration of St-Nle-VIP intranasally to AF64A-treated animals significantly improved their performance in the Morris water maze in comparison to animals treated with AF64A alone. AF64A animals treated with the VIP analogue (10 μg or 70 μg per day per animal) learned to reach the submerged platform after five days of training ($p < 0.05$), whereas AF64A animals treated with vehicle exhibited similar behavior only after seven days of training. These results indicated an increase in learning and memory capacities associated with St-Nle-VIP. Similar results were obtained in other models exhibiting learning impairments associated with aged rats or animals treated with VIP antagonists. Animals treated with AF64A and St-Nle-VIP exhibited an identical behavioral pattern to vehicle-treated controls. Treatment with St-Nle-VIP alone in control animals did not result in improvement of learning and memory. After nine days of training and testing, the platform was removed and on day 10 the animals were subjected to swimming in a pool without the platform. The time spent in the area, which includes the area where the platform used to be, was recorded (these measurements were designated as a probe trial). It was apparent from the probe trial that there was a dose-dependent effect of St-Nle-VIP on the spatial memory, because the time spent by the rats in the quadrant of the pool where the platform had previously been was significantly increased in the VIP analogue-treated rats. The effect of St-Nle-VIP on the spatial memory was evident both in the control rates ($p < 0.05$) and in the AF64-treated animals ($p < 0.007$). To evaluate for possible motor deficits that could impair performance unrelated to memory, on day 11 of testing the animals were allowed to swim to an exposed platform, and no significant differences were observed among the various treatment groups.

CONCLUSIONS AND FUTURE STUDIES

A novel strategy for Alzheimer's drug treatment was suggested using lipophilic neurotrophic VIP analogues. St-Nle-VIP had the remarkable property that it entered the brain intact after intranasal administration.[21] This property greatly enhances St-Nle-VIP therapeutic potential, because a major obstacle in the use of any neurotrophic substance is the challenge of crossing the blood–brain barrier. These data are consistent with previous observations indicating that topically applied St-Nle-VIP exhibits stability and biological activity in the noninvasive treatment of impotence.[24] Other neuropeptide analogues (agonists and antagonists) have been suggested, in the past, to alleviate learning deficits, for example, vasopressin[27] and galanin.[28] The possibility of a combinatorial drug treatment is intriguing as well as the possibility of lipophilization of other model peptides, which should facilitate their brain entry.

The mechanism through which St-Nle-VIP exerted neuroprotective effects is not yet clear. St-Nle-VIP has been shown to differentiate between two classes of VIP receptors in the central nervous system,[25] preferring the high-affinity site and not recognizing the receptor associated with adenylate cyclase activation. The neurotrophic activity associated with VIP is apparently mediated through high-affinity receptors on glial cells[29] that are linked to the mobilization of calcium[30] and the release of survival-promoting substances.[11,13,31–33] Our working hypothesis is that the neuroprotective action of VIP apparent in the Alzheimer's models presented herein is mediated through these glia-derived molecules. We further speculate that interference with the action of these endogenous molecules is a part of the etiology of Alzheimer's disease. The behavioral model that included the cholinotoxin AF64A is known to induce neuronal cell loss, and specific neuronal cell loss has been associated with learning impairments. Taken together with the ability of St-Nle-VIP to protect against neurotoxicity, one mechanism by which VIP analogues protect against learning and memory deficits may be through the inhibition of accelerated neuronal cell death. Our future studies are aimed at further characterization of the receptor involved in maintaining neuronal survival, as well as in delineating active sites on the VIP molecule. Because VIP exerts its neuroprotection via glial cells, one major breakthrough would be the molecular structure of novel neuroprotective "factors" secreted from glial cells in the presence of VIP.

ACKNOWLEDGMENT

We thank Dr. Haim Leder for the initial sample of AF64A.

REFERENCES

1. GOZES, I. & D. E. BRENNEMAN. 1993. Neuropeptides as growth and differentiation factors in general and VIP in particular. J. Mol. Neurosci. **4:** 1–9.
2. GOZES, I., D. E. BRENNEMAN, G. LILLING, A. DAVIDSON & T. W. MOODY. 1994. Neuropeptide regulation of mitosis. Ann. N.Y. Acad. Sci. **739:** 253–261.
3. PALKOVITS, M. 1995. Neuropeptide messenger plasticity in the CNS neurons following axotomy. Mol. Neurobiol. **10:** 91–103.
4. CARNAHAN, J. & H. NAWA. 1995. Regulation of neuropeptide expression in the brain by neurotrophins. Mol. Neurobiol. **10:** 135–149.
5. HOKFELT, T., X. ZHANG, & Z. WEISENFELD-HALLIN. 1994. Messenger plasticity in primary sensory neurons following axotomy and its functional implications. Trends Neurosci. **17:** 22–30.
6. SCHWARTZ, J. P. & T. TANIWAKI. 1994. Heterogeneity of expression of neuropeptide genes by astrocytes: Functional implications. Perspect. Dev. Neurobiol. **2:** 251–257.
7. MATTSON, M. P. 1994. Calcium and neuronal injury in Alzheimer's disease. Contributions of beta amyloid precursor protein metabolism, free radicals and metabolic compromise. Ann. N.Y. Acad. Sci. **747:** 50–76.
8. YAMAMOTO, K., T. MIYOSHI, T. YAE, K. KAWASHIMA, H. ARAKI, K. HANADA, D. A. OTERO, J. M. ROCH & T. SAITOH. 1994. The survival of rat cerebral cortical neurons in the presence of trophic APP peptides. J. Neurobiol. **25:** 585–594.
9. ROBINSON, L. J., W. LEITNER, B. DRAZNIN & K. A. HEIDENREICH. 1994. Evidence that p21ras mediates the neurotrophic effects of insulin and insulin-like growth factor I in chick forebrain neurons. Endocrinology **135:** 2568–2573.
10. VAUGHAN, P. J., C. J. PIKE, C. W. COTMAN & D. D. CUNNINGHAM. 1995. Thrombin receptor activation protects neurons and astrocytes from cell death produced by environmental insults. J. Neurosci. **15:** 5389–5401.
11. FESTOF, B., P. G. NELSON & D. E. BRENNEMAN. 1996. Prevention of activity-dependent neuronal death: Vasoactive intestinal polypeptide stimulates astrocytes to secrete the thrombin-inhibiting, neurotrophic serpin, protease nexin I. J. Neurobiol. **30:** 255–266.
12. SHEN, Y., J. A. HALPERIN & C. M. LEE. 1995. Complement-mediated neurotoxicity is regulated by homologous restriction. Brain Res. **671:** 282–292.
13. BRENNEMAN, D. E., M. SCHULTZBERG, T. BARTFAI & I. GOZES. 1992. Cytokine regulation of neuronal survival. J. Neurochem. **58:** 454–460.
14. BRENNEMAN, D. E., & L. E. EIDEN. 1986. Vasoactive intestinal peptide and electrical activity influence neuronal survival. Proc. Natl. Acad. Sci. USA **83:** 1159–1162.
15. GOZES, I. & D. E. BRENNEMAN. 1989. VIP molecular biology and neurobiological function. Mol. Neurobiol. **3:** 201–236.
16. GRESSENS, P., J. M. HILL, I. GOZES, M. FRIDKIN & D. E. BRENNEMAN. 1993. Growth

factor function of vasoactive intestinal peptide in whole cultured mouse embryos. Nature **362:** 155–158.

17. BALDINO, F., JR., S. FITZPATRICK-MCELLIGOTT, I. GOZES & J. P. CARD. 1989. Localization of VIP and PHI-27 messenger RNA in rat thalamic and cortical neurons. J. Mol. Neurosci. **1:** 199–207.

18. GLOWA, J. R., L. V. PANLILIO, D. E. BRENNEMAN, I. GOZES, M. FRIDKIN & J. M. HILL. 1992. Learning impairment following intracerebral administration of the HIV envelope protein gp120 or a VIP antagonist. Brain Res. **570:** 49–53.

19. GOZES, I., J. GLOWA, D. E. BRENNEMAN, S. K. MCCUNE, E. LEE & H. WESTPHAL. 1993. Learning and sexual deficiencies in transgenic mice carrying a chimeric vasoactive intestinal peptide gene. J. Mol. Neurosci. **4:** 185–193.

20. BRENNEMAN, D. E., G. L. WESTBROOK, S. P. FITZGERALD, D. L. ENNIST, K. L. ELKINS, M. R. RUFF & C. B. PERT. 1988. Neuronal cell killing by the envelope protein of HIV and its prevention by vasoactive intestinal peptide. Nature **335:** 639–642.

21. GOZES, I., A. BARDEA, A. RESHEF, R. ZAMOSTIATNO, S. ZHUKOVSKY, S. RUBINRAUT, M. FRIDKIN & D. E. BRENNEMAN. 1996. Novel neuroprotective strategy for Alzheimer's disease: Inhalation of a fatty neuropeptide. Proc. Natl. Acad. Sci. USA **93:** 427–432.

22. YANKNER, B. A., I. K. DUFFY & D. A. KIRSCHNER. 1990. Neurotrophic and neurotoxic effects of amyloid beta protein: Reversal by tachykinin neuropeptides. Science **250:** 279–282.

23. PIKE, C. J., D. BURDICK, A. J. WALENCEWICZ, C. G. GLABE & C. W. COTMAN. 1993. Neurodegeneration induced by beta amyloid peptides in vitro: The role of peptide assembly state. J. Neurosci. **13:** 1676–1687.

24. GOZES, I., A. RESHEF, D. SALAH, S. RUBINRAUT & M. FRIDKIN. 1994. Stearyl-norleucine-VIP, a novel VIP analogue for non-invasive impotence treatment. Endocrinology **134:** 2121–2125.

25. GOZES, I., G. LILLING, R. GLAZER, A. TICHER, I. E. ASHKENAZI, A. DAVIDSON, S. RUBINRAUT, M. FRIDKIN & D. E. BRENNEMAN. 1995. Superactive lipophilic peptides discriminate multiple vasoactive intestinal peptide receptors. J. Pharmacol. Exp. Ther. **273:** 161–167.

26. FISHER, A., R. BRANDEIS, Z. PITTEL, I. KARTON, M. SAPIR, S. DACHIR, A. LEVY & E. HELDMAN. 1989. (+−)-cis-2-methyl-spiro (1.3-oxathiolane-5.3′) quinuclidine (AF102B): A new M1 agonist attenuates cognitive dysfunctions in AF64A-treated rats. Neurosci. Lett. **102:** 325–331.

27. PITMAN, R. K., S. P. ORR & N. B. LASKO. 1993. Effects of intranasal vasopressin and oxytocin on physiologic responding during personal combat imagery in Vietnam veterans with posttraumatic stress disorder. Psychiatry Res. **48:** 107–117.

28. CRAWLEY, J. N. 1993. Functional interactions of galanin and acetylcholine: Relevance to memory and Alzheimer's disease. Behav. Brain Res. **57:** 133–141.

29. GOZES, I., S. K. MCCUNE, L. JACOBSON, D. WARREN, T. W. MOODY, M. FRIDKIN & D. E. BRENNEMAN. 1991. An antagonist to vasoactive intestinal peptide: Effects on cellular functions in the central nervous system. J. Pharmacol. Exp. Ther. **257:** 959–966.

30. FATATIS, A., L. A. HOLTZCLAW, R. AVIDOR, D. E. BRENNEMAN & J. T. RUSSELL. 1994. Vasoactive intestinal peptide increases intracellular calcium in astroglia: Synergism with alpha-adrenergic receptors. Proc. Natl. Acad. Sci. USA **91:** 2036–2040.

31. BRENNEMAN, D. E., E. A. NEALE, G. A. FOSTER, S. W. D'AUTREMONT & G. L. WESTBROOK. 1987. Non-neuronal cells mediate neurotrophic action of VIP. J. Cell Biol. **104:** 1603–1610.

32. BRENNEMAN, D. E., T. NICOL, D. WARREN & L. M. BOWERS. 1990. Vasoactive intestinal peptide: A neurotrophic releasing agent and an astroglial mitogen. J. Neurosci. Res. **25:** 386–394.

33. BRENNEMAN, D. E. & I. GOZES. 1996. A femtomolar-acting neuroprotective peptide. J. Clin. Invest. **97:** 2299–2307.

Identity of Neurotrophic Molecules Released from Astroglia by Vasoactive Intestinal Peptide[a]

DOUGLAS E. BRENNEMAN,[b,f] TERRY M. PHILLIPS,[c]
BARRY W. FESTOFF,[d] AND ILLANA GOZES[e]

[b]Section on Developmental and Molecular Pharmacology
National Institute of Child Health and Human Development
National Institutes of Health
Bethesda, Maryland 20892

[c]Department of Medicine
George Washington University School of Medicine
Washington, D.C.

[d]Department of Neurology
University of Kansas Medical Center
Kansas City, Kansas

[e]Department of Clinical Biochemistry
Sackler School of Medicine
Tel Aviv University
Tel Aviv, Israel

INTRODUCTION

Vasoactive intestinal peptide (VIP), a 28 amino acid neuropeptide, has neurotrophic actions on developing neurons in the central nervous system (CNS). Previous studies have shown that 1–100 pM VIP increases the survival of electrically blocked spinal cord neurons.[1] Further substantiating a neurotrophic role for VIP are studies that show that blocking the action of VIP with neutralizing antisera to VIP or with VIP receptor antagonists[1,2] results in neuronal cell death for a subpopulation of CNS neurons in spinal cord, hippocampus or cerebral cortex. However, VIP does not act directly on neurons to increase their survival. Rather, the survival-enhancing action of VIP is contingent on the presence of astroglia.[3,4] The mode of action for neurotrophism therefore resides in VIP's secretagogue action: releasing other substances from these CNS support cells. The identification of some of these secreted substances has been achieved. The purpose of this paper is to summarize our knowledge of these VIP-related molecules and to propose how these substances may play important roles in neurodevelopment and provide potential lead compounds for drug design.

[a] This research was supported in part by the U.S.–Israel Binational Science Foundation. This paper was written while Prof. Illana Gozes was a scholar-in-residence at the Fogarty International Center for Advanced Studies in the Health Sciences, National Institutes of Health, Bethesda, Maryland.
[f] Address correspondence to Dr. Douglas E. Brenneman, Chief, SDMP/LDN/NICHD, Building 49, Room 5A38, National Institutes of Health, Bethesda, Maryland 20892.

VIP AND INTERLEUKIN-1α

Cytokines are now recognized as key regulators of development.[5,6] Indeed, the distinction between cytokines and the classical growth factors is becoming increasingly one of historical rather than functional significance. The neurotrophic actions of cytokines are distinct from those involved in immune and inflammatory responses of the mature nervous system. Many of the cytokines are found in early embryonic development, and a growing number of these molecules have mitogenic, survival-promoting, and differentiating actions on the developing nervous system.[7–9]

Previous studies from our laboratory indicated that interleukin-1α (IL-1α) had a survival-promoting action in developing spinal cord cultures derived from fetal mice.[10] Like VIP, the requirement of some neurons for IL-1α was only apparent when the action of this cytokine was interrupted. Three observations supported the conclusion that IL-1α was essential for the survival of some CNS neurons. First, if neutralizing antiserum to IL-1α was added to developing spinal cord cultures, about 30% of the neurons died within five days.[10] Preimmune serum had no detectable effect on neuron viability. Furthermore, the cell death produced by anti-IL-1α could be prevented by co-treatment with recombinant IL-1α. Treatment of one-month-old spinal cord cultures with IL-1α antisera produced no detectable effect on neuronal survival, demonstrating the developmental specificity of this action. The second observation supporting a neurotrophic role of IL-1α was the killing action of antisera to the type 1 IL-1 receptor.[11] Antiserum to the IL-1 receptor produced neuronal cell death in about 30% of the neurons, an effect that also was dependent on the age of culture. A third observation supporting a neurotrophic action of IL-1α was its action in electrically blocked spinal cord cultures. Treatment of spinal cord cultures with tetrodotoxin blocks the spontaneous action potentials that all these neurons exhibit in cell culture.[12] Importantly, when the synaptic activity is blocked, the release of neurotransmitters and neuropeptides, including VIP, is also inhibited.[13] Thus, by blocking electrical activity, one effectively prevents the release of the IL-1 secretagogue, VIP. Neuronal cell death produced by electrical blockade in spinal cord cultures was prevented by picomolar concentrations of IL-1α. Treatment of electrically active spinal cord cultures with IL-1α produced no detectable effect on neuronal survival. The fundamental concept that emerged is that IL-1α, similar to other glia-derived substances, is present in effective concentrations in electrically active cultures, with the addition of more cytokine resulting in no further survival effects. The obligatory presence of IL-1α is demonstrated only upon pharmacological interference with its biological action. Although IL-1α has these trophic effects on developing neurons, this cytokine has an opposite biological action in one-month-old cultures. Treatment with recombinant IL-1α in mature cultures produces a dose-dependent decrease in neuronal survival, indicating that the biological action of this cytokine is highly dependent on the stage of neurodevelopment.[11]

Neuronal cell death produced by anti-IL-1α was prevented by co-treatment with VIP.[10] This protective effect of VIP was hypothesized to be a result of VIP-induced secretion of a IL-1α-like substance. As shown in FIGURE 1, VIP can release IL-1α from astrocytes derived from rat cerebral cortical astrocyte cultures.[11] This release is rapid, with significant increases observed within 10 min of treatment with 0.1 nM VIP. Concentration-effect studies indicate that physiologically relevant amounts of VIP (1–100 pM) release significant amounts of IL-1α.[11] These data are consistent with the model that IL-1α is among the neurotrophic molecules secreted from astroglia by VIP.

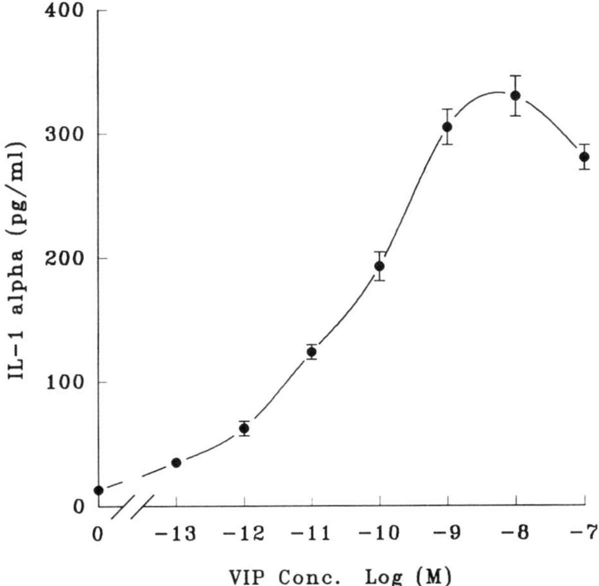

FIGURE 1. Release of interleukin-1α (IL-1α) immunoreactivity into the medium of VIP-stimulated astroglial cultures. Cells (80% confluent) were incubated in phosphate-buffered saline for one hour in the presence of 2 μg/mL leupeptin. The medium was centrifuged and then analyzed for IL-1α by capillary electrophoresis followed by chemiluminescence-based ELISA.[30] (Figure redrawn from Brenneman *et al.*[11])

VIP AND PROTEASE NEXIN I

A protease inhibitor (cathepsin G inhibitor, α_1-antichymotrypsin) of the serpine superfamily has been reported to increase the survival of cultured neurons.[14] Several observations prompted our examination of a possible relationship between VIP and another serine protease inhibitor, protease nexin I (PNI), as a potential component of VIP-induced neurotrophic milieu produced from glia. Motor neuron survival was increased in the presence of minute amounts from an extract of adult chicken sciatic nerve.[15] PNI was shown to be increased after sciatic nerve crush.[16] Preliminary studies indicated that PNI was present in glial cultures and that a PNI-like substance was present in the conditioned medium of VIP-stimulated astroglia.[17] Although PNI has been shown to stimulate neurite outgrowth on hippocampal neurons, previous investigations indicated PNI had no apparent effect on neuronal survival.[18] As described above for IL-1α, our approach for assessing the role of PNI in VIP neurotrophism was to assume that its importance would be apparent only upon interference with PNI action in a biologically relevant system.

Treatment of electrically blocked spinal cord cultures with PNI resulted in an increase in neuronal survival (FIG. 2). This effect of PNI was apparent over a very narrow range of concentrations, and like VIP, exhibited an attenuation of the survival-promoting response at higher concentrations.[19] Similar to IL-1α, treatment of electrically active cultures with PNI resulted in no significant increases

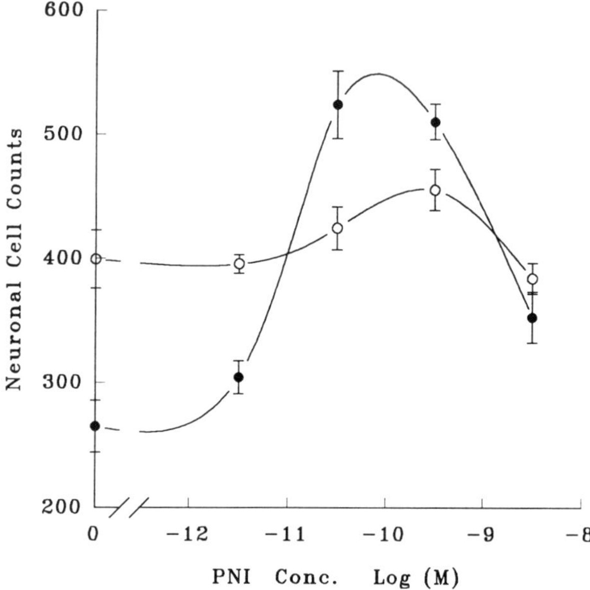

FIGURE 2. Protease nexin I (PNI) increases neuronal survival in spinal cord–dorsal root ganglion cultures blocked with tetrodotoxin. Purified PNI was incubated for five days in cultures blocked with 1 μM tetrodotoxin (*closed circles*) or in electrically active cultures (*open circles*). Significant increases in neuronal cell counts were observed at concentrations of PNI between 0.3 and 3 nM in electrically blocked cultures ($p < 0.05$). No significant changes in cell counts were found with PNI treatment of electrically active cultures. Each point is the mean of 3–5 determinations $\pm$ the standard error. (Redrawn from Festoff, Nelson & Brenneman.[19])

in neuronal survival (FIG. 2). Further studies indicated that anti-PNI immunoprecipitated a 47-kDa PNI-like substance from conditioned medium of VIP-stimulated astroglia cultures.[19] The 47-kDa protein was not detectable in the medium of control astroglia cultures. The survival-promoting action of PNI is probably due to the inhibition of thrombin-like activity. In this regard, a recent study indicated that thrombin produced apoptosis in neurons and glia at nanomolar concentrations.[20] Together, these data indicate that VIP can release PNI from astroglia and that this serine protease inhibitor is another "passive glial factor" that is necessary for the survival of a subpopulation of neurons in the CNS.

VIP AND ACTIVITY-DEPENDENT NEUROTROPHIC FACTOR

In addition to the recognized substances that contribute to the VIP-induced neurotrophic milieu from glia, novel proteins were sought that could mediate VIP's effect on neuronal survival.[21,22] For these neurotrophic factor isolation experiments, proteins released from astroglia were collected after three hour incubations with 0.1 nM VIP. Sequential chromatographic separations (including

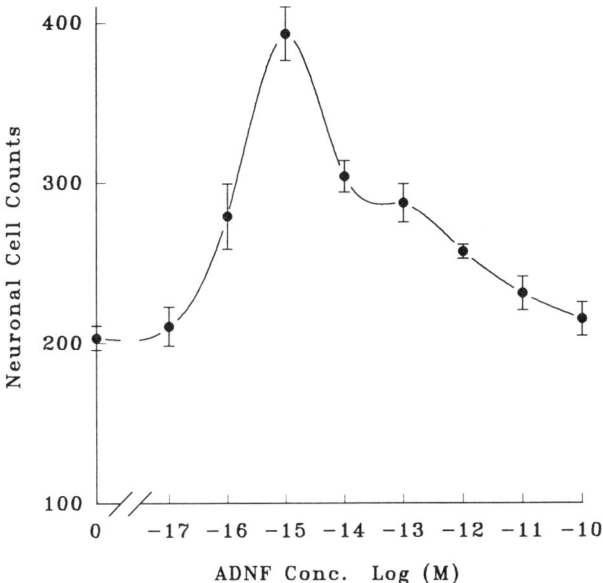

FIGURE 3. Purified activity-dependent neurotrophic factor (ADNF) increases survival of electrically blocked cerebral cortical neurons in dissociated cultures. ADNF was incubated with cerebral cortical cultures for five days in the presence of 1 μM tetrodotoxin. Neuronal cell counts released significant increases at concentrations $>10^{-16}$ M to 10^{-12} M ($p < 0.05$). Each point is the mean of 3–5 determinations $\pm$ the standard error. (Redrawn from Brenneman and Gozes.[22])

anionic exchange, size exclusion, and hydrophobic interaction) resulted in a purified protein, as assessed by SDS polyacrylamide gel electrophoresis. The isolated protein was named activity-dependent neurotrophic factor (ADNF), because its biological activity was detected in spinal cord cultures that were electrically blocked with tetrodotoxin. ADNF has an apparent molecular mass of 14 kDa and a pI of 8.1 + 0.25. It appears to be a monomeric protein that exhibits extraordinary potency in protecting neurons from cell death associated with electrical blockade (FIG. 3). In cerebral cortical cultures, the EC50 for the survival-promoting action was in subfemtomolar concentrations, amounts that are at least 100 times more potent than the biological activities of previously described growth factors or cytokines.[22] Upon sequencing, ADNF exhibited N-terminal blockade. Analysis of protease digests of ADNF has indicated amino acid sequence homology with one of the stress proteins: heat shock protein 60. These data suggest that ADNF is an extracellular stress protein that has remarkable neuroprotective properties for a subpopulation of CNS neurons. Furthermore, these data indicate that ADNF, like IL-1α and PNI, is an important component of the neurotrophic environment produced by astrocytes stimulated with VIP. Although all three molecules are apparently essential to the survival of a group of neurons, ADNF is the only substance known to work at femtomolar concentrations, making it the most potent neuroprotective factor thus far reported.

SUMMARY

Subnanomolar concentrations of VIP elicit a survival-producing action in CNS cultures composed of both astroglia and neurons. This neurotrophic action is mediated by a complex array of substances released by VIP from astrocytes. Included in this glial protein mixture is a cytokine (interleukin-1α), a serine protease inhibitor (protease nexin I), and an extracellular stress protein (activity-dependent neurotrophic factor). In dissociated spinal cord cultures, all of these substances exhibit neuroprotection from neuronal cell death produced by electrical blockade with tetrodotoxin. All these substances produce neuronal cell death when test cultures are treated with neutralizing antisera to any one of them. They are all apparently necessary for the survival of a subpopulation of neurons that are dependent on spontaneous, excitatory neurotransmission. Our view is that these substances are components of a glia-derived environment that regulates, together with target-derived growth factors, the survival fate of developing neurons. In addition, it is our belief that some of these glia-derived substances will be found to have active roles in the injury response to the CNS or in the regulation of VIP-mediated growth in other tissues. Drugs based on these substances may provide therapeutic agents for the treatment of neurodegeneration[22–25] and tumor growth.[26–29]

REFERENCES

1. BRENNEMAN, D. E. & L. E. EIDEN. 1986. Vasoactive intestinal peptide and electrical activity influence neuronal survival. Proc. Natl. Acad. Sci. USA **83:** 1159–1162.
2. GOZES, I., S. K. MCCUNE, L. JACOBSON, D. WARREN, T. W. MOODY, M. FRIDKIN & D. E. BRENNEMAN. 1991. An antagonist to vasoactive intestinal peptide affects cellular functions in the central nervous system. J. Pharmacol. Exp. Ther. **257:** 959–966.
3. BRENNEMAN, D. E., E. A. NEALE, G. A. FOSTER, S. W. D'AUTREMONT & G. L. WESTBROOK. 1987. Nonneuronal cells mediate neurotrophic action of vasoactive intestinal peptide. J. Cell. Biol. **104:** 1603–1610.
4. BRENNEMAN, D. E., T. NICOL, D. WARREN & L. M. BOWERS. 1990. Vasoactive intestinal peptide: A neurotrophic releasing agent and an astroglial mitogen. J. Neurosci. Res. **25:** 386–394.
5. ROBERTSON, S. A. & R. F. SEAMARK. 1992. Granulocyte-macrophage colony stimulating factor (GM-CSF): One of a family of epithelial cell-derived cytokines in the preimplantation uterus. Reprod. Fertil. Dev. **4:** 435–448.
6. OTERO, G. C. & J. E. MERRILL. 1994. Cytokine receptors on glial cells. Glia **11(2):** 117–128.
7. KUSHIMA, Y., T. HAMA & H. HATANAKA. 1992. Interleukin-6 as a neurotrophic factor for promoting the survival of cultured catecholaminergic neurons in a chemically defined medium from fetal and postnatal rat midbrains. Neurosci. Res. **13:** 267–280.
8. RICHARDSON, P. M. 1994. Ciliary neurotrophic factor: A review. Pharmacol. Ther. **63:** 187–198.
9. MORONI, S. C. & A. ROSSI. 1995. Enhanced survival and differentiation in vitro of different neuronal populations by some interleukins. Int. J. Dev. Neurosci. **13(1):** 41–49.
10. BRENNEMAN, D. E., M. SCHULTZBERG, T. BARTFAI & I. GOZES. 1992. Cytokine regulation of neuronal survival. J. Neurochem. **58:** 454–460.
11. BRENNEMAN, D., J. M. HILL, G. W. GLAZNER, I. GOZES & T. M. PHILLIPS. 1995. Interleukin-1 alpha and vasoactive intestinal peptide: Enigmatic regulation of neuronal survival. Int. J. Dev. Neurosci. **13:** 187–200.

12. JACKSON, M. B., H. LECAR, D. E. BRENNEMAN, S. FITZGERALD & P. G. NELSON. 1982. Electrical development in spinal cord culture. J. Neurosci. **2:** 1052–1061.

13. BRENNEMAN, D. E., L. E. EIDEN & R. E. SIEGEL. 1985. Neurotrophic action of VIP on spinal cord cultures. Peptides **6(Suppl. 2):** 35–39.

14. MIZUGUCHI, M. & S. U. KIM. 1991. Alpha 1-antichymotrypsin supports short-term survival of cerebral neurons in culture. Neurosci. Lett. **124:** 166–168.

15. POPIELA, H., T. PORTER, R. L. BEACH & B. W. FESTOFF. 1984. Peripheral nerve extract promotes long-term survival and neurite outgrowth in cultured spinal cord neurons. Cell. Mol. Neurobiol. **4:** 67–77.

16. SMIRNOVA, I. V., G. J. HO, J. W. I. FENTON & B. W. FESTOFF. 1994. Extravascular proteolysis and the nervous system: Serine protease/serpin balance. Semin. Thromb. Hemostasis **20:** 426–432.

17. FESTOFF, B. W., J. S. RAO & D. E. BRENNEMAN. 1990. Vascoactive intestinal polypeptide (VIP) is a secretatogue for protease nexin I (PNI) release from astrocytes. Neurosci. Abstr. **17:** 909.

18. MONARD, D. D., K. STOCKEL, R. GOODMAN & H. THOENEN. 1975. Distinction between nerve growth factors and glial factors. Nature **258:** 444–445.

19. FESTOFF, B. W., P. G. NELSON & D. E. BRENNEMAN. 1996. Inhibition of activity-dependent neuronal death: Vasoactive intestinal polypeptide stimulates secretion of the neurotrophic serpin. J. Neurobiol. **30:** 255–266.

20. VAUGHAN, P. J., C. J. PIKE, C. W. COTMAN & D. D. CUNNINGHAM. 1994. The protease thrombin induces morphological degeneration and cell death in cultures of astrocytes and neurons. Soc. Neurosci. Abstr. **20:** 641.

21. BRENNEMAN, D. E. & I. GOZES. 1992. Vasoactive intestinal peptide: A secretagogue for permissive glia-derived growth factors. Int. Soc. Dev. Neurosci. **10(Suppl. 1):** 44.

22. BRENNEMAN, D. E. & I. GOZES. 1996. A femtomolar-acting neuroprotective peptide. J. Clin. Invest. **97:** 2299–2307.

23. GOZES, I., G. LILLING, R. GLAZER, A. TICHER, I. E. ASHEKNAZI, A. DAVIDSON, S. RUBINRAUT, M. FRIDKIN & D. E. BRENNEMAN. 1995. Superactive lipophilic peptides discriminate multiple VIP receptors. J. Pharmacol. Exp. Ther. **273:** 161–167.

24. BRENNEMAN, D. E. & I. GOZES. 1993. Prevention of gp120-associated neuronal cell death by activity-dependent neurotrophic factor. Soc. Neurosci. Abstr. **19:** 659.

25. GOZES, I., A. BARDEA, A. RESHEF, R. ZAMOSTIANO, S. ZHUKOVSKY, S. RUBINRAUT, M. FRIDKIN & D. E. BRENNEMAN. 1996. Novel neuroprotective strategy for Alzheimer's disease: Inhalation of a fatty neuropeptide. Proc. Natl. Acad. Sci. USA **93:** 427–432.

26. MOODY, T. W., F. ZIA, M. DRAOUI, D. E. BRENNEMAN, M. FRIDKIN, A. DAVIDSON & I. GOZES. 1993. A novel VIP antagonist inhibits non-small cell lung cancer growth. Proc. Natl. Acad. Sci. USA **90:** 4345–4349.

27. BRENNEMAN, D. E., G. L. WESTBROOK, S. P. FITZGERALD, D. L. ENNIST, K. L. ELKINS, M. R. RUFF & C. B. PERT. 1988. Neuronal cell killing by the envelope protein of HIV and its prevention by vasoactive intestinal peptide. Nature **335:** 639–642.

28. MOODY, T. W., F. ZIA, A. L. GOLDSTEIN, P. H. NAYLOR, E. SARIN, D. E. BRENNEMAN, A. M. C. KOROS, J. C. REUBI, L. Y. KORMAN, M. FRIDKIN & I. GOZES. 1992. VIP analogues inhibit small cell lung cancer growth. Biomedical Res. **13(Suppl. 2):** 131–135.

29. LILLING, G., Y. WOLLMAN, M. N. GOLDSTEIN, S. RUBINRAUT, M. FRIDKIN, D. E. BRENNEMAN & I. GOZES. 1995. Inhibition of human neuroblastoma growth by a specific VIP antagonist. J. Mol. Neurosci. **5:** 231–239.

30. PHILLIPS, T. M. & P. I. KIMMELL. 1994. High-performance capillary electrophoretic analysis of inflammatory cytokines in human biopsies. J. Chromatogr. **656:** 259–266.

Growth of the Early Postimplantation Embryo

Regulation by High-Affinity, GTP-Insensitive VIP Receptors

JOANNA M. HILL,[a,c] PIERRE GRESSENS,[b] AND
DOUGLAS E. BRENNEMAN[a]

[a]Section on Developmental and Molecular Pharmacology
Laboratory of Developmental Neurobiology
National Institutes of Health
Bethesda, Maryland 20892

[b]Laboratoire de Neurologie du Développement, and
Service Neuropédiatrie
Hôpital Robert-Debri
Paris, France

INTRODUCTION

Vasoactive intestinal peptide (VIP) is a 28 amino acid neuropeptide with a wide variety of functions both in the central and peripheral nervous systems. Many of the actions of VIP are indirect, occurring through the VIP-stimulated release of other factors.[1] VIP is known to function through cAMP-mediated mechanisms, and adenylate-cyclase linked VIP receptors have been cloned[2,3] and were reported to be widespread throughout the central nervous system.[3] However, evidence exists that VIP acts through a site linked to an alternate second messenger system. GTP sensitivity, resulting in the separation of ligand from receptor, is characteristic of receptors linked to adenylate cyclase;[4] however, GTP-insensitive VIP binding has been described in brain and several other tissues.[5,6] In addition, in astrocyte cultures, a high-affinity non-cyclic-AMP-linked VIP binding site has been identified which is associated with glial mitosis as well as the secretory and neurotrophic actions of VIP.[7–9] Subnanomolar concentrations of VIP have been shown to stimulate the accumulation of intracellular calcium, increase inositol phosphates in astrocyte cultures,[10] and produce translocation of selected protein kinase C isozymes,[11] suggesting that VIP may act through a phospholipid/PKC signal transduction mechanism.

Recent studies have shown that VIP regulates growth of the early postimplantation rodent embryo. VIP has been shown to dramatically stimulate the growth of E9.5 whole-cultured mouse embryos.[12] Bromodeoxyuridine incorporation showed an increase in the number of neuroepithelial cells in S-phase in VIP-treated embryos, indicating that VIP-induced growth occurred through a stimulation of mitosis. Treatment of pregnant mice during days E9–E11 with a VIP antagonist resulted

[c]Address correspondence to Dr. Joanna M. Hill, LDN/NICHD, Building 49, Room 5A38, 9000 Rockville Pike, Bethesda, MD 20892.

in growth retardation accompanied by microcephaly.[13] Treatment after E11 did not inhibit growth. These studies suggested that VIP was necessary for normal embryonic growth and that the period of its growth-regulating action in the rodent was limited to a three-day early postimplantation period.

Although VIP receptors were present in the embryonic nervous system during this period, neither *in situ* hybridization histochemistry nor reverse transcriptase–polymerase chain reaction revealed the mRNA for VIP in the early embryo.[14] However, examination of maternal serum during pregnancy revealed that levels of VIP were raised 6 to 10 times baseline levels during the early postimplantation period in the rat (E10–E12), a time coinciding with the period during which embryonic growth was shown to be regulated by VIP.[14] (The developmental events occurring during E9–E11 in the mouse occur during E10–E12 in the rat embryo.) Further studies showed that the E11 rat embryo had four times the concentration of VIP found in the E17 embryo. In addition, radiolabeled VIP injected into the uterine vasculature of the pregnant mouse appeared in the embryo with little degradation.[14] These data suggest that maternal VIP acts on embryonic receptors to regulate growth.

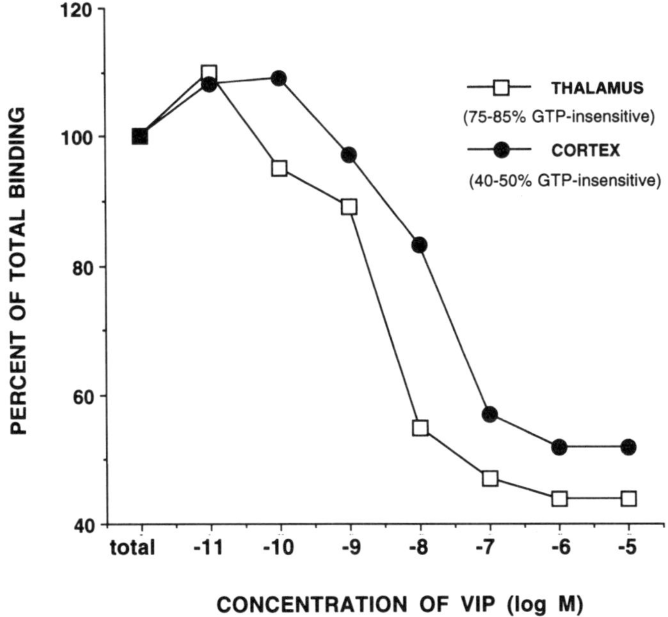

FIGURE 1. Comparison of displacement of [^{125}I]VIP from adjacent adult rat brain sections with a 10^{-11} M to 10^{-5} M nonradiolabeled VIP in a brain region with moderate GTP-insensitive binding (cortex) with that of a brain region with high levels of GTP-insensitive binding (thalamus). VIP binding was performed as previously described.[5,6,14] The region with the greater number of GTP-insensitive binding sites (thalamus) expressed the higher affinity for VIP, suggesting that GTP-insensitive binding sites had a higher affinity for VIP than the GTP-sensitive sites. Each point is the mean of six measurements from two different animals. The standard errors were between 2 and 7% of the mean.

TABLE 1. Physiological Parameters Characterizing VIP-sensitive Period of Embryonic Growth Compared with Later Development

VIP-dependent Embryonic Growth (Mouse E9–E11; Rat E10–E12)	VIP-independent Embryonic Growth (after Mouse E12; after Rat E13)
• Elevated maternal serum VIP	• Baseline maternal VIP
• Elevated embryonic VIP	• Lower levels of embryonic VIP
• Absence of embryonic VIP mRNA	• VIP mRNA in embryonic sphenopalatine ganglia, aorta, and intestine[6]
• VIP binding sites only in central nervous system	• Appearance of peripheral VIP binding sites
• Low to moderate VIP binding	• High VIP binding in central nervous system
• GTP-insensitive binding sites	• Both GTP-sensitive and GTP-insensitive binding sites

GTP-SENSITIVE AND GTP-INSENSITIVE VIP BINDING

Both GTP-sensitive and GTP-insensitive VIP binding sites have been revealed by *in vitro* autoradiography with [125I]VIP in the presence of a nonhydrolyzable guanosine 5′-triphosphate analogue, guanylyl-imidodiphosphate (GMP-PNP, 10^{-5} M).[5,6] In most brain regions, including the neocortex, GMP-PNP reduced VIP binding between 40 and 60%, showing that both receptor types were about equally abundant. However, in some regions, such as the supraoptic nucleus, locus coeruleus, interpeduncular nucleus, olfactory tubercle, and periventricular hypothalamic nucleus, the GTP-sensitive type was most abundant, and 80% or more of VIP binding was inhibited. In other brain regions, including the medial geniculate, olfactory bulb, and ventral thalamic nuclei, GTP had little effect on VIP binding.[5] These studies showed that the ratio of GTP-sensitive to GTP-insensitive binding sites differed from region to region in the brain, with the thalamus and olfactory bulb being especially enriched in the GTP-insensitive receptor. *In vitro* autoradiography with [125I]VIP was performed on adjacent adult rat brain sections in the presence of increasing concentrations of nonradiolabeled VIP and revealed differences in the affinity of VIP in different brain regions. The affinity of VIP for the binding site was related to the enrichment of the brain region in GTP-insensitive sites, with the GTP-insensitive enriched regions having a higher affinity for VIP than those with a greater proportion of GTP-sensitive sites (FIG. 1). For example, in the thalamus, 75–85% of the binding sites are GTP-insensitive, and the IC_{50} was approximately 5×10^{-8} M VIP, whereas in the cortex where only 40–50% for the binding is GTP-insensitive, the IC_{50} was approximately 5×10^{-7} M (FIG. 1). These data suggest that GTP-insensitive sites have a higher affinity for VIP than the GTP-sensitive receptors.

FIGURE 2. Autoradiography of [125I]VIP binding in E11 **(A)** and E12 **(B)** rat embryos. VIP binding was performed as previously described[5,6,14] with incubation in [125I]VIP (total) occurring with or without 10^{-7} M nonradiolabeled VIP (specific binding) or 10^{-5} M GTP analogue (GMP-PNP) (displacing VIP at GTP-sensitive VIP binding sites). Bar = 1.0 mm.

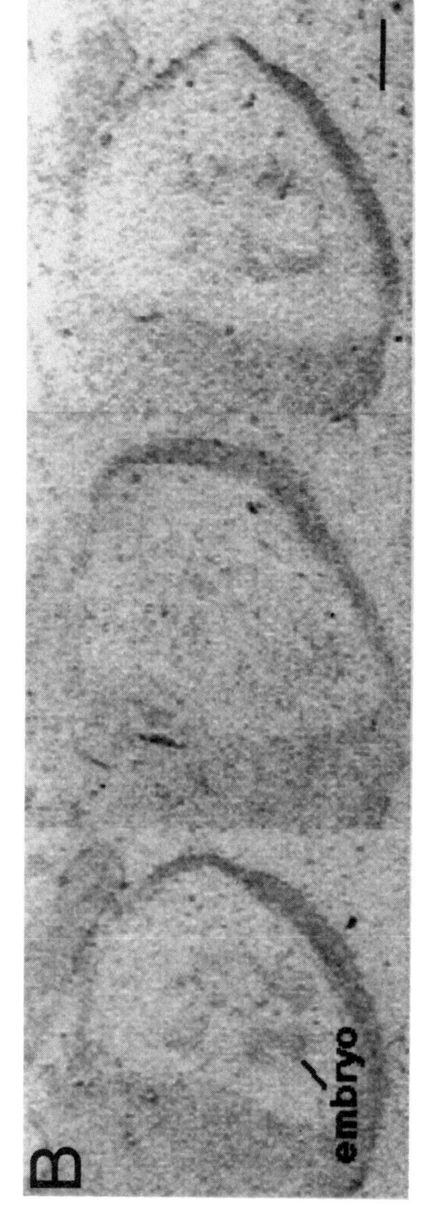

FIGURE 2

VIP BINDING SITES DURING EMBRYONIC DEVELOPMENT

During development of the rat from embryonic day 14 (E14) to adult, both GTP-sensitive and GTP-insensitive binding sites were shown to be abundant, to exhibit distribution patterns related to ontogenic events, and, early in development, to be primarily localized to the central nervous system.[6]

In further studies, *in vitro* autoradiography with [125I]VIP was used to assess the VIP binding sites on rat embryos both during the period of VIP-regulated growth, during which maternal serum levels of VIP were significantly elevated (E10–E12), and after this period (E13–E14).[14] An accurate assessment of VIP binding could not be made from autoradiographs of E10 embryos (3 to 8 somites) because of their small size. In the older embryos, [125I]VIP bound to receptors throughout the brain and spinal cord (FIGS. 2, 3, and 4, left panels) and the addition of 1 μM unlabeled VIP to the incubation medium reduced [125I]VIP binding to almost background levels (FIGS. 2, 3, and 4, center panels). In the E11 and E12 embryos, low to moderately dense specific VIP binding was seen only in the brain and spinal cord (FIG. 2). This binding was GTP-insensitive, because none was displaced with GTP (FIG. 2, right panel). In the E13 embryo (FIG. 3), VIP binding sites were also limited to the nervous system; however, binding sites were approximately triple the density seen at E11 and E12. Within the possible optical density range of 0 to 255, the E13 spinal cord had an optical density of 99 $\pm$ 4.2 (mean $\pm$ SEM) compared with a density of 38 $\pm$ 4.5 for E11 spinal cord, ANOVA ($p < 0.001$). About 50% of the binding in the E13 spinal cord was displaced with GTP (FIG. 3, right panel). As previously described,[6] this pattern continued in the E14 spinal cord (FIG. 4), where most VIP binding was GTP-sensitive and displaced with GTP. The GTP-insensitive binding was localized to the floor and roof plates (FIG. 4, right panel). These studies revealed that during the early postimplantation period, only GTP-insensitive binding sites were present, but that immediately after the fall of VIP in maternal serum, VIP binding sites became abundant and both GTP-sensitive and GTP-insensitive binding sites were present.

We have found a combination of distinct physiological events that characterize the early postimplantation period during which VIP regulates embryonic growth (TABLE 1). Following this period, change is seen in all of these parameters. Although some of them have been only assessed in the mouse and others only in the rat, if these measures are typical of developing rodents, these data indicate that VIP plays an important role in the regulation of growth during a short ontogenic period. Further, the data indicate that VIP-induced growth occurs through high-affinity, GTP-insensitve binding sites which are localized to the embryonic neuroepithelium. Many VIP functions have been

---

FIGURE 3. Autoradiography of [125I]VIP binding in E13 rat embryos. VIP binding was performed as previously described[5,6,14] with incubation in [125I]VIP (total) occurring with or without 10^{-7} M nonradiolabeled VIP (specific binding) or 10^{-5} M GTP analogue (GMP-PNP) (displacing VIP at GTP-sensitive VIP binding sites). Bar = 1.6 mm.

FIGURE 4. Autoradiography of [125I]VIP binding in E14 rat embryos. VIP binding was performed as previously described[5,6,14] with incubation in [125I]VIP (total) occurring with or without 10^{-7} M nonradiolabeled VIP (specific binding) or 10^{-5} M GTP analogue (GMP-PNP) (displacing VIP at GTP-sensitive VIP binding sites). Bar = 1.4 mm.

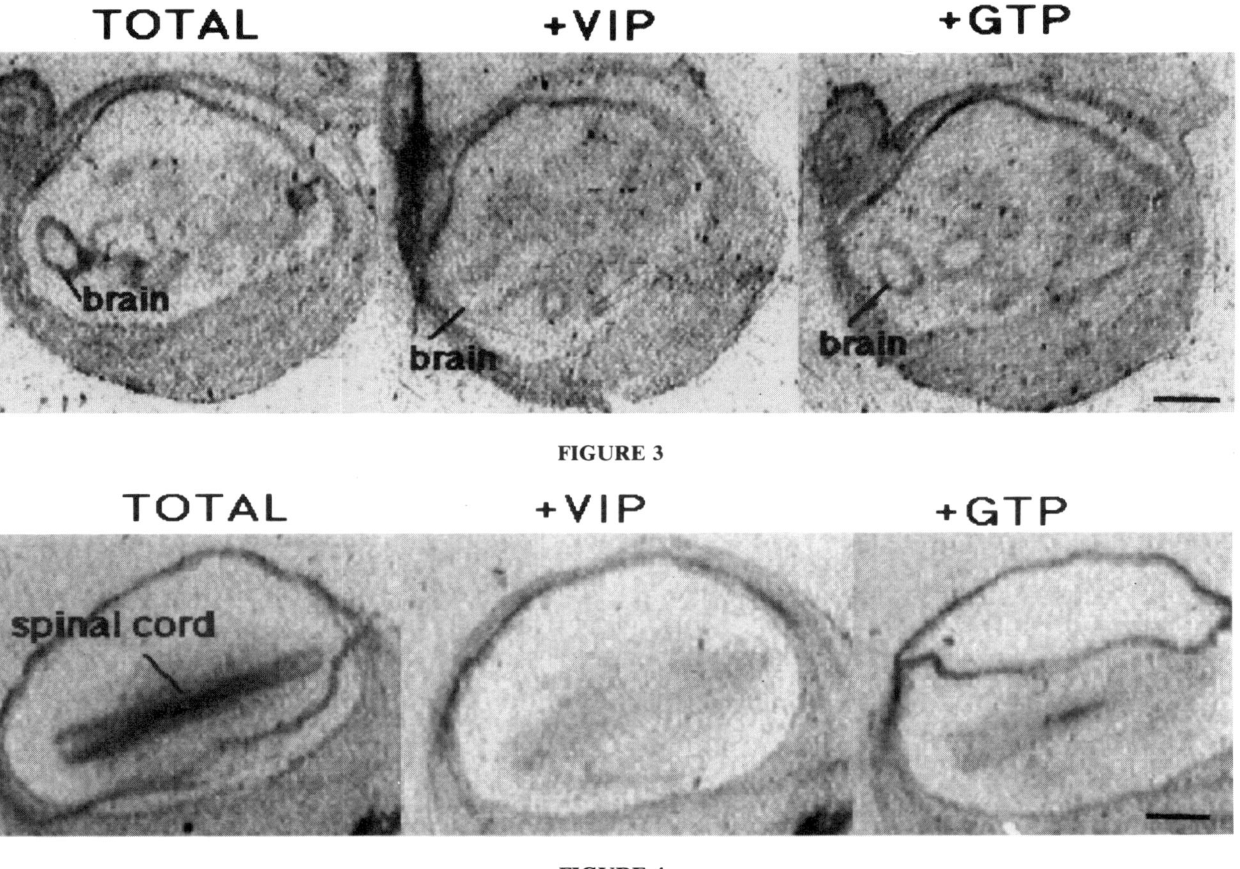

FIGURE 3

FIGURE 4

shown to be indirect, occurring through VIP-stimulated release of other factors.[1] Thus, the coordinated growth of both brain and body resulting from VIP stimulation indicates that VIP acts through high-affinity GTP-insensitive sites in the embryonic central nervous system to stimulate the release of additional growth-promoting factor(s) which, in turn, stimulate body growth.

REFERENCES

1. BRENNEMAN, D. E., T. NICOL, D. WARREN & L. M. BOWERS. 1990. Vasoactive intestinal peptide: A neurotrophic releasing agent and an astroglial mitogen. J. Neurosci. Res. **25:** 386–394.
2. SREEDHARAN, S. P., A. ROBICHON, K. E. PETERSON & E. J. GOETZL. 1991. Cloning and expression of the human vasoactive intestinal peptide receptor. Proc. Natl. Acad. Sci. USA **88:** 4986–4990.
3. ISHIHARA, T., R. SHIGEMOTO, K. MORI, K. TAKAHASHI & S. NAGATA. 1992. Functional expression and tissue distribution of a novel receptor for vasoactive intestinal polypeptide. Neuron **8:** 811–819.
4. MOODY, T. W., D. P. TAYLOR & C. B. PERT. 1981. Effects of guanine nucleotides on CNS neuropeptide receptors. J. Supramol. Struct. **15:** 153–159.
5. HILL, J. M., A. HARRIS & D. I. HILTON-CLARKE. 1992. Regional distribution of guanine nucleotide-sensitive and guanine nucleotide-insensitive vasoactive intestinal peptide receptors in rat brain. Neuroscience **48:** 925–932.
6. HILL, J. M., D. V. AGOSTON, P. GRESSENS & S. K. McCUNE. 1994. Distribution of VIP mRNA and two distinct VIP binding sites in the developing rat brain: Relation to ontogenic events. J. Comp. Neurol. **342:** 186–205.
7. BRENNEMAN, D. E., L. E. EIDEN & R. E. SIEGEL. 1985. Neurotrophic action of VIP on spinal cord cultures. Peptides **6(Suppl. 2):** 35–39.
8. BRENNEMAN, D. E. & L. E. EIDEN. 1986. Vasoactive intestinal peptide and electrical activity influence neuronal survival. Proc. Natl. Acad. Sci. USA **83:** 1159–1162.
9. BRENNEMAN, D. E., E. A. NEALE, G. A. FOSTER, S. d'AUTREMONT & G. L. WESTBROOK. 1987. Nonneuronal cells mediate neurotrophic action of vasoactive intestinal peptide. J. Cell Biol. **104:** 1603–1610.
10. FATATIS, A., L. A. HOLTZCLAW, R. AVIDOR, D. E. BRENNEMAN & J. T. RUSSELL. 1994. Vasoactive intestinal peptide increases intracellular calcium in astroglia: Synergism with alpha-adrenergic receptors. Proc. Natl. Acad. Sci. USA **91:** 2036–2040.
11. OLAH, Z., C. LEHEL, W. B. ANDERSON, D. E. BRENNEMAN & D. V. AGOSTON. 1994. Subnanomolar concentration of VIP induces the nuclear translocation of protein kinase C in neonatal rat cortical astrocytes. J. Neurosci. Res. **39:** 355–363.
12. GRESSENS, P., J. M. HILL, I. GOZES, M. FRIDKIN & D. E. BRENNEMAN. 1993. Growth factor function of vasoactive intestinal peptide in whole cultured mouse embryos. Nature (Lond.) **362:** 155–158.
13. GRESSENS, P., J. M. HILL, B. PAINDAVEINE, I. GOZES, M. FRIDKIN & D. E. BRENNEMAN. 1994. Severe microcephaly induced by blockade of vasoactive intestinal peptide function in the primitive neuroepithelium of the mouse. J. Clin. Invest. **94:** 2020–2027.
14. HILL, J. M., S. K. McCUNE, R. J. ALVERO, G. W. GLAZNER, K. A. HENINS, S. F. STANZIALE, J. R. KEIMOWITZ & D. E. BRENNEMAN. 1996. Maternal vasoactive intestinal peptide and the regulation of embryonic growth in the rodent. J. Clin. Invest. **97:** 202–208.

Regulation of Neuropeptide Expression in Sympathetic Neurons

Paracrine and Retrograde Influences[a]

R. E. ZIGMOND[b] AND Y. SUN

Department of Neurosciences
Case Western Reserve University
School of Medicine
10900 Euclid Avenue
Cleveland, Ohio 44106-4975

INTRODUCTION

A great deal of information has accumulated during the past two decades concerning the distribution and function of neuropeptides in the peripheral nervous system.[1–3] The neuropeptide phenotype of a particular peripheral neuron is now cited as one of its distinguishing features. For example, in mammalian species, a large number of sympathetic neurons contain neuropeptide Y (NPY); few, if any, contain substance P. On the other hand, few sensory neurons contain NPY, whereas a distinct subpopulation contains substance P. A number of *in vivo* and *in vitro* studies have examined the mechanisms involved in the regulation of peptide expression in these neurons. One manipulation that produces a profound change in peptide expression in a variety of peripheral neurons is axotomy. For example, after axotomy, sympathetic neurons decrease their expression of NPY and begin to express substance P,[4,5] whereas sensory neurons decrease their expression of substance P and begin to express NPY.[6–10] Recent studies on these and other axotomy-induced changes in neuropeptide expression have begun to explain how peptide expression is altered after nerve injury; they also have revealed important mechanisms involved in maintaining normal peptide expression under control conditions.[11] In this brief review, emphasis will be placed on studies in the rat superior cervical sympathetic ganglia (SCG).

NEUROPEPTIDE EXPRESSION IN SYMPATHETIC NEURONS IN THE SCG UNDER NORMAL CONDITIONS AND AFTER POSTGANGLIONIC AXOTOMY

NPY is estimated to be present in 60% of the principal neurons in the rat SCG.[12–14] Immunoreactivities (IR) for methionine-enkephalin and a C-terminal extended enkephalin have also been observed in about 10–20% of the neurons.[15–17]

[a] This work was supported by grants from the National Institute of Neurological Disease and Stroke (NS17512 and NS12651) and the National Institute of Mental Health (MH00162).
[b] E-mail: REZ@po.cwru.edu

Interestingly, the enkephalin-positive neurons were NPY-negative, indicating a high degree of cellular specificity of peptide expression even with a single ganglion.[17] One hypothesis that is commonly proposed to explain such specificity is that it depends, at least in part, on the particular target cells a neuron innervates. For example, in sympathetic target tissues, NPY is a more common component of neurons that innervate the vasculature than in those innervating parenchymal cells.[18] It is interesting that heterogeneity is less apparent when the distribution of mRNA for these two peptides are localized by *in situ* hybridization.[19,20] This difference quite likely reflects important regulation of neuropeptide expression at the posttranscriptional level.

In contrast to NPY and enkephalin, very few neurons in the SCG contain detectable substance P, vasoactive intestinal peptide (VIP) or galanin. Nevertheless, all three of these peptides and their mRNA are clearly detectable within overlapping subpopulations of neurons 48 hours after the major postganglionic trunks of the ganglion are transected (FIG. 1).[5,21–23] The peptide that is induced to the greatest extent is galanin. Galanin mRNA and galanin-like immunoreactivity were detectable in about 80 and 50% of the neurons, respectively.[22,23] The finding that, in such cases, there is an increase in both the peptides and their corresponding mRNA argues against the traditional idea that axotomy-induced increases in peptide content in cell bodies is simply due to accumulation of the peptides resulting from the blockade of anterograde axonal transport.[1,24] Increases in VIP mRNA expression have also been found after administration of colchicine[25] or 6-hydroxydopamine.[26] One of the features these two pharmacological treatments have in common with postganglionic nerve transection is that all three interfere with trophic communication between autonomic target tissues and the sympathetic neurons that innervate them (see below).

Similar changes in peptide expression occur when sympathetic ganglia are placed in cell and in organ culture, thus providing *in vitro* systems for studies on the mechanisms underlying these changes.[22,27–29] Because culturing SCG involves both axotomy and deafferentation of the sympathetic neurons, the effects of cutting the predominantly preganglionic cervical sympathetic trunk, the nerve trunk that contains the preganglionic input to the SCG, were also examined. Although changes in peptide expression are seen after decentralization, they are at least an order of magnitude smaller than those seen after explantation or axotomy.[21,22,30,31] Furthermore, we have proposed that at least some of the peptide changes that occur in the SCG after cutting the cervical sympathetic trunk take place in a small population of neurons that are axotomized.[23,32] A second possible mechanism by which decentralization or explantation might affect peptide expression has been suggested, namely, the resulting decrease in synaptic stimulation.[30,33] In our studies, rather than blocking peptide inductions as the hypothesis would predict,[33] continuous depolarization of sympathetic ganglia in culture for 24 hours led to even larger increases than those seen with explantation alone.[34] Higher concentrations of depolarization agents (i.e., elevated potassium and veratridine) led to a blockade of peptide induction, but also to considerable neuronal cell death.

Since one of the features of the general neuronal cell body reaction to axotomy is an increase in RNA and protein synthesis,[35] one question that arises is how specific these increases are in peptide expression. In sympathetic ganglia, the most dramatic evidence for specificity comes from studies on NPY, which as already noted, is the most abundant neuropeptide in the SCG under normal conditions. Postganglionic axotomy led to a decrease in NPY mRNA levels.[4] As discussed below, the expression of a large number of other proteins have also been shown to decrease after axotomy.

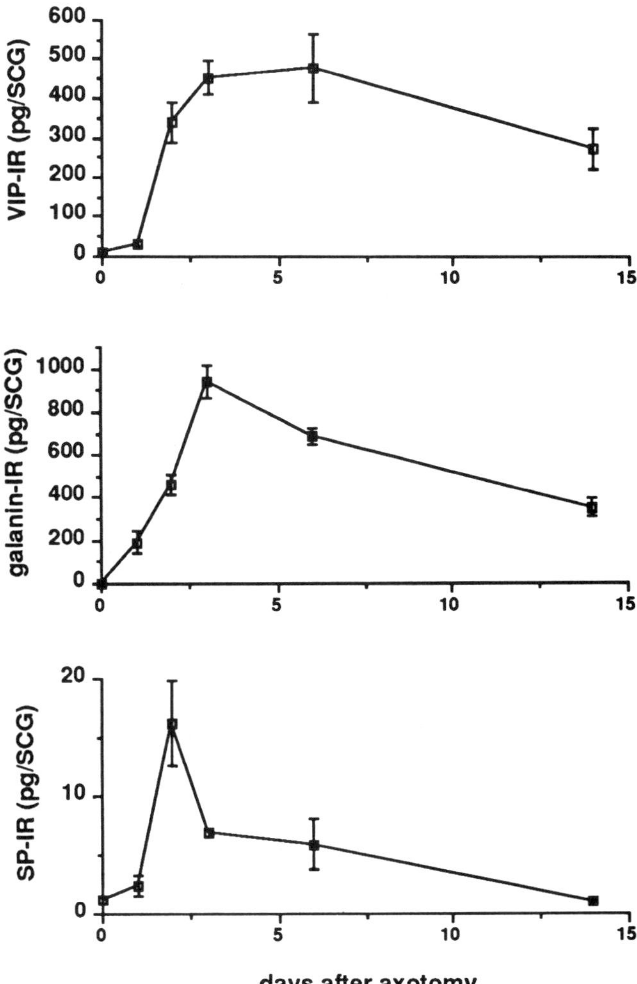

FIGURE 1. Time course of induction of VIP, substance P, and galanin in the superior cervical ganglia (SCG) *in vivo* after axotomy. The two major postganglionic trunks of the SCG, the internal and external carotid nerves, were transected near the SCG and the various peptide-immunoreactivities (IR) were measured by radioimmunoassay at various times from 1 to 14 days later. No significant changes in peptide levels occurred following a sham operation in which the ganglion was exposed but the nerve trunks were not transected. The data here and in subsequent figures are presented as mean values plus or minus the standard error of the mean.

A KEY SIGNAL IN THE AXOTOMY-INDUCED CHANGES IN NEUROPEPTIDE EXPRESSION: INDUCTION AND RELEASE OF THE CYTOKINE LEUKEMIA INHIBITORY FACTOR BY NONNEURONAL CELLS

Our initial hypothesis was that the increases in peptide expression after axotomy were the result of the removal of a target-derived differentiation factor. The first clue that this could not be the whole story came from experiments with dissociated SCG cells. By preplating these cells on uncoated plastic petri dishes, it was possible to separate a neuron-enriched fraction (which does not stick to the dish) and a nonneuronal cell fraction (which does). The induction of VIP was compared in mixed cell cultures and in neuron-enriched cultures. The idea behind the experiment was to determine whether nonneuronal cells played any role in the regulation of VIP expression or whether the only important variable was the removal of the influence of the target tissue. VIP induction was significantly higher in the mixed cultures than in the neuron-enriched cultures (FIG. 2A).[29] Furthermore, medium conditioned for 48 hours by cultures of nonneuronal cells contained an activity that stimulated VIP induction in neuron-enriched cultures (FIG. 2B).[29]

Two plausible candidates for the VIP-inducing factor in the conditioned medium were ciliary neurotrophic factor (CNTF) and leukemia inhibiting factor (LIF), because both of these factors had been shown to increase VIP expression when added to neonatal sympathetic cultures.[36,37] These two molecules belong to a family of cytokines, which, though differing substantially in their primary structure, share a similar secondary and tertiary structure.[38] Other members of this family include oncostatin M, cardiotrophin-1, interleukin (IL)-6, and IL-11.[39,40] CNTF was known to be produced by Schwann cells, and LIF by fibroblasts, both cell types that are present in our nonneuronal cell cultures.[29] The possible contribution of either of these two molecules to VIP induction was assessed by immunoprecipitating the conditioned medium, prior to its exposure to the neuronal cultures, with specific antisera raised against them. The LIF antiserum substantially blocked the VIP induction produced by the conditioned medium, whereas the antiserum to CNTF had no effect (FIG. 3).[29] Furthermore, when CNTF activity was bioassayed in conditioned medium obtained from SCG nonneuronal cells, no activity was detectable.[29]

Dissociating ganglia prior to cell culture could lead to changes in gene expression due to the use of proteolytic enzymes and the disruption of cell–cell interactions. It was, therefore, important to determine whether LIF was also involved in VIP induction in organ culture and *in vivo* after axotomy. Fortunately, the LIF antiserum used in the immunoprecipitation experiments was capable of blocking rat LIF bioactivity. When this antiserum was added to the organ culture medium, the increases in VIP levels normally seen were reduced by 60% (FIG. 4).[29] That this inhibition was not due to a toxic effect of the antiserum was shown by adding to the medium human recombinant LIF, which is not recognized by the LIF antiserum, and obtaining an increase in the level of VIP (FIG. 4).[29] Further evidence for the involvement of LIF in VIP induction in explants came from studies using transgenic mice that had their LIF gene "knocked out" by homologous recombination (LIF⁻ mice). When ganglia from these animals were explanted, no increase in VIP expression was observed (FIG. 5A),[41] although large increases were seen in ganglia from wild-type mice. The availability of these transgenic mice also allowed an investigation of the role of LIF *in vivo*. Forty-eight hours after axotomy a much smaller increase in VIP levels was seen in the LIF⁻ mice than in wild-type controls (FIG. 5B).[41] The results suggested that both LIF-dependent and LIF-independent mechanisms for VIP induction exist *in vivo*.

Having shown the involvement of LIF in VIP induction, we performed a more limited set of experiments to determine whether the same cytokine also underlies the increases in galanin and substance P expression seen in explant cultures and *in vivo* after axotomy. The LIF antiserum diminishes the increases in galanin and substance P expression seen in explants by 40 and 99%, respectively.[42] In mice, although no detectable increase in substance P levels was seen in wild-type animals, an increase in substance K could be detected. Because both tachykinins are products of the same precursor protein, β-preprotachykinin, we attribute this apparent difference in regulation to a difference in sensitivity of the antibodies used for detecting the two peptides. The increase in both galanin and substance K seen in wild-type mice after postganlionic axotomy was substantially reduced in the LIF$^-$ mice.[41]

Interestingly, it has been shown that LIF can decrease NPY expression in neonatal SCG neurons in dissociated culture.[37] To determine whether, in addition to being involved in the induction of certain peptides after axotomy, LIF might also be involved in the suppression of others, the effects of axotomy on NPY mRNA were examined in wild-type and LIF$^-$ mice. Although the difference was small, the reduction in NPY was significantly larger in the wild-type than in the transgenic mice.[4]

IN VIVO REGULATION OF LIF EXPRESSION

The lack of expression of VIP, substance P, and galanin in sympathetic neurons prior to axotomy raises the question of whether LIF itself is induced by axotomy. Unfortunately, it is not possible at present to measure levels of LIF in an SCG or a small number of SCG, due to the lack of a sensitive enough assay. We have, therefore, examined the levels of LIF mRNA. In normal ganglia, LIF mRNA is barely detectable; however, its level increases dramatically after explantation or after axotomy *in vivo*.[29,43] LIF mRNA also increases when ganglionic nonneuronal cells are cultured by themselves under the conditions we used to produce conditioned medium.[29] Although the cell type expressing LIF mRNA in the SCG has not been determined definitively, current evidence suggests that it is ganglionic Schwann cells and/or satellite cells.[44,45] Other studies have indicated independently that these cells do react dramatically to axotomy both by an increased expression of glial fibrillary acidic protein[46,47] and by an increase in cell proliferation.[48] Concerning a possible involvement of CNTF in the changes that occur after axotomy, it is noteworthy that CNTF mRNA does not increase when ganglia are explanted (H. Hyatt-Sachs and R. E. Zigmond, unpublished data).

The satellite/Schwann cells within the SCG, however, may not be the only source of LIF relevant to the cell body reaction. In the sciatic nerve, after a transection some distance from the lumbar sympathetic ganglion, dorsal root ganglia and spinal cord, LIF mRNA increased in the distal nerve stump, again probably within Schwann cells.[43–45] In *in vivo* studies, [^{125}I]LIF was shown to be retrogradely transported by sensory and motor neurons, particularly after injury, though not by sympathetic neurons.[45,49] On the other hand, retrograde transport of LIF has been demonstrated in sympathetic neurons maintained *in vitro* in a compartmentalized culture system.[50] These findings raise the possibility that LIF synthesized near the site of injury is transported back to neurons in the SCG and

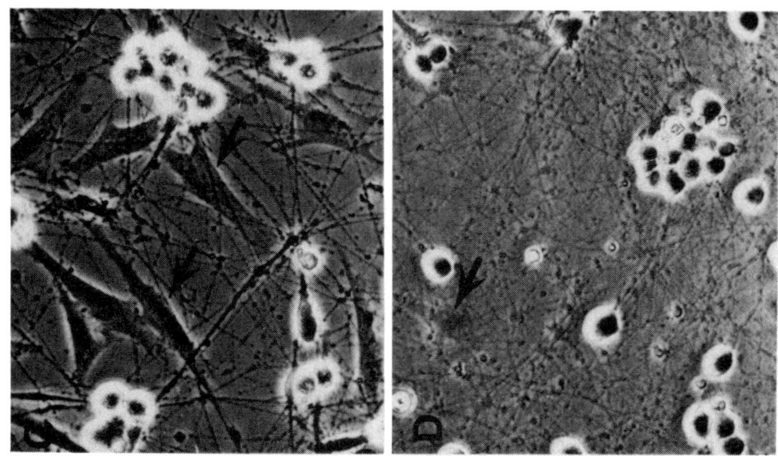

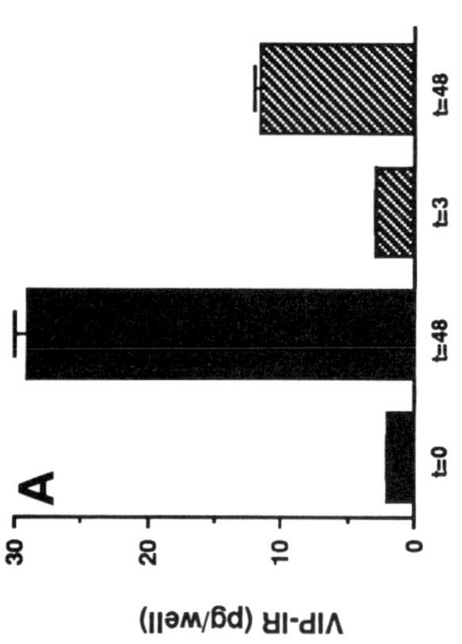

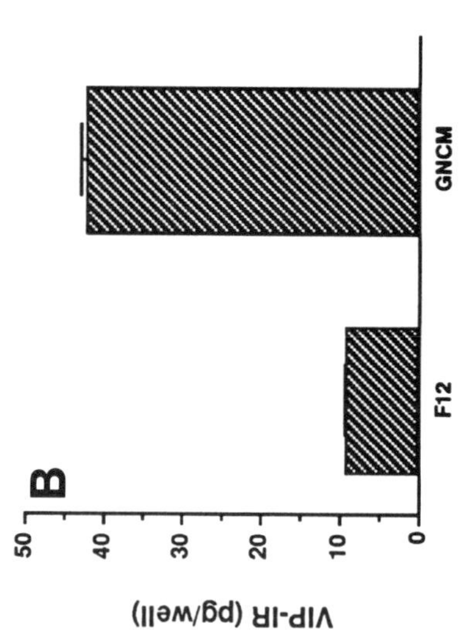

FIGURE 2

trigger aspects of the cell body reaction. Two issues need to be resolved, however, before retrogradely transported LIF can be considered an important source of LIF with respect to changes in peptide expression in the SCG. The first is the determination of whether sympathetic neurons transport LIF *in vivo*. The second is that it must be demonstrated that LIF synthesized at the site of injury actually affects gene expression in the neuronal cell body. Parenthetically, this second requirement also must be satisfied in other cases where growth/differentiation factors are expressed by Schwann cells in the distal nerve stumps after axotomy and have been proposed to participate in the regeneration response (see below).

POSSIBLE MECHANISMS BY WHICH LIF CHANGES NEUROPEPTIDE EXPRESSION

The intracellular signaling events mediating the effect of LIF on neuropeptide expression are being elucidated.[51,52] The functional LIF receptor complex contains a heterodimer of two subunits (LIFβ and gp130). This complex also forms the receptor or part of the receptor for other members of the LIF cytokine family such as CNTF and oncostatin M. After ligand binding, both LIFβ and gp130 are phosphorylated on tyrosine residues and become associated with members of the nonreceptor, JAK tyrosine kinase family. These kinases, once associated with the receptor complex, are themselves phosphorylated and subsequently phosphorylate the transcription factors, STAT 1 and STAT 3. After their phosphorylation, these STAT proteins translocate to the nucleus, form homo- and/or heterodimers, and bind to specific DNA response elements that have the sequence -TTC(C/T)NG(G/T)AA-.[52–54]

Several neuropeptides and neurotransmitter-related proteins have been found to contain such response elements, which are referred to as cytokine response elements (CyRE) or acute phase response elements (APRE). These include the genes for human VIP, human choline acetyltransferase, and rat substance P. Through promoter analysis of the VIP promoter, this response element has been shown to be essential for induction of VIP expression by LIF and CNTF in a neuroblastoma cell line (NBFL).[54] A recent short communication in which a gel shift assay was used indicates that, after axotomy, STAT 1 and 3 proteins capable of binding to a CyRE are found in the SCG.[55]

The mechanism by which these cytokines might decrease the expression of other genes such as that for NPY remains to be determined. LIF may suppress these genes directly through a CyRE or perhaps an as-yet-unidentified response element in the genes. Alternatively, the inhibitory effect of LIF on gene expression could be indirect and might be mediated by other transcription factors.

FIGURE 2. (A) VIP induction in SCG cultures with (*solid bars*) or without (*hatched bars*) nonneuronal cells. The cells were cultured for 48 h ($t = 48$). Peptide levels were also measured in ganglia immediately after being removed from animals ($t = 0$) or in a neuron-enriched fraction immediately after nonneuronal cells were removed by preplating ($t = 3$). **(B)** VIP-inducing activity of nonneuronal cell-conditioned medium. Neuron-enriched cultures of the SCG were incubated for 48 h with defined medium (F12) or medium conditioned by ganglionic nonneuronal cells (GNCM). The phase contrast micrographs are of 48 h mixed cultures **(top)**, which contain neurons and nonneuronal cells, and of neuron-enriched cultures **(bottom)**. (Data have been replotted from Sun *et al.*[29])

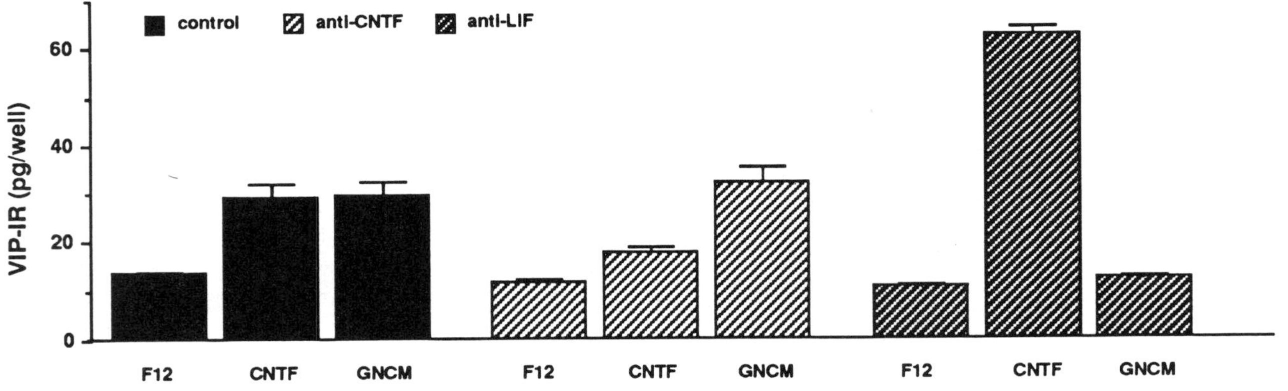

FIGURE 3. Ability of leukemia inhibiting factor (LIF) and ciliary neurotrophic factor (CNTF) antisera to immunoprecipitate the VIP-inducing activity present in nonneuronal cell-conditioned medium. Defined medium (F12), F12 medium containing 10 ng of recombinant CNTF, and medium conditioned for 48 h from ganglionic nonneuronal cells (GNCM) were treated similarly except that some were incubated with an antiserum to CNTF (anti-CNTF) or LIF (anti-LIF). VIP-IR was measured after neuron-enriched cultures were exposed to the various media for 48 h. Although the CNTF antiserum was not able to decrease the VIP-inducing activity of conditioned medium, it was able to block the effects of exogenous CNTF. (Data have been replotted from Sun *et al.*[29])

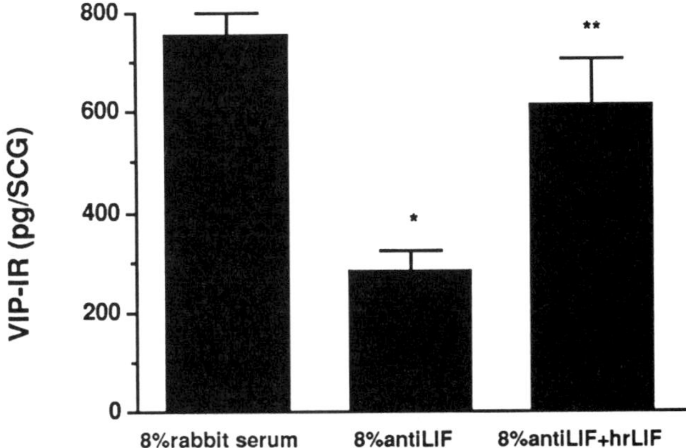

FIGURE 4. Blocking effect of a LIF antiserum on VIP induction in explant cultures. Adult SCG were maintained in culture for 48 h in medium containing 8% normal rabbit serum or antiserum raised against LIF. In one group, in addition to receiving the LIF antiserum, 100 ng/mL of human recombinant LIF was also added to the medium. * indicates a significant difference compared to 8% rabbit serum. ** indicates a significant difference compared to 8% antiLIF. (Data reproduced, with permission, from the *Journal of Neurobiology*.[29])

POSSIBLE ROLE THAT DEPRIVATION OF TARGET-DERIVED NGF MAY PLAY IN CHANGES IN NEUROPEPTIDE EXPRESSION AFTER AXOTOMY

As noted in the experiments on axotomy *in vivo* in LIF⁻ mice, even in the total absence of LIF expression a small increase in the expression of VIP, galanin, and substance K is found. One possible factor that might mediate these changes is NGF. Previous studies have indicated that NGF can increase substance P expression in sensory neurons in dorsal root ganglia[56–58] and NPY expression in PC12 cells.[59] Axotomy has two effects on NGF in the sympathetic system. First, there is almost a complete loss of NGF from the SCG, presumably because of the interference with retrograde transport of NGF from sympathetic target tissues.[60,61] Similar decreases in NGF in the SCG are produced after 6-hydroxydopamine or colchicine administration.[62] Second, axotomy leads to an increased expression of NGF by Schwann cells,[63] at least in the distal stump of the transected sciatic nerve, and NGF can be retrogradely transported by sympathetic and sensory neurons.[64] Whether, however, NGF from Schwann cells affects metabolic changes in axotomized neurons has not yet been demonstrated.

Studies in explanted SCG gave the first direct evidence that the loss of target-derived NGF after axotomy may participate in the induction of galanin. Thus, NGF, when added to adult explants for 48 hours, decreased the expression of galanin though it did not alter the expression of VIP.[28,42] *In vivo*, application of exogenous NGF directly to the axotomized SCG for 14 days decreased the expression of both galanin and VIP.[65] Finally, injection of rats with an antiserum against NGF, in the absence of postganglionic nerve transection, led to a significant increase in galanin and VIP expression, although these effects were relatively small compared to those observed after axotomy.[66] NGF does not appear to play a significant role in the induction of substance P after axotomy, because NGF

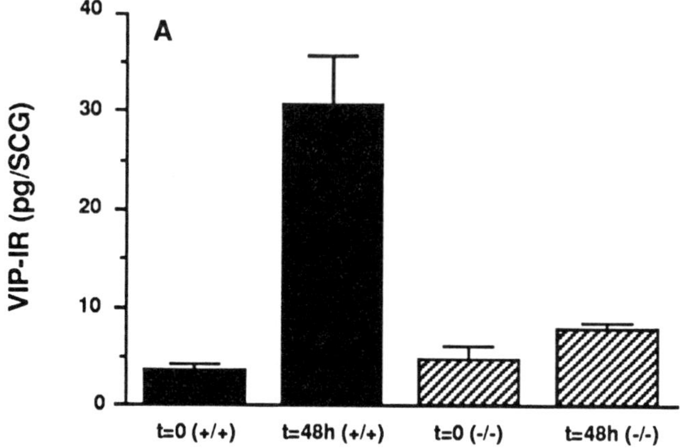

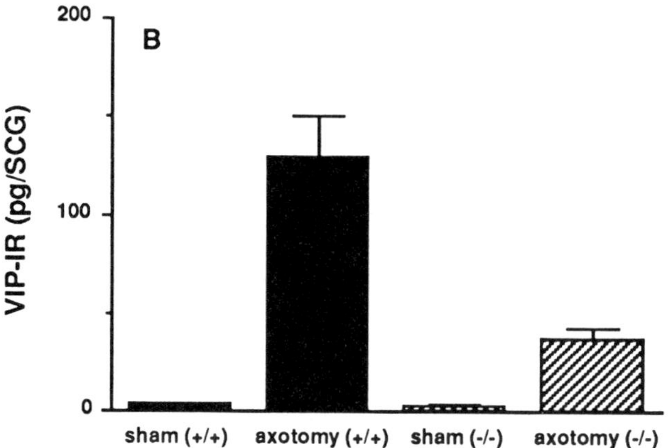

FIGURE 5. VIP induction in SCG from wild type ($+/+$) and LIF$^-$ transgenic ($-/-$) mice. **(A)** VIP levels were measured in ganglia immediately after their removal from animals ($t = 0$) or after 48 h in explant culture ($t = 48$). **(B)** In addition, VIP levels were measured in ganglia 48 h after a sham operation or after transection of the ganglia's major postganglionic trunks. (Data are reproduced, with permission, from *Neuron*.[41])

stimulates substance P and NGF antiserum inhibits substance P expression in the SCG,[42,65,66] as they do in sensory neurons.

CHANGES IN PEPTIDE EXPRESSION AFTER AXOTOMY VIEWED AS PART OF THE "CELL BODY REACTION"

These changes in peptide expression are best viewed in light of other biochemical and morphological changes previously shown to take place in axotomized peripheral neurons and, in general, are referred to as the "cell body reaction." Among the earliest changes noted were general increases in RNA and protein synthesis[35] and the chromatolytic reaction, which involves a dissolution of the rough endoplasmic reticulum.[67,67b] As changes in specific proteins and mRNA began to be examined, increases were found in the expression of several cytoskeletal proteins (and their mRNA), such as α- and β-tubulin,[68–70] actin,[68] and peripherin.[71] Also, expression of the growth-associated protein, GAP-43, increased,[72] as did the activity in certain metabolic pathways.[73] Concurrent with these changes decreases were found in a variety of proteins associated with synaptic transmission. For example, in sympathetic neurons a marked decrease is found in the expression of tyrosine hydroxylase, the enzyme that catalyzes the rate-limiting step in norepinephrine biosynthesis,[69,74] in acetylcholinesterase,[75] and in cholinergic receptor expression.[73,76]

NGF is known to be involved in some of these aspects of the cell body response to axotomy. Thus, the chromatolysis, depression of synaptic transmission, and loss of TH activity that normally occurs in axotomized sympathetic neurons is prevented by administering NGF.[77–79] Furthermore, synaptic depression can be produced by injecting normal animals with antiserum to NGF.[78]

These changes have been traditionally seen as reflecting a shift in the "focus" of the neuron from synaptic transmission to regeneration.[74,80] The morphological switch away from rough endoplasmic reticulum toward more free ribosomes has also been viewed as a shift away from the synthesis of proteins for secretion to the synthesis of proteins for internal use.[67b] Do the changes in neuropeptides described earlier in this paper fit with this view of the cell body reaction to axotomy? The decreases in neuropeptide expression, such as the decrease in NPY expression in sympathetic neurons and the decrease in substance P expression in small diameter sensory neurons, fit in rather well, because these peptides function as neurotransmitters in these neurons under normal conditions. Whether the increases in expression of other neuropeptides not normally synthesized by these neurons also fit with this hypothesis will remain unclear until the functions of these neuropeptides after axonal damage are determined. Some evidence exists in other systems that VIP and substance P may play a role in neural regeneration. Exogenous VIP increases proliferation, survival, and neurite extension in sympathetic neuroblasts,[81] reduces neurite retraction and cell death in cultured SCG neurons deprived of NGF,[82] and acts as an autocrine growth factor for neuroblastoma cells.[83] In the central nervous system, substance P can counteract the degenerative effects of 6-hydroxydopamine on noradrenergic neurons in neonates[84] and enhance regeneration in adults.[85]

The target cells for the effects of neuropeptides after axotomy could be either sympathetic neurons, satellite/Schwann cells, or macrophages.[86–88] Both VIP and galanin are transported by axotomized neurons down to the site of injury and thus the possibility exists for these peptides to act not only within the ganglion but also at the growing nerve tip.[89,90]

SUMMARY

Sympathetic neurons and other peripheral neurons exhibit a great deal of plasticity in their neuropeptide phenotype in adulthood. In this review, two phenotypes have been described in detail: that of normal sympathetic neurons and that of axotomized neurons. Two factors produced by nonneuronal cells, LIF and

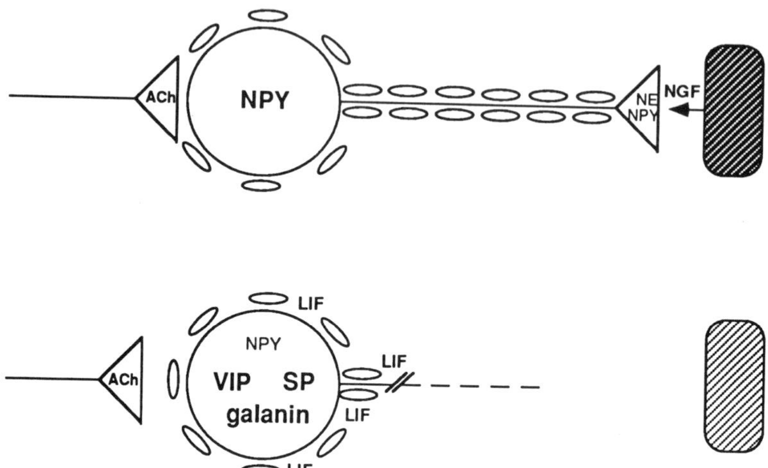

FIGURE 6. A working model of the factors triggering the cell body response to axotomy. Sympathetic neurons innervating their targets take up NGF and transport it back to the neuronal cell body. NGF maintains the normal neuropeptide phenotype of these neurons, promoting the expression of NPY in the majority of the neurons and inhibiting the expression of VIP and galanin. Sympathetic neurons do not normally express substance P (SP) even though NGF stimulates SP expression after axotomy. Postganglionic axotomy removes the influence of target-derived NGF and causes the production of LIF in nonneuronal cells both within the ganglion and at the site of injury. LIF promotes the "axotomy phenotype" in which NPY expression is decreased and VIP, SP, and galanin expressions are increased.

NGF, determine which of these phenotypes is expressed. Under normal conditions, the neurons receive NGF primarily, if not exclusively, from the target tissues they innervate. Prior to surgery the nonneuronal cells within the ganglion and nerve tract express little, if any, LIF. This milieu favors the expression of NPY and suppresses the expression of VIP, galanin, and substance P (FIG. 6). After axotomy, however, this situation is reversed. The neuronal cell bodies are deprived of target-derived NGF and are exposed to LIF both within the ganglion and at the site of the injury (FIG. 6). Both the removal of NGF and the exposure to LIF inhibit NPY expression, while promoting the expression of VIP and galanin. Expression of substance P after axotomy occurs primarily, if not entirely, because of the effects of LIF, with the removal of NGF playing no obvious role in the regulation of this peptide.

ACKNOWLEDGMENTS

We acknowledge the many contributions to this work made by Robert Mohney, Hillary Hyatt-Sachs, Rebecca Schreiber, Dr. Annette Shadiack, and Stacey Vaccariello.

REFERENCES

1. HÖKFELT, T., O. JOHANSSON, Å. LJUNGDAHL, J. M. LUNDBERG & M. SCHULTZBERG. 1980. Peptidergic neurones. Nature **284:** 515–521.

2. HEYM, C. & R. LANG. 1986. Transmitters in sympathetic postganglionic neurons. *In* Neurohistochemistry: Modern Methods and Applications. P. Panula, H. Päivärinta, & S. Soinila, Eds.: 493–525. Alan R. Liss. New York.

3. ELFVIN, L-G., B. LINDH & T. HÖKFELT. 1993. The chemical neuroanatomy of sympathetic ganglia. Annu. Rev. Neurosci. **16:** 471–507.

4. SUN, Y. & R. E. ZIGMOND. 1996. Involvement of leukemia inhibitory factor in the increases in galanin and vasoactive intestinal peptide mRNA and the decreases in neuropeptide and tyrosine hydroxylase mRNA after axotomy of sympathetic neurons. J. Neurochem. **67:** 1751–1760.

5. RAO, M. S., Y. SUN, U. VAIDYANATHAN, S. C. LANDIS & R. E. ZIGMOND. 1993. Regulation of substance P is similar to that of vasoactive intestinal peptide after axotomy or explantation of the rat superior cervical ganglion. J. Neurobiol. **24:** 571–580.

6. JESSELL, T., A. TSUNOO, I. KANAZAWA & M. OTSUKA. 1979. Substance P: Depletion in the dorsal horn of rat spinal cord after section of the peripheral processes of primary sensory neurons. Brain Res. **168:** 247–259.

7. NIELSCH, U. & P. KEEN. 1989. Reciprocal regulation of tachykinin- and vasoactive intestinal peptide–gene expression in rat sensory neurones following cut and crush injury. Brain Res. **481:** 25–30.

8. NOGUCHI, K., E. SENBA, Y. MORITA, M. SATO & M. TOHYAMA. 1989. Prepro-VIP and preprotachykinin mRNAs in the rat dorsal root ganglion cells following peripheral axotomy. Mol. Brain Res. **6:** 327–330.

9. WAKISAKA, S., K. C. KAJANDER & G. J. BENNETT. 1991. Increased neuropeptide Y (NPY)-like immunoreactivity in rat sensory neurons following peripheral axotomy. Neurosci. Lett. **124:** 200–203.

10. NOGUCHI, K., M. DE LEON, R. L. NAHIN, E. SENBA & M. A. RUDA. 1993. Quantification of axotomy-induced alteration of neuropeptide mRNAs in dorsal root ganglion neurons with special reference to neuropeptide Y mRNA and the effects of neonatal capsaicin treatment. J. Neurosci. Res. **35:** 54–66.

11. ZIGMOND, R. E. 1995. Retrograde and paracrine influences on neuropeptide expression in sympathetic neurons after axonal injury. *In* Cytokines and the CNS: Development, Defenses and Disease. R. M. Ransohoff & E. N. Benveniste, Eds.: 169–186. CRC Press. Boca Raton, FL.

12. LUNDBERG, J., L. TERENIUS, T. HÖKFELT & K. TATEMOTO. 1984. Comparative immunohistochemical and biochemical analysis of pancreatic polypeptide-like peptides with special reference to presence of neuropeptide Y in central and peripheral neurons. J. Neurosci. **4:** 2376–2386.

13. JÄRVI, R., P. HELÉN, M. PELTO-HUIKKO & A. HERVONEN. 1986. Neuropeptide Y (NPY)-like immunoreactivity in rat sympathetic neurons and small granule-containing cells. Neurosci. Lett. **67:** 223–227.

14. TYRRELL, S. & S. C. LANDIS. 1994. The appearance of NPY and VIP in sympathetic neuroblasts and subsequent alterations in their expression. J. Neurosci. **14:** 4529–4547.

15. SCHULTZBERG, M., T. HÖKFELT, L. TERENIUS, L-G. ELFVIN, J. M. LUNDBERG, J. BRANDT, R. P. ELDE & M. GOLDSTEIN. 1979. Enkephalin immunoreactive nerve fibers and cell bodies in sympathetic ganglia of the guinea-pig and rat. Neuroscience **4:** 249–270.

16. HÄPPÖLÄ, O., S. SOINILA, H. PÄIVÄRINTA & P. PANULA. 1987. [Met^5]enkephalin-Arg^6-Phe^7 and [met^5]enkephalin-Arg^6-Gly^7-Leu^8 immunoreactive nerve fibers and neurons in the superior cervical ganglion of the rat. Neuroscience **21:** 283–295.

17. DOMEIJ, S., A. DAHLQVIST & S. FORSGREN. 1991. Enkephalin-like immunoreactivity in ganglionic cells in the larynx and superior cervical ganglion of the rat. Regul. Pept. **32:** 95–107.

18. LUNDBERG, J., T. HÖKFELT, A. ÄNGGÅRD, L. TERENIUS, R. ELDE, K. MARKEY, M. GOLDSTEIN & J. KIMMEL. 1977. Organizational principles in the peripheral sympathetic nervous system: Subdivision by co-existing peptides (somatostatin-, avian pancreatic polypeptide-, and vasoactive intestinal polypeptide-like immunoreactive material). Proc. Natl. Acad. Sci. USA **79:** 1303–1307.

19. GREIF, K. F. 1994. Expression of preproenkephalin mRNA in rat superior cervical ganglion during postnatal development. Neurosci. Lett. **180:** 203–208.

20. TYRRELL, S. & S. C. LANDIS. 1994. Distribution of target interactions prevents the development of enkephalin immunoreactivity in sympathetic neurons. J. Neurosci. **14:** 5708–5721.

21. HYATT-SACHS, H., R. C. SCHREIBER, T. A. BENNETT & R. E. ZIGMOND. 1993. Phenotypic plasticity in adult sympathetic ganglia in vivo: Efects of deafferentiation and axotomy on the expression of vasoactive intestinal peptide. J. Neurosci. **13:** 1642–1653.

22. SCHREIBER, R. C., H. HYATT-SACHS, T. A. BENNETT & R. E. ZIGMOND. 1994. Galanin expression increases in adult rat sympathetic neurons after axotomy. Neuroscience **60:** 17–27.

23. MOHNEY, R. P., R. E. SIEGEL & R. E. ZIGMOND. 1994. Galanin and vasoactive intestinal peptide messenger RNA expression increase following axotomy of the adult rat SCG. J. Neurobiol. **25:** 108–118.

24. MOORE, R. Y. 1989. Cranial motor neurons contain either galanin or calcitonin gene-related peptide-like immunoreactivity. J. Comp. Neurol. **282:** 512–522.

25. ZIGMOND, R. E., H. HYATT-SACHS, R. P. MOHNEY, R. C. SCHREIBER, A. M. SHADIACK, Y. SUN & S. A. VACCARRIELLO. 1997. Changes in neuropeptide phenotype after axotomy of adult peripheral neurons and the role of leukemia inhibitory factor. Perspec. Dev. Neurobiol. In press.

26. HYATT-SACHS, H., M. BACHOO, R. C. SCHREIBER & R. E. ZIGMOND. 1993. Chemical sympathectomy with 6-hydroxydopamine (6-HDA) alters peptide expression in the superior cervical ganglion (SCG). Soc. Neurosci. Abstr. **19:** 663.

27. ADLER, J. E. & I. B. BLACK. 1984. Plasticity of substance P in mature and aged sympathetic neurons in culture. Science **225:** 1499–1500.

28. ZIGMOND, R. E., H. HYATT-SACHS, C. BALDWIN, X. M. QU, Y. SUN, T. W. McKEON, R. C. SCHREIBER & U. VAIDYANATHAN. 1992. Phenotypic plasticity in adult sympathetic neurons: Changes in neuropeptide expression in organ culture. Proc. Natl. Acad. Sci. USA **89:** 1507–1511.

29. SUN, Y., M. S. RAO, R. E. ZIGMOND & S. C. LANDIS. 1994. Regulation of vasoactive intestinal peptide expression in sympathetic neurons in culture: The role of cholinergic differentiation factor/leukemia inhibitory factor. J. Neurobiol. **25:** 415–430.

30. KESSLER, J. A. & I. B. BLACK. 1982. Regulation of substance P in adult rat sympathetic ganglia. Brain Res. **234:** 182–187.

31. ZIGMOND, R. E. 1994. Axotomy changes peptide expression. Trends Neurosci. **17:** 297–298.

32. BOWERS, C. W. & R. E. ZIGMOND. 1979. Localization of neurons in the rat superior cervical ganglion that project into different postganglionic trunks. J. Comp. Neurol. **185:** 381–392.

33. KESSLER, J. A., J. E. ADLER, M. BOHN & I. B. BLACK. 1981. Substance P in principal sympathetic neurons: Regulation by impulse activity. Science **214:** 335–338.

34. SUN, Y., M. S. RAO, S. C. LANDIS & R. E. ZIGMOND. 1992. Depolarization increases vasoactive intestinal peptide- and substance P-like immunoreactivities in cultured neonatal and adult sympathetic neurons. J. Neurosci. **12:** 3717–3728.

35. LIEBERMAN, A. R. 1971. The axon reaction: A review of the principal features of perikaryal responses to axon injury. Int. Rev. Neurobiol. **14:** 49–124.

36. ERNSBERGER, U., M. SENDTNER & H. ROHRER. 1989. Proliferation and differentiation of embryonic chick sympathetic neurons: Effects of ciliary neurotrophic factor. Neuron **2:** 1275–1284.

37. NAWA, H., S. NAKANISHI & P. PATTERSON. 1991. Recombinant cholinergic differentiation factor (leukemia inhibitory factor) regulates sympathetic neuron phenotype by

alterations in the size and amounts of neuropeptide mRNAs. J. Neurochem. **56:** 2147–2150.

38. BAZAN, J. F. 1991. Neuropoietic cytokines in the hematopoietic fold. Neuron **7:** 197–208.

39. PATTERSON, P. H. 1992. The emerging neuropoietic cytokine family: First CDF/LIF, CNTF and IL-6; next ONC, MGF, GCSF? Curr. Opin. Neurobiol. **2:** 94–97.

40. PENNICA, D., K. J. SHAW, T. A. SWANSON, M. W. MOORE, D. L. SHELTON, K. A. ZIONCHECK, A. ROSENTHAL, T. TAGA, N. F. PAONI & W. I. WOOD. 1995. Cardiotrophin-1. Biological activities and binding to the leukemia inhibitory factor receptor/gp130 signaling complex. J. Biol. Chem. **270:** 10915–10922.

41. RAO, M. S., Y. SUN, J. L. ESCARY, J. PERREAU, P. H. PATTERSON, R. E. ZIGMOND, P. BRULET & S. C. LANDIS. 1993. Leukemia inhibitory factor mediates an injury response but not a developmental transmitter switch in sympathetic neurons. Neuron **11:** 1175–1185.

42. SUN, Y., A. M. SHADIACK, S. C. LANDIS & R. E. ZIGMOND. 1993. Differential effects of cholinergic differentiation factor/leukemia inhibitory factor (CDF/LIF) and NGF on peptide expression in adult rat superior cervical ganglion (SCG). Soc. Neurosci. Abstr. **19:** 136.

43. SUN, Y., S. C. LANDIS & R. E. ZIGMOND. 1996. Signals triggering the induction of leukemia inhibitory factor in sympathetic superior cervical ganglia and their nerve trunks after axonal injury. Mol. Cell. Neurosci. **7:** 152–163.

44. BANNER, L. R. & P. H. PATTERSON. 1994. Major changes in the expression of the mRNAs for cholinergic differentiation factor/leukemia inhibitory factor and its receptor after injury to adult peripheral nerves and ganglia. Proc. Natl. Acad. Sci. USA **91:** 7109–7113.

45. CURTIS, R., S. S. SCHERER, R. SOMOGYI, K. M. ADRYAN, N. Y. IP, Y. ZHU, R. M. LINDSAY, & P. S. DI STEFANO. 1994. Retrograde axonal transport of LIF is increased by peripheral nerve injury: Correlation with increased LIF expression in distal nerve. Neuron **12:** 191–204.

46. ELFVIN, L-G., H. BJÖRKLUND, D. DAHL & Å SEIGER. 1987. Neurofilament-like and glial fibrillary acidic protein-like immunoreactivities in rat and guinea-pig sympathetic ganglia in situ and after perturbation. Cell Tissue Res. **250:** 79–86.

47. VAIDYANATHAN, U., H. HYATT-SACHS & R. E. ZIGMOND. 1992. Increases in glial fibrillary acidic protein (GFAP)-like immunoreactivity (IR) in the rat superior cervical ganglion (SCG) after decentralization, axotomy, and explanation. Soc. Neurosci. Abstr. **18:** 624.

48. BACHOO, M., A. HALL, C. POLOSA & R. ZIGMOND. 1992. Postganglionic axotomy and 6-hydroxydopamine (6-HDA) produce a burst of satellite cell proliferation in adult rat superior cervical ganglion (SCG). Soc. Neurosci. Abstr. **18:** 624.

49. HENDRY, I. A., M. MURPHY, D. J. HILTON, N. A. NICOLA & P. F. BARTLETT. 1992. Binding and retrograde transport of leukemia inhibitory factor by the sensory nervous system. J. Neurosci. **12:** 3427–3434.

50. URE, D. R. & R. B. CAMPENOT. 1994. Leukemia inhibitory factor and nerve growth factor are retrogradely transported and processed by cultured rat sympathetic neurons. Dev. Biol. **162:** 339–347.

51. STAHL, N. & G. D. YANCOPOULOS. 1994. The alphas, betas, and kinases of cytokine receptor complexes. Cell **74:** 587–590.

52. BONNI, A., D. A. FRANK, C. SCHINDLER & M. E. GREENBERG. 1993. Characterization of a pathway for ciliary neurotrophic factor signaling to the nucleus. Science **262:** 1575–1579.

53. SHUAI, K., A. ZIEMIECKI, A. F. WILKS, A. G. HARPUR, M. Z. GILMAN & J. E. DARNELL. 1993. Polypeptide signaling to the nucleus through tyrosine phosphorylation of Jak and Stat proteins. Nature **366:** 580–583.

54. SYMES, A., S. LEWIS, L. CORPUS, P. RAJAN, S. E. HYMAN & J. S. FINK. 1994. STAT proteins participate in the regulation of the vasoactive intestinal peptide gene by the ciliary neurotrophic factor family of cytokines. Mol. Endocrinol. **8:** 1750–1763.

55. RAJAN, R., C. L. STEWART & J. S. FINK. 1995. LIF-mediated activation of STAT proteins after neuronal injury *in vivo*. Neuroreport **6:** 2240–2244.

56. SCHWARTZ, J. P., J. PEARSON & E. M. JOHNSON. 1982. Effect of exposure to anti-NGF on sensory neurons of adult rats and guinea pigs. Brain Res. **244:** 378–381.
57. FITZGERALD, M., P. D. WALL, M. GOEDERT & P. C. EMSON. 1989. Nerve growth factor counteracts the neurophysiological and neurochemical effects of chronic sciatic nerve section. Brain Res. **332:** 131–141.
58. WONG, J. & M. M. OBLINGER. 1991. NGF rescues substance P expression but not neurofilament or tubulin gene expression in axotomized sensory neurons. J. Neurosci. **11:** 543–552.
59. ALLEN, J. M., J. B. MARTIN & G. HEINRICH. 1987. Neuropeptide Y gene expression in PC12 cells and its regulation by nerve growth factor: A model for developmental regulation. Mol. Brain Res. **3:** 39–43.
60. NAGATA, Y., M. ANDO, K. TAKAHAMA, M. IWATA & K. KATO. 1987. Retrograde transport of endogenous nerve growth factor in superior cervical ganglia of adult rats. J. Neurochem. **49:** 296–302.
61. ZHOU, Z-F., C. ZETTLER & R. A. RUSH. 1994. An improved procedure for the immuno-histochemical localization of nerve growth factor-like immunoreactivity. J. Neurosci. Methods **54:** 95–102.
62. KORSCHING, S. & H. THOENEN. 1985. Treatment with 6-hydroxydopamine and colchicine decreases nerve growth factor levels in sympathetic ganglia and increases them in the corresponding target tissues. J. Neurosci. **5:** 1058–1061.
63. HEUMANN, R., S. KORSCHING, C. BANDTLOW & H. THOENEN. 1987. Changes of nerve growth factor synthesis in nonneuronal cells in response to sciatic transection. J. Cell Biol. **104:** 1623–1631.
64. THOENEN, H., C. BANDTLOW, R. HEUMANN. D. LINDHOLM, M. MEYER & H. ROHRER. 1988. Nerve growth factor: Cellular localization and regulation of synthesis. Cell Mol. Neurobiol. **8:** 35–40.
65. SUN, Y. & R. E. ZIGMOND. 1995. Differential effects of long-term and short-term NGF treatment on neuropeptide expression in axotomized superior cervical ganglia (SCG) *in vivo*. Soc. Neurosci. Abstr. **21:** 1052.
66. ZIGMOND, R. E., A. M. SHADIACK & Y. SUN. 1995. Antiserum to nerve growth factor (NGF) alters neuropeptide expression in the superior cervical ganglion (SCG) and dorsal root ganglion (DRG) in vivo. Soc. Neurosci. Abstr. **21:** 1025.
67. CRAGG, B. G. 1970. What is the signal for chromatolysis? Brain Res. **23:** 1–21.
67b. MATTHEWS, M. R. & G. RAISMAN. 1972. A light and electron microscopic study of the cellular response to axonal injury in the superior cervical ganglion of the rat. Proc. R. Soc. Lond. **181:** 43–79.
68. HOFFMAN, P. N., D. W. CLEVELAND, J. W. GRIFFIN, P. W. LANDES, N. J. COWAN & D. L. PRICE. 1987. Neurofilament gene expression: A major determinant of axonal caliber. Proc. Natl. Acad. Sci. USA **84:** 3472–3476.
69. KOO, E. H., R. N. HOFFMAN & D. L. PRICE. 1988. Levels of neurotransmitter and cytoskeletal protein mRNAs during nerve regeneration in sympathetic ganglia. Brain Res. **449:** 361–363.
70. MATHEW, T. C. & F. D. MILLER. 1990. Increased expression of $T\alpha$ 1-α tubulin mRNA during collateral and NGF-induced sprouting of sympathetic neurons. Dev. Biol. **141:** 84–92.
71. OBLINGER, M. M., J. WONG, & L. M. PARYSEK. 1989. Axotomy-induced changes in the expression of a type III neuronal intermediate filament gene. J. Neurosci. **9:** 3766–3775.
72. SKENE, J. H. P. 1989. Axonal growth-associated proteins. Annu. Rev. Neurosci. **12:** 127–156.
73. SINICROPI, D. V., F. C. KAUFFMAN & D. R. BURT. 1979. Axotomy in rat sympathetic ganglia: Reciprocal effects on muscarinic receptor binding and 6-phosphogluconate dehydrogenase activity. Brain Res. **161:** 560–565.
74. CHEAH, T. B. & L. B. GEFFEN. 1973. Effects of axonal injury on norepinephrine, tyrosine hydroxylase, and monoamine oxidase levels in sympathetic ganglia. J. Neurobiol. **4:** 443–452.
75. VIANA, G. B. & F. C. KAUFFMAN. 1984. Cholinesterase activity in the rat superior cervical ganglion: Effect of denervation and axotomy. Brain Res. **304:** 37–45.

76. FUMAGALLI, L. & G. DE RENZIS. 1980. α-Bungarotoxin binding sites in the rat superior cervical ganglion are influenced by postganglionic axotomy. Neuroscience **5:** 611–616.

77. HENDRY, I. A. 1992. Response of autonomic neurones to target deprivation: Axotomy and regeneration. *In* Development, Regeneration and Plasticity of the Autonomic Nervous System. I. A. Hendry & C. E. Hill, Eds.: 415–462. Harwood Academic Publishers. Chur, Switzerland.

78. PURVES, D. E. & J. W. LICHTMAN. 1975. Formation and maintenance of synaptic connections in autonomic ganglia. Physiol. Rev. **58:** 821–862.

79. FEDEROFF, H. J., M. D. GESCHWIN, A. I. GELLER & J. A. KESSLER. 1992. Expression of nerve growth factor *in vivo* from a defective herpes simplex virus 1 vector prevents effects of axotomy on sympathetic ganglia. Proc. Natl. Acad. Sci. USA **89:** 1636–1640.

80. GRAFSTEIN, B. & I. G. McQUARRIE. 1978. Role of the nerve cell body in axonal regeneration. *In* Neuronal Plasticity. C. W. Cotman, Ed.: 155–196. Raven Press. New York.

81. PINCUS, D. W., E. M. DiCICCO-BLOOM & I. B. BLACK. 1990. Vasoactive intestinal peptide regulates mitosis, differentiation and survival of cultured sympathetic neuroblasts. Nature **343:** 564–567.

82. TANAKA, S. & T. KOIKE. 1994. Vasoactive intestinal peptide suppresses neuronal cell death induced by nerve growth factor deprivation in rat sympathetic ganglion cells in vitro. Neuropeptides **26:** 103–111.

83. O'DORISIO, M. S., D. J. FLESHMAN, S. J. QUALMAN & T. M. O'DORISIO. 1992. Vasoactive intestinal peptide: Autocrine growth factor in neuroblastoma. Regul. Pept. **37:** 213–326.

84. JONSSON, G. & H. HALLMAN. 1982. Modulation of 6-hydroxydopamine induced alteration of the postnatal development of central noradrenaline neurons. Brain Res. Bull. **9:** 635–640.

85. NAKAI, K. & T. KASAMATSU. 1984. Accelerated regeneration of central catecholamine fibers in cat occipital cortex: Effects of substance P. Brain Res. **323:** 374–379.

86. IP, N. Y., C. BALDWIN & R. E. ZIGMOND. 1985. Regulation of the concentration of adenosine $3',5'$-cyclic monophosphate and the activity of tyrosine hydroxylase in the rat superior cervical ganglion by three neuropeptides of the secretin family. J. Neurosci. **5:** 1947–1954.

87. YASUDA, T., G. SOBUE, T. MITSUMA & A. TAKAHASHI. 1988. Peptidergic and adrenergic regulation of the intracellular $3',5'$-cyclic adenosine monophosphate content in cultured rat Schwann cells. J. Neurol. Sci. **88:** 315–325.

88. HARTUNG, H-P. 1988. Activation of macrophages by neuropeptides. Brain Behav. Immun. **2:** 275–281.

89. SHADIACK, A. M., R. P. MOHNEY & R. E. ZIGMOND. 1994. Galanin expression in sympathetic ganglia after partial axotomy. Soc. Neurosci. Abstr. **20:** 1499.

90. SHADIACK, A. M. & R. E. ZIGMOND. 1995. The axotomy-induced neuropeptide galanin is anterogradely transported in axotomized sympathetic neurons. Soc. Neurosci. Abstr. **21:** 1075.

Target Regulation of VIP Expression in Sympathetic Neurons[a]

BETH A. HABECKER,[b] STEVEN E. ASMUS,
NICOLE FRANCIS, AND STORY C. LANDIS

Department of Neurosciences
Case Western Reserve University School of Medicine
10900 Euclid Avenue
Cleveland, Ohio 44106-4975

INTRODUCTION

After neural crest cells coalesce to form sympathetic ganglia along the dorsal aorta, they acquire a noradrenergic phenotype, characterized by catecholamine histofluorescence and immunoreactivity for tyrosine hydroxylase, the rate-limiting enzyme in catecholamine biosynthesis.[1-3] The neurotransmitter and neuropeptide phentoype expressed by the precursors and the neurons that they give rise to can change during the course of development. These changes include the coexpression of one or more neuropeptides with norepinephrine, or the loss of noradrenergic properties coincident with the acquisition of cholinergic and peptidergic properties. As a class, sympathetic neurons produce a number of neuropeptides including neuropeptide Y (NPY), the enkephalins, CGRP, and vasoactive intestinal peptide (VIP),[4] which exhibit different developmental patterns.[3,5] NPY and VIP are expressed widely during early neuronal development; NPY is evident several days before VIP, and VIP is more limited in its expression than NPY. Subsequently, NPY and VIP become restricted distinct subsets of sympathetic neurons; in mature animals many, but not all, noradrenergic sympathetic neurons coexpress NPY, while the much smaller population of cholinergic sympathetic neurons coexpress VIP (FIG. 1).[6,7] In contrast to the early expression of NPY and VIP, the enkephalin peptides do not appear in sympathetic ganglia until birth.[5,8] Enkephalin expression peaks at approximately postnatal day 21 (P21), then decreases significantly to reach steady-state adult levels.

The mechanisms that control neuropeptide expression during development are not well understood. The observation that sympathetic neurons alter their expression of neuropeptides during development indicates that these neurons are plastic with regard to neurochemical properties, and raises the possibility that extracellular signals contribute to alterations in phenotype. A number of investigators have noted that the transmitter repertoire expressed by mature neurons is often correlated with their particular peripheral target,[7,9-13] and suggested that target tissues instruct the innervating neurons to synthesize the specific neurotransmitters and neuropeptides appropriate for the target, or suppress the synthesis of unnecessary neurotransmitters and neuropeptides.

[a] This work was supported by National Institutes of Health grants NS-023678 (S.C.L.), NS-09336 (B.A.H.) and NS-09709 (S.E.A.). S.C.L. is a McKnight Senior Neuroscience Investigator.
[b] E-mail: habecker@codon.nih.gov

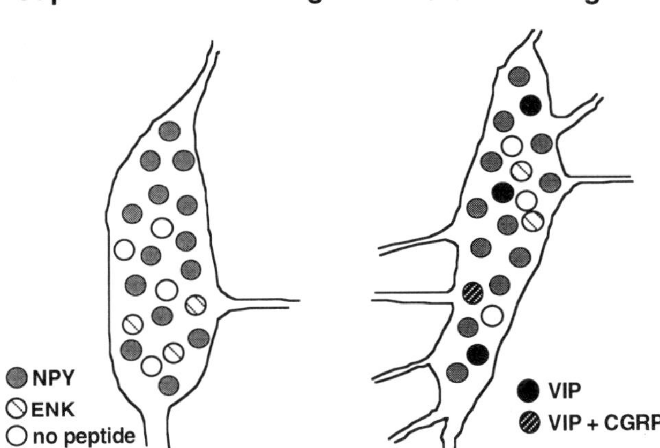

FIGURE 1. Neuropeptide expression differs between ganglia. In mature rats, the superior cervical ganglion is composed of neurons that are noradrenergic and contain no neuropeptides, as well as noradrenergic neurons that coexpress either NPY or enkephalin. In addition to these cell types, the stellate ganglion contains neurons that are VIP-IR, and a small number of neurons that express both VIP and CGRP.

SWEAT GLANDS INDUCE VIP EXPRESSION IN SYMPATHETIC NEURONS

As part of our effort to elucidate the developmental mechanisms regulating the neurotransmitters and neuropeptide phenotype of sympathetic neurons *in vivo*, we examined the sympathetic innervation of rodent sweat glands. During normal maturation, the sweat gland innervation undergoes a striking change in the expression of small molecule transmitters. Initially, the fibers display only noradrenergic properties, but as development proceeds catecholamine histofluorescence disappears and immunoreactivity for the catecholamine synthetic enzymes decreases as cholinergic properties, including choline acetyltransferase (ChAT) and acetylcholinesterase, appear.[14,15] Transplantation of target tissues in early postnatal rats indicates that this switch in neurotransmitter phenotype is induced by interactions with the sweat gland target tissue.[15-17]

In addition to the change in small molecule neurotransmitter expression, the sweat gland innervation also acquires immunoreactivity (IR) for VIP. This peptide is not detectable at P4 when sympathetic axons reach the developing glands, but first appears around P10 (FIG. 2).[12,18] Destruction of developing noradrenergic sympathetic neurons with the neurotoxin 6-hydroxydopamine prevents the appearance of VIP,[19] indicating that after noradrenergic axons innervate the developing sweat glands, they become VIP immunoreactive.

Several lines of evidence suggest that sweat glands retrogradely specify the production of VIP in neurons that innervate them, just as they induce a cholinergic phenotype. When sweat gland primordia are replaced in early postnatal rats with parotid gland, a target which receives noradrenergic sympathetic innervation that lacks VIP-IR,[20,21] the sympathetic fibers that would have innervated sweat glands

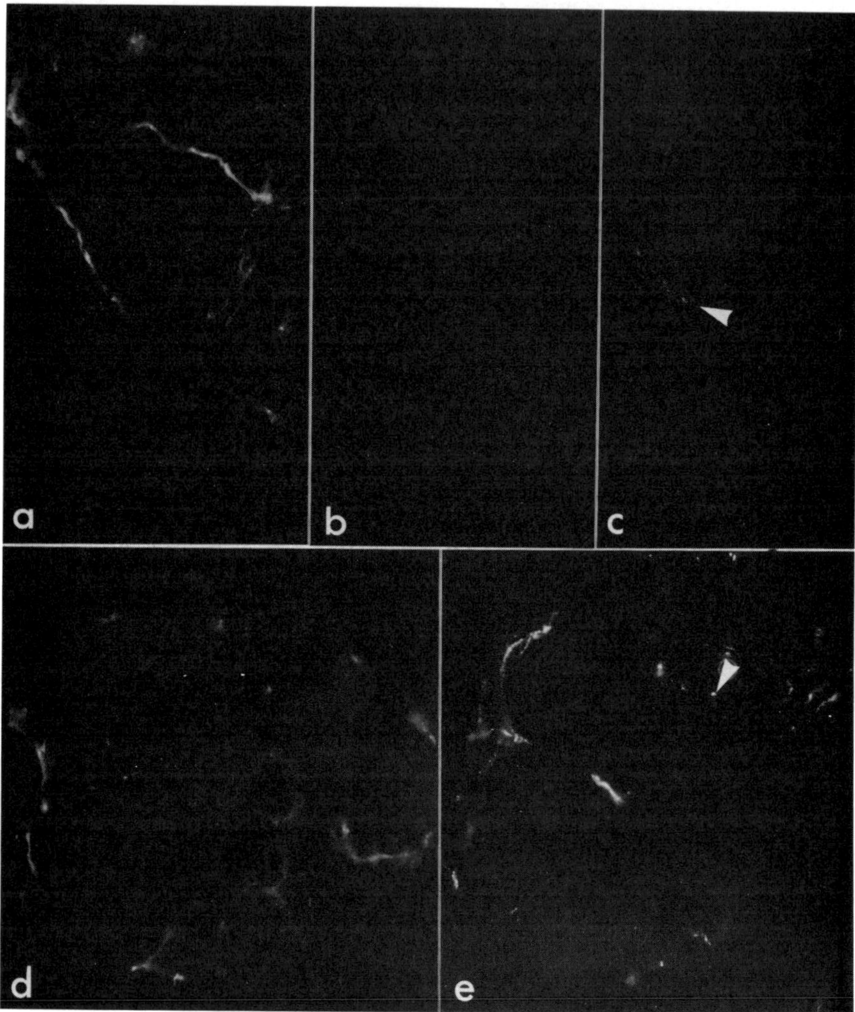

FIGURE 2. VIP immunoreactivity appears in the sweat gland sympathetic innervation on postnatal day 10. (**a**, **b**) At P7, tyrosine hydroxylase (TH) immunoreactive fibers (**a**) surround the sweat glands, while VIP immunoreactivity (**b**) is absent. (**c**) By P10, some sympathetic fibers are VIP immunoreactive. (**d**, **e**) At P15, the intensity of TH immunoreactivity (**d**) has decreased in the sweat gland innervation, while many brightly fluorescent VIP immunoreactive (**e**) fibers are visible. Sections (10 μm) from footpads of paraformaldehyde-perfused rats were incubated with primary antisera raised against either tyrosine hydroxylase (TH; 1 : 1000 dilution) or VIP (1 : 250 dilution), followed by the addition of Texas red– or tetramethylrhodamine isothiocyanate–conjugated secondary antisera.

and acquired VIP-IR innervate the transplanted parotid and do not develop VIP-IR or cholinergic markers, but retain noradrenergic properties instead. In addition, sympathetic fibers that pathfind successfully to the footpads of *Tabby* mutant mice, which lack sweat glands, do not possess VIP-IR.[22] Thus, in the absence of sweat glands the presumptive sweat gland innervation does not produce VIP or alter its small molecule transmitter expression. Conversely, when sweat gland-containing footpad skin is transplanted in place of hairy skin, a target that receives noradrenergic innervation lacking VIP, the sympathetic innervation of the transplanted glands acquires VIP-IR as well as cholinergic properties.[15,16] It is of interest that although catecholamine fluorescence disappears from all sweat glands in these transplants, presumably reflecting the acquisition of cholinergic function, not all innervated glands contain VIP-immunoreactive innervation. This finding raises the possibility that sympathetic neurons are differentially responsive to the target-derived differentiation factor with respect to the induction of cholinergic and peptidergic properties.

The induction of VIP-IR observed in the sweat gland innervation *in vivo* is reproduced *in vitro*. Extracts from sweat gland-containing footpads induce the production of VIP as well as ChAT in cultured sympathetic neurons, which in the absence of extracts produce little or no VIP and ChAT.[23–25] In contrast to extracts from footpads of wild-type mice, extracts from *Tabby* footpads do not induce phenotypic changes in cultured sympathetic neurons, suggesting that the extracted activity is derived from sweat glands rather than other footpad tissues.[26,27] To facilitate examination of the cholinergic differentiation activity produced by the glands, we established primary cultures of sweat gland cells. These cultures contain the two cell types found in rodent sweat glands *in vivo*, secretory cells with a much smaller population of myoepithelial cells.[12,26,28]

PRODUCTION OF SWEAT GLAND–DERIVED DIFFERENTIATION ACTIVITY IS INNERVATION DEPENDENT

Although extracts of sweat gland-containing footpads induce production of VIP in cultured sympathetic neurons, neither conditioned medium nor extracts from primary cultures of sweat gland cells increase ChAT activity or VIP in sympathetic neuron cultures. This indicates that sweat gland cells grown alone do not produce or secrete the differentiation activity present in footpad extracts, and suggests that they lack an instructive signal that normally induces differentiation factor production.[26] Denervation of developing footpads on P14 by sciatic nerve section results in a significant decrease in the ability of footpad extracts to induce VIP and ChAT in cultured sympathetic neurons.[25] These observations, coupled with the data suggesting that most of the cholinergic differentiation activity is gland-associated, raise the possibility that innervation is critical for production of the sweat gland-derived differentiation factor, and that the lack of differentiation activity in sweat gland cultures reflects the absence of innervation. Consistent with this possibility, co-culture of sympathetic neurons with sweat gland cells increases ChAT activity and VIP in the neurons and decreases catecholamine content.[26] Sweat gland cells co-cultured with sympathetic neurons, therefore, reproduce three of the phenotypic changes observed in the sweat gland innervation during normal development.

The phenotypic switch in sympathetic neurons co-cultured with sweat gland cells can be blocked by the addition of adrenergic antagonists, suggesting that

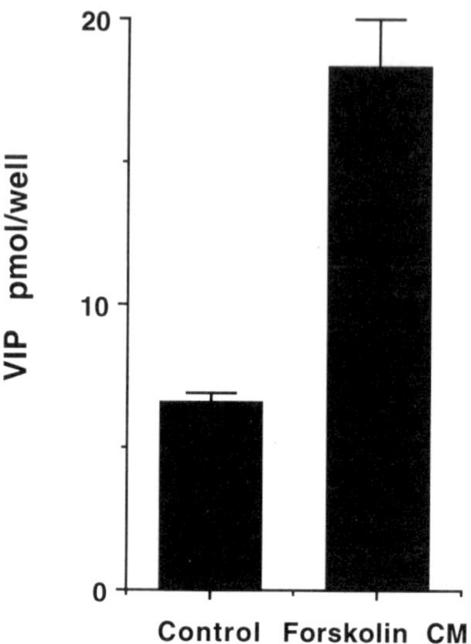

FIGURE 3. Conditioned medium (CM) from forskolin-stimulated sweat gland cells induces VIP production in cultured sympathetic neurons. Sweat gland and sympathetic neuron cultures were established as described,[26] and gland cells were grown for 24 h in the presence of 10 μM forskolin, a direct activator of adenylyl cyclase. Conditioned medium was collected, concentrated, and added onto sympathetic neurons. Neurons were grown in either control medium or conditioned medium for one week, and VIP content measured by RIA as described.[26] Neurons grown in medium from forskolin-stimulated sweat gland cells contained significantly ($p < 0.5$) more VIP than control neurons. The data shown represent the mean of duplicate samples $\pm$ SD, and are representative of results obtained in five independent experiments.

catecholamines released by sympathetic neurons induce the production and secretion of differentiation activity in the sweat gland cells. Similarly, neurons grown in medium conditioned by sweat gland cells treated with forskolin, which mimics noradrenergic activation of β-adrenergic receptors, contained significantly higher levels of ChAT and VIP than control neurons[29] (FIG. 3). In contrast, co-culture of sweat gland cells with sensory neurons, which do not produce catecholamines, does not cause the release of differentiation activity into the medium.[29] Sympathetic innervation is necessary for the production of differentiation activity in developing sweat glands *in vivo* as well. Treatment of newborn rats with the neurotoxin 6-hydroxydopamine destroys developing sympathetic neurons, preventing sympathetic axons from innervating the sweat glands, while leaving sensory innervation to footpads intact.[19] Although footpad extracts from vehicle-injected control rats induce a cholinergic phenotype in sympathetic neuron cultures, extracts from sympathectomized animals do not,[29] indicating that developing sweat glands deprived of noradrenergic input do not produce differentiation activity.

Developing sweat glands require noradrenergic innervation in order to produce differentiation activity, but the sweat glands of adult rats continue to synthesize this activity while their innervation is no longer noradrenergic. There are at least two possible explanations for this apparent discrepancy. First, the maintenance of differentiation factor production may become innervation independent once gland cells have begun to make the factor. Alternatively, the actions of acetylcholine and VIP, mediated by muscarinic acetylcholine and VIP receptors in sweat glands,[30,31] might substitute for the earlier activation of α- and β-adrenergic receptors by catecholamines.[30] To determine if differentiation factor production becomes innervation independent, or if the mature innervation[18,32,33] is required to maintain its expression, we examined whether the differentiation activity in adult rat footpads decreased following denervation. Footpads from adults rats were denervated by unilateral sciatic nerve transection, and the control and denervated footpads were collected one or two weeks later for preparation of extracts. Addition of extracts from control and denervated footpads onto cultured sympathetic neurons indicated that the loss of innervation did not significantly alter the ability of footpad extracts to induce cholinergic function in cultured sympathetic neurons (FIG. 4). This suggests that, in contrast to the induction of differentiation factor expression, the maintenance of factor production is innervation independent.

SWEAT GLAND DIFFERENTIATION ACTIVITY FUNCTIONALLY RESEMBLES NEUROPOIETIC CYTOKINE ACTIVITY

Although the expression of the sweat gland-derived differentiation activity is regulated in an unusual manner, its effects on sympathetic neuron phenotype are

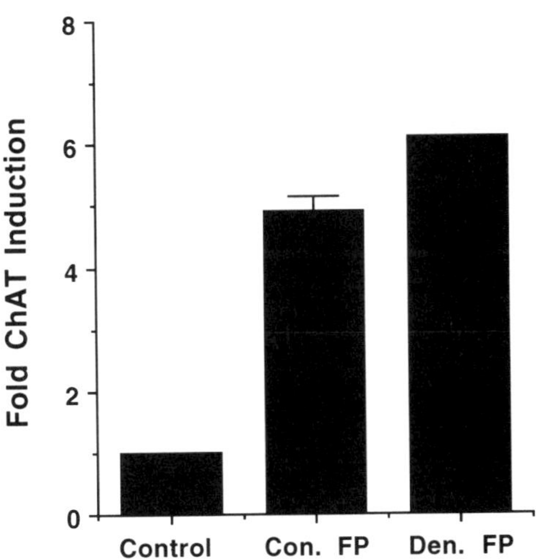

FIGURE 4. Denervated sweat glands maintain production of differentiation activity. Rat footpads were denervated by unilateral sciatic nerve lesion. Extracts were prepared from control and denervated footpads one week after surgery and were added at a concentration of 100 μg/mL to cultures of dissociated sympathetic neurons. Seven days after the addition of extract, triplicate samples of treated and control neurons were homogenized and assayed for choline acetyltransferase (ChAT) activity. Results are expressed as the mean fold induction of ChAT activity ± SEM compared to control neurons, and are representative of five experiments.

similar to those of a family of molecules known as neuropoietic cytokines. These molecules, which include leukemia inhibitory factor (LIF), ciliary neurotrophic factor (CNTF), and cardiotrophin-1 (CT-1), share limited sequence homology, but utilize common receptor subunits and signal transduction pathways.[34–43] Each of these cytokines induce both VIP and ChAT in cultured sympathetic neurons, while decreasing catecholamine production. These findings indicate that a single factor can cause all the changes in transmitter and peptide phenotype associated with the sweat gland-derived activity.[26,40,42–44] Although the sweat gland factor resembles the neuropoietic cytokines functionally and LIF, CNTF, and CT-1 transcripts can be detected in developing rat footpads with RT-PCR, several lines of evidence indicate that the known members of this family do not account for the differentiation activity present in the glands. The most direct evidence comes from the analysis of mice lacking CNTF, LIF, or both CNTF and LIF. If either or both of these molecules were required for the induction of VIP in the sweat gland sympathetic innervation, then the loss of these proteins would result in mice lacking VIP-IR sympathetic innervation in the footpads. In each line of transgenic mice, however, the sweat gland sympathetic innervation acquires cholinergic properties and VIP-IR during development,[42,43,45] indicating that neither LIF nor CNTF is responsible for the differentiation activity present in sweat glands. CT-1, a newly identified member of this family, also induces VIP expression and ChAT activity in cultured sympathetic neurons.[39] Although mice lacking CT-1 are not available, an antiserum has been raised against CT-1 that blocks the ability of recombinant CT-1 to induce cholinergic differentiation in cultured sympathetic neurons. This CT-1 antiserum does not, however, inhibit the ability of sweat gland cells or footpad extracts to induce cholinergic function in sympathetic neurons, making it unlikely that CT-1 is responsible for the cholinergic differentiation activity present in sweat glands.[39] Thus, the sweat gland-derived activity remains a novel differentiation factor for the induction of VIP and acetylcholine during sympathetic neuron development.

Target-induced transmitter differentiation of sympathetic neurons is not limited to sweat glands. Other peripheral targets, including the periosteum, receive cholinergic and peptidergic sympathetic innervation as well. Periosteum, the connective tissue covering of bone, initially receives noradrenergic sympathetic innervation which does not contain VIP,[46] whereas the mature sympathetic innervation is VIP-IR and cholinergic (FIG. 5). Transplant studies and analysis of periosteal cells in culture indicate that periosteum induces cholinergic properties and VIP-IR in sympathetic neurons, both *in vitro* and *in vivo*, and that this is due to the release of a secreted differentiation activity.[46,47] The differentiation factor present in periosteum is therefore functionally similar to the neuropoietic cytokines and the sweat gland-derived factor. LIF cannot account for the differentiation activity in periosteum, because the developmental appearance of VIP-IR in the periosteum sympathetic innervation is not altered in transgenic mice lacking LIF;[47] preliminary data suggest that neither CNTF nor CT-1 is responsible for the differentiation activity.

A minority population of mature sympathetic neurons produce VIP and coexpress the neurotransmitter acetylcholine,[6,7] although sympathetic neurons express noradrenergic properties during ganglion formation[1,2] and, in the absence of environmental influences, will continue to produce norepinephrine. Two peripheral targets that receive cholinergic/peptidergic sympathetic innervation, periosteum and sweat glands, induce the acquisition of VIP and acetylcholine in their innervation. Although the majority of sympathetic neurons do not become cholinergic, many noradrenergic sympathetic neurons express one or more neuropeptides at

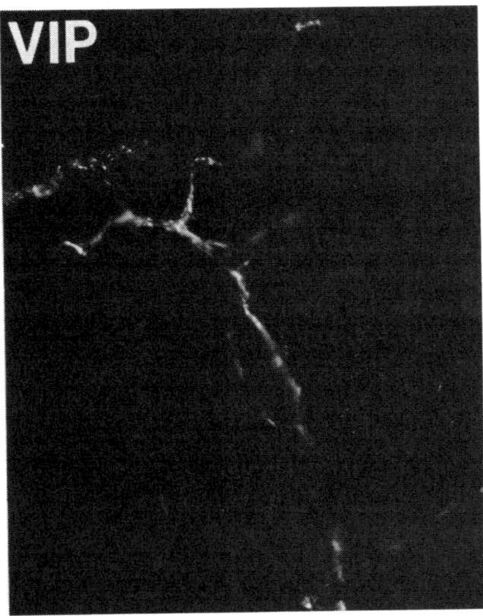

FIGURE 5. The sympathetic innervation of periosteum contains VIP. Brightly fluorescent VIP immunoreactive fibers are present in periosteum from an adult rat. Periosteum was removed from the sternum of a paraformaldehyde-perfused adult rat, and 10-μm sections were incubated with primary antisera raised against VIP (1:250 dilution), followed by the addition of tetramethylrhodamine isothiocyanate-conjugated secondary antisera.

some time throughout development. The ability of target-derived differentiation factors to induce the expression of VIP along with acetylcholine in noradrenergic sympathetic neurons raises the possibility that target interactions may serve as a general mechanism to influence peptide expression in neurons that remain noradrenergic as well. Consistent with this notion is the observation that disruption of target interactions prevents the production of enkephalins in noradrenergic neurons that would ordinarily coexpress enkephalins with catecholamines during postnatal development.[5] Although the mechanisms that control neuropeptide expression during development are incompletely understood, target-derived differentiation factors are one influence regulating neuropeptide production during sympathetic neuron development.

ACKNOWLEDGMENT

The authors thank Dr. Sophia Tyrrell for contributions to the production of FIGURE 1.

REFERENCES

1. COCHARD, P. M., M. GOLDSTEIN & I. B. BLACK. 1979. Initial development of the noradrenergic phenotype in autonomic neuroblasts of the rat embryo in vivo. Dev. Biol. **71:** 100–114.

2. TEITELMAN, G., H. BAKER, T. H. JOH & D. J. REIS. 1979. Appearance of catecholamine synthesizing enzymes during development of the rat nervous system: Possible role of tissue environment. Proc. Natl. Acad. Sci. USA **76:** 509–513.

3. TYRRELL, S. & S. C. LANDIS. 1994. NPY and VIP expression in sympathetic neuroblasts and subsequent regulation of neuropeptide expression. J. Neurosci. **14:** 4529–4547.

4. ELFVIN, L., B. LINDH & T. HOKFELT. 1993. The chemical neuroanatomy of sympathetic ganglia. Annu. Rev. Neurosci. **16:** 471–507.

5. TYRRELL, S. & S. C. LANDIS. 1994. Disruption of target interactions prevents the development of enkephalin immunoreactivity in sympathetic neurons. J. Neurosci. **14:** 5708–5721.

6. LUNDBERG, J. M., A. ANGAARD, J. FAHRENKRUG, T. HOKFELT & V. MUTT. 1980. Vasoactive intestinal polypeptide in cholinergic neurons of exocrine glands: Functional significance of coexisting transmitters for vasodilation and secretion. Proc. Natl. Acad. Sci. USA **77:** 1651–1655.

7. LINDH, B., J. LUNDBERG & T. HOKFELT. 1989. NPY-, galanin-, VIP/PHI-, CGRP- and substance P-immunoreactive neuronal subpopulations in cat autonomic and sensory ganglia and their projections. Cell Tissue Res. **256:** 259–273.

8. KONDO, H., M. YAMAMOTO, N. YANAIHARA & I. NAGATSU. 1988. Transient involvement of enkephalins in both sympathetic and parasympathetic innervations of the submandibular gland of rats. Cell Tissue Res. **253:** 529–537.

9. MORRIS, J. L. & I. L. GIBBINS. 1989. Co-localization and plasticity of transmitters in peripheral autonomic and sensory neurons. Int. J. Dev. Neurosci. **7:** 521–531.

10. LEBLANC, G. G. & S. C. LANDIS. 1988. Target specificity of neuropeptide Y-immunoreactive cranial parasympathetic neurons. J. Neurosci. **8:** 146–155.

11. LUNDBERG, J. M., T. HOKFELT, A. ANGAARD, L. TERENIUS, R. ELDE, K. MARKEY & M. GOLDSTEIN. 1982. Organizational principles in the peripheral nervous system: Subdivisions by coexisting peptides (somatostatin, avian pancreatic polypeptide and vasoactive intestinal peptide-like materials). Proc. Natl. Acad. Sci. USA **79:** 1303–1307.

12. LANDIS, S. C. & D. KEEFE. 1983. Evidence for neurotransmitter plasticity in vivo: Developmental changes in the properties of cholinergic sympathetic neurons. Dev. Biol. **98:** 349–372.

13. COSTA, M., J. B. FURNESS & I. L. GIBBINS. 1986. Chemical coding of enteric neurons. Prog. Brain Res. **68:** 217–239.

14. LANDIS, S. C. 1990. Target regulation of neurotransmitter phenotype. TINS **13:** 344–350.

15. SCHOTZINGER, R., X. YIN & S. LANDIS. 1994. Target determination of neurotransmitter phenotype in sympathetic neurons. J. Neurobiol. **25:** 620–639.

16. SCHOTZINGER, R. & S. C. LANDIS. 1988. Cholinergic phenotype developed by noradrenergic sympathetic neurons after innervation of a novel cholinergic target in vivo. Nature **335:** 637–639.

17. SCHOTZINGER, R. & S. C. LANDIS. 1990. Acquisition of cholinergic and peptidergic properties by the sympathetic innervation of rat sweat glands requires interaction with normal target. Neuron **5:** 91–100.

18. LANDIS, S. C., M. SCHWAB & R. E. SIEGEL. 1988. Evidence for neurotransmitter plasticity in vivo: II. Immunocytochemical studies of rat sweat gland innervation during development. Dev. Biol. **126:** 129–138.

19. YODLOWSKI, M. L., J. R. FREDIEU & S. C. LANDIS. 1984. Neonatal 6-hydroxydopamine treatment eliminates cholinergic sympathetic innervation and induces sensory sprouting in rat sweat glands. J. Neurosci. **4:** 1535–1548.

20. NORBERG, K. & L. OLSON. 1965. Adrenergic innervation of the salivary glands in the rat. Z. Zellforsch. **68:** 183–189.

21. HAND, A. R. 1972. Adrenergic and cholinergic nerve terminals in the rat parotid gland. Electron microscopic observations on permanganate-fixed glands. Anat. Rec. **173:** 131–140.

22. GUIDRY, G. & S. C. LANDIS. 1995. Sympathetic axons pathfind successfully in the absence of target. J. Neurosci. **15:** 7565–7574.

23. RAO, M. & S. LANDIS. 1990. Characterization of a target-derived neuronal cholinergic differentiation factor. Neuron **5:** 899–910.

24. RAO, M. S., P. H. PATTERSON & S. C. LANDIS. 1992. Multiple cholinergic differentiation factors are present in footpad extracts: Comparison with known cholinergic factors. Development **116:** 731–744.

25. ROHRER, H. 1992. Cholinergic neuronal differentiation factors: Evidence for the presence of both CNTF-like and non-CNTF-like factors in developing footpad. Development **114:** 689–698.

26. HABECKER, B. A., S. J. TRESSER, M. S. RAO & S. C. LANDIS. 1995. Production of sweat gland cholinergic differentiation factor depends on innervation. Dev. Biol. **167:** 307–316.

27. RAO, M. S., E. JASZCZAK & S. C. LANDIS. 1994. Innervation of footpads of normal and mutant mice lacking sweat glands. J. Comp. Neurol. **346:** 613–625.

28. QUICK, D. C., W. R. KENNEDY & K. S. YOON. 1984. Ultrastructure of the secretory epithelium, nerve fibers and capillaries on the mouse sweat gland. Anat. Rec. **208:** 491–499.

29. HABECKER, B. A. & S. C. LANDIS. 1994. Noradrenergic regulation of cholinergic differentiation. Science **264:** 1602–1604.

30. HABECKER, B. A., N. M. MALEC & S. C. LANDIS. 1996. Differential regulation of adrenergic receptor development by sympathetic innervation. J. Neurosci. **16:** 229–237.

31. GRANT, M. P. & S. C. LANDIS. 1991. Developmental expression of muscarinic cholinergic receptors and coupling to phospholipase C on rat sweat glands are independent of innervation. J. Neurosci. **11:** 3772–3782.

32. LEBLANC, G. & S. C. LANDIS. 1986. Development of choline acetyltransferase activity in the cholinergic sympathetic innervation of sweat glands. J. Neurosci. **6:** 260–265.

33. STEVENS, L. M. & S. C. LANDIS. 1987. Development and properties of the secretory response in rat sweat glands: Relationship to the induction of cholinergic function in sweat gland innervation. Dev. Biol. **123:** 179–190.

34. STAHL, N. & G. YANCOPOULOS. 1993. The alphas, betas and kinases of cytokine receptor complexes. Cell **74:** 587–590.

35. STAHL, N. & G. D. YANCOPOULOS. 1994. The tripartite CNTF receptor complex: Activation and signaling involves components shared with other cytokines. J. Neurobiol. **25:** 1454–1466.

36. STAHL, N., T. G. BOULTON, T. FARRUGGELLA, N. Y. IP, S. DAVIS, B. A. WITTHUHN, F. W. QUELLE, O. SILVENNOINEN, G. BARBIERI, S. PELLEGRINI, J. N. IHLE & G. D. YANCOPOULOS. 1994. Association and activation of Jak-Tyk kinases by CNTF-LIF-OSM-IL6β receptor components. Science **263:** 92–95.

37. DAVIS, S., T. ALDRICH, N. STAHL, L. PAN, T. TAGA, T. KISHIMOTO, N. IP & G. YANCOPOULOS. 1993. LIFRβ and gp130 as heterodimerizing signal transducers of the tripartite CNTF receptor. Science **260:** 1805–1808.

38. IP, Y., S. NYE, T. BOULTON, S. DAVIS, T. TAGA, Y. LI, S. BIRREN, K. YASUKAWA, T. KISHIMOTO, D. ANDERSON, N. STAHL & G. YANCOPOULOS. 1992. CNTF and LIF act on neuronal cells via shared signalling pathways that involve the IL-6 signal transducing receptor component gp130. Cell **69:** 1121–1132.

39. HABECKER, B. A., D. PENNICA & S. C. LANDIS. 1995. Cardiotrophin-1 is not the sweat gland-derived differentiation factor. NeuroReport **7:** 41–44.

40. PENNICA, D., K. L. KING, K. J. SHAW, E. LUIS, J. RULLAMUS, S.-M. LUOH, W. C. DARBONNE, D. S. KNUTZON, R. YEN & K. R. CHIEN. 1995. Expression cloning of cardiotrophin-1, a cytokine that induces cardiac myocyte hypertrophy. Proc. Natl. Acad. Sci. USA **92:** 1142–1146.

41. RAO, M. S., S. TYRRELL, S. C. LANDIS & P. H. PATTERSON. 1992. Effects of ciliary neurotrophic factor (CNTF) and depolarization on neuropeptide expression in cultured sympathetic neurons. Dev. Biol. **150:** 2811–293.

42. RAO, M. S., J. ESCARY, Y. SUN, J. PERREAU, P. H. PATTERSON, R. E. ZIGMOND, P. BRULET & S. C. LANDIS. 1993. Leukemia inhibitory factor mediates an injury

response but not a target-mediated developmental transmitter switch in sympathetic neurons. Neuron **11:** 1175–1185.

43. MASU, Y., E. WOLF, B. HOLTMANN, M. SENDTER, G. BREM & H. THOENEN. 1993. Disruption of the CNTF gene results in motor neuron degeneration. Nature **365:** 27–32.

44. PENNICA, D., K. J. SHAW, T. A. SWANSON, M. W. MOORE, D. L. SHELTON, K. A. ZIONCHECK, A. ROSENTHAL, T. TAGA, N. F. PAONI & W. I. WOOD. 1995. Cardiotrophin-1: Biological activities and binding to the leukemia inhibitory factor/gp130 signaling complex. J. Biol. Chem. **270:** 10915–10922.

45. FRANCIS, N., B. HABECKER, D. PENNICA & S. LANDIS. 1995. The sweat gland-derived cholinergic differentiation factor is distinct from LIF, CNTF, and CT-1. Soc. Neurosci. Abstr. **21:** 1544.

46. ASMUS, S. E., R. J. SCHOTZINGER & S. C. LANDIS. 1994. Periosteum alters the transmitter phenotype of sympathetic neurons. Soc. Neurosci. Abstr. **20:** 1690.

47. ASMUS, S. E. & S. C. LANDIS. 1995. Periosteal cells in culture induce choline acetyltransferase in cultured sympathetic neurons. Soc. Neurosci. Abstr. **21:** 1052.

Target Tissue Influence on Somatostatin Expression in the Avian Ciliary Ganglion[a]

JAMES N. COULOMBE AND KSENIJA KOS

Department of Anatomy and Cell Biology
Uniformed Services University of the Health Sciences
4301 Jones Bridge Road
Bethesda, Maryland 20814

INTRODUCTION

Formation of the nervous system involves the establishment of a vast number of synaptic connections between neurons and their neuronal or muscle target cells. For these synaptic contacts to function appropriately each neuron must contain the specific neurotransmitters that induce suitable responses in the target cells it innervates. Relatively little is known about the mechanisms by which this specificity of neurotransmitter expression is achieved during neuronal differentiation. One plausible explanation is that target cells are able to regulate neurotransmitter expression in the neurons.

Evidence for this view has accumulated in recent years, with observations which suggest that both classical and neuropeptide transmitter phenotypes may be regulated by interactions between neurons and the targets that they innervate. Neurotransmitter expression can be specifically altered by a variety of treatments in cultured neurons of both peripheral and central nervous system origin.[1-10] Moreover, experimental manipulations of neuronal targets *in vivo* have also shown that in many cases transmitter phenotype can be altered by target influences.[11-14]

The nature of these target-derived influences is unclear. One likely mechanism envisions target cell release of factors at synaptic contacts to regulate neurotransmitter expression in a manner analogous to that hypothesized for the control of neuronal apoptosis by the release of target-derived "trophic" factors (briefly reviewed by Oppenheim[15]). The identification of molecules that might serve as these target-derived neurodifferentiation factors is an important step toward understanding how cell-to-cell interactions might coordinate neurotransmitter expression.

Our studies have focused on the expression of the neuropeptide transmitter somatostatin in the avian ciliary ganglion (CG). In birds, the CG contains two distinct populations of neurons: ciliary neurons, which innervate striated muscle of the iris and ciliary body, and choroid neurons, which innervate vascular smooth muscle in the choroid layer of the eye.[16] Both types of neurons utilize acetylcholine as a small molecule neurotransmitter;[17,18] but only those neurons that innervate the vascular choroid layer, the choroid neurons, use somatostatin as a peptide co-transmitter.[18,19] Within the choroid layer, somatostatin functions as a neuro-

[a] This work is supported by the National Science Foundation IBN 9309932 and the Uniformed Services University of the Health Sciences RO70DU.

modulatory agent inhibiting the release of acetylcholine.[19] Both ciliary and choroid neurons share a common neural crest origin,[20] have the same source of preganglionic input,[21] and develop within the same ganglionic environment. In our studies we have therefore concentrated on examining whether choroidal vascular smooth muscle target cells are influential in regulating somatostatin expression in the CG.

In this paper we summarize the evidence that target cell influences regulate somatostatin expression in CG neurons and that the chemical mediator of this target-derived influence is a molecule known as activin A. We hypothesize that it is the availability of target-derived activin A at nerve terminals that controls expression of somatostatin in CG neurons.

EVIDENCE FOR TARGET INFLUENCE ON SOMATOSTATIN EXPRESSION

Epstein *et al.*[18] first observed somatostatin-like immunoreactivity (Som-IR) in sections of CG from newly hatched chickens and quail. In the chick, Som-IR was located in a population of small neurons located around the periphery of the ganglion. The size, location, and presence of substance P-like immunoreactive boutons around these somatostatin-containing neurons led to the conclusion that the neurons with Som-IR correspond to the population of choroid neurons. This conclusion has subsequently been corroborated on the basis of ultrastructural characteristics,[22] confirming the observation that neurons innervating the choroidal layer express Som-IR, whereas the larger more centrally located ciliary neurons do not express Som-IR, *in vivo*.

We were primarily interested in what developmental influences control this differential expression of somatostatin. Because CG neurons that express somatostatin contact smooth muscle cells in the choroid layer, we began our studies by examining whether choroid smooth muscle cells could influence somatostatin expression.

Developmental Timing of Somatostatin Expression

We initially examined when in development Som-IR is first expressed in CG neurons. A finding that neurons contain Som-IR before contact with target cells would indicate that earlier developmental influences have determined which neurons express somatostatin. We examined anti-somatostatin-immunostained sections of CG prepared from staged[23] CG. At each developmental stage sections from at least 20 different embryos were examined. Neurons with Som-IR were found within a few ganglia as early as stage 29 (embryonic day 6; E6). By stage 30 (E6.5) all ganglia examined contained a few (less than 1%) neurons with Som-IR.[24] Because neurite extension to targets in the choroid layer of the eye is first observed at E6,[25] the timing of this initial appearance of Som-IR is consistent with the idea that an interaction with target cells in the choroid layer induces somatostatin expression.

Target Influence on Somatostatin Induction in Co-Cultures

As a test of whether somatostatin expression could be induced by interaction with target choroid tissue, we co-cultured CG with dissociated choroid cells, skeletal muscle myotubes, or with ganglionic nonneuronal cells. Multiple cultures of each co-culture type were established and sister cultures were fixed at daily intervals and immunostained with an anti-somatostatin antiserum. In all these culture situations, nonneuronal cells from the ganglion were present, and the medium was supplemented with eye extract (1% v/v) in order to support the survival, growth, and differentiation of the neurons in culture.[26] We chose E8 ciliary ganglia, an age when less than 10% of the neurons contain Som-IR *in vivo*, for these cultures because information about the survival and cholinergic development of neurons from this age in culture was already available.[26,27]

In these co-cultures, increased expression of Som-IR was only observed when CG neurons were co-cultured with choroid cells, but not when co-cultured with striated muscle myotubes, or ganglionic nonneuronal cells. Over a 6-day culture period, the percentage of CG neurons containing Som-IR increased in cultures containing choroid cells. In contrast, the percentage of CG neurons with Som-IR remained low (< 3%) in cultures with pectoral muscle or with CG nonneuronal cells.[24] A possible explanation of this result is that choroid cells selectively support survival of somatostatinergic neurons. However, no significant differences were found in the number of neurons present with time in culture and no apparent differences that would easily explain the differences in somatostatin expression.

In addition to cell survival, the possibility exists that differences in neuronal cell health or differences in their ability to continue neuronal differentiation might explain the apparent induction of somatostatin in the presence of choroid cells. As a measure of these parameters we measured choline acetyltransferase (ChAT) activity in sister cultures. As expected from earlier studies[26,27] ChAT activity increased throughout the culture period and no significant differences in ChAT activity were found between the different co-culture conditions.[24] This observation suggests that somatostatin expression is regulated independently of ChAT expression and that in these culture conditions the neurons are capable of continued neuronal differentiation. Thus, neither neuronal cell health nor selective survival can account for the differences in the percentages of neurons expressing somatostatin in these different co-culture regimes.[24] Instead it appears that embryonic CG neurons require an interaction with cells from the choroid layer for the initial expression of Som-IR in cell culture.

Target Influence Maintaining Somatostatin Expression in Co-Cultures

We also tested whether choroid cells were necessary to maintain the expression of Som-IR in more mature CG neurons. We used E13–E14 CG to examine the maintenance of somatostatin expression. By this age, the choroid neurons, which constitute approximately one-half of the CG neurons, already contain Som-IR.[22,24,28] In a co-culture paradigm, similar to that described above, we observed that the percentage of Som-IR neurons after the first day of culture was approximately 50% in all three culture conditions. In cultures with striated muscle or ganglionic nonneuronal cells, the proportion of neurons containing Som-IR declined to less than 10% by five days after plating. In contrast, approximately 50% of the CG neurons co-cultured with choroid cells contained Som-IR throughout the 7-day culture period. No evident differences were found in either survival or ChAT expression during the five days of declining Som-IR in the absence of choroid cells.[24] Thus, it appears that CG neurons which have already begun to express Som-IR *in vivo* still require an interaction with choroid cells to maintain expression in culture.

The observation that somatostatin expression begins at a developmental age when neurites first contact the choroid layer and the apparent requirement for choroid cells to support somatostatin expression in cultured CG neurons both strongly suggest that a target-derived influence from the choroid cells is necessary for the expression of somatostatin in CG neurons.

SOLUBLE FACTORS PRODUCED AND SECRETED BY CULTURED CHOROID CELLS STIMULATE SOMATOSTATIN IN CULTURED CG NEURONS

As a first step toward understanding the mechanism by which this target-derived influence is able to stimulate expression of Som-IR, we asked whether the effect was mediated by soluble molecules or required direct cell contact. We developed methods for producing cultures of dissociated choroid cells on a collagen gel substrate. In these cultures the dissociated choroid cells first dedifferentiate, multiply, and then form two superimposed layers of elongated confluent cells. At least one-half of these cultured choroid cells subsequently redifferentiate as smooth muscle cells as shown by the presence of immunoreactivity for a smooth muscle cell specific form of actin (FIG. 1a). Medium was collected from these choroid cell cultures (choroid cell conditioned medium, ChCM) and fed to cultures of CG neurons resulting in a dose-dependent increase in the percentage of Som-IR neurons[24] (FIG. 1b). Thus, the choroid cell stimulation of somatostatin expression in cultured CG neurons could be mediated by diffusible molecules. We named this biological activity of ChCM somatostatin stimulating activity (SSA).

We hypothesized that SSA might represent a neurodifferentiation factor for CG neurons that would be released by healthy choroid target cells *in vivo*. Therefore, we examined whether SSA was present in ChCM due to secretion or whether it was the product of cell lysis. Choroid cells were allowed to condition medium for 24 hours before we removed the ChCM and lysed the cells. As a measure of cell lysis, we compared the amount of lactate dehydrogenase (LDH) activity present in ChCM and in the cell lysate[29] (FIG. 2c). Because LDH is a ubiquitous cytoplasmic enzyme that is not secreted,[30] our finding that ChCM contained less than 5% of the LDH present in the cell lysate implied that very little lysis had occurred during the conditioning period. SSA was only found in ChCM and was not detectable in the cell lysate suggesting that it is a secreted factor.

It thus appears that the SSA from choroid cells is due to diffusible molecules that are secreted by the cultured choroid cells. This observation is consistent with the idea that release of a target-derived neurodifferentiation factor controls somatostatin expression.

BOTH CHOROID AND CILIARY NEURONS CAN EXPRESS SOMATOSTATIN

Although there appears to be a requirement for a target influence for CG neurons to express somatostatin, this target influence might be a permissive factor allowing choroid neurons to express somatostatin. Conversely, in the context of CG neuron development this influence might be instructive, stimulating the expression of somatostatin by either ciliary or choroid CG neurons.

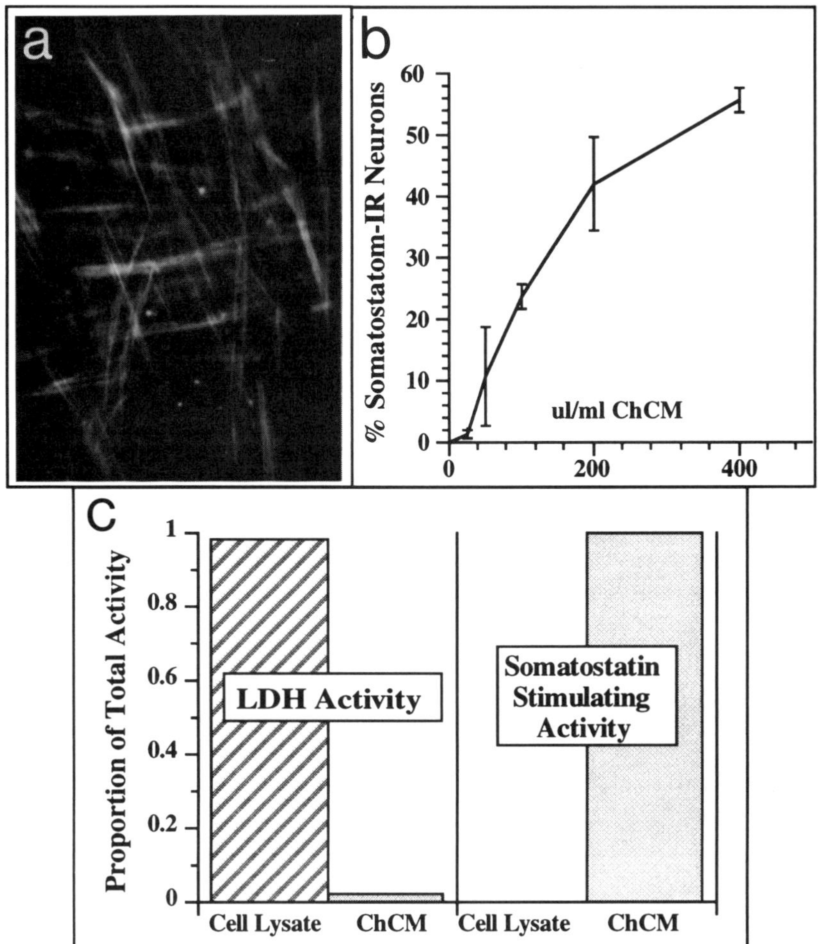

FIGURE 1. Medium from cultured choroid cells stimulates somatostatin expression in cultured ciliary neurons. (a) Fluorescence micrograph of cultured choroid cells immuno-stained with a smooth muscle-specific antibody. Choroid cell cultures were established and stained with an anti-smooth muscle-specific actin antibody as previously described.[24] Choroid cells in culture form two confluent layers; approximately half of the cells have smooth muscle-specific actin-like immunoreactivity. (b) Culture of E8 CG neurons in ChCM results in a dose-responsive increase in the percentage of Som-IR neurons. Choroid cultures were allowed to condition a serum-free medium for 24 h as described previously.[33] E8 CG were grown in the presence of varying amounts of this ChCM. After 3 days, the neuron cultures were processed and assayed for Som-IR as previously described.[24] Values presented are the average of at least four different cultures ± SD. (c) Comparison of LDH and SSA. Choroid cell cultures were allowed to condition a serum-free medium for 24 h. Following a rinse with serum-free medium the cells were lysed with distilled water, and the amounts of LDH and SSA were determined for the ChCM and cell lysates. ChCM contained less than 5% of the LDH activity present in the lysate indicating that very little lysis occurs during the conditioning period.[29] All of the SSA was found in the ChCM suggesting that this activity is secreted.[29]

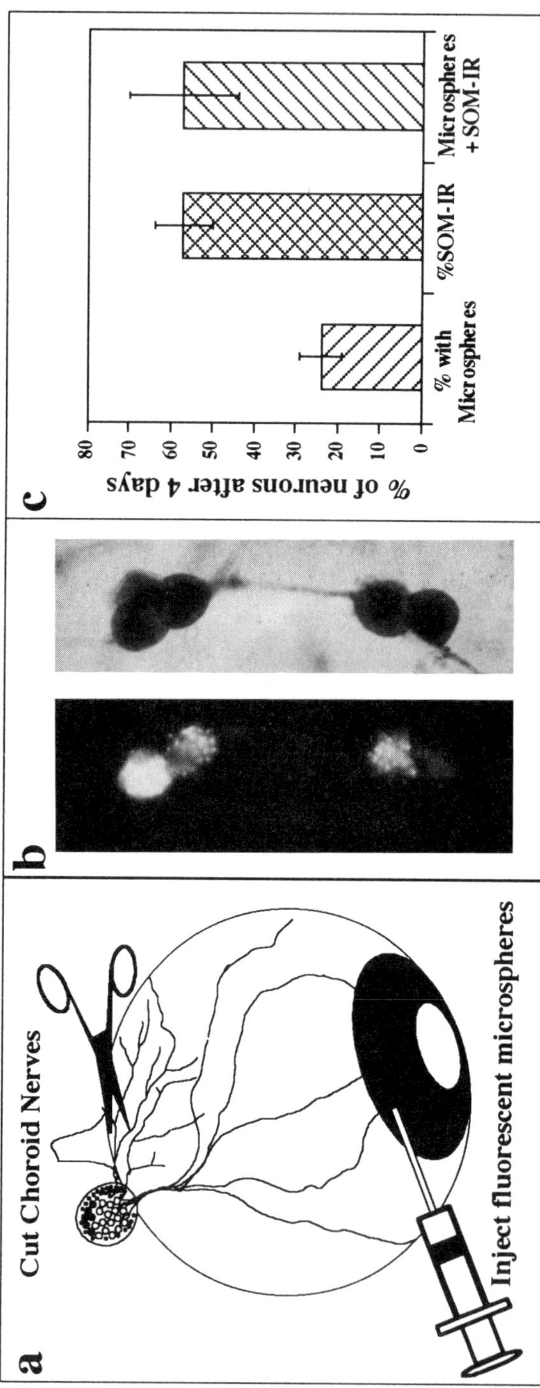

FIGURE 2. Ciliary neurons, which would not express somatostatin *in vivo*, contain Som-IR when cultured with ChCM. **(a)** Eyes, with the CG in place, were removed from E8 embryos, the choroid nerves were severed and fluorescent latex microspheres injected into the iris/ciliary body region. After incubation overnight to allow retrograde transport of the microspheres to the cell bodies of the ciliary neurons, the CG were placed into culture. **(b)** After culture for 4 days in the presence of ChCM, many retrogradely labeled neurons that had projected to the iris/ciliary body region (ciliary neurons) contained Som-IR.[24] **(c)** Counts of Som-IR neurons, retrogradely labeled neurons, and Som-IR retrogradely labeled neurons. Retrogradely labeled neurons contained Som-IR in the same proportion as the total neuron population.[24] Values presented are averages ± SD.

In vivo the neurons of the CG are divided equally between ciliary and choroid neurons. Only half of the neurons, those that contact the choroid layer, normally express somatostatin.[18,22] Interestingly, when CG neurons are cultured in the presence of ChCM, more than 90% of the neurons contain Som-IR nine days after plating.[24] Neuron cell counts showed that this effect could not be accounted for by selective death, suggesting that ciliary neurons which normally do not contain Som-IR *in vivo* could be induced to express somatostatin when cultured with SSA.[24]

As a direct test of whether ciliary neurons could express somatostatin in the presence of SSA, we retrogradely labeled ciliary neurons with fluorescent latex microspheres[31,32] prior to culturing them in ChCM. Specificity of the labeling paradigm was insured by severing the choroid nerves prior to injection of the microspheres into the iris/ciliary body region (FIG. 2a). After allowing time for retrograde transport, the ganglia were removed, dissociated, and cultured for four days in the presence of ChCM. In these cultures 24% of the neurons were labeled with fluorescent latex microspheres. Of the labeled neurons 57% also contained Som-IR (FIG. 2b and c). This percentage of retrogradely labeled neurons expressing Som-IR was identical to the total percentage of Som-IR neurons in the cultures.[24] Thus, ChCM can induce Som-IR in retrogradely labeled ciliary neurons that would not normally express somatostatin.

These two observations suggest that for CG neurons the choroid cell influence plays an instructive role in specifying somatostatin neuropeptide expression.

THE SOLUBLE FACTOR IS ACTIVIN A

Stimulation of Somatostatin Expression in Culture by Activin

In order to identify SSA we began to biochemically characterize its activity from ChCM. SSA appeared to be a macromolecule because it could be concentrated behind a Centriprep™ (Amicon) filter that retains components larger than 10 kDa in apparent size.[24] This factor appeared to be a protein because it was ammonium sulfate precipitable and, although extremely heat resistant, was destroyed by heating to 100 °C for more than 10 min. When ChCM was fractionated by gel filtration chromatography a peak of activity for SSA was eluted with an apparent size of 30–40 kDa.[33]

Before attempting to purify this molecule from ChCM, we tested whether SSA could be due to a previously characterized growth factor. The apparent size of 30–40 kDa for SSA is considerably larger than that known for mitogenic factors such as epidermal growth factor (EGF), insulin-like growth factor (IGF), and the fibroblast growth factors (FGFs), but suggested that SSA could be due to growth factors in the platelet-derived growth factor (PDGF), or the transforming growth factor-β (TGF-β) family. We therefore obtained PDGF-BB, TGF-β, and two other TGF-β superfamily members—activin and inhibin—and tested them for their ability to stimulate expression of Som-IR in cultured CG neurons. Exposure of CG neurons to either PDGF-BB, TGF-β, or inhibin A, which shares one subunit in common with activin A, did not result in any neurons with detectable Som-IR. In contrast, the response of CG neurons to activin A was dosage dependent, with a half-maximal response at a concentration of approximately 2.5 ng/mL and a plateau above 10 ng/mL.[33] Basic FGF, rat CNTF, and chicken GPA—three factors which are known to support survival of CG neurons in culture—were also tested but did not result in any neurons with detectable Som-IR. Of the growth factors tested, only activin was able to mimic ChCM in stimulating somatostatin expression.[33]

Activin-like Biological Activity in Choroid-conditioned Medium

To determine whether ChCM contains activin, we used a standard bioassay for activin-like biological activity—the induction of hemoglobin expression by a human erythroleukemic cell line, K562.[34] In these experiments, K562 cells express hemoglobin both in the presence of activin A and ChCM but not in response to the control medium, indicating that ChCM contains an activin-like activity.[33]

Inhibin, a heterodimer (α, β) that shares one of the subunits of activin ($\beta\beta$), can inhibit many of activin's known biological activities including the induction of hemoglobin in K562 cells.[35] Inhibin inhibited the hemoglobin-inducing activities of both activin A and ChCM in a dosage-dependent manner that was maximal at 50 ng/mL.[33] This result further confirms the activin-like nature of the hemoglobin-inducing activity in ChCM.

Activin-like Immunoreactivity in Choroid-conditioned Medium

As a further means of confirming the presence of an activin-like molecule in ChCM, we tested antisera raised against human activin A ($\beta A\beta A$) and B ($\beta B\beta B$) in western blots of samples of ChCM and control medium. With antisera against activin A, a single band with an apparent size of 24 kDa (consistent with the apparent sze of chicken activin in SDS PAGE[36]) was observed in the lane loaded with ChCM. In contrast, no bands were detected with antisera against activin B although an appropriately sized band was found in a positive control lane spiked with human recombinant activin B.[33] Thus, antisera that are able to distinguish between human activin A and B, which share 85% amino acid identity, are able to specifically recognize a protein similar to activin A in ChCM.

Activin Messenger RNA in Cultured Choroid Cells

As another means of confirming the presence of activin in ChCM, we examined whether cultured choroid cells express mRNA coding for activin. In Northern blot analysis, using an antisense cDNA coding for human activin A, a single band of polyadenylated RNA from choroid cultures hybridized with the activin probe. The same amount of polyadenylated RNA from muscle cultures, which produce little or no SSA, did not hybridize. A control blot hybridized with a probe for cyclophilin indicating that the amount of RNA loaded was similar in both choroid and muscle samples.[33] Thus, choroid cells that produce SSA also contain mRNA coding for activin.

The results described above indicate that activin A can stimulate somatostatin expression in E8 CG neurons mimicking the effects of ChCM. Additionally, ChCM contains a substance that is immunologically and biologically identical to activin A. Moreover, mRNA for activin A is present in cultured choroid cells. From these observations we surmise that activin is at least one of the components in ChCM with SSA. Is all of the SSA in ChCM due to activin or might another factor with SSA be present as well?

Inhibition of Somatostatin Expression by an Activin-binding Protein, Follistatin

In order to examine whether the activin-like molecule that induced hemoglobin synthesis in K562 cells was the agent responsible for SSA, we sought to block or deplete ChCM of activin to determine if SSA was also eliminated. We used another protein present in ovarian fluid, follistatin, a highly specific activin-binding protein that inhibits all known biological activities of activin.[37] In this experiment duplicate samples of medium containing either activin A or one of three different preparations of ChCM were prepared. To one of each preparation, 100 ng/mL of follistatin was added, prior to assaying the samples for induction of hemoglobin in K562 cells and SSA in E8 CG neurons. In all cases, follistatin reduced the expression of hemoglobin to or below background levels. Similarly, in the presence of follistatin very few neurons (one Som-IR neuron in one sample of ChCM) contained Som-IR.[33] Thus, follistatin effectively eliminated the ability of ChCM to induce hemoglobin expression and to stimulate somatostatin expression, which suggests that both activities are due to the presence of activin in ChCM.

It appears that when choroid cells, the target cells of somatostatinergic neurons in the CG, are placed into cell culture they produce and secrete activin A, which can stimulate somatostatin expression in both choroid and ciliary CG neurons in culture.

EVIDENCE FOR ACTIVIN A AS A NEURODIFFERENTIATION FACTOR *IN VIVO*

The results described in culture are consistent with the developmental and spacial expression of somatostatin *in vivo*. Nevertheless, it is possible that the expression of activin A by choroid cells or the response of CG neurons to activin are artifacts of the cell culture environment. To substantiate the putative role of activin A as a target-derived neurodifferentiation factor, it is necessary to show that activin A is available in the choroid layer *in vivo* and that the CG neurons are capable of responding to activin *in vivo*.

Activin Is Produced by Choroid Cells **in Vivo**

In order to examine whether mRNA for activin A was present in the choroid layer of the chicken eye, we designed primers to amplify a fragment of activin A mRNA with the reverse transcription–polymerase chain reaction (RT-PCR). Because the full sequence of chicken activin was not yet known, degenerate primers were designed so as to hybridize to regions of the preproactivin sequence that were highly conserved between reported sequences for activin A from the human, rat, mouse, pig, and sheep. To avoid amplification of genomic DNA the primers were chosen so that the region to be amplified flanked an intron present in the activin A gene.

When these activin A-specific primers were used with cDNA reverse transcribed from microdissected choroid tissue, they amplified a single band of 478

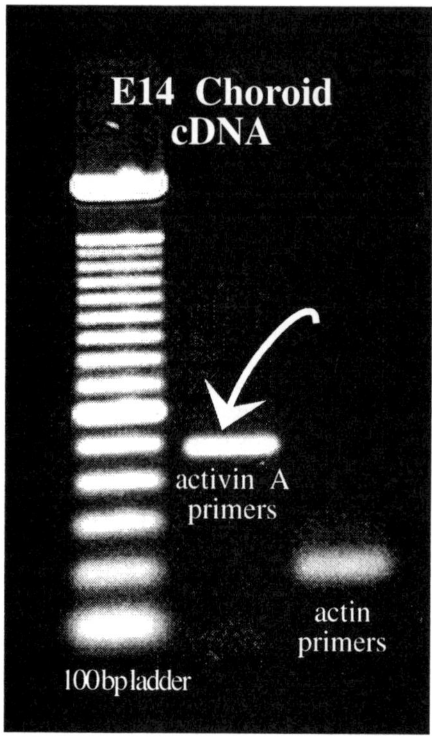

FIGURE 3. *In vivo*, the choroid layer contains activin A mRNA. Activin A-specific primers used in RT-PCR with E15 choroid layer RNA amplify a band of the predicted size (473 bp) which, when subcloned, was found to match the sequences of chicken activin A.

bp (FIG. 3). This RT-PCR product was subcloned and sequenced, confirming its nucleotide sequence identity (99%) with the recently published sequence for chicken activin A.[38]

It thus appears that mRNA for activin A is present *in vivo* in the choroid tissues, consistent with the hypothesis that activin is a neurodifferentiation factor for somatostatinergic CG neurons. However, using this RT-PCR based technique, we were also able to detect activin A mRNA in a variety of other tissues from embryonic chicken eyes including the iris. This observation is problematic in light of our earlier finding that both ciliary and choroid neurons can respond to activin in culture,[24] yet only choroid neurons and not ciliary neurons express somatostatin *in vivo*. Somatostatin, *in vivo*, is used only by CG neurons innervating the choroid although both choroid and ciliary neurons are capable of expressing somatostatin in culture. Although this *in vivo* selectivity might be due to limiting amounts of activin produced exclusively in the eye by choroid cells, there may also be an inhibitory factor within the iris that ensures selective expression of this neuropeptide in choroid neurons.

Recently, Darland *et al.*[39] reported the detection and quantification of activin A and follistatin mRNAs in choroid and iris tissues using an RNase protection assay. These authors used a multiprobe RNase protection assay to simultaneously measure mRNAs for activin A, follistatin, and a constitutively expressed gene, chick ribosomal binding protein S17 (CHRPS).[40] Choroid and iris tissues at several embryonic ages from E9 to E16 were assayed, and the measurements were normalized to the levels of CHRPS transcript. In the iris, low and approximately equal levels of follistatin and activin A transcripts were present at E9, and these levels increased in parallel through E16. In the choroid, relatively high levels of activin A transcripts were present at E9 through E14 and then substantially declined at E16. Follistatin transcripts were detectable but at low levels from E9 through E16.

Immunocytochemistry was also used to determine whether the activin and

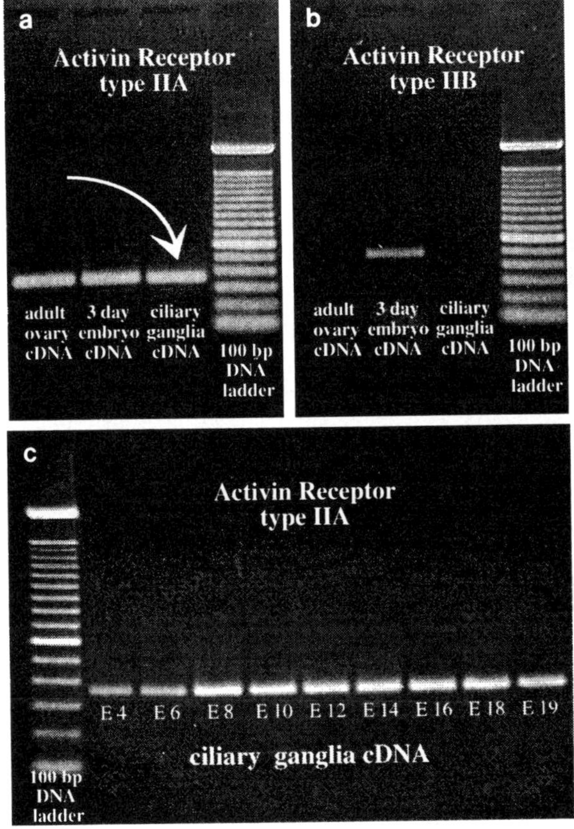

FIGURE 4. Messenger RNA for cActR-IIA is present in embryonic CG throughout development. **(a)** Primers specific for cActR-IIA used in RT-PCR with E15 choroid layer RNA amplify a band of the predicted size (473 bp) which, when subcloned, was found to match the sequence of chicken activin A. A corresponding amplified band was also present when RNA from E3 embryos and adult ovaries was used.[42] **(b)** Primers specific for cActR-IIB used in RT-PCR with E3 whole embryo RNA amplify a band of the predicted size which, when subcloned, was found to match the sequence of activin B. No corresponding amplified band was present when RNA from CG or adult ovaries was used.[42] **(c)** Primers specific for cActR-IIA used in RT-PCR with RNAs from CG of varying embryonic age from E4–E19 amplify bands of the appropriate size. This indicates that the mRNA for cActR-IIA is expressed from early in CG development up to a time shortly before hatching.

follistatin transcripts were translated into detectable levels of protein. Immunocytochemistry staining for activin A was found in the choroid at E12, and in the iris and cilary body at E11. Follistatin-like immunoreactivity was found in the iris and ciliary body at E11 but not in the choroid layer, nor was it detectable in the choroid layer at E12. These observations suggest that activin A is expressed in the choroid, iris, and ciliary body but that the activin inhibitor follistatin is expressed in significant levels only in the iris and ciliary body.

Thus, *in vivo*, the differential expression of somatostatin by choroid neurons can be explained not by a limited expression of activin A alone but instead by the regional expression of the inhibitor follistatin in the iris and ciliary body. Because of the presence of follistatin, activin is unlikely to be available to nerve terminals in the iris and ciliary body. In contrast, activin is likely to be available to nerve terminals innervating the choroid, thereby supporting somatostatin expression in the choroid neurons.

CILIARY GANGLION NEURONS EXPRESS THE ACTIVIN RECEPTOR TYPE IIA *IN VIVO*

Although both choroid and ciliary neurons can respond to activin when dissociated and placed in culture, this response to activin might be induced after the neurons are dissociated and placed into an artificial culture environment.

To examine whether CG neurons, *in vivo*, have the ability to respond to activin, and as a first step toward understanding the signal transduction mechanisms employed by CG neurons in responding to activin, we sought to determine whether any of the known activin receptors are expressed by developing CG neurons *in vivo*.

We used RT-PCR to detect chicken activin receptor type IIA (cAct-RIIA) and type IIB (cAct-RIIB) mRNAs obtained from CG. Design of the PCR primers was based on the published sequences for these receptors.[41] cActR-IIA specific primers amplified cDNA transcribed from CG RNA yielding a 358 bp DNA product (FIG. 4a). This RT-PCR product was subcloned and sequenced confirming its nucleotide sequence identity (99%) with the published sequence coding for chicken cActR-IIA.[41] In addition, RT-PCR with cActR-IIA specific primer pairs amplified cDNA derived from adult chicken ovaries and E3 whole embryos (FIG. 4a). In contrast, cActR-IIB specific primers did not amplify a product from CG cDNA. E3 embryo cDNA was used as a positive control template to assure the efficacy of the cActR-IIB PCR primers. Amplification of this cDNA yielded the expected 466 bp PCR product with 99% nucleotide identity with the chicken cActR-IIB sequence (FIG. 4b). Transcripts for the cActR-IIA were detected from the earliest time (E4) that the CG could be isolated through E19, shortly before hatching[42] (FIG. 4c). It thus appears that mRNA for the cActR-IIA, but not the cActR-IIB, is expressed in the CG. The cActR-IIA appears to be expressed throughout embryonic development of the chicken CG.

We used *in situ* hybridization on cryostat sections of CG to characterize which cell types contain mRNA for cActR-IIA. Digoxigenin-labeled riboprobes were synthesized from the cloned cActR-IIA fragment and *in situ* hybridization was performed according to a modification of the procedure described by Harland.[43] Antisense probes for cActR-IIA hybridized to cells within these sections (FIG. 5). No hybridization was detectable with sense strand riboprobes. Hybridization of the antisense probe was detectable in sections of CG from E4–E19. From

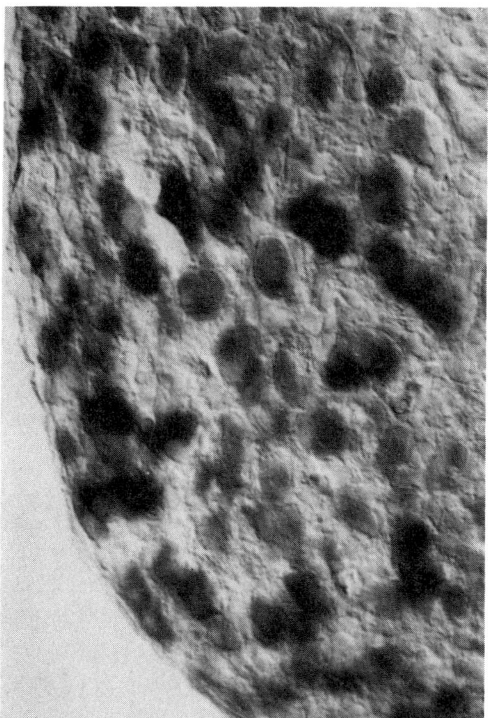

FIGURE 5. *In situ* hybridization of a cActR-IIA probe to CG neurons. The subcloned fragment of the cActR-IIA was used to produce digoxigenin-labeled riboprobes for *in situ* hybridization. The antisense strand cActR-IIA riboprobe hybridized to neurons within cryostat sections of CG.[42] In this differential interference contrast micrograph of a cryostat-sectioned E15 CG, the neurons contain a dark alkaline phosphatase reaction product indicating the sites of hybridization of the digoxigenin-labeled cActR-IIA sense riboprobe. Hybridization to both ciliary and choroid neurons is observed, suggesting that *in vivo* expression of the cActR-IIA occurs in both populations of neurons in the CG.

the earliest ages at which neurons could be morphologically distinguished (E7), hybridization of the cActR-IIA probe appears to be neuron specific.[42] Both ciliary and choroid neurons appear to hybridize with the cActR-IIA probe and do not significantly differ in their expression of mRNA for this activin receptor. It appears that, *in vivo*, both choroid and ciliary CG neurons express mRNA transcripts coding for an activin receptor. Consistent with our earlier findings in culture, both these types of CG neurons are thus likely to be capable of responding to activin, *in vivo*.

SUMMARY

Activin as a neurodifferentiation factor. Our studies of neurotransmitter expression have focused on the expression of neuropeptide transmitters in the avian ciliary ganglion (CG) and have examined the influence of choroidal vascular smooth muscle cells in regulating the differential expression of somatostatin in the CG. In these studies we have identified activin A as a potential target-derived neurodifferentiation factor that can stimulate somatostatin expression in cultured CG neurons. In cultured CG neurons, activin can stimulate the expression of somatostatin in choroid neurons, the pattern of neurotransmitter expression found *in vivo*, and in the ciliary neurons that would normally not express somatostatin. *In vivo*, mRNA transcripts of the cActR-IIA appear to be expressed by both choroid and ciliary CG neurons. This suggests that activin might serve as an instructive factor in controlling neuropeptide phenotype.

For activin to serve as an instructive factor requires that activin be produced by choroid smooth-muscle target cells. Indeed, activin mRNA and activin-like immunoreactivity are found in choroid cells, *in vivo*.[39] However, the lack of somatostatin expression by ciliary neurons suggests that activin is not produced by their targets, the iris and ciliary body. This simple view is countered by the observation that activin A mRNA is also present in the iris and activin-like immunoreactivity is detectable in the iris and ciliary body.[39] Instead, the production of the specific activin inhibitor follistatin in the iris and ciliary body is likely to limit the availability of activin to only those neurites innervating the choroid layer, thus accounting for the differential expression of somatostatin in only the choroid CG neurons. This somewhat more complicated arrangement is similar to the mechanism thought to be employed for primary induction during frog embryogenesis.[44,45]

The observations reviewed here are all consistent with the hypothesized role for activin as a molecule whose availability to neurites in the target regulates neurotransmitter expression. Additional *in vivo* perturbation experiments are needed to further examine this hypothesis; nevertheless, activin appears as a strong candidate for a target-derived neurotransmitter differentiation factor.

Activin's potential roles in differentiation. A wide variety of biological effects have been ascribed to activin. Initially identified and purified as a gonadal hormone stimulating the production and release of FSH from the pituitary,[46,47] activin is also implicated in the stimulation of erythroid differentiation,[48] as a modulator of follicular granulosa cell differentiation,[49,50] as a mesodermalizing factor in both amphibian[51] and avian[52] early development, and as a component in establishing left-right axial patterning in the chicken embryo.[53] Activin has also been found to be a survival factor for several neuronal cell lines and for rat embryonic neural retina cells in culture.[54] However, activin is not a survival factor for chicken CG neurons in culture.[54] Our observation that activin may play a function in target-derived control of neuropeptide expression adds yet another aspect to the list of its potential biological functions. In addition, activin shares regions of amino acid sequence identity with members of the TGF-β superfamily, which includes the TGF-βs, Mullerian inhibitory substance, Drosophila decapentaplegic gene product, dorsalin, bone morphogenetic proteins, inhibin, and glial-derived neurotrophic factor. Interestingly, these are all factors that have effects upon cellular differentiation.

Effects of activin on other neurons. Activin A—as well as two other TGF-β superfamily members, BMP-2 and BMP-6—has been shown to induce expression of mRNAs for several neuropeptides in cultured rat sympathetic neurons.[9,10] In addition, activin A induces ChAT mRNA in cultured sympathetic neurons.[9] In preliminary *in situ* hybridization experiments we found that embryonic chicken dorsal root ganglion neurons express mRNA for the cActR-IIA, suggesting the possibility that these sensory neurons may respond to activin as well. Thus, in addition to the role hypothesized for activin in CG development, activin and other TGF-β superfamily members may serve as neurodifferentiation factors controlling neuropeptide expression in other regions of the developing nervous system.

ACKNOWLEDGMENTS

We thank Walter Debnam and Tatjana Odineca for their excellent technical assistance.

REFERENCES

1. PATTERSON, P. H. & L. Y. CHUN. 1977. The induction of acetylcholine synthesis in primary cultures of dissociated rat sympathetic neurons. I. Effects of conditioned medium. Dev. Biol. **56:** 263–280.
2. KESSLER, J. A., J. E. ADLER, G. M. JONAKAIT & I. B. BLACK. 1984. Target organ regulation of substance P in sympathetic neurons in culture. Dev. Biol. **103:** 71–79.
3. POTTER, D. D., S. C. LANDIS, S. G. MATSUMOTO & E. J. FURSHPAN. 1986. Synaptic functions in rat sympathetic neurons in microcultures. II. Adrenergic/cholinergic dual status and plasticity. J. Neurosci. **6:** 1080–1090.
4. IACOVITTI, L., M. J. EVINGER, T. H. JOH & D. J. REIS. 1989. A muscle derived factor(s) induces expression of a catecholamine phenotype in neurons of cultured rat cerebral cortex. J. Neurosci. **9:** 3529–3537.
5. SCHOENEN, J., P. DELTREE, P. LEPRINCE & G. MOONEN. 1989. Neurotransmitter phenotype plasticity in cultured dissociated adult rat dorsal root ganglia: An immunological study. J. Neurosci. Res. **22:** 473.
6. LANDIS, S. C. 1990. Target regulation of neurotransmitter phenotype. TINS **13:** 344–350.
7. NAWA, H. & D. W. Y. SAH. 1990. Different biological activities in conditioned media control the expression of a variety of neuropeptides in cultured sympathetic neurons. Neuron **4:** 279–287.
8. NAWA, H., Y. BESSHO, J. CANAHAN, S. NAKANISHI & K. MIZUNO. 1993. Regulation of neuropeptide expression in cultured cerebral cortical neurons by brain-derived neurotrophic factor. J. Neurochem. **60:** 772–775.
9. FANN, M. J. & P. H. PATTERSON. 1994. Neuropoietic cytokines and activin A differentially regulate the phenotype of cultured sympathetic neurons. Proc. Natl. Acad. Sci. USA **91:** 43–47.
10. FANN, M. J. & P. H. PATTERSON. 1994. Depolarization differentially regulates the effects of bone morphogenetic protein (BMP)-2, BMP-6, and activin A on sympathetic neuronal phenotype. J. Neurochem. **63:** 2074–2079.
11. McMAHON, S. B. & S. GIBSON. 1987. Peptide expression is altered when afferent nerve reinnervates inappropriate tissue. Neurosci. Lett. **73:** 9–15.
12. SCHOTZINGER, R. J. & S. C. LANDIS. 1988. Cholinergic phenotype developed by noradrenergic sympathetic neurons after innervation of a novel cholinergic target tissue *in vivo*. Nature **335:** 637–639.
13. STEVENS, L. M. & S. C. LANDIS. 1987. Development and properties of the secretory response in rat sweat glands: Relationship to the induction of cholinergic function in sweat gland innervation. Dev. Biol. **123:** 179–190.
14. HORGAN, K. & D. VAN DER KOY. 1992. Visceral targets specify calcitonin gene-related peptide and substance P enrichment in trigeminal afferent projection. J. Neurosci. **12(4):** 1135–1143.
15. OPPENHEIM, R. W. 1989. The neurotrophic theory and naturally occurring motoneuron death. TINS **12(7):** 252–255.
16. MARWITT, R., G. PILAR & J. WEAKLY. 1971. Characterization of two ganglion cell populations in avian ciliary ganglia. Brain Res. **25:** 317–334.

17. MERINEY, S. D. & G. PILAR. 1987. Cholinergic innervation of the smooth muscle cells in the choroid coat of the chick eye and its development. J. Neurosci. **7:** 3827–3839.
18. EPSTEIN, M. L., J. P. DAVIS, L. E. GELMAN, J. R. LAMB & J. L. DAHL. 1988. Cholinergic neurons of the chicken ciliary ganglion contain somatostatin. Neuroscience **25:** 1053–1060.
19. GRAY, D. B., D. ZELAZNY, N. MANTHAY & G. PILAR. 1990. Endogenous modulation of ACh release by somatostatin and the differential roles of Ca^{2+} channels. J. Neurosci. **10:** 2687–2698.
20. NARAYANAN, C. H. & Y. NARAYANAN. 1978. On the origin of the ciliary ganglion in birds studied by the method of interspecific transplantation of embryonic brain regions between quail and chick. J. Embryol. Exp. Morphol. **46:** 137–148.
21. NARAYANAN, C. H. & Y. NARAYANAN. 1976. An experimental injury into the central source of preganglionic fibers to the chick ciliary ganglion. J. Comp. Neurol. **166:** 101–110.
22. DE STEFANO, M. E., A. CIOFI LUZZATTO & E. MUGNAINI. 1993. Neuronal ultrastructure and somatostatin immunolocalization in the ciliary ganglion of chicken and quail. J. Neurocytol. **22:** 868–892.
23. HAMBURGER, V. & H. L. HAMILTON. 1951. A series of normal stages in the development of the chick embryo. J. Morphol. **88:** 49–92.
24. COULOMBE, J. N. & R. NISHI. 1991. Stimulation of somatostatin expression in developing ciliary ganglion neurons by cells of the choroid layer. J. Neurosci. **11(2):** 553–562.
25. MERINEY, S. D. & G. PILAR. 1987. Cholinergic innervation of the smooth muscle cells in the choroid coat of the chick eye and its development. J. Neurosci. **7:** 3827–3839.
26. NISHI, R. & D. K. BERG. 1981. Two components from eye tissue that differentially stimulate the growth and development of ciliary ganglion neurons in cell culture. J. Neurosci. **1:** 505–513.
27. NISHI, R. & D. K. BERG. 1977. Dissociated ciliary ganglion neurons in vitro: Survival and synapse formation. Proc. Natl. Acad. Sci. USA **74:** 5171–5175.
28. SMET, P. J. & R. A. RUSH. 1993. Effect of ciliary neuronotrophic factor on somatostatin expression in chick ciliary ganglion neurons. Brain Res. **609:** 351–356.
29. COULOMBE, J., R. NISHI & F. ECKENSTEIN. 1991. Evidence for target cell secretion of survival and differentiation factors for ciliary ganglion neurons. Soc. Neurosci. Abstr. **17(2):** 1122.
30. KORZENIEWSKE, C. & D. M. CALLEWAERT. 1983. An enzyme-release assay for natural cytotoxicity. J. Immunol. Methods **664:** 313–320.
31. KATZ, L. C., A. BURKHALTER & W. J. DREYER. 1984. Fluorescent latex microspheres as a retrograde neuronal marker for *in vivo* studies of visual cortex. Nature **310:** 786–789.
32. COULOMBE, J. & M. BRONNER-FRASER. 1990. Development of cholinergic traits in the quail ciliary ganglion: Expression of choline acetyltransferase-like immunoreactivity. Neuroscience **37(1):** 259–270.
33. COULOMBE, J. N., R. SCHWALL, A. S. PARENT, F. ECKENSTEIN & R. NISHI. 1993. Induction of somatostatin immunoreactivity in cultured ciliary ganglion neurons by activin in choroid cell-conditioned medium. Neuron **10:** 899–906.
34. SCHWALL, R. H. & C. LAI. 1991. Erythroid differentiation bioassays for activin. Methods Enzymol. **198:** 340–346.
35. YU, J., L-E. SHAO, J. VAUGHAN, W. VALE & A. L. YU. 1989. Characterization of the potentiation effect of activin on human erythroid colony formation in vitro. Blood **73:** 952–960.
36. KOKAN-MOORE, N. P., D. BOLENDER & J. LOUGH. 1991. Secretion of inhibin βA by endoderm cultured from early embryonic chicken. Dev. Biol. **146:** 242–245.
37. NAKAMURA, T., K. TAKIO, Y. ETO, H. SHIBAI, K. TITANI & H. SUGINO. 1990. Activin-binding protein from rat ovary is follistatin. Science **247:** 836–838.
38. CHEN, C.-C. & P. A. JOHNSON. 1996. Molecular cloning of inhibin activin beta(A)-subunit complementary deoxyribonucleic acid and expression of inhibin activin alpha- and beta(A)-subunits in the domestic hen. Biol. Reprod. **54(2):** 429–435.
39. DARLAND, D. C., B. A. LINK & R. NISHI. 1996. Activin A and follistatin expression in developing targets of ciliary ganglion neurons suggests a role in regulating neurotransmitter phenotype. Neuron **15:** 857–866.

40. TRUEB, B., T. SCHREIER, K. H. WINTERHALTER & E. STREHLER. 1988. Sequence of a cDNA clone encoding chicken ribosomal protein S17. Nucleic Acids Res. **16:** 4723.
41. STERN, C. D., R. T. YU, A. KAKIZUKA, C. R. KINTNER, L. S. MATHEWS, W. W. VALE, R. M. EVANS & K. UMESONO. 1995. Activin and its receptors during gastrulation and the later phases of mesoderm development in the chick embryo. Dev. Biol. **172(1):** 192–205.
42. KOS, K. & J. COULOMBE. 1995. Activin type IIA receptor mRNA expression by neurons of the avian ciliary ganglion. Soc. Neurosci. Abstr. **21(3):** 1783.
43. HARLAND, R. M. 1991. In situ hybridization: An improved whole-mount method for Xenopus embryos. Methods Cell Biol. **36:** 685–694.
44. HEMMATI-BRINVALOU, A. & D. A. MELTON. 1992. Inhibition of activin receptor signaling promotes neuralization in Xenopus. Cell **77:** 273–281.
45. HEMMATI-BRIVANLOU, A., O. G. KELLY & D. A. MELTON. 1992. Follistatin, an antagonist of activin, is expressed in the Spemann organizer and displays direct neutralizing activity. Cell **77:** 283–295.
46. VALE, W., A. HSUEH, C. RIVIER & J. YU. 1990. The inhibin/activin family of hormones and growth factors. *In* Peptide Growth Factors and Their Receptors. M. B. Sporn & A. B. Roberts, Eds.: 211–248. Springer-Verlag. Berlin.
47. WOODRUFF, T., R. LYON, S. HANSEN, G. RICE & J. MATHER. 1990. Inhibin and activin locally regulate rat ovarian folliculogenesis. Endocrinology **127(6):** 3196–3205.
48. YU, J., L-E. SHAO, J. VAUGHAN, W. VALE & A. L. YU. 1989. Characterization of the potentiation effect of activin on human erythroid colony formation in vitro. Blood **73:** 952–960.
49. SUGINO, H., T. NAKAMURA, Y. HASEGAWA, K. MIYAMOTO, M. IGARASHI, Y. ETO, H. SHIBAI & K. TITANI. 1988. Identification of a specific receptor for erythroid differentiation factor on follicular granulosa cell. J. Biol. Chem. **263(30):** 15249–15252.
50. HUTCHINSON, L., J. FINDLAY, F. DE VOS & D. ROBERTSON. 1987. Effects of bovine inhibin, transforming growth factor-beta and bovine activin-A on granulosa cell differentiation. Biochem. Biophys. Res. Commun. **146(4):** 1405–1412.
51. SMITH, J., K. VAN NIMMEN & D. HUYLEBROECK. 1990. Identification of a potent Xenopus mesoderm-inducing factor as a homologue of activin A. Nature **345:** 729–731.
52. MITRANI, E., T. ZIV, G. THOMSEN, Y. SHIMONI, D. MELTON & A. BRIL. 1990. Activin can induce the formation of axial structures and is expressed in the hypoblast of the chick. Cell **63:** 495–501.
53. LEVIN, M., R. L. JOHNSON, C. D. STERN, M. KUEHN & C. TABIN. 1995. A molecular pathway determining left-right asymmetry in chick embryogenesis. Cell **82:** 803–814.
54. SCHUBERT, D., H. KIMURA, M. LACORBIERE, J. VAUGHAN, D. KARR & W. FISCHER. 1990. Activin is a nerve cell survival molecule. Nature **344:** 868–870.

Neonatal ACTH Administration Elicits Long-term Changes in Forebrain Monoamine Innervation

Subsequent Disruptions in Hypothalamic-Pituitary-Adrenal and Gonadal Function[a]

S. E. ALVES,[b] H. M. AKBARI,[c] G. M. ANDERSON,[c]
E. C. AZMITIA,[d] B. C. McEWEN,[b] AND F. L. STRAND[d]

[b]Laboratory of Neuroendocrinology, Box 165
The Rockefeller University
1230 York Avenue
New York, New York 10021

[c]The Yale Child Study Center
Yale University School of Medicine
New Haven, Connecticut 06520

[d]Department of Biology and Center for Neural Science
New York University
New York, New York 10003

INTRODUCTION

Perinatal manipulation of the hypothalamic-pituitary-adrenal (HPA) axis, either by stress or by the administration of pituitary/adrenal stress hormones, alters the development of central monoamine neurons.[1–7] Such changes in the monoaminergic systems are believed to be involved in neuroendocrine and behavioral alterations observed in perinatally "stressed" rodents. For example, the relatively small stress of handling neonatal rats for the first 7 days of life can promote the development of a permanently more efficient HPA system, manifested by an increase in hippocampal type II glucocorticoid receptor levels, thereby facilitating the termination of a stress response.[8] This alteration in HPA function in turn appears to attenuate some of the changes that occur in the aging central nervous system (CNS).[9] These apparently permanent and wide-reaching neuroendocrine effects are believed to involve the serotonergic innervation of the hippocampus during this critical developmental period.[10]

Our laboratory has been interested in the effects of stress hormones on the sexual differentiation of the rat brain, and the role of the monoaminergic systems in this process. Segarra and colleagues[5,11] demonstrated the late-gestational treatment of adrenocorticotropic hormone (ACTH) decreases the expression of masculine sexual behavior in young adult male rats, a behavioral change linked to an enhanced inhibitory serotonergic input within the medial preoptic area, a primary regulatory site for male sexual behavior. Subsequently, we investigated the mono-

[a] This work was supported by The Council for Tobacco Research.

226

amine innervation of the developing female hypothalamus following ACTH treatment during the first postnatal week, the period believed to be critical for the feminization of the female rat brain,[12,13] and normally a time of relative HPA quiescence. Postnatal ACTH administration, which greatly increased endogenous corticosterone (CORT) release from the neonatal adrenal glands, significantly increased hypothalamic serotonin (5-HT) and dopamine (DA) innervation, as measured by high-affinity specific uptake, in both neonatal (day 7) and young adult (days 80–90) female rats.[7] Associated with these neural changes, hormone-treated animals displayed a disruption in reproductive physiology, manifested by a delay in the onset of puberty and some deficits in female sexual behavior as intact young adults.[7]

The apparent trophic action of stress hormones on the developing 5-HT system is in agreement with a study by Azmitia and de Kloet[14] which demonstrated that ACTH and its various neurotrophic peptide fragments enhance neurite outgrowth of cultured fetal 5-HT neurons. Other, earlier studies have shown that the adrenal corticosteroids enhance rodent brain tryptophan hydroxylase activity,[15-17] and more recently its synthesis,[18] as well as tyrosine hydroxylase activity,[19,20] the rate-limiting enzymes in the production of 5-HT and the catecholamines, respectively. It should be pointed out that 5-HT and the catecholamines themselves appear to act as regulatory factors during brain development, assisting in cellular differentiation, proliferation, and synaptogenesis.[21-24] Taken together, these data suggest that elevated levels of ACTH/CORT during development can have profound effects on the maturation and subsequent function of numerous central neuronal circuits.

Although many studies have focused on the effects of prenatal stress or stress hormone treatment on the central monoamine systems, few have looked at the effects of early postnatal stress hormone increases on the development of these neural systems, and, to our knowledge, no one has investigated possible long-term consequences. Indeed, findings from studies of postnatal effects of ACTH could have important clinical relevance because ACTH has been widely used to treat infantile spasms, a neonatal/early childhood neurological disorder with a suspected serotonergic etiology.[25-28] Although peripherally administered ACTH has rapid and potent antimyoclonic and antiepileptic effects, its mechanism of action remains unknown.[27,28] Despite the acute therapeutic benefits of ACTH, several clinical studies using computed tomography (CT) scan have suggested that ACTH treatment induces cerebral atrophy in human patients, probably via adrenocortical secretions.[29-31]

The apparent plasticity of the neonatal HPA axis and the marked effects of ACTH on CORT secretion in the neonatal rat, as well as the reported *in vitro* and *in vivo* effects of monoaminergic neurons and the possible long-term deleterious effects of ACTH administration to human neonates, have prompted us to study this area further. In the present study, we more extensively investigate postnatal effects of ACTH on the development and maintenance of the forebrain monoamine systems, as well as on the functioning of the HPA and gonadal axes. Specifically, monoamine content and activity are assessed in the hypothalamus, as well as two other behaviorally and functionally relevant forebrain regions that receive extensive monoaminergic projections, the striatum and the cortex, over the course of one year. Furthermore, we investigate whether ACTH-induced stimulation of the neonatal adrenal cortex persistently alters basal and/or post-stress adrenocortical function beyond the end of the treatment period. Finally, we attempt to elucidate a physiological basis underlying the postnatal ACTH-related deficits in female sexual behavior that we previously reported among young, intact virgin animals.

EXPERIMENTAL DESIGN AND METHODS

Animals and Treatment

Pregnant Sprague-Dawley rats (gestation day 14) from Hilltop Laboratories
were housed individually; water and Purina rat chow were provided ad libitum.
Animals were kept in a 12L:12D reversed photoperiod with lights off at 12 noon.
On the day of birth (day 1), the pups were weighed and each litter culled to five
females and three males to maintain the same sex ratio. Because our model is the
female rat, only female offspring were used in this study. Pups were weighed and
injected subcutaneously (sc) from day 1 (day of birth) to day 7 with either ACTH
(1-24) (0.5 mg/kg) donated by Organon B.V., or 0.9% saline, once daily between 10
A.M. and 12 noon. The peptide dosage was consistent with our previous studies.[5,7]
ACTH 1-24 was dissolved in saline and administered in a volume of 0.1 mL/10 g
body weight. This peptide contains the full steroidogenic properties of the whole
molecule, ACTH 1-39, and will be referred to as ACTH throughout the rest of
the paper.

Plasma Stress Hormones

Neonatal Treatment Period

Plasma ACTH and CORT levels were measured at the end of the treatment
period, on postnatal day 7, to determine the effects of saline or ACTH injection
on circulating ACTH levels, and the adrenal response to these treatments. Neonates from each treatment group ($n = 5$) were rapidly decapitated 45 minutes
following their final injection. Trunk blood from five animals per group was collected in EDTA and pooled and adrenal glands were dissected out and weighed.

Juvenile to Adult, Peak Basal Plasma CORT

Plasma CORT levels were measured at several points during development
(days 15, 32, 65, 90, and 1 year, 5 to 7 rats/treatment/age) at the diurnal "peak,"
just prior to darkness, to determine whether neonatal ACTH adrenal stimulation
would produce a long-term effect on subsequent basal adrenocortical activity.
Although measurement at this one time point does not give a full profile of adrenocortical secretion throughout the circadian cycle, it does allow an assessment of
"maximum" basal pituitary/adrenal activity.

Adult Post-Stress Plasma ACTH and CORT

Young adult virgin animals (90 days) were exposed to the handling manipulation
of vaginal smearing for the first time, approximately three hours prior to darkness.
Approximately 45 minutes after, blood was collected from animals found to be
in proestrus (basal bodies are the predominant cell type in the smear; $n = 6$ saline,
$n = 9$ ACTH).
    Middle-aged, ovariectomized animals (1 year old; $n = 6$ per group) were ex-
posed to ether vapors for one minute by placing them in a sealed glass desiccator
containing paper napkins dampened with ether below the porous bottom. Animals

were removed to the home cages, the time was recorded, and then the animals were sacrificed one hour later.

Animals were anesthetized with sodium pentobarbital (60 mg/kg) and blood was collected via cardiac puncture with a 10-mL syringe containing 1 mL of EDTA, approximately one hour before the onset of darkness. Following extraction, all blood was immediately placed on ice, and then centrifuged in a Beckman TJ-6 refrigerated centrifuge at 5,000 $\times$ g for 25 min. The plasma was placed in vials and stored at -70 °C until the day the assays were performed.

CORT radioimmunoassay. Plasma CORT levels were determined using the Coat-A-Count rat CORT radioimmunoassay kit from Diagnostic Products Corporation (Los Angeles, CA) with [^{125}I]CORT as a tracer. The CORT antiserum is highly specific for rat CORT with minimal cross-reactivity with deoxycorticosterone ($\sim$2%) and less than 1% for other steroid hormones present in samples. Determination of hormonal levels was interpolated from a standard curve prepared in triplicate. Plasma samples (50 μL) were run in duplicate (triplicate for pooled samples). The data reduction was calculated by linear regression and logit-log representation with the aid of a computer program. The intra- and interassay coefficients of variance were 4.3 and 5.8%, respectively. The assay had a sensitivity of up to 5.7 ng/mL.

ACTH radioimmunoassay. Determination of plasma ACTH levels was performed using the ACTH Double Antibody RIA kit from Diagnostic Products Corporation. The antiserum is highly specific for ACTH (including 100% reactivity with ACTH 1-24) with very low ($<$1.0%) cross-reactivity to other compounds that might be present in the samples. Plasma samples (100 μL) were incubated with ACTH antiserum for one hour at room temperature and then with [^{125}I]ACTH overnight at 4 °C. Goat anti-rabbit gamma globulin (GARGG) was used to separate bound from free fractions. All samples were assayed in duplicate (triplicate for pooled samples). Concentrations were interpolated from a standard curve prepared in triplicate. The data reduction was calculated by linear regression and logit-log representation with the aid of a computer program. The sensitivity of the assay was 8 pg/mL, and the intra- and interassay coefficients of variance were 8.9 and 6.4%, respectively.

Forebrain Monoamine Levels and Metabolism

Neurochemical concentrations were determined using high-performance liquid chromatography (HPLC) following a previously described technique.[32,33] Animals at days 7, 15, 90, and 1 year ($n = 6$/treatment/age) were rapidly decapitated under sodium pentobarbital anesthesia (60 mg/kg) and the brains were removed, immediately frozen in powdered dry ice, and stored at -70 °C. On the days the samples were prepared, brains were partially thawed, and the hypothalamus, striatum, and cortex were dissected and weighed. Ice cold pH 6.5, 0.1 M MOPS buffer (10/1 v/wt) containing 0.1% EDTA, 0.1% ascorbic acid, and internal standard [3,4 dihydroxybenzylamine (DHBA, 200 ng/mL), or *N*-methylserotonin (NMS, 200 ng/mL)] was added to the brain tissue. The samples were then sonicated on ice using a Branson Sonic Power model 185 sonifier for three cycles of 10 s on/10 s off. The homogenate was then divided, a portion immediately frozen, and a portion deproteinized by adding 10% v/v 3.2 M $HClO_4$ and centrifuging for 5 min at 12,000 $\times$ g. All samples were then frozen at -70 °C until analyses were performed.

For 5-HT, 5-hydroxyindoleacetic acid (5-HIAA), and homovanillic acid (HVA) measurements, NMS was used as an internal standard. Determination was made by directly injecting the supernatant of the deproteinized sample and analyzing it by HPLC with serial fluorometric and electrochemical detection.

DHBA was used as an internal standard for norepinephrine (NE), DA, and 3,4-dihydroxyphenylacetic acid (DOPAC, measured only in the striatum). Determinations were made by HPLC electrochemical detection after alumina extraction. Neurochemical concentrations were expressed as nanograms per gram of wet tissue weight. Monoamine turnover rates (for 5-HT and DA) were estimated by the ratio of metabolite to neurotransmitter levels.

Reproductive Function and Behavior

Ovarian Hormone Profile

Intact, young adult (90 days) animals were smeared beginning three hours before the onset of darkness. Approximately 45 minutes following vaginal smearing, animals found to be in proestrus were sacrificed, and blood was collected as described above. Plasma estradiol and progesterone levels were measured via specific radioimmunoassays (saline = 9; ACTH = 12).

Radioimmunoassays

Plasma hormone levels were determined using Coat-a-Count estradiol and progesterone RIA kits from Diagnostic Products Corporation. The antisera used in each of these assays are highly specific for the hormone measured (estradiol or progesterone, respectively) with very little cross-reactivity to other naturally occurring steroid hormones (e.g., cross-reactivity of the progesterone antiserum with 11-deoxycorticosterone is 1.7%). The sample size required for both assays was 100 μL. The data reduction was calculated by the conventional RIA techniques of linear regression and logit-log representation with standard calibrators and quality control serum pools (assayed in triplicate), and the aid of a computer program. All samples were assayed in duplicate. The estradiol assay had a sensitivity of 8 pg/mL, and intra- and interassay coefficients of variance of 5.8 and 7.4%, respectively. The progesterone assay had a sensitivity of 0.05 ng/mL and intra- and interassay coefficients of variance of 5.1 and 8.8%, respectively.

Female Sexual Behavior—Estradiol Dose Response

Young adult female rats (60 days old) were ovariectomized (saline, $n = 7$; ACTH, $n = 9$) under ketamine (0.075 mL/100 g) and xylazine (0.025 mL/100 g) anesthesia and allowed one week to recover. These animals were then tested for female sexual behavior during three separate trials, each spaced approximately one week apart, starting with the initial "virgin" trial (day 70) and two subsequent "experienced" trials (days 80–90). Estradiol benzoate (EB) was administered sc, approximately 52 hours and 28 hours prior to testing. Dosages of EB administered were (1) 1 μg for the virgin trial, (2) 2.5 μg for the second trial, and (3) 5 μg for the third trial. Approximately five hours before each trial, animals were injected sc with 500 μg of progesterone. The lordosis quotient (LQ) and the lordotic quality score (LQS), a quantitative and a qualitative measure of female sexual behavior, respectively,[34] were recorded for each animal in response to a mounting male as we previously described.[7]

Female Sexual Behavior—Intact, Experienced Animals

Intact, sexually experienced female rats, approximately one year of age, were tested for female sexual behavior to determine whether neonatal ACTH treatment would produce long-lasting deficits in this behavior.

"Male Typical" Sexual Behavior Test

Ovariectomized female rats were tested for male typical sexual behavior under the influence of testosterone to observe whether neonatal ACTH administration would have altered the sexual differentiation of the brain in the male direction. Animals received a 3-cm silastic tube filled with testosterone propionate implanted in the dorsal neck region. One week later, these "experimental females" were tested for male sexual behavior against "stimulus" females (intact females that received a 2-cm silastic tube filled with EB one week prior and 500 μg of progesterone four hours before testing). All animals were allowed one minute to acclimate to the test environment. Each "experimental" female was tested for male copulatory behavior as expressed in interest toward the "stimulus" females (number of genital sniffs), number of mounts, and intromission/ejaculation patterns within a 5-minute test period ($n = 6$ per treatment).

Statistical Analysis

All data were analyzed by the two-tailed Student's *t* test. Significance was determined at $p < 0.05$.

RESULTS AND DISCUSSION

Body Weight

ACTH treatment (PD 1–7) produced a small but significant decrease in body weight, detectable by postnatal day 3. This significant decrease in weight continued to day 15, but by day 25 no significant change was evident. By adulthood, at 90 days of age, body weights of hormone- and saline-treated animals were nearly identical (FIG. 1).

Hypothalamic-Pituitary-Adrenal Profile: Plasma Stress Hormones and Adrenal Weights

Neonatal Treatment Period

Plasma ACTH levels measured 45 minutes after the last injection, on day 7, were nearly 10 times greater in ACTH-treated neonates compared to those measured in the saline control group. The mean plasma ACTH concentration was 889.1 ± 6.2 pg/mL among hormone-treated animals versus 99.2 ± 7.8 pg/mL in saline-treated controls ($n = 5$, pooled samples; $p < 0.001$; $\pm$ SEM). Simultaneously, ACTH-treated animals exhibited extremely high plasma CORT levels compared to those measured in control animals ($p < 0.001$, FIG. 2). In contrast to 4-day-old pups that had nondetectable plasma CORT following injection of vehicle,[7] 7-day-old saline-treated animals did have low levels of plasma CORT, as would be expected in this neonatal stress hyporesponsive period. In addition, ACTH treatment induced adrenal hypertrophy by the end of the treatment period on day 7. The mean weight of paired adrenals from ACTH-treated animals was 2.320 ± 0.243 mg, compared to 1.180 ± 0.095 mg for saline-treated controls ($p < 0.002$).

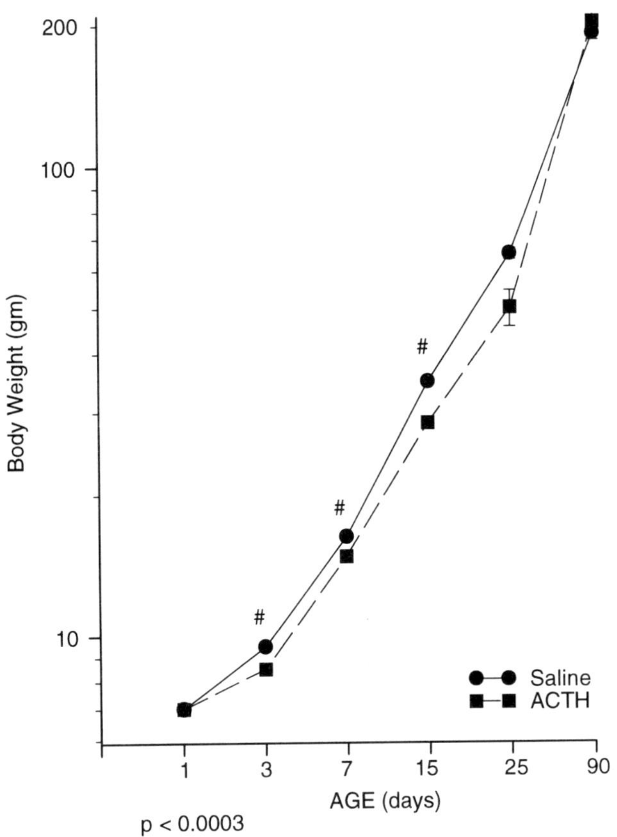

FIGURE 1. Growth curve showing change in body weight from day 1 through several developmental ages and adulthood of female rats treated with either saline or ACTH (0.5 mg/kg/day) for the first 7 days postnatal. Values are expressed as weight $\pm$ SEM (days 1–15, $n = 12$; days 25 and 90, $n = 7$). Note that weight (grams) on the Y-axis is on a log scale to more clearly demonstrate the changes in the lower range.

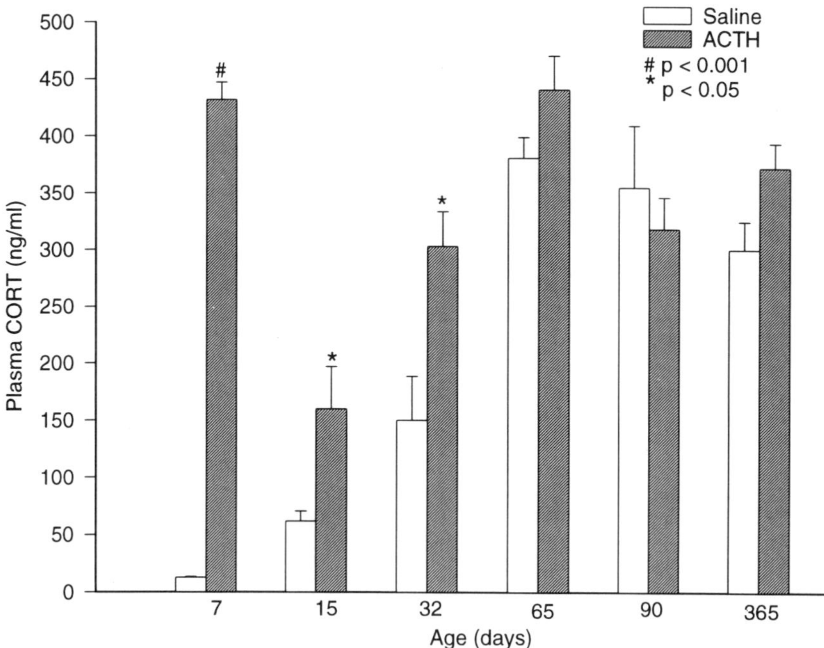

FIGURE 2. Plasma corticosterone (CORT) levels following the last sc injection with either saline or ACTH (0.5 mg/kg/day) on day 7, and then "peak" basal levels at various ages thereafter. Values are expressed as ng/mL ± SEM (day 7, $n = 5$ [pooled]; days 15–365, $n = 5–7$).

Juvenile to Adult, Peak Basal CORT Levels

Daily ACTH treatment from day 1 to day 7 significantly elevated peak basal CORT levels on days 15 and 32, indicating that this neonatal HPA manipulation produces a prolonged effect on basal adrenocortical activity (FIG. 2). Thus, brains of peptide-treated animals were exposed to greatly elevated circulating corticosteroid levels from the neonatal period at least through to pubescence. Although plasma CORT levels of ACTH-treated animals tended to be higher than control levels at 65 days, the difference was not significant. No significant difference in peak basal CORT levels was seen between saline and ACTH treatment groups at 90 days or 1 year (FIG. 2). The decreased body weight exhibited by peptide-treated animals was concurrent with significantly elevated basal CORT levels; as CORT is largely catabolic, chronic exposure to elevated levels of this steroid can decrease body mass.

Adult, Post-Stress Plasma ACTH and CORT

Young adult animals (90 days, proestrus) that had been treated neonatally with ACTH exhibited significantly increased plasma CORT levels compared to control animals, approximately 45 minutes following vaginal smearing for the first time (TABLE 1). Considering that the stressful manipulations involved in vaginal smearing (removing animals from their cages, handling, and obtaining the vaginal lavage) are relatively mild compared to other experimental "stressors" often employed, and that the procedure is executed rather quickly (approximately 1–2 min), one would expect that the stress response would be short in duration. Thus, a 30% increase in plasma CORT levels among ACTH-treated animals, 45 minutes after the stressor, suggests a hyperactive or prolonged adrenocortical response in these animals compared to controls.

Interestingly, this increase in plasma CORT was correlated with a significant decrease in circulating ACTH levels (TABLE 1). Perhaps the elevated circulating CORT levels among hormone-treated animals acted more rapidly, either centrally (hippocampal type II receptors) and/or at the pituitary to inhibit further ACTH release. However, a rapid and thus more efficient shutoff of the stress response is usually correlated with a decrease in both pituitary and adrenal stress hormones. Alternatively, perhaps developmental exposure to greatly elevated ACTH/CORT concentrations could have disrupted pituitary cell (corticotrope) differentiation, thereby altering the activity and/or responsiveness of the pituitary to stress. It is apparent that adrenal cells were affected by this treatment, as indicated by the observed adrenal weight difference at seven days. A hypertrophy/accelerated maturation of the steroidogenic adrenocortical cells could have persistently heightened their responsiveness to ACTH stimulation.

It was not surprising that middle-aged animals (1 year old) that had been ovariectomized as young adults exhibited rather low post-stress plasma CORT levels, because the higher basal and post-stress activity of the HPA axis—and thus plasma CORT levels typical of post-pubescent female rats—result from ovarian stimulation. Despite ovariectomy, ACTH-treated animals exhibited significantly elevated post-stress plasma CORT compared to saline-treated animals, measured one hour after exposure to ether (TABLE 2). Although peptide-treated animals also tended to have higher post-stress ACTH levels, these differences did not attain statistical significance due to variability within both groups (TABLE 2).

These findings demonstrate that, in addition to a developmental increase in basal plasma CORT levels, animals treated neonatally with ACTH continue to exhibit a significantly elevated adrenocortical response to stress through adulthood. An important determinant for the normal functioning of the HPA system seems to be a quick and efficient turning on and shutting off of the adrenocortical stress response.[35] Persistent elevation of circulating CORT following chronic stress decreases the number of cytosolic type II corticosteroid receptor sites within the

TABLE 1. Plasma ACTH and Corticosterone Levels at 90 Days (Proestrus), Approximately 45 Minutes after Vaginal Smearing

Treatment	ACTH (pg/mL)	Corticosterone (ng/mL)
Saline (n = 6)	113.0 ± 10.9	504.3 ± 41.8
ACTH (0.5 mg/kg) (n = 9)	78.6 ± 5.3[a]	647.8 ± 32.9[a]

Values are the mean ± SEM.
[a] $p < 0.02$.

TABLE 2. Plasma ACTH and Corticosterone Levels at One Year (Ovariectomized), Approximately One Hour after Ether Exposure

Treatment	ACTH (pg/mL)	Corticosterone (ng/mL)
Saline ($n = 6$)	57.0 ± 10.3	268.0 ± 7.0
ACTH (0.5 mg/kg) ($n = 6$)	71.3 ± 10.2	396.4 ± 25.1[a]

Values are the mean ± SEM.
[a] $p < 0.02$.

hippocampus[36,37] which can ultimately lead to a reduction in the stress shutoff response.[38] Whether the intense stimulation of the neonatal adrenal cortex evoked in the present study resulted in a change in hippocampal corticosteroid receptors is not certain. Preliminary observations with Nissl staining indicate that hippocampal morphology is altered by this neonatal exposure to ACTH/CORT (Alves, unpublished results). Further investigation into the duration of elevated CORT levels following a stress response in these ACTH-treated animals, and how the hippocampus is affected by this treatment, is needed before making any definite conclusions on these stress hormone-induced HPA alterations. Since the developing serotonergic system is believed to be directly involved in the neonatal handling-induced enhancement of HPA feedback at the level of the hippocampus,[10] it is possible that the ACTH/CORT-induced changes in this neural system among peptide-treated animals may have been involved in this disruption of HPA feedback (see below).

Forebrain Monoamine Levels and Metabolism

Early postnatal ACTH treatment and subsequent adrenocortical steroid secretion appeared to elicit a trophic or stimulatory effect on forebrain monoamine growth, followed by alterations in monoamine metabolism at adulthood and, eventually, a decrease in monoamine levels later in life, at one year. Along with these ontogenic differences, we observed region-specific changes as well as differential effects on each monoamine. Actual neurochemical values are presented in TABLES 3–5, while FIGURE 3 summarizes the neurochemical changes measured in ACTH-treated animals based on percent of control. At 7 days, ACTH-treated animals showed significantly elevated levels of the catecholamines in both the hypothalamus and striatum (TABLES 3 and 5). Hypothalamic 5-HT values were 13% above control values but did not achieve statistical difference. In contrast to the subcortical regions studied, cortical monoamine levels at 7 days were decreased following ACTH treatment (cortical DA levels were too low to measure in either treatment group at 7 days) (TABLE 4). Of the three forebrain regions studied, the cortex is the furthest target for these neurons to innervate. A shift in monoamine innervation may have occurred, with the stress hormones inducing a "hyperinnervation" of the subcortical regions, the hypothalamus and striatum, but a delayed development of the cortex by the end of the treatment period at 7 days. Changes in the monoamine metabolite levels paralleled the changes found for the respective monoamines; an increase in striatal DOPAC was observed at 7 days as was a decrease

TABLE 3. Hypothalamic Monoamine and Monoamine Metabolite Concentrations of Female Rats Treated with either Saline or ACTH 1-24 during the First Postnatal Week[a]

Age	Treatment	5-HT	5-HIAA	DA	HVA	NE
7 Days	Saline	211.8 ± 12.2	193.6 ± 31.0	231.0 ± 9.1	138.2 ± 5.1	488.0 ± 6.8
	ACTH 1-24	239.0 ± 12.0	196.8 ± 22.4	297.7 ± 10.0^b	164.3 ± 13.7	583.3 ± 34.4^c
15 Days	Saline	351.4 ± 3.3	803.2 ± 64.3	296.8 ± 19.8	146.8 ± 6.6	849.8 ± 48.2
	ACTH 1-24	440.8 ± 6.2^b	917.6 ± 19.1	327.0 ± 21.6	113.3 ± 8.4^d	890.0 ± 44.2
90 Days	Saline	675.8 ± 40.3	763.8 ± 34.4	303.6 ± 11.5	n.m.	1443.6 ± 81.2
	ACTH 1-24	754.5 ± 20.5	727.0 ± 21.0	325.3 ± 16.8	n.m.	1415.3 ± 52.3
1 Year	Saline	807.7 ± 38.9	454.8 ± 12.5	417.1 ± 23.9	n.m.	1758.8 ± 56.8
	ACTH 1-24	705.7 ± 48.4	450.2 ± 20.7	349.2 ± 16.9^c	n.m.	1591.7 ± 41.4^c

[a] Dosage of ACTH 1-24 = 0.5 mg/kg/day. Values are expressed as nanograms per gram of wet tissue weight $\pm$ SEM (n = 6 per treatment, per age; n.m. = not measured).
[b] $p < 0.001$.
[c] $p < 0.05$.
[d] $p < 0.02$.

TABLE 4. Cortical Monoamine and Monoamine Metabolite Concentrations of Female Rats Treated with either Saline or ACTH 1-24 during the First Postnatal Week[a]

Age	Treatment	5-HT	5-HIAA	DA	HVA	NE
7 Days	Saline	75.8 ± 9.4	205.5 ± 10.3	n.d.	74.1 ± 13.5	101.8 ± 11.9
	ACTH 1-24	56.3 ± 6.6	113.0 ± 7.6[b]	n.d.	76.4 ± 7.3	91.4 ± 3.3
15 Days	Saline	66.7 ± 3.2	226.8 ± 14.3	45.6 ± 4.0	23.9 ± 1.9	81.1 ± 4.7
	ACTH 1-24	89.6 ± 2.7[b]	272.6 ± 11.4[c]	37.1 ± 6.9	26.2 ± 6.6	102.0 ± 6.5[c]
90 Days	Saline	213.8 ± 12.9	372.2 ± 15.6	79.8 ± 7.4	30.0 ± 4.8	297.0 ± 16.8
	ACTH 1-24	235.0 ± 12.8	324.3 ± 21.8	81.6 ± 12.8	31.0 ± 4.5	277.3 ± 9.7
1 Year	Saline	484.4 ± 20.0	216.2 ± 9.6	228.0 ± 49.7	54.7 ± 3.3	403.5 ± 10.5
	ACTH 1-24	390.4 ± 24.5[d]	236.8 ± 16.7	170.2 ± 19.3	49.2 ± 7.7	399.0 ± 16.8

[a] Dosage of ACTH 1-24 = 0.5 mg/kg/day. Values are expressed as nanograms per gram of wet tissue weight ± SEM (n = 6 per treatment, per age; n.d. = not detectable).
[b] $p < 0.001$.
[c] $p < 0.05$.
[d] $p < 0.02$.

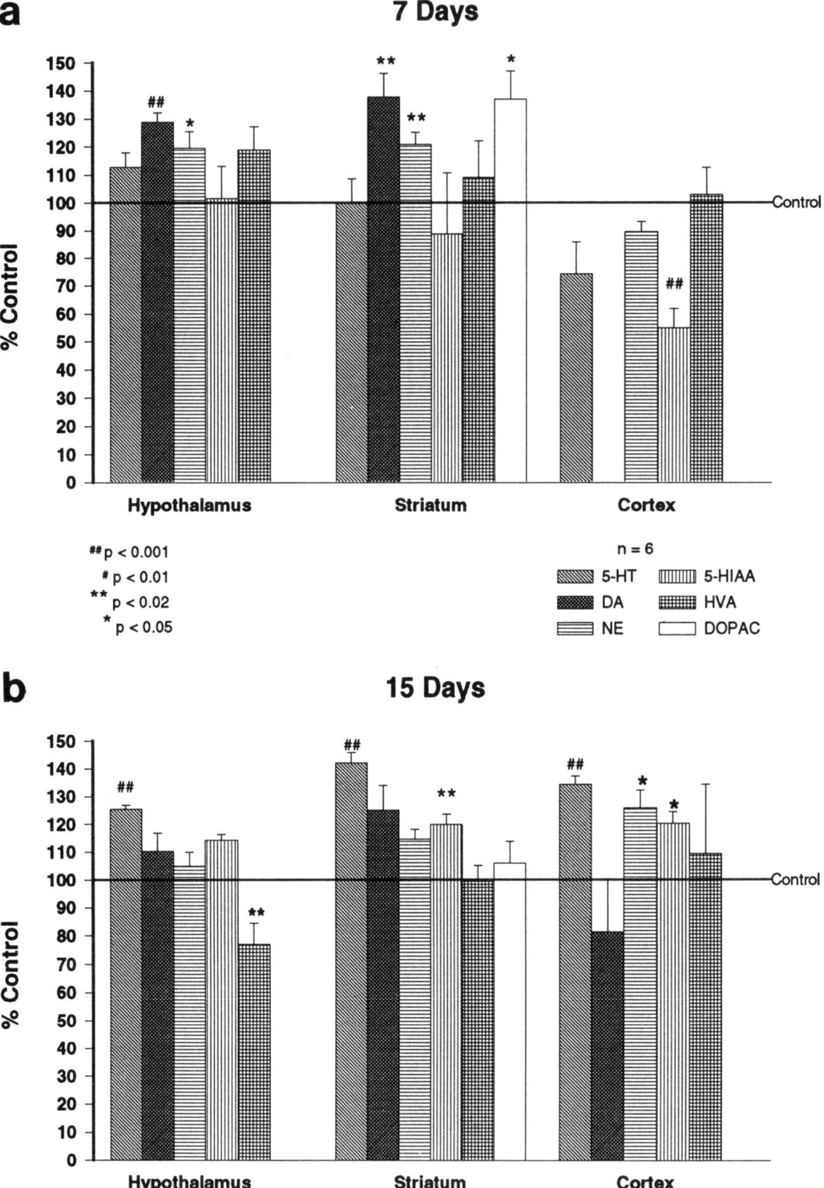

FIGURE 3. Changes in monoamine and metabolite levels at days 7 **(a)**, 15 **(b)**, 90 **(c)**, and one year **(d)**, within the hypothalamus, striatum, and cortex of animals treated with ACTH (0.5 mg/kg/day) during the first week postnatal. Values are expressed as percent of saline control ± SEM ($n = 6$).

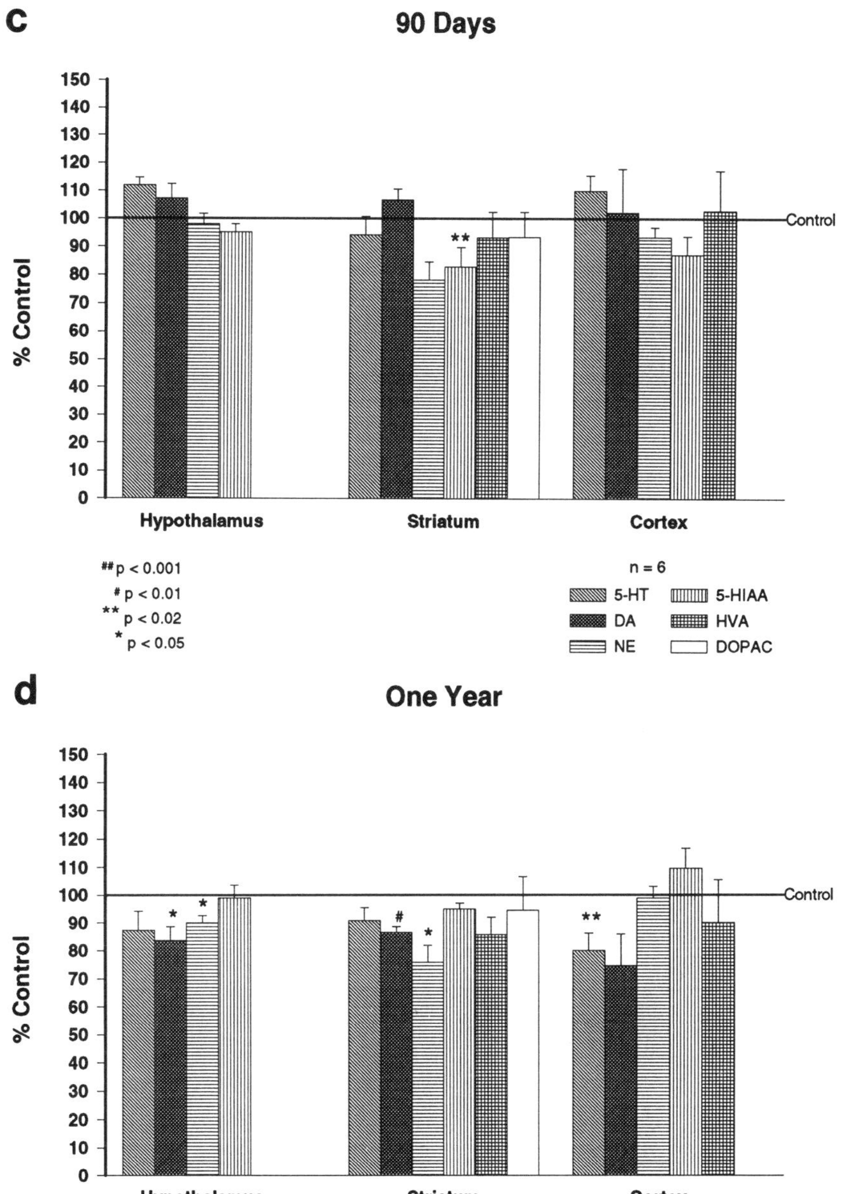

FIGURE 3. (*Continued*)

TABLE 5. Striatal Monoamine and Monoamine Metabolite Concentrations of Female Rats Treated with either Saline or ACTH 1-24 during the First Postnatal Week[a]

Age	Treatment	5-HT	5-HIAA	DA	HVA	DOPAC	NE
7 Days	Saline	75.0 ± 2.4	62.7 ± 1.6	1065.0 ± 55.6	179.2 ± 10.8	107.2 ± 8.5	110.2 ± 5.5
	ACTH 1-24	75.0 ± 6.5	55.8 ± 12.1	1466.6 ± 123.8[b]	195.6 ± 25.2	146.8 ± 14.8[c]	133.2 ± 5.6[b]
15 Days	Saline	167.8 ± 9.5	319.6 ± 14.8	2333.6 ± 356.7	543.2 ± 61.4	113.8 ± 22.6	83.9 ± 6.5
	ACTH 1-24	238.2 ± 8.8[d]	382.8 ± 14.4[b]	2914.6 ± 261.0	540.4 ± 30.6	120.7 ± 9.3	96.2 ± 3.4
90 Days	Saline	493.6 ± 16.8	634.2 ± 9.9	12142.6 ± 334.8	1182.8 ± 108.6	293.6 ± 36.5	110.8 ± 9.7
	ACTH 1-24	465.2 ± 30.5	525.8 ± 36.1[b]	12961.4 ± 518.8	1104.0 ± 100.7	274.2 ± 24.5	86.8 ± 5.4
1 Year	Saline	503.4 ± 25.0	738.2 ± 22.4	11722.3 ± 402.8	1030.3 ± 98.1	266.3 ± 22.4	183.3 ± 14.8
	ACTH 1-24	457.1 ± 21.2	701.6 ± 14.2	10149.3 ± 211.5[e]	883.4 ± 54.7	251.8 ± 30.7	138.9 ± 8.4[c]

[a] Dosage of ACTH 1-24 = 0.5 mg/kg/day. Values are expressed as nanograms per gram of wet tissue weight ± SEM (n = 6 per treatment, per age).
[b] $p < 0.02$.
[c] $p < 0.05$.
[d] $p < 0.001$.
[e] $p < 0.01$.

in cortical 5-HIAA. No significant differences in turnover rates were observed at 7 days (TABLE 6).

By 15 days of age, the 5-HT system was most markedly affected (TABLES 3–5; FIG. 3). All three brain regions studied showed highly significant ($p < 0.001$) increases in 5-HT levels. Levels of the 5-HT metabolite 5-HIAA also were increased in both the striatum and the cortex, and because there were no changes in 5-HT turnover, as assessed by the 5-HIAA/5-HT ratio (TABLE 6), these increases suggest an elevated innervation or accelerated maturation of this neurotransmitter system. A marked effect of ACTH and CORT on the catecholamines was found at 7 days; however, significant differences at 15 days included only an increase in cortical NE levels and a decrease in hypothalamic HVA, correlated with a decrease in the DA turnover index, the HVA/DA ratio (TABLE 6).

At 90 days, most neurochemical measures were similar between treated and control animals. The differences observed among ACTH-treated animals included a decrease in striatal 5-HIAA (TABLE 5) along with decreases in 5-HT turnover within the striatum and the cortex (TABLE 6), which suggest a decrease in serotonergic activity within these two brain regions at this time.

Forebrain monoamine and metabolite levels were generally decreased among ACTH-treated animals at one year (TABLES 3–5; FIG. 3). The catecholamines were significantly decreased in the hypothalamus and the striatum, whereas in the cortex a prolonged effect was observed on the serotonergic system. A decrease in cortical 5-HT was correlated with a large increase in the 5-HIAA/5-HT ratio (TABLE 6), suggesting a much higher cortical 5-HT turnover rate one year after ACTH treatment. The simultaneous decrease in striatal DA was not accompanied by a change in DA turnover, because no differences in striatal HVA/DA or DOPAC/DA ratios were found, indicating a general decline in the striatal DA system.

Many studies have shown that a decline in central monoamine systems is associated with the aging CNS. A decline in the striatal DA system of the aged rat has been demonstrated,[39–42] and a decrease in striatal DA innervation is a prominent age-related change in the "normal" human brain and, to a much greater extent, in the Parkinsonian brain.[43] Although the decline in striatal DA among ACTH-treated animals was not severe (-13.4%, $p < 0.01$), a far greater decrease in striatal DA was seen among ACTH-treated animals from early adulthood (90 days) to 1 year of age (-22%), compared to practically no change (-3%) observed in striatal DA levels among saline-treated animals during the same time span. In addition to this change in DA, several studies have reported increased 5-HIAA/5-HT ratios in the aged rat CNS,[40,44,45] similar to the 50% increase in the cortical 5-HIAA/5-HT ratio found in the present study. Thus, it is tempting to speculate that these significant "age-related" changes observed in middle-aged animals exposed to elevated ACTH and the chronic hyperactive adrenocortical activity during a very sensitive developmental period may lead to premature aging of the monoamine systems. Thus, in contrast to a "mild" manipulation of the HPA axis neonatally, such as handling, which appears to attenuate some neuronal changes that occur during aging, this extreme activation of the stress response system during development may accelerate not only the maturation of the systems, but also their decline during adulthood.

Two groups have previously investigated possible effects of postnatal ACTH treatment on the central monoamine systems in the rat;[26,46] however, the dosages, treatment periods, ages studied, and brain regions examined differed considerably from the present investigation. They reported either no changes in monoamine levels in midsagittally sectioned half brains at the two periods examined (30 and

TABLE 6. Estimated Monoamine Turnover within the Hypothalamus, Striatum, and Cortex of Female Rats Treated with either Saline or ACTH 1-24 during the First Postnatal Week[a]

Age	Treatment	Hypo 5-HIAA/5-HT	Hypo HVA/DA	Striatum 5-HIAA/5-HT	Striatum HVA/DA	Striatum DOPAC/DA	Cortex 5-HIAA/5-HT	Cortex HVA/DA
7 Days	Saline	0.891 ± 0.105	0.600 ± 0.016	0.839 ± 0.027	0.170 ± 0.013	0.101 ± 0.007	2.836 ± 0.295	n.d.
	ACTH	0.798 ± 0.055	0.557 ± 0.055	0.780 ± 0.090	0.152 ± 0.016	0.104 ± 0.016	2.128 ± 0.193	n.d.
15 Days	Saline	2.282 ± 0.168	0.503 ± 0.032	1.937 ± 0.151	0.249 ± 0.029	0.048 ± 0.004	3.421 ± 0.215	0.600 ± 0.073
	ACTH	2.083 ± 0.045	0.352 ± 0.030[b]	1.610 ± 0.043	0.190 ± 0.014	0.042 ± 0.004	3.052 ± 0.154	0.685 ± 0.116
90 Days	Saline	1.126 ± 0.071	n.m.	1.290 ± 0.038	0.098 ± 0.010	0.024 ± 0.003	1.765 ± 0.106	0.374 ± 0.043
	ACTH	0.970 ± 0.045	n.m.	1.132 ± 0.038[c]	0.085 ± 0.008	0.021 ± 0.002	1.393 ± 0.105[d]	0.375 ± 0.049
1 Year	Saline	0.571 ± 0.035	n.m.	1.487 ± 0.087	0.087 ± 0.007	0.023 ± 0.002	0.431 ± 0.021	0.295 ± 0.057
	ACTH	0.653 ± 0.046	n.m.	1.530 ± 0.101	0.087 ± 0.006	0.025 ± 0.003	0.650 ± 0.075[d]	0.298 ± 0.033

[a] Dosage of ACTH 1-24 = 0.5 mg/kg/day. Values are expressed as metabolite/neurotransmitter ratios ± SEM (n = 6 per treatment, per age; n.d. = not detectable; n.m. = not measured). Hypo, hypothalamus.
[b] $p < 0.01$.
[c] $p < 0.02$.
[d] $p < 0.05$.

90 days),[46] or significant decreases in neocortical 5-HT$_2$ receptors (-13%) and tryptophan levels (-29%) at the one age studied (41 days).[26] As we have observed in the present study, both ontogenic-specific and region-specific changes were found in transmitter and/or metabolite levels, as well as turnover rates following early postnatal ACTH administration.

There are several means by which the pituitary-adrenal stress hormones could have exerted their effects on the forebrain monoamine systems we examined. ACTH, a melanocortin, has potent neurotrophic properties in both developing and regenerating neural systems.[47–49] Although this hormone is a peptide and thus subject to enzymatic degradation in the periphery, ample studies have demonstrated that peptides do cross the blood–brain barrier (BBB),[50,51] particularly during development when there appears to be greater permeability of this barrier. The recent isolation of melanocortin receptors[52] in all of the major monoaminergic nuclei and many of the terminal fields[53] provides evidence for a direct modulatory role of ACTH peptides on these systems. Although the exact mechanisms by which ACTH exerts its neurotrophic effects are not clear, the neuroactive portion of this peptide appears to be contained within the short peptide fragment ACTH 4-10.[49] A recent study by Lee and colleagues[54] has demonstrated immunoreactivity for a trophic ACTH 4-10-like peptide throughout the neonatal rat brain, but only in discrete hypothalamic regions in the adult brain, suggesting an early postnatal developmental role for this ACTH peptide in the CNS. Interestingly, this peptide fragment appears to be re-expressed centrally in the adult CNS during periods of "regrowth," following nerve damage.[47,55]

It is likely that many of the effects on the monoamines observed in this study are due to the corticosteroids. These adrenal steroids have profound effects on the brain throughout life, and the roles that they play differ drastically during development, at adulthood and through senescence. Depending upon the systems involved, they can have facilitative developmental effects as well as reversible modulatory actions in mature neurotransmitter systems.[35,37] Eventually, they may play a role in neuronal degeneration and cell death in the aging CNS.[35,37,56,57] Our findings from the present study support these steroidal actions on the central monoamines.

The corticosteroids may modify the growth of monoaminergic neurons by altering the metabolism of these neurons. In addition to a stimulatory effect of CORT on the rate-limiting enzymes in monoamine synthesis previously mentioned,[15–17,19,20] monoamine turnover rate is also modulated by glucocorticoids.[58–60] A current hypothesis states that 5-HT may act as a differential signal for raphe neurons and their target cells, via neuronal and glial 5-HT$_{1A}$ receptor mediated mechanisms.[24,61] Therefore, a CORT-induced augmented synthesis of 5-HT during development could result in an increase in the activation of glial 5-HT$_{1A}$ receptors, thereby stimulating the release of S-100β, a potent neurotrophic factor for 5-HT, cortical, and motor neurons.[61] This glucocorticoid-induced, 5-HT-mediated stimulation of neuronal growth represents a means by which the stress hormones can exert their effects on and through neurotransmitter systems.

A recent study by Azmitia and colleagues[18] has shown that dexamethasone (DEX) treatment of ADX rats induces cellular hypertrophy among dorsal raphe and supralemniscal neurons, suggesting a type II receptor-mediated role of the corticosteroids in 5-HT cell growth. Furthermore, chronic stress-induced increases in CORT alter the expression of hippocampal neurotrophin-3 mRNA and brain-derived neurotrophic factor (BNDF) mRNA,[62] indicating that both of these neurotrophic factors are products of stress-responsive genes. Alterations in the concentrations of these and other similarly regulated growth factors may be involved in the stress hormone-induced neural changes during development.

Hypothalamic-Pituitary-Gonadal Function: Ovarian Hormones and Sexual Behavior

Plasma Ovarian Hormones

Proestrous plasma estradiol levels of ACTH-treated animals were found to be significantly decreased by approximately 44%, compared to saline-treated animals at young adulthood (TABLE 7). Progesterone levels did not differ between animals of either treatment group. Although not quantified over a long period of time, intact ACTH-treated animals displayed normal 4–5 day estrous cycle lengths, no different from saline-treated animals, assessed during daily vaginal smearing prior to blood collection. No animals were found to be acyclic, and no significant difference in ovary weight or the ovary/body weight ratio was found. The mean ovary/body weight ratio of ACTH-treated animals was $6.941 \pm 0.228 \times 10^{-4}$ versus $7.531 \pm 0.261 \times 10^{-4}$ of saline-treated animals. Thus, if an alteration in the estrogen-producing capacity of the ovaries occurred following stress hormone exposure, it was not sufficient to disrupt the cyclicity of these animals.

We previously reported no significant difference in proestrous plasma estradiol between young adult animals of either treatment group, although a trend toward a decrease among ACTH animals was evident.[7] A higher "n" in the current study (9–12 per group vs. 4 per group in the previous study) exposed the decrease as statistically significant. Although a decrease in circulating estradiol could possibly explain deficits in receptivity, results from the current OVX + EB dose response experiment suggest that there may be something more than simply a decrease in plasma estradiol responsible for behavioral deficits among intact ACTH-treated animals that we previously observed.[7]

Sexual Behavior Tests

Under equal gonadal hormone conditions, ACTH-treated animals displayed a significantly decreased number of lordosis responses to a mounting male in the

TABLE 7. Plasma Ovarian Steroid Hormone Levels during Proestrus, Approximately Three Hours Prior to Darkness at 90 Days of Age[a]

Treatment	Estradiol (pg/mL)	Progesterone (ng/mL)
Saline ($n = 9$)	94.0 ± 11.8	23.5 ± 7.1
ACTH (0.5 mg/kg) ($n = 12$)	52.9 ± 6.7^{b}	21.4 ± 3.9

[a] Values are the mean ± SEM.
[b] $p < 0.01$.

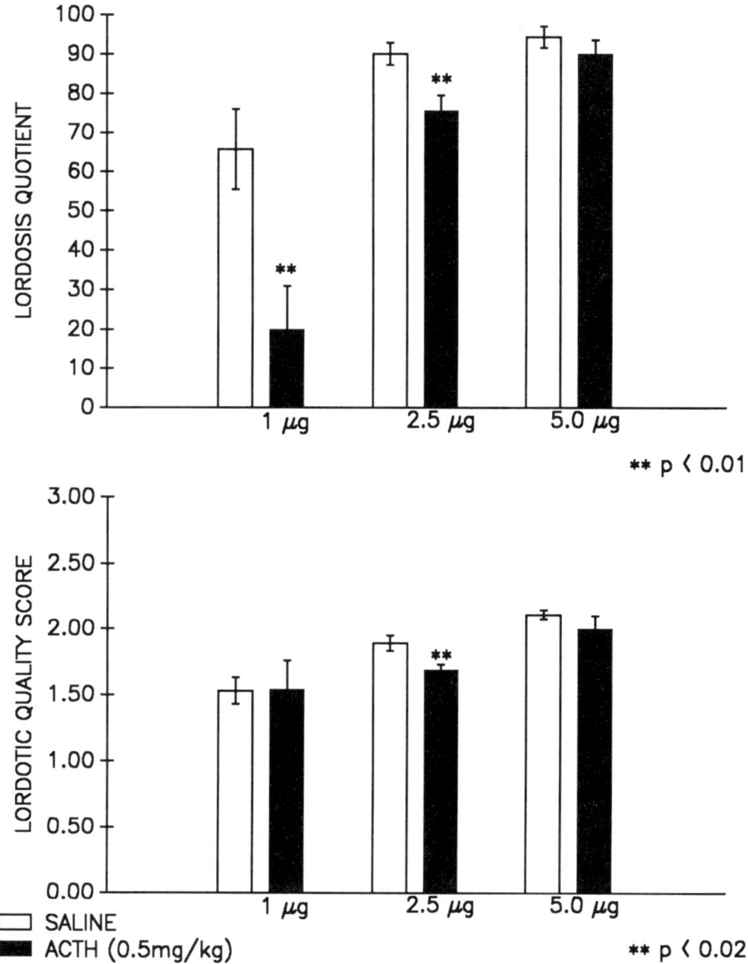

FIGURE 4. Estradiol dose response. Mean lordosis quotients (*upper graph*) and lordotic quality scores (*lower graph*) ± SEM of ovariectomized rats treated neonatally with either saline or ACTH. All animals (70–90 days of age) were administered estradiol benzoate in sesame oil, sc, approximately 52 and 28 hours prior to behavior testing, at a dosage of 1.0 μg for the first trial, 2.5 μg for the second trial, and 5.0 μg for the third trial. Approximately 5 hours before each trial, all animals were injected sc with 500 μg of progesterone (saline, $n = 7$; ACTH, $n = 9$).

first two trials of this study (FIG. 4). Following a 1 μg EB regimen, animals from the ACTH treatment group displayed severe deficits in the LQ compared to saline-treated animals, where four out of nine animals failed to lordose in response to a mounting male, compared to only one of seven saline-treated animals (FIG. 4). Simultaneously, ACTH animals exhibited high levels of anxious, rejection behavior as we previously observed in intact, virgin ACTH females.[7] However, those animals that did lordose from either treatment group displayed similar quality scores (FIG. 4). Approximately one week later, following a regimen of 2.5 μg of EB, ACTH-treated animals continued to show significant deficits in the LQ, as well as a small, but significant, decrease in the LQS compared to saline-treated animals (FIG. 4). Following a regimen of 5 μg of EB, ACTH-treated animals exhibited high levels of receptivity, comparable to controls, with very little anxious/defensive behavior evident (FIG. 4).

The decreased responsiveness to EB observed among neonatally manipulated female rats during the first two trials may suggest a disruption in their neurocircuitry regulating female reproductive behavior. Estrogen "priming" of specific neural circuits controlling female sexual behavior is necessary for the induction of full receptivity, and it occurs only in the female brain. This estrogen-mediated process involves the up-regulation of progestin receptors within the regulatory circuitry.[63] Could developmental ACTH and excessive adrenocortical stimulation have resulted in a defeminization and/or masculinization of the circuitry regulating sexual behavior?

Findings from the male typical sexual behavior test indicate that ACTH treatment did not cause a masculinization of the female brain, since animals of this group tended to actually display less male-typical behavior than saline controls under the influence of testosterone (TABLE 8). In fact, no ACTH animals exhibited intromission/ejaculatory patterns, whereas several control animals displayed this behavior quite vigorously under the influence of testosterone. Furthermore, as demonstrated by the high levels of receptivity eventually displayed by sexually experienced, intact animals (TABLE 9), as well as OVX and hormone-replaced animals (FIG. 4), it appears that the "required" feminine neurocircuitry is present among these animals. Perhaps the threshold to a novel stimulus, such as a mounting male, was somehow elevated by neonatal stress hormone exposure, and that experience, together with the increased dosage of estradiol, had an additive effect to "reset" or lower the threshold.

Prenatal exposure to stress can alter the behavior of the adult animal during times of stress, particularly in response to a novel stimulus.[6,64-66] In addition, direct alterations to the monoaminergic systems can lead to similar behavioral changes. Jarzab and colleagues[67] demonstrated that elevating central 5-HT levels of neonatal female rats by the administration of 1-tryptophan later led to a significant decrease in lordotic behavior among these animals, similar in magnitude to the decreases observed in the present study. Decreased sexual behavior was also reported following neonatal clonidine treatment, an α_2-adrenergic agonist.[67] Thus, it is plausible that the ACTH/CORT-induced stimulation of the monoamines early

TABLE 8. Male Sexual Behavior Parameters among Ovariectomized Female Rats That Had Been Treated with Testosterone and Placed with Receptive Females for the First Time[a]

Treatment	Genital Sniffs	Mounts	Intromission/Ejaculatory Patterns
Saline	9.8 ± 1.1	3.4 ± 1.7	2.2 ± 1.2
ACTH (0.5 mg/kg)	9.0 ± 1.1	1.7 ± 0.9	0

[a] Testosterone: 3 cm silastic implants, sc. Values are expressed as the mean ± SEM; $n = 6$ per treatment.

TABLE 9. Lordosis Quotients (LQ) and Lordotic Quality Scores (LQS) of Intact Animals during the Virgin Trial at Early Adulthood and Then as Sexually Experienced Adults at One Year of Age[a]

Treatment	Virgin Trial[b] (60–90 Days)		Experienced Trials (1 Year)	
	LQ	LQS	LQ	LQS
Saline	95.3 ± 1.5	2.35 ± 0.07	91.7 ± 2.8	1.95 ± 0.05
ACTH (0.5 mg/kg)	78.2 ± 3.7[c]	1.80 ± 0.11[c]	93.8 ± 4.7	2.13 ± 0.13

[a] Values represent the mean ± SEM: virgin trial, $n = 17$; experienced trial, $n = 6$ saline, 8 ACTH.
[b] Data from Alves *et al.*[7]
[c] $p < 0.05$.

in development is involved in the observed differences in the lordosis response at adulthood. Because few differences were detected in the monoaminergic systems roughly at the age of behavior testing (90 days), it is likely that these neurochemicals exerted an effect during synaptogenesis.

Interestingly, depletion of 5-HT levels by the administration of para-chlorophenylalanine (pCPA) during the second postnatal week can increase reactivity to environmental and social cues, and decrease anxiety in adult male rats.[68] Perhaps by significantly increasing 5-HT levels during this time, as occurred following neonatal ACTH treatment (5-HT levels were highly and significantly increased in all three forebrain regions studied at 15 days of age), the reverse may occur, producing animals with a lower capability for social interaction and/or increased expression of anxiety. As discussed above, female rats treated neonatally with ACTH appear to have a permanently altered hormonal response to stress.

COMPLEXITY OF MEDIATORS

Manipulation of circulating stress hormone levels during early postnatal life can affect numerous physiological systems, in addition to those examined within this study. For example, functioning of the hypothalamic-pituitary-thyroid (HPT) system is influenced by the adrenal steroids, because these hormones act to suppress pituitary thryoid-stimulating hormone secretion.[69] Considering that manipulation of thyroid hormone levels during development has been linked to numerous biochemical and behavioral abnormalities, such as drastic learning and memory deficits which last into adulthood,[70,71] some of the observed alterations in the present study may have also involved pertubations in HPT functioning. This is one example of the complexity of possible mediators of the findings reported within this paper.

SUMMARY

The findings from this study demonstrate that the manipulation of the HPA system resulting from ACTH administration during neonatal development produces long-term, differential effects, not only on adrenocortical activity, but also on the activity and integrity of the forebrain monoamine systems. Increased concentrations of the monoamines within the forebrain regions studied at days 7 and 15, suggest a hastened maturation of these neural systems in animals neonatally treated with ACTH. The observed neurochemical alterations in these animals at one year are suggestive of an accelerated aging in the monoamine systems. A further consequence of these disturbances during development is an altered functioning of the HPG axis, as demonstrated by a delayed onset of puberty as previously reported,[7] as well as significantly decreased proestrus plasma estradiol. Although deficits in sexual behavior also existed, it seems probable that these behavioral changes are a manifestation of altered neural systems regulating the ability to cope with a novel stimulus or situation, rather than a disruption of the "feminization" of the brain during sexual differentiation. This is in contrast to the male rat which exhibits permanent deficits in male typical sexual behavior following developmental ACTH treatment.[5]

The clinical relevance of these findings may be extensive. Perinatal exposure to events or agents that markedly increase ACTH and the corticosteroids may cause significant immediate and long-term changes in central monoamine functioning. These changes may constitute some of the most deleterious effects of stress exposure in infants and children. The alterations may be especially devastating in individuals with predispositions to stress-sensitive disorders such as anxiety, depression, and Tourette's syndrome. Finally, the use of ACTH in the treatment of infantile spasms may need to be reassessed in light of the possible long-term effects of ACTH on central monoamine functioning.

ACKNOWLEDGMENTS

The authors thank Laura Hall for her time and assistance in HPLC analysis, and Organon B. V. for the generous donation of the ACTH 1-24 peptide.

REFERENCES

1. HUTTUNEN, M. O. 1971. Persistent alteration of turnover of brain noradrenaline in the offspring of rats subjected to stress during pregnancy. Nature **230:** 53–55.
2. MOYER, J. A., L. R. HERRENKOHL & D. M. JACOBOWITZ. 1978. Effects of stress during pregnancy on catecholamines in discrete brain regions. Brain Res. **121:** 385–393.
3. ENDRÖCZI, E. 1991. Stress and Adaptation. Akademiai Kiado. Budapest.
4. KING, J. A., M. DAVILA-GARCIA, E. C. AZMITIA & F. L. STRAND. 1991. Differential effects of prenatal and postnatal ACTH or nicotine exposure on 5-HT high affinity uptake in the neonatal rat brain. Int. J. Dev. Neurosci. **9(3):** 281–286.
5. SEGARRA, A. C., V. N. LUINE & F. L. STRAND. 1991. Sexual behavior of male rats is differentially affected by timing of perinatal ACTH administration. Physiol. Behav. **50:** 689–697.
6. TAKAHASHI, L. K., J. G. TURNER & N. H. KALIN. 1992. Prenatal stress alters brain catecholaminergic activity and potentiates stress-induced behavior in adult rats. Brain Res. **574:** 131–137.
7. ALVES, S. E., H. M. AKBARI, E. C. AZMITIA & F. L. STRAND. 1993. Neonatal ACTH and corticosterone alter hypothalamic monoamine innervation and reproductive parameters in the female rat. Peptides **14:** 379–384.
8. MEANEY, M. J. & D. H. AIKEN. 1985. The effects of early postnatal handling on the development of hippocampal glucocorticoid receptors: Temporal parameters. Dev. Brain Res. **22:** 301–304.

9. MEANEY, M. J., D. H. AITKEN, S. BHATNAGAR, C. VAN BERKEL & R. M. SAPOLSKY. 1988. Postnatal handling attenuates neuroendocrine, anatomical, and cognitive impairments related to the aged hippocampus. Science **238:** 766–768.

10. MEANEY, M. J., J. DIORIO, D. FRANCIS, S. LAROCQUE, D. O'DONNELL, J. W. SMYTHE, S. SHARMA & B. TANNENBAUM. 1994. Environmental regulation of the development of glucocorticoid receptor systems in the rat forebrain: The role of serotonin. Ann. N. Y. Acad. Sci. **746:** 260–273.

11. SEGARRA, A. C., S. D. MENDELSON, F. L. STRAND & B. S. MCEWEN. 1990. Decreased sexual behavior induced by prenatal ACTH is correlated with increased 5-HT$_{1B}$ receptors in the MPN of male rats (Abstr.). Neuroendocrinol. Lett. **12(4):** 301.

12. MCEWEN, B. S., V. N. LUINE & C. T. FISCHETTE. 1988. Developmental action of hormones: From receptors to function. *In* From Message to Mind. S. S. Easter, Jr., K. F. Barald & B. M. Carlson, Eds.: 272–287. Sinauer Associated. Sunderland, MA.

13. MCCARTHY, M. M., E. H. SCHLENKER & D. W. PFAFF. 1993. Enduring consequences of neonatal treatment with antisense oligodeoxynucleotides to estrogen receptor mRNA on sexual differentiation of rat brain. Endocrinology **133:** 433–439.

14. AZMITIA, E. C. & E. DE KLOET. 1987. ACTH neuropeptide stimulation of serotonergic neuronal maturation in tissue culture: Modulation by hippocampal cells. Prog. Brain Res. **72:** 311–318.

15. AZMITIA, E. C. & B. S. MCEWEN. 1969. Corticosterone regulation of tryptophan hydroxylase in midbrain of the rat. Science **166:** 1274–1276.

16. AZMITIA, E. C. & B. S. MCEWEN. 1976. Early response of rat brain tryptophan hydroxylase activity to cycloheximide, puromycin and corticosterone. J. Neurochem. **27:** 773–778.

17. SINGH, V. B., K. C. CORLEY, H. PHAN & M. C. BOADLE-BIDER. 1990. Increases in the activity of tryptophan hydroxylase from rat cortex and midbrain in response to acute or repeated sound stress are blocked by adrenalectomy and restored by dexamethasone treatment. Brain Res. **516:** 66–76.

18. AZMITIA, E. C., B. LIAO & Y. S. CHEN. 1993. Increase of tryptophan hydroxylase enzyme protein by dexamethasone in adrenalectomized rat midbrain. J. Neurosci. **13:** 5041–5055.

19. MARKEY, K. A., A. C. TOWLE & P. Y. SZE. 1982. Glucocorticoid influence on tyrosine hydroxylase activity in mouse locus coeruleus during postnatal development. Endocrinology **111:** 1519–1523.

20. SHEN, J.-T. & W. F. GANONG. 1976. Effect of variations in pituitary-adrenal activity on dopamine-β-hydroxylase activity in various regions of rat brain. Neuroendocrinology **20:** 311–318.

21. CAVINESS, V. S. & M. G. KORDE. 1981. Monoaminergic afferents to the neocortex: A developmental histofluorescence study in normal and reeler mouse embryos. Brain Res. **209:** 1–9.

22. KÖNIG, N., G. ROCH & R. MARTY. 1975. The onset of synaptogenesis in rat temporal cortex. Anat. Embryol. **148:** 73–87.

23. LAUDER, J. M. 1990. Ontogeny of the serotonergic system in the rat: Serotonin as a developmental signal. Ann. N. Y. Acad. Sci. **600:** 297–314.

24. WHITAKER-AZMITIA, P. M. 1991. Role of serotonin and other neurotransmitter receptors in brain development: Basis for developmental pharmacology. Pharmacol. Rev. **43(4):** 553–561.

25. HRACHOVY, R. A., J. D. FROST, P. KELLAWAY & T. ZION. 1980. A controlled study of ACTH therapy in infantile spasms. Epilepsia **21:** 631–636.

26. PRANZATELLI, M. R. 1989. In vivo and in vitro effects of adrenocorticotropic hormone on serotonin receptors in neonatal rat brain. Dev. Pharmacol. Ther. **12:** 49–56.

27. PRANZATELLI, M. R. 1994. On the molecular mechanism of adrenocorticotropic hormone in the CNS—Neurotransmitters and receptors. Exp. Neurol. **125:** 142–161.

28. PRANZATELLI, M. R. 1994. Putative neurotransmitter abnormalities in infantile spasms—Cerebrospinal-fluid neurochemistry and drug effects. J. Child Neurol. **9:** 119–129.

29. LYEN, L. R., I. M. HOLLAND & Y. C. LYEN. 1979. Reversible cerebral atrophy in infantile spasms caused by corticotropin. Lancet **2:** 37–38.

30. OKUNO, T., M. ITO, Y. KONISHI, M. YOSHIOKA & Y. NAKANO. 1980. Cerebral atrophy following ACTH therapy. J. Comput. Assist. Tomogr. **4:** 20–23.

31. ITO, M., T. TAKAO, T. OKUNO & H. MIKAWA. 1983. Sequential CT studies of 24 children with infantile spasms on ACTH therapy. Dev. Med. Child Neurol. **25:** 475–480.

32. ANDERSON, G. M., J. G. YOUNG, D. K. BATTER, S. N. YOUNG, D. J. COHEN & B. A. SHAYWITZ. 1981. Determination of indoles and catechols in rat brain and pineal using liquid chromatography with fluorometric and amperometric detection. J. Chromatogr. **223:** 315–320.

33. ANDERSON, G. M., F. C. FEIBAL & D. J. COHEN. 1987. Determination of serotonin in whole blood, PRP, PPP and plasma ultrafiltrate. Life Sci. **40:** 1063–1070.

34. HARDY, D. F. & J. F. DEBOLD. 1971. The relationship between levels of exogenous hormones and the display of lordosis by the female rat. Horm. Behav. **2:** 287–297.

35. MCEWEN, B. S., E. R. DE KLOET & W. ROSTENE. 1986. Adrenal steroid receptors and actions in the nervous system. Physiol. Rev. **66(4):** 1121–1188.

36. SAPOLSKY, R. M., L. C. KREY & B. S. MCEWEN. 1984. Stress down-regulates corticosterone receptors in a site-specific manner. Endocrinology **114:** 287–292.

37. SAPOLSKY, R. M., L. C. KREY & B. S. MCEWEN. 1985. Prolonged glucocorticoid exposure reduced hippocampal neuron number: Implications for aging. J. Neurosci. **5:** 1222–1227.

38. SAPOLSKY, R. M., L. C. KREY & B. S. MCEWEN. 1984. Glucocorticoid-sensitive hippocampal neurons are involved in terminating the adrenocortical stress response. Proc. Natl. Acad. Sci. USA **81:** 6174–6177.

39. MAY, T. & M. SUGAWA. 1993. Altered dopamine receptor mediated signal transmission in the striatum of aged rats. Brain Res. **604:** 106–111.

40. MORETTI, A., N. CARFAGNA & F. TRUNZO. 1987. Effects of aging on monoamines and their metabolites in the rat brain. Neurochem. Res. **12:** 1035–1039.

41. SIMPKINS, J. W. 1984. Regional changes in monoamine metabolism in the aging constant estrous rat. Neurobiol. Aging **5:** 309–313.

42. WATANABE, H. 1987. Differential decrease in the rate of dopamine synthesis in several dopaminergic-neurons of aged rat brain. Exp. Gerontol. **22:** 17–25.

43. PALMER, A. M. & S. T. DEKOSKY. 1993. Monoamine neurons in aging and Alzheimer's disease. J. Neural Transm. **91:** 135–159.

44. GODEFROY, F., M. H. BASSANT, J. WEIL-FUGAZZA & Y. LAMOUR. 1990. Age-related changes in dopaminergic and serotonergic indices in the rat forebrain. Neurobiol. Aging **10:** 187–190.

45. GOZLAN, H., G. DAVAL, D. VERGE, U. SPAMPINATO, C. M. FATTACHINI, M. C. GALLISSOT, S. EL MESTIKAWY & M. HAMON. 1990. Aging associated changes in serotonergic and dopaminergic pre- and postsynaptic neurochemical markers in the rat brain. Neurobiol. Aging **11:** 437–449.

46. ITO, M., V. W. YONG, T. L. PERRY & V. K. SINGH. 1985. Chronic treatment with ACTH$_{1\text{-}24}$ does not produce permanent damage to the developing rat brain. Dev. Brain Res. **19:** 315–317.

47. ANTONAWICH, F. J., E. C. AZMITIA, H. K. KRAMER & F. L. STRAND. 1994. Specificity versus redundancy of melanocortins in nerve regeneration. Ann. N. Y. Acad. Sci. **739:** 60–73.

48. STRAND, F. L., K. J. ROSE, L. A. ZUCCARELLI, J. KUME, S. E. ALVES, F. J. ANTONAWICH & L. Y. GARRETT. 1991. Neuropeptide hormones as neurotrophic factors. Physiol. Rev. **71:** 1017–1046.

49. STRAND, F. L., S. J. LEE, T. S. LEE, L. A. ZUCCARELLI, F. A. ANTONAWICH, J. KUME & K. A. WILLIAMS. 193. Non-corticotropic ACTH peptides modulate nerve development and regeneration. Rev. Neurosci. **4:** 321–364.

50. DE WIED, D. 1969. Effects of peptide hormones on behavior. *In* Frontiers in Neuroendocrinology. W. F. & L. Martini, Eds.: 97–140. Oxford University Press. London.

51. BANKS, W. & A. KASTIN. 1985. Permeability of the blood-brain-barrier to neuropeptides: The case for penetration. Psychoneuroendocrinology **10:** 385–399.

52. TATRO, J. B. 1990. Melanotropin receptors in the brain are differentially distributed and recognize both corticotropin and α-melanocyte stimulating hormone. Brain Res. **536:** 124–132.

53. TATRO, J. B. & M. L. ENTWHISTLE. 1994. Distribution of melanocortin receptors in the lower brainstem of the rat. Ann. N. Y. Acad. Sci. **739:** 311–314.

54. LEE, S. J., C. AOKI & F. L. STRAND. 1997. Immunocytochemical localization of an ACTH 4-10-like peptide in the developing rat brain: A light and electron microscopic study. Submitted.

55. LEE, S. J., T. S. LEE, S. E. ALVES & F. L. STRAND. 1994. Immunocytochemical localization of ACTH 4-10 in the rat spinal cord following peripheral nerve trauma. Ann. N. Y. Acad. Sci. **739:** 320–323.

56. DE KLOET, E. R. 1992. Corticosteroids, stress and aging. Ann. N. Y. Acad. Sci. **663:** 357–371.

57. JOËLS, M. & E. R. DE KLOET. 1995. Corticosteroid hormones: Endocrine messengers in the brain. NIPS **10:** 71–76.

58. VAN LOON, G. R., A. SHUM & M. J. SOLE. 1981. Decreased brain serotonin turnover after short-term (two-hour) adrenalectomy in rats: A comparison of four turnover methods. Endocrinology **108:** 1392–1402.

59. DE KLOET, E. R., G. L. KOVACS, G. SZABO, G. TELEGDY, B. BOHUS & D. H. G. VERSTEEG. 1982. Decreased serotonin turnover in the dorsal hippocampus of rat brain shortly after adrenalectomy: Selective normalization after corticosterone substitution. Brain Res. **239:** 659–663.

60. VERSTEEG, D. H., I. VAN ZOEST & E. R. DE KLOET. 1984. Acute changes in dopamine metabolism in the medio-basal hypothalamus following adrenalectomy. Experientia **40:** 112–114.

61. WHITAKER-AZMITIA, P. M., R. MURPHY & E. C. AZMITIA. 1990. S-100 protein is released from astroglial cells by stimulation of 5-HT$_{1A}$ receptors and regulates development of serotonin neurons. Brain Res. **528:** 155–162.

62. SMITH, M. A., S. MAKINO, R. KVETNANSKY & R. M. POST. 1995. Stress and glucocorticoids affect the expression of brain-derived neurotrophic factor and neurotrophin-3 mRNAs in the hippocampus. J. Neurosci. **15(3):** 1768–1777.

63. PARSONS, B., M. Y. MCGINNIS & B. S. MCEWEN. 1981. Sequential inhibition by progesterone: Effects on sexual receptivity and associated changes in brain cytosol progestin binding in the female rat. Brain Res. **221:** 149–160.

64. FRIDE, E., Y. DAN, J. FELDON, G. HALEVY & M. WEINSTOCK. 1986. Effects of prenatal stress on vulnerability to stress in prepubertal and adult rats. Physiol. Behav. **37:** 681–687.

65. FRIDE, E. & M. WEINSTOCK. 1988. Prenatal stress increases anxiety-related behavior and alters cerebral lateralization of dopamine activity. Life Sci. **42:** 1059–1065.

66. WEINBERG, J. 1988. Hyperresponsiveness to stress: Differential effects of prenatal ethanol on males and females. Alcohol. Clin. Exp. Res. **12:** 647–652.

67. JARZAB, B., P. M. SICKMOLLER, D. KOKOCINSKA, M. KAMINSKI, E. GUBALA, W. ACHTLIK, J. WAGIEL & K. D. DOHLER. 1990. Influence of neurotransmitters on sexual differentiation of the brain. *In* Neuroendocrinology: New Frontiers. D. Gupta, H. A. Wollmann & M. B. Ranke, Eds.: 87–96. Brain Research Promotion. Tübingen.

68. FARABOLLINI, F., D. R. HOLE & C. A. WILSON. 1988. Behavioral effects at adulthood of serotonin depletion by p-chlorophenylalanine given neonatally to male rats. Int. J. Neurosci. **41:** 187–199.

69. LARSEN, P. R. & S. H. INGBAR. 1992. The thyroid gland. *In* Textbook of Endocrinology. J. D. Wilson & D. W. Foster, Eds.: 357–487. W. B. Saunders Co. Philadelphia, PA.

70. MACFAUL, R., S. DORNER, E. M. BRETT & D. B. GRANT. 1978. Neurological abnormalities in patients treated with hypothyroidism from early life. Arch. Dis. Child. **53:** 611–619.

71. GOULD, E., C. S. WOOLLEY & B. S. MCEWEN. 1991. The hippocampal formation: Morphological changes induced by thyroid, gonadal and adrenal hormones. Psychoneuroendocrinology **16:** 67–84.

Developmental Neurobiology of the Stress Response: Multilevel Regulation of Corticotropin-Releasing Hormone Function[a]

TALLIE Z. BARAM,[b–e] SUJIN YI,[c]
SARIT AVISHAI-ELINER,[b] AND LINDA SCHULTZ[b]

*Departments of Anatomy & Neurobiology,[b] Pediatrics,[c]
and Neurology[d]
University of California, Irvine
Irvine, California 92697-4475*

INTRODUCTION

In the mature animal, a variety of stressors activate the hypothalamic-pituitary-adrenal (HPA) axis:[1–3] Increased secretion of corticotropin-releasing hormone (CRH) from the hypothalamic paraventricular nucleus (PVN) induces ACTH release from the pituitary and increases plasma corticosterone (CORT). The depletion of hypothalamic CRH is followed by a "compensatory" up-regulation of CRH gene expression.[2–4] Limbic input from hippocampus and amygdala, mainly the central nucleus (ACE), further modulates stress-induced alteration of CRH gene expression.[5,6]

The presence of most of these components of the "stress response" in the neonatal (first postnatal week) and infant (second postnatal week) rat has been debated.[7,8] Elevation of plasma glucocorticoids in response to stressors and the relative roles of CRH and vasopressin have been subjects of investigation.[4,7–10] A "stress-hypo-responsive" period with diminished or immature hormonal responses has been described during the first two postnatal weeks.

The goal of this report is to present data confirming the presence of robust stress-induced elevation of plasma glucocorticoids (GC) and the involvement of CRH in this response. Evidence for immature regulation of CRH gene expression by GC and stress will be presented, as well as the role of the cyclic AMP cascade in the regulation of the CRH gene promoter.

MATERIALS AND METHODS

Animals

Pups were products of time-pregnant Sprague-Dawley rats (Zivic-Miller, Zeliehople, PA), kept on a 12-h light/dark cycle and given access to unlimited lab chow

[a] This work was supported in part by NINDS NS 28912 (T.Z.B.) and by USC-CHLA BRSG awards (S.J.Y. and S.A.E.).

[e] Address correspondence to Tallie Z. Baram, M.D., Ph.D., Med. Sci. I, 4475, University of California, Irvine, Irvine, CA 92697-4475.

and water. Delivery was verified at 12-h intervals, and the date of birth was considered day 0. Litters were culled to 12 pups, and undisturbed for 48 h prior to experiments. All experiments were started at 9 A.M. to avoid circadian variation in HPA axis components and tone.[1,11]

Materials

Dibutyryl-cyclic-AMP (db-cAMP), isobutyl methylxanthine (IBMX), dexamethasone (DEX), forskolin, and GTP were obtained from Sigma Chemical Co. (St. Louis, MO). DEX was dissolved in 100% ethanol, and forskolin was dissolved in DMSO. Stock solutions of both drugs were diluted with 10^2 to 10^6 volumes of buffer immediately prior to use. CRH antiserum (CRH-AS) produced in sheep was a gracious gift of Dr. W. W. Vale (pool No. 238-293). Normal sheep serum (NSS) was purchased from Calbiochem (San Diego, CA). CRH-antagonist, (9-41)-alpha-helical CRH, was purchased from BACHEM (Torrance, CA).

Stress Paradigm and Hormonal Studies

Acute cold-separation was chosen as a potent, age-specific and painless stress.[4,10] Pups were separated from their mothers and placed individually in glass jars in a cold room (4 °C). Preliminary experiments defined maximally tolerated cold exposure, as defined by the development of rigor and little response to tactile stimulus (core temperature averaged 9.8 °C). Thus, cold exposure lasted an average of 30 min in 4–6-day-old pups and 60 min in 11–16-day-old pups. Stressed pups were rewarmed on a heating pad under a heat lamp as a group. Rats were sacrificed by decapitation and trunk blood was collected before the onset of stress and at 0, 20, 40 (or 30), 60, 150, and 240 min after its termination. Control rats were left in home cages and sacrificed within 45 s of disturbance. Plasma CORT levels were determined by radioimmunoassay (ICN, Irvine, CA). Assay sensitivity was 0.5 μg/dL; interassay variability was determined by two dilutions of adult rat plasma, and averaged 15%.[4,10]

CRH-Antiserum Effect on Stress-Induced Plasma CORT Elevation

Rat pups were divided into the following experimental groups: undisturbed controls, injection controls, cold exposure, cold and NSS-injection and cold and CRH-AS injection. Rats were injected with CRH-AS or NSS (100 μL, ip) at 9:30 A.M. One hour later, rats were exposed to cold for 25 min (6-day-old) or 40 min (9–10-day-old), rewarmed for 50 min and sacrificed. Undisturbed controls were sacrificed at 9 A.M. and injection controls (not exposed to cold) immediately prior to the corresponding cold-exposed groups.

Time-Course of and Age Dependence of Stress-Induced Up-Regulation of CRH-mRNA Abundance

In a preliminary time-course experiment, 16-day-old-rats were divided into six groups ($n = 4$). Groups 1–3 were subjected to cold-separation stress as above, and sacrificed at 1, 4, or 28 h after termination of stress. Groups 4–6 served as controls for both separation-cold stress and diurnal variability in CRH-mRNA.[11] Controls were undisturbed in home cages until sacrifice at the corresponding hours. Based on the results of this experiment,[4] 6-, 9- and 16-day-old rats were subjected to cold-separation stress as above, and decapitated 4 or 28 h after the onset of cold-stress (the 28-h group was returned to home cages at 4 h and sacrificed 24 h later).

Effect of db-cAMP on Stress-Induced CRH Synthesis in the 6-Day-Old Rat

Six-day-old rats were assigned to one of the following groups: (1) Undisturbed controls; (2) cold-stressed controls; (3) cannulated, saline-infused, stressed controls; (4) cannulated, db-cAMP-infused, stressed; and (5) cannulated, stressed, with db-cAMP infused into the striatum. Cold was used as both stressor and anesthetic. Rats were placed in an ice-bath and transferred to a stereotaxic apparatus when motionless. Stainless steel cannulae were implanted unilaterally, with the tip directed immediately above the PVN.[12,13] We have previously shown that in the neonatal rat, a unilateral infusion of a number of compounds results in diffusion to the contralateral PVN (refs. 12 and 13 and unpublished observations). Coordinates for striatal cannulae were A: 0.95, V: 4.0, L: 3.4 in reference to bregma. Saline or db-cAMP (1 μmole in 1 μL) was infused using a Hamilton syringe. Cannula placement was verified in each animal using 20-μm coronal sections stained with cresyl violet.

Methodology of cAMP Analysis

Rats were sacrificed at 9 A.M., and brains were removed within 45 s of initial disturbance. Anterior hypothalami were dissected on ice as follows: The brain was placed, hypothalamus up, on an ice-chilled petri dish. With a razor blade, a coronal cut was made 2 mm (for immature) or 2.5 mm (for adult rats) anterior to the mammillary bodies. Diagonal cuts separated the hypothalamus, including the medial preoptic area, from the telencephalon. A horizontal cut, 3 mm above the inferior surface of the hypothalamus separated it from the thalamus. The resultant tissue-block, as determined in preliminary experiments, consisted of the PVN, periventricular and medial preoptic nuclei, and the anterior hypothalamic area. Dissection was achieved in 20 s or less, and the weight of the excised block averaged 19.8 $\pm$ 0.8 mg. Dissected tissues were immediately transferred to ice-chilled buffer, pooled, and minced to a thickness of 300–400 μm. Minced tissue was suspended in 50 vol of freshly oxygenated Krebs buffer (NaCl; 119, KCl; 2.5, $MgSO_4$; 1.3, $CaCl_2$; 2.5, KH_2PO_4; 1.0, $NaHCO_3$; 26.2, glucose; 11 mM, pH 7.4) containing 0.5 mM IBMX and 0.1 μM GTP, and incubated for 30 min at 37 °C. During preincubation, medium was replaced with equal volume of freshly oxygenated buffer. Tissue aliquots were portioned into test tubes containing fresh buffer with various concentrations of DEX for an initial incubation of 20 min at 37 °C, and then pipetted into chambers containing fresh buffer and desired concentrations of forskolin and DEX for a final 15 min. Incubation was terminated by transferring chambers onto ice and pipetting the tissue into test tubes containing 500 μL of ice-cold 10% trichloroacetic acid. Homogenization (on ice) and centrifugation (1,000 $\times$ g for 10 min) were followed by extraction of supernatants with 5 vol of diethyl ether (3$\times$). Supernatants were lyophilized and stored at -20 °C until used for cAMP assay. Pellets were resuspended in 50 mM sodium acetate buffer and assayed for protein. cAMP radioimmunoassay was performed using commercial cAMP tracer (Biomedical Technologies Inc., Stoughton, MA) and cAMP antibody obtained from the National Institutes of Health. Assay sensitivity was 0.05 pmol/mL.

Tissue Processing and in Situ Hybridization

In all experiments, brains were rapidly removed onto powdered dry ice and stored at -80 °C. Brains were cut into 20-μm coronal sections in a cryostat and mounted on gelatin-coated slides.[14,15] Preparation of oligonucleotides for CRH and CRF_1, and details of ISH and image analysis have been described.[4,10,12,14,15] Briefly, prior to *in situ* hybridization (ISH), slides were brought to room temperature, air-dried and fixed in buffered paraformaldehyde. Following a graded ethanol treatment, sections were exposed to acetic anhydride-triethanolamine, then dehydrated through 100% ethanol. Sections were prehybridized for 1 h, then hybridized for 20 h at 40 °C in a humidity chamber. Serial washes ($2\times$ SSC for 15 min $\times 4$ at 40 °C; $1\times$ and $0.3\times$ SSC for 30 min each at room temperature) were followed by dehydration and apposition to film (Hyperfilm B-Max, Amersham). Selected sections were subsequently dipped in emulsion (NTB-2; Kodak), and developed as previously described, with the exception that the emulsion was not diluted.

Analysis

Quantitation and statistical analysis were described previously.[4,12,14,15] Briefly, optical density (OD) was determined over areas of interest, using the MCID software image analysis system (Imaging Research, Ont., Canada). Each point was derived from 6–12 sections from a minimum of four individual rats. Statistical significance between groups was determined using two-way analysis of variance, followed by Duncan's multiple range tests.

RESULTS

Cold-Separation Stress Elevates Plasma CORT throughout the First Two Postnatal Weeks

The time course and magnitude of cold-induced CORT elevation are shown in FIGURE 1. Clear, stress-induced peaks of plasma CORT were evident on postnatal days (PND) 4–6, 9, and 11–13. Maximal cold-stress induced plasma CORT values were 7.57 ± 0.86, 6.48 ± 0.26, and 6.64 ± 1.31 μg/dL on PND 6, 9, and 11–13, respectively.

Cold-Stress-Induced Elevation of Plasma CORT Is Abolished by Antiserum to CRH

Injection of either AS or NSS resulted in increased plasma CORT, presumably because of handling and pain. Cold-separation stress induced a significant ($p < 0.01$) further increase in plasma CORT levels at both days 6 and 9 (FIG. 2A

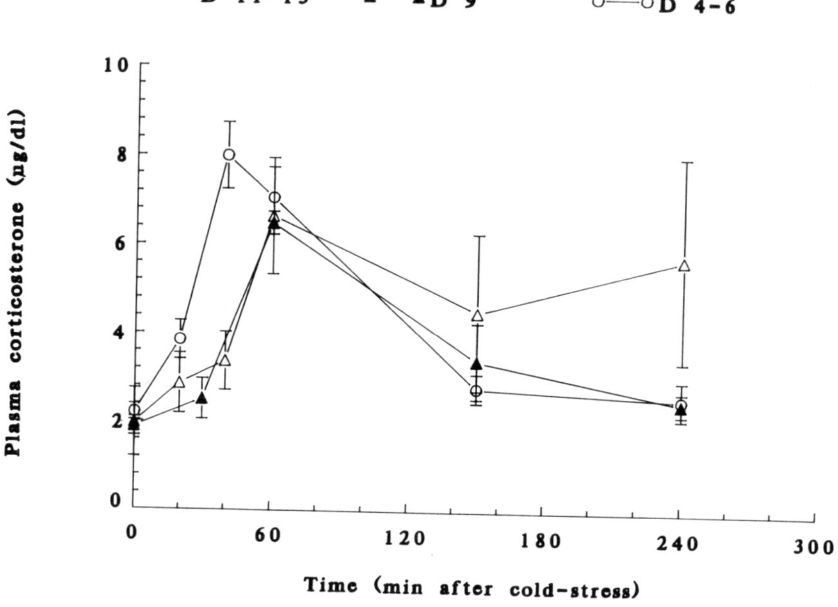

FIGURE 1. Plasma CORT time courses in response to cold-separation stress in 4- to 6-, 9-, and 11- to 13-day-old rats. Values are mean ± SEM of 6–8 rats/group. For all age groups, 60-min values are significantly different from prestress ones ($p < 0.05$). On postnatal day 6, the 40-min value is significantly elevated as well ($p < 0.05$).

and 2B). At both ages, administration of CRH-AS abolished cold-induced plasma CORT ($p < 0.01$, cold + NSS vs. cold + AS).

Stress Induces CRH Gene Expression Only in 9-Day, or Older, Rats

A significant increase of CRH-mRNA abundance in the PVN of 16-day-old rats was observed at 4 h (but not at 1 h), and persisted at 28 h subsequent to cold-separation stress. Therefore, these time points were studied in younger rats. In 9-day-old rats, cold-separation stress enhanced CRH-mRNA abundance at 4 h, but not at 28 h. No change in the CRH-message was found in 6-day-old rats (FIG. 3).

Infusion of a Cyclic AMP Analogue Enables the Up-Regulation of CRH Synthesis by Stress in the 6-Day-Old Rat

As is evident from FIGURE 4, infusion of db-cAMP permitted cold stress to up-regulate CRH gene expression in response to cold stress. This was evident upon infusion of the cAMP analogue to the PVN. Striatal infusion of db-cAMP was ineffective (not shown). db-cAMP was thus able to "reconstitute" a missing regulatory factor(s) acting on the CRH gene promoter.

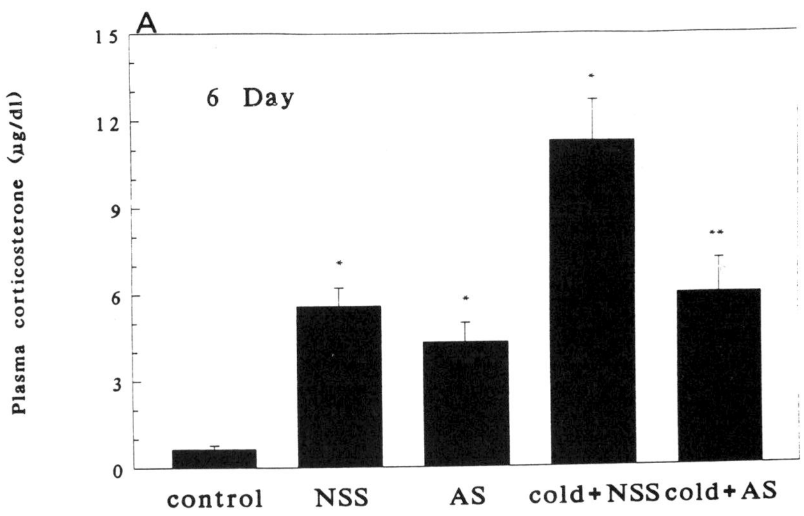

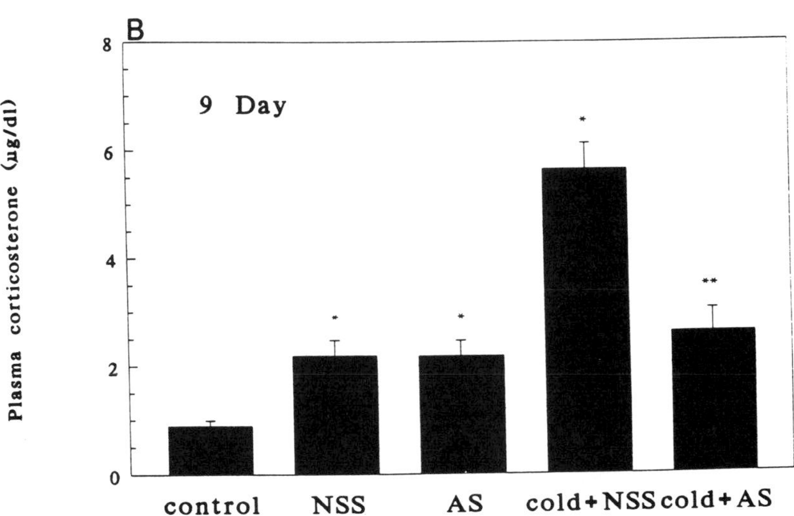

FIGURE 2. Effect of passive immunization against CRH on cold-separation stress-induced plasma CORT. **Panel A:** On postnatal day 6. **Panel B:** On postnatal day 9. NSS, normal sheep serum; AS, antiserum directed against CRH. $n = 6$ per group. Bars indicate standard errors. *Significantly different from control ($p < 0.05$); **significantly different from cold + NSS ($p < 0.05$).

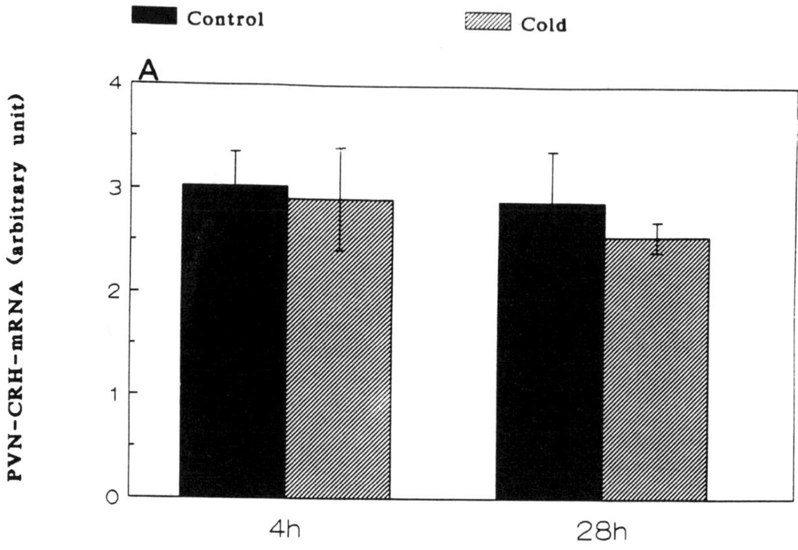

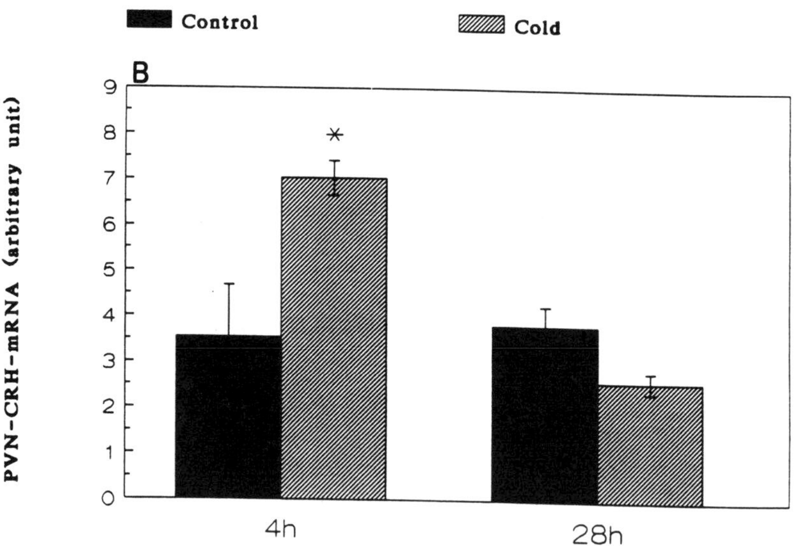

FIGURE 3. Effect of cold-separation stress on CRH-mRNA in the paraventricular nucleus of 6- **(panel A)** and 9- **(panel B)** day-old rats. Pups were subjected to age-appropriate maximal tolerated cold-stress. CRH-mRNA was determined using *in situ* hybridization. Values (mean ± SEM) were derived as detailed in the METHODS section. *Significantly different from control ($p < 0.05$).

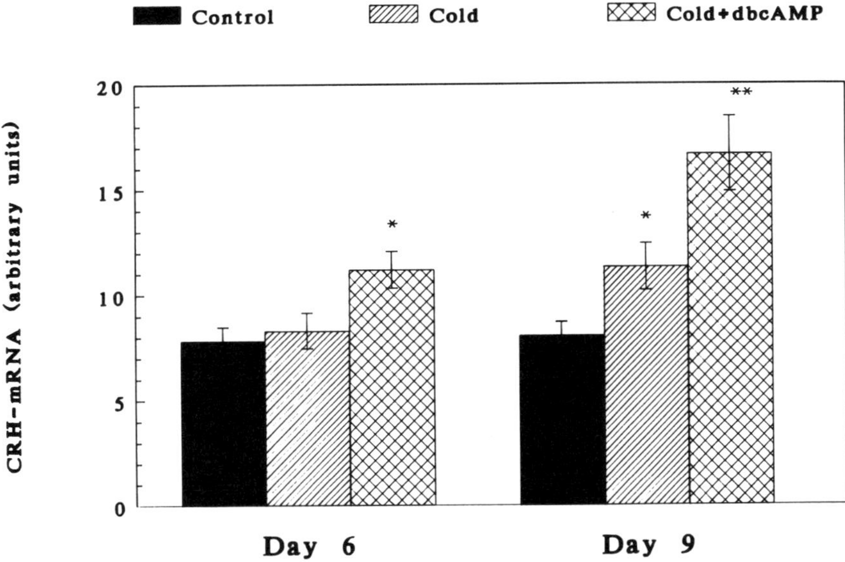

FIGURE 4. The effect of db-cAMP infusion, combined with cold + surgical stress, on CRH-mRNA abundance in the paraventricular nucleus of the 6- and 9-day-old rats. In the 6-day-old, stress alone did not result in "compensatory" increase of CRH-mRNA abundance at 4 h. db-cAMP enabled this effect, seen normally in the 9-day-old. In the 9-day-old rat, db-cAMP further increased stress-induced elevation of steady-state CRH-mRNA levels. *Significantly different from control ($p < 0.05$); **different from cold stress.

The Regulation of cAMP by Glucocorticoids Is Deficient in Immature Rat Hypothalamus

Anterior hypothalami derived from adult rats contained less basal cAMP (pmole/mg protein) than those of neonatal rats (FIG. 5A). This was probably due to higher protein concentrations in the adult rat. DEX alone (0.01 μM to 10 μM) did not significantly affect cAMP content at any age (FIG. 5A). Forskolin (10 μM) stimulated adenylate cyclase in the anterior hypothalamus significantly in all age groups (FIG. 5B). A 10-fold increase in cAMP production was found in adult rats, and a 3.2- to 6-fold increase in neonatal rats (3- to 13-day-old, $p < 0.05$).

DEX inhibited the stimulatory effect of forskolin on adenylate cyclase in the anterior hypothalamus of adult rats in a concentration-dependent manner, with maximal effect (a 60% reduction) observed with 1 μM. This inhibitory effect of dex was considerably reduced in the anterior hypothalamus of 9- to 13-day-old rats, and was minimal in 3- to 4-day-old rats (FIG. 5C). Data analysis revealed a significant effect of age ($p < 0.05$). IC$_{50}$'s of the inhibition curves for the adult and 9- to 13-day-old rats were $1.87 \pm 0.4 \times 10^{-8}$ M and $1.68 \pm 0.32 \times 10^{-7}$ M, respectively.

CRH Receptor Is Highly Abundant in Limbic Structures of the Immature Rat

FIGURE 6 is a schematic representation of the CRH receptor (CRF$_1$) messenger RNA distribution in hippocampus, amygdala, and cortex of the developing rat.

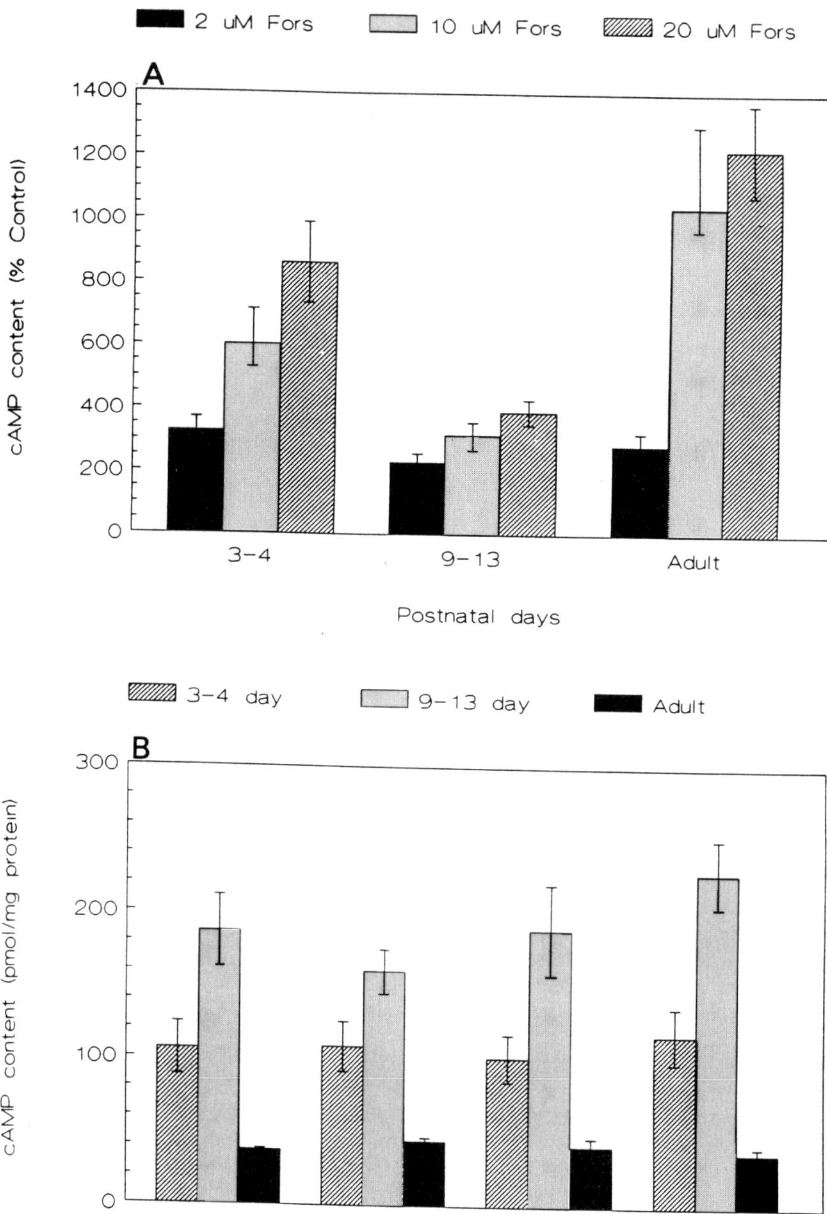

FIGURE 5. **(A)** Basal cAMP content (pmole/mg protein) in the anterior hypothalamus of neonatal and adult rats. Dexamethasone alone did not significantly affect cAMP accumulation at any age. Values represent the mean ± SEM of 6–7 experimental pools (10–12 neonatal or 2 adult hypothalami/pool). **(B)** Forskolin-stimulated adenylate cyclase in the anterior hypothalamus of neonatal and adult rats. Forskolin (2–20 μM) caused a robust increase at all ages. Values are the mean ± SEM of 5–6 pooled experiments. **(C)** Effect of dexamethasone on forskolin-induced increase in adenylate cyclase in anterior hypothalamus of neonatal and adult rats. *$p < 0.05$ vs. forskolin alone. Values obtained from 5–6 pooled experiments. Fors, forskolin; Dex, dexamethasone.

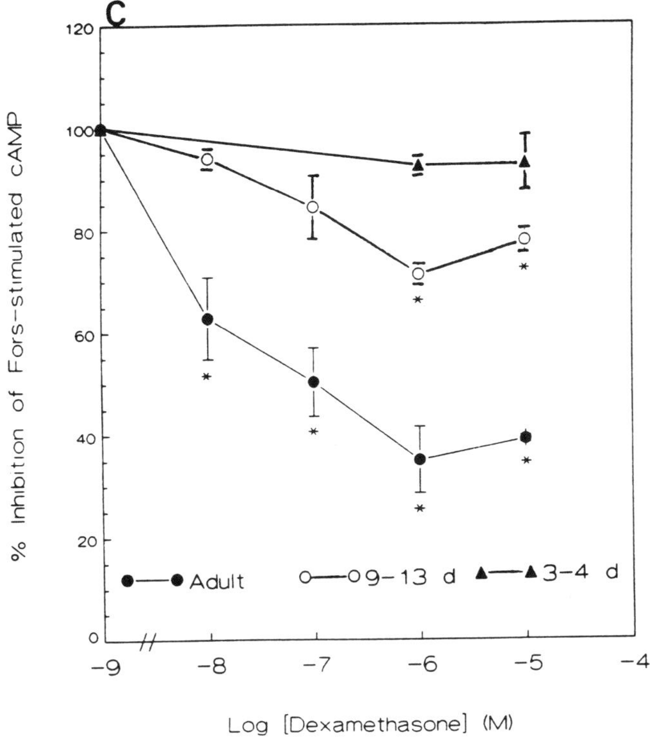

FIGURE 5. (*Continued*)

Hippocampal values are a composite of CA1–CA2 and CA3 values.[16] Amygdala values represent primarily the basolateral and central nuclei. Cortex was sampled at the frontoparietal convexity. Maximal receptor levels in the hippocampus are evident on PND 6 and peak in amygdala on PND 9.

DISCUSSION

The hormonal response to stressful cues is complex. The final common pathway consists of a transient, CRH-induced elevation of plasma ACTH and CORT. Negative feedback effects of glucocorticoids, which help terminate the response, up-regulation of CRH gene expression after stress-induced "depletion" of hypothalamic stores, and limbic input impinging on the hypothalamic PVN are only some of the molecular events contributing to this complex adaptive mechanism.[1–6] During the neonatal and infant developmental periods in the rat, "immaturity" of this hormonal stress response implies quantitative or qualitative alterations of one or more components of this chain of events.[4,7–10] The specific control points at which the immature animal differs from the adult are unclear. Obviously,

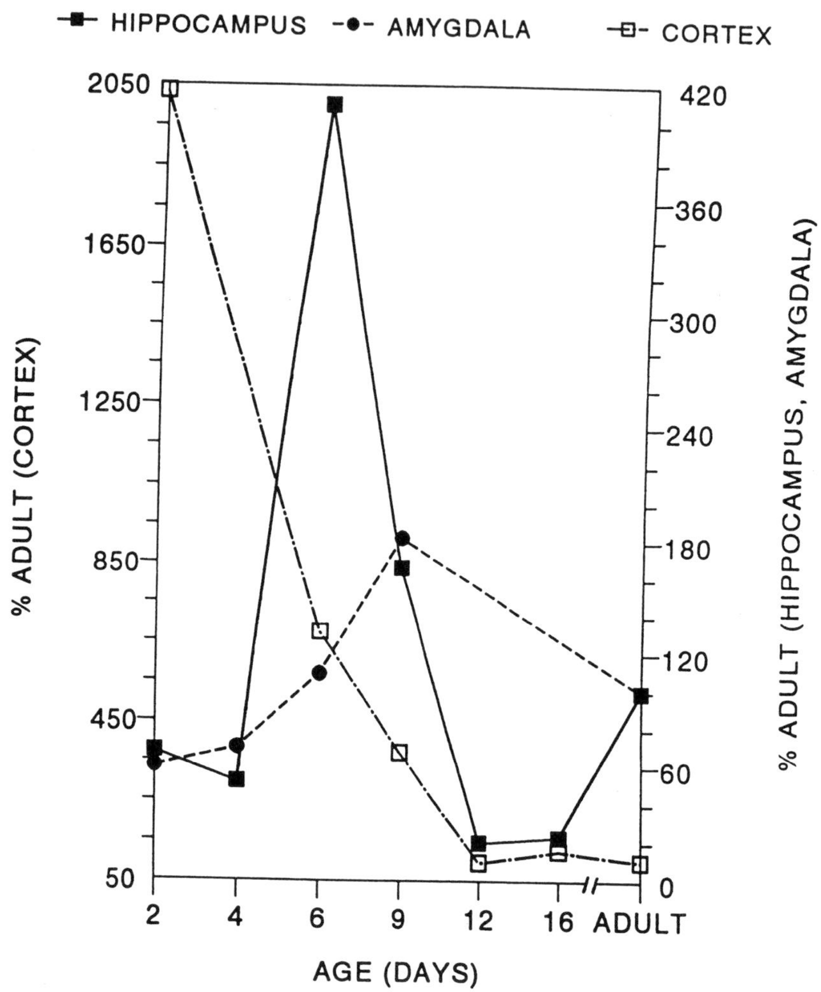

FIGURE 6. Schematic of the developmental profile of corticotropin-releasing hormone receptor (CRF_1) in the amygdala, hippocampus, and cortex of the rat. See reference 16 for methodology.

perturbation of regulation of CRH gene expression, for example, should result in significant alteration of "downstream" components such as plasma CORT.

The studies described affirm that cold, a defined, age-specific acute stress, induces plasma CORT throughout the neonatal and infant period. Further, this elevation of plasma CORT depends on CRH in that it is abolished by antiserum to the peptide. As such, no major qualitative difference exists between the neonatal and adult rat in the "effector" arm of the CRH-CORT cascade.

Stress-induced up-regulation of CRH synthesis does not occur to any significant degree prior to the second postnatal week. This is consistent with previous findings regarding "insensitivity" of the CRH gene promoter, *in vivo*, in the neonatal rat.[12] Neither the elimination of negative glucocorticoid feedback, by blocking of GC receptors in the PVN, nor the chronic stress of surgical cannula implantation resulted in increased steady-state CRH-mRNA prior to the ninth day of life.[12] Both maneuvers increased CRH gene expression in the PVN of older rats.[12]

Candidate mechanisms for this insensitivity of the CRH gene promoter in the neonatal rat *in vivo* are suggested in FIGURES 4 and 5. First, infusion of a cAMP analogue, dibutyryl-cAMP, into the PVN during cold-surgery stress on PND 6, permitted enhancement of CRH-mRNA abundance, an effect normally observed on PND 9 or later. The effect was specific to db-cAMP and was not observed upon saline infusion. Further, neither db-cAMP nor saline was effective upon infusion into the striatum, suggesting anatomic specificity to the hypothalamus. A similar approach, implicating cAMP in the regulation of neuropeptide Y has recently been reported.[17] The mechanism by which db-cAMP, in conjunction with acute stress, enhanced CRH-mRNA abundance on PND 6 remains speculative. The rationale for infusing the cAMP analogue derives from the fact that the CRH gene promoter contains a cAMP response element (CRE).[18–20] Forskolin enhances the human CRH gene promoter, and mutation of the CRH-CRE sequence completely abolishes the stimulation by cAMP.[19] We speculated that the cAMP-dependent signal transduction pathway might not be fully functional on PND 6. An excess of exogenous cAMP enabled this "mature response" of the CRH promoter on PND 6, consistent with a role for cAMP in the activation of the CRH gene promoter.

Additionally, the interaction between GC and the cAMP second messenger cascade is altered in the developing hypothalamus (FIG. 5A–C).[21] A consensus GC response element sequence is not found at, or upstream from, the CRH gene promoter site, suggesting other mechanisms for the observed down-regulation of CRH synthesis by GC in the adult.[20] We find that GC decreases cAMP production in mature hypothalamus.[21] The response curve is shifted markedly to the right in the neonatal or infant rat (FIG. 5C).

An additional determinant of the magnitude of the CRH effect is the presence and abundance of receptors mediating the peptide's effects.[22–24] In the immature rat, distinct temporal and spatial profiles of one of these receptor types, CRF_1, is clearly evident. For example, on the sixth postnatal day, the hippocampal/ amygdala receptor ratio is higher than on the ninth postnatal day or in the adult. Whether this may result in augmented inhibitory input from hippocampus to PVN at this age remains speculative.[16]

SUMMARY

The ability to respond to adverse environmental cues is present in the neonatal and infant rat, although in an immature form: A number of laboratories have demonstrated stress-induced elevations of plasma glucocorticoids during the first two postnatal weeks. The limbic and hypothalamic mechanisms controlling the hormonal stress-response during this period are not fully understood and are, therefore, the focus of this report.

Both hypothalamic corticotropin-releasing hormone (CRH) and vasopressin contribute to the release of ACTH from the pituitary in the adult. The relative roles of these two peptides during the neonatal (first week) and infant (second week) developmental period, are controversial. Evidence is presented that argues strongly for a major role for CRH.

Up-regulation of hypothalamic CRH synthesis is a major component in the mature stress response. CRH-mRNA levels in the hypothalamic PVN are increased with cold stress by the ninth postnatal day, but not during the first postnatal week. Further, down-regulation of CRH gene expression by glucocorticoids (GC) constitutes a critical "shut-down" mechanism for the hormonal stress response. *In vivo* and *in vitro* experiments supporting the "immaturity" of GC feedback on CRH synthesis during the first postnatal week are described.

CRH-mediated neurotransmission, in both the endocrine and neuronal effector arms of the response to stress may be modeulated via alteration of receptor number. The first member of the CRH receptor family, CRF_1, probably mediates the neuroendocrine effects of CRH. The developmental profile of CRF_1-mRNA reveals several distinctive spatial and temporal patterns. In the hippocampal CA1, CA2, and CA3a, peak (300–600% adult values) CRF_1-mRNA is found on postnatal day 6. In the amygdala, CRH receptor mRNA levels are maximal on the ninth postnatal day (at 180% of adult values). In cortex, a steady decline from high postnatal day 2 levels results in adult levels by day 12. These findings demonstrate distinct, regional, age-specific control of the synthesis of CRF_1. Receptor expression profile may provide important information regarding modulation of the age-specific roles of CRH in different regions. For example, a high ratio of hippocampus/amygdala receptors may preferentially activate negative hippocampal input to the hypothalamus during the neonatal period. Additionally, increased CRH receptor mRNA in the infant compared with the adult provides a mechanism for enhanced excitatory effect of the peptide at this age.

In conclusion, increasing evidence exists for multiple control points of the early postnatal response and adaptation to stress. CRH synthesis in hypothalamus and amygdala, its sensitivity to GC feedback, and the abundance and distribution of at least two distinct CRH receptors in the limbic central nervous system and the pituitary are developmentally regulated. All serve as control points permitting an effective endocrine, autonomic, and behavioral response to stressful environmental cues.

REFERENCES

1. DALLMAN, M. F., S. F. AKANA, C. S. CASCIO, D. N. DARLINGTON, L. JACOBSON & N. LEVIN. 1987. Regulation of ACTH: Variations on a theme of B. Recent Prog. Horm. Res. **43:** 113–131.
2. LIGHTMAN, S. L. & M. S. HARBUZ. 1993. Expression of corticotropin releasing factor mRNA in response to stress. *In* Corticotropin Releasing Factor. CIBA Symposium. Wiley. Chichester. pp. 173–198.
3. HARBUZ, M. S. & S. L. LIGHTMAN. 1989. Responses of hypothalamic and pituitary mRNA to physical and psychological stress in the rat. J. Endocrinol. **122:** 705–711.
4. YI, S.-J. & T. Z. BARAM. 1994. Corticotropin releasing factor mediates the response to cold stress in the neonatal rat, without compensatory enhancement of the peptide's gene expression. Endocrinology **135:** 2364–2368.
5. JACOBSON, L. & R. SAPOLSKY. 1991. The role of the hippocampus in feedback regulation of the HPA axis. Endocr. Rev. **12:** 118–134.
6. GRAY, T. S., C. CARNEY & D. J. MAGNUSON. 1989. Direct projections from the central amygdaloid nucleus to the hypothalamic PVN: Possible role in stress-induced ACTH release. Neuroendocrinology **50:** 433–446.
7. WALKER, C. D., K. A. SCRIBNER, C. S. CASCIO & M. F. DALLMAN. 1991. The pituitary-adrenocortical systems of neonatal rats is responsive to stress throughout development in a time dependent and stressor-specific fashion. Endocrinology **128:** 1385–1395.

8. SUCHEKI, D., D. MOZAFFARIAN, G. GRAZIELLA, P. ROSENFELD & S. LEVINE. 1993. Effects of maternal deprivation on the ACTH stress response in the infant rat. Neuroendocrinology **57:** 204–212.

9. PAULMYER-LACROIX, O., G. ANGLADE & M. GRINO. 1994. Stress regulates differently the AVP-containing and AVP-deficient CRF synthesizing cell bodies in the hypothalamic paraventricular nucleus of the developing rat. Endocrine **2:** 1037–1043.

10. AVISHAI-ELINER, S., S. J. YI, C. L. NEWTH & T. Z. BARAM. 1995. Effects of maternal and sibling deprivation on basal and stress-induced HPA components in the infant rat. Neurosci. Lett. **192:** 1–4.

11. WATTS, A. G. & L. W. SWANSON. 1989. Diurnal variation in the content of preprocorticotropin releasing hormone mRNA in the hypothalamic paraventricular nucleus of rats of both sexes as measured by in situ hybridization. Endocrinology **125:** 1734–1738.

12. YI, S.-J., J. N. MASTERS & T. Z. BARAM. 1993. The effect of RU-38486 on CRH gene expression in the neonatal rat hypothalamus. Dev. Brain Res. **73:** 253–259.

13. YI, S.-J. & T. Z. BARAM. 1993. Methods for implanting steroid-containing cannulae into the paraventricular nucleus of neonatal rats. J. Pharmacol. Toxicol. Methods **30:** 97–102.

14. BARAM, T. Z. & S. P. LERNER. 1991. Corticotropin releasing hormone: Ontogeny of gene expression in rat hypothalamus. Int. J. Dev. Neurosci. **9:** 473–478.

15. BARAM, T. Z. & L. SCHULTZ. 1992. CRH gene expression in the fetal rat is not increased after pharmacological adrenalectomy. Neurosci. Lett. **142:** 215–218.

16. AVISHAI-ELINER, S., S. J. YI & T. Z. BARAM. 1996. Developmental profile of CRH-receptor messenger RNA in the rat limbic system. Dev. Brain Res. **91:** 159–163.

17. AKABAYASHI, A., C. T. B. V. ZAIA, S. M. GABRIEL, I. SILVA, W. K. CHEUNG & S. F. LEIBOWITZ. 1994. Intracerebroventricular injection of dibutyryl cyclic adenosine 3'5'-monophosphate increases hypothalamic levels of neuropeptide Y. Brain Res. **660(2):** 323–328.

18. SEASHOLTZ, A. F., R. C. THOMPSON & I. O. DOUGLASS. 1988. Identification of a cyclic AMP-responsive element in the rat corticotropin-releasing hormone gene. Mol. Endocrinol. **2:** 1311–1318.

19. SPENGLER, D., R. RUPPRECHT, L. P. VAN & F. HOLSBOER. 1992. Identification and characterization of 3',5'-cyclic adenosine monophosphate responsive element in the human corticotropin releasing hormone gene promoter. Mol. Endocrinol. **6:** 1931–1941.

20. MAJZOUB, J. A., R. EMANUEL, G. K. ADLER, C. MARTINEZ, B. ROBINSON & G. WITTERT. 1993. Second messenger regulation of mRNA for CRF. *In* Corticotropin Releasing Factor. CIBA Symposium. Wiley. Chichester. pp. 30–43.

21. YI, S.-J. & T. Z. BARAM. 1993. Ontogeny of forskolin-stimulated cAMP-dependent protein kinase and of dexamethasone suppression of forskolin-stimulated cAMP production in rat hypothalamus. Soc. Neurosci. Abstr. **23:** 1186.

22. DE SOUZA, E. B. 1987. Corticotropin-releasing factor receptors in the rat central nervous system: Characterization and regional distribution. J. Neurosci. **7:** 88–100.

23. CHANG, C.-P., R. V. PEARSE, S. O'CONNELL & M. G. ROSENFELD. 1993. Identification of a seven transmembrane helix receptor for corticotropin-releasing factor and sauvagine in mammalian brain. Neuron **11:** 1187–1195.

24. PERRIN, M. H., C. J. DONALDSON, R. CHEN, K. A. LEWIS & W. W. VALE. 1993. Cloning and functional expression of a rat brain corticotropin-factor (CRF) receptor. Endocrinology **133:** 3058–3061.

25. LOVENBERG, T. W., C. W. LIAW, D. E. GRIGORIADIS, W. CLEVENGER, D. T. CHALMERS, E. B. DE SOUZA & T. OLTERSDORF. 1995. Cloning and characterization of functionally distinct corticotropin-releasing factor receptor subtype from rat brain. Proc. Natl. Acad. Sci. USA **92:** 836–840.

Maternal Stress, HPA Activity, and Fetal/Infant Outcome[a]

CURT A. SANDMAN,[b,e] PATHIK D. WADHWA,[c]
ALEKSANDRA CHICZ-DeMET,[b]
CHRISTINE DUNKEL-SCHETTER,[d] AND MANUEL PORTO[c]

Departments of Psychiatry,[b] and Obstetrics and Gynecology[c]
University of California, Irvine
Irvine, California

[d]Department of Psychology
University of California, Los Angeles
Los Angeles, California

The consequences of perinatal stress and/or exposure to neuropeptides on the developing organism can include permanent influences on the brain and behavior. In human subjects, infant and early childhood separation stress is associated with a significant increase in adult psychopathology and permanent elevation in plasma β-endorphin (βE) and cortisol.[1] A study of over 3,000 patients[2] reported that the type of ante- or intrapartum stress significantly predicted the category of self-destructive behavior exhibited among adults. For instance, perinatal hypoxia was associated with significant tendencies to attempt suicide by asphyxic means such as suffocation. Intrapartum "mechanical" trauma (e.g., use of forceps) was associated with significant tendencies to self-destruct with instruments such as guns. Some less severe forms of self-injurious behavior related to developmental compromise observed among mentally retarded and autistic individuals can be attenuated by centrally acting drugs.[3–6] In these individuals the best predictor of a positive response to drugs is a history of perinatal stress.[7,8] These studies in clinical populations suggest that perinatal stress is strongly associated with developmental patterns and with influences on the central nervous system.

Controlled observations in animals provide evidence that early exposure to stressful events known to release peptides from the hypothalamic-pituitary-adrenal (HPA) axis have persisting influences on the brain and behavior. Restraint stress in pregnant rats results in decreased binding of μ-opiate receptors in the striatum of the offspring at 42 days of age.[9] Hypothalamic levels of βE are elevated in 10-day-old rats prenatally stressed during the first or second trimester.[10] The finding that maternal pretreatment with opiate blockers protects rabbit fetuses from the damaging influence of stress and trauma[11] supports the pivotal role of the HPA in mediating the influence of stress on development.

Administration of peptides from the HPA axis also produces long-term influences on the nervous system and behavior. Rats exposed as fetuses directly to βE during the first and second trimester had "permanent" decreases (subsensitivity) in

[a] This work was supported by grants HD28413 and HD28202 from the National Institute of Child Health and Human Development.

[e] Address correspondence to Curt A. Sandman, Fairview Developmental Center, 5A, 2501 Harbor Boulevard, Costa Mesa, California 92626.

the density of striatal, dopamine (D2) receptors,[12] and "permanent" alterations in behavior.[13] Perinatal exposure to MSH/ACTH and its analogues permanently altered learning, memory, and growth,[14,15] decreased central monoaminergic neurons, and disrupted the HPA axis during stress.[16] Offspring of rats administered corticotrophin-releasing hormone (CRH) during the third trimester weighed less at birth than controls and emitted more stress-related vocalizations.[17] Male offspring from the CRH-treated group had shorter anogenital distances at birth. These findings in the CRH-treated group are identical to findings observed in offspring of stressed mothers.

The primary objective of our research is to characterize the influence of stress during human pregnancy on the HPA axis, birth outcome, and fetal behavior. Understanding the effects of stress during human pregnancy is complicated by the development of the placenta as a significant endocrine, perhaps "stress-sensitive," organ.[18,19] All HPA-axis peptides increase during human gestation, but the dramatic elevations of placental CRH in maternal plasma during the course of pregnancy reach levels observed only in the hypothalamic portal system during physiological stress.[20] In addition to the increase in levels of CRH as pregnancy advances, a sharp increase is found in the availability of bioactive CRH in the peripheral circulation because of the reduction of CRH-binding protein during the last two to four weeks of gestation.[21,22] For these reasons, the role of CRH among the HPA products is of special interest.

Placental CRH is identical to hypothalamic CRH in structure, immunoreactivity, and bioactivity;[23,24] however, in contrast to the negative control on hypothalamic CRH, glucocorticoids stimulate the expression of hCRH mRNA in the placenta. This establishes a positive feedback loop resulting in parallel increases of CRH, ACTH, and cortisol over the course of gestation.[25] Thus, during the course of pregnancy, the CRH system becomes increasingly activated and reaches the peak of bioactivity just before term. The consequences of increased CRH for maternal stress are not known, but several possibilities are suggested by the model in FIGURE 1. One possibility is that hypophyseal corticotrophs may become desensitized as pregnancy develops. If this is true, the threshold is increased for stress-induced, pituitary release of ACTH and βE by hypothalamic CRH. The normal function of this feature of the model may effectively *immunize* pregnant women approaching term from the effects of CRH-modulated environmental stress. As the concentration of placental CRH rises in maternal circulation, the ability decreases for hypothalamic CRH to stimulate pituitary βE and ACTH and, in turn, for ACTH to stimulate adrenal cortisol. The classical HPA response to environmental stress may be greatly lessened as pregnancy proceeds.

A second possibility illustrated in FIGURE 1 is that the placenta may *amplify* the stress signal that does get transmitted. This possibility exists because placental CRH production is stimulated by adrenal cortisol. Thus, in this down-regulated system, the stress-induced release of ACTH from the pituitary eventually stimulates the release of adrenal cortisol, which is turn evokes the synthesis and release of placental CRH. In this model, even though the system is dampened, stress will ultimately result in an increase in CRH. The increase in CRH after stress in pregnant women may trigger a cascade of other events that contribute to specific birth outcomes. For instance, a precocious rise in CRH is associated with greater risk of preterm delivery.[21]

The purpose of our studies is to determine the relationship between psychosocial stress and the HPA axis and to characterize their joint and independent contribution to fetal behavior and birth outcome. Results are reported that (1) quantify the relationship between psychosocial stress and birth outcome, (2) illus-

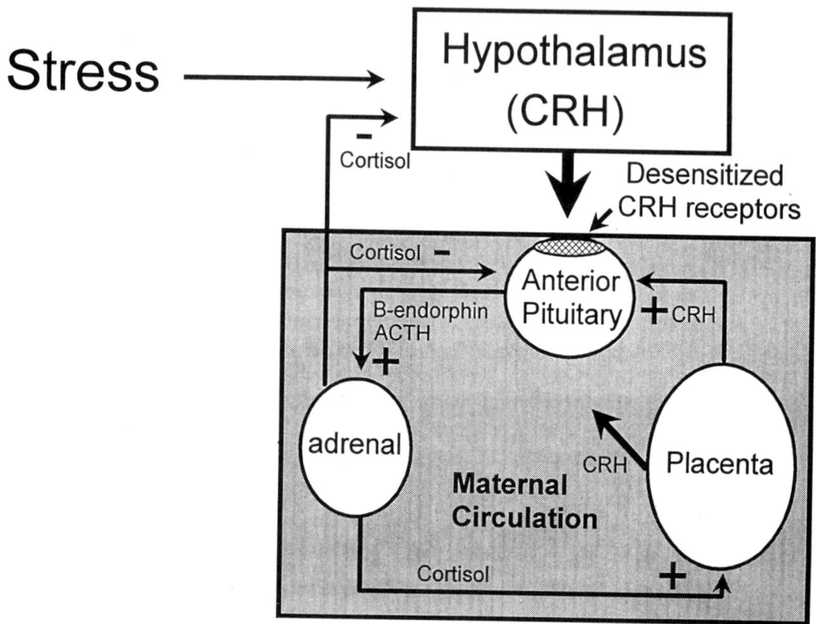

FIGURE 1. Model illustrating relationships among HPA-placental axis. Placental CRH is stimulated by adrenal cortisol resulting in (1) possible amplification of stress response in the placenta and (2) decreased sensitivity of anterior pituitary to effects of hypothalamic CRH.

trate the association between psychosocial stress and the HPA axis during pregnancy, (3) demonstrate the consequences of elevated maternal CRH on birth outcome, and (4) evaluate the effects of maternal HPA activity on fetal behavior.

RESEARCH PROTOCOL

As illustrated in FIGURE 2, women are recruited during weeks 20–24 of pregnancy, and demographic information, obstetric history, and antepartum risk are obtained for each subject. During weeks 30–32, measures of prenatal psychosocial stress, sociodemographic factors, and health practices are collected by interview and questionnaire, 20 mL of plasma are drawn and assayed for CRH, ACTH, βE, and cortisol, and a comprehensive fetal evaluation is performed including an experimental fetal heart rate (FHR) habituation procedure to measure learning.

Transabdominal transducers for measuring FHR and uterine contractions were attached in 33 women during weeks 31–32 to test fetal habituation and dishabituation patterns to a vibroacoustic stimulus. As reported elsewhere,[5] a series of 15 vibroacoustic stimuli (S1) were presented on the mother's abdomen, over the fetal head, for two seconds with pseudorandom intervals between stimuli of 20 and 45 s. On the sixteenth stimulus, a novel (dishabituating) stimulus (change in db) was presented. Trials 17–31 repeated the S1 series. The difference between prestimulus HR during each trial was subtracted from each respective average FHR value.

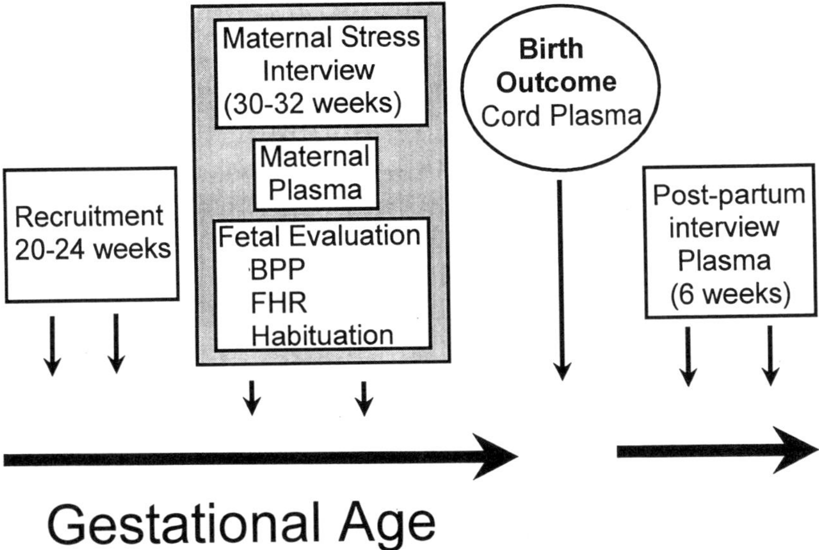

FIGURE 2. Protocol for procedures followed in the study of the effects of stress on the fetus and birth outcome.

The influence of the novel stimulus was tested by computing the slope of the dFHR index for the last four trials of the first series of S1 (immediately before the novel S2) and the slope of the first four trials of the second series of S1 (immediately after the novel S2). The results in a larger sample ($n = 93$) indicated that FHR followed a classical habituation and dishabituation pattern and that uterine contractions did not contribute to the patterns.[26]

Blood was collected at the time of the FHR assessment, at birth and at six weeks postpartum. At birth, intrapartum complications and indices of birth outcome were recorded. The 6-week sample provides a baseline of nonpregnancy so that peptide change during pregnancy can be estimated. At six weeks postpartum, maternal plasma was drawn and assayed from peptide concentration, and an abbreviated stress interview was conducted.

RESULTS AND DISCUSSION

Stress and Birth Outcomes

In a subsample of 90 women, life event stress and pregnancy anxiety significantly predicted infant birth weight and gestational age at birth independent of obstetric risk.[27] Each unit increase of maternal life event stress during the third trimester was associated with a 55.03 g decrease in infant birth weight. Each unit increase of maternal pregnancy anxiety during the third trimester was associated with a 3-day decrease in gestational age at birth.

Previous research by members of our group[28] supports the role of stress in preterm labor and delivery. Prenatal stress scores (a combination of the undesirability of life events, perceived chronic stress, and state anxiety aggregated over the course of pregnancy) predicted gestational age at delivery controlling for infant birth weight, parity, and medical risk (including substance use) in 130 low-income women of diverse ethnicity, who were interviewed on multiple occasions during the second trimester. These findings are consistent with studies from other groups. A study of 5,459 women in Denmark[29] concluded that generalized distress (anxiety and depression) in the 30th week of pregnancy was associated with risk of preterm delivery, controlling for smoking, education, parity, previous preterm delivery, maternal height, and prepregnancy weight. A study of 1,545 women[30] in Alabama indicated that the risk of having a fetal growth-retarded infant was significantly higher if an elevated psychosocial risk profile was present (stress, depression, and anxiety measured at 24 to 26 weeks or at 30 to 32 weeks). An interesting and rigorous study[31] concluded that change (increase) in life event scores from the second to the third trimester (but not high scores per se) was significantly associated with low birth weight. These selected studies reflect the general consensus that stress influences birth outcome. A number of unresolved issues include the effects of the timing and intensity of stress on outcome. Findings that change in stress levels during the course of pregnancy is more predictive than absolute levels of stress are consistent with HPA results from our research program as discussed below.

Stress and the Maternal HPA Axis

The relationship between psychosocial stress and HPA peptides was examined to a subsample of 54 women.[32] Plasma levels of ACTH, βE, and cortisol measured in maternal blood drawn at 28 weeks gestation were compared with measures of prenatal stress and social support administered at 28 and 30 weeks gestation. Elevated psychosocial stress was associated with higher plasma levels of ACTH and cortisol. A combination of the maternal psychosocial and sociodemographic factors during pregnancy accounted for 36% of the variance in ACTH, 13% of the variance in cortisol, and 3% of the variance in βE.

Physiological processes including neuroendocrine function are known to mediate the relationship between psychological stress and behavior;[33–35] however, virtually no human studies have systematically assessed this relationship during human pregnancy. Pregnancy is a unique condition for this relationship because, as reviewed above, neuroendocrine processes are altered significantly during pregnancy by the evolution of the placenta. Not only does the placenta contribute significant concentrations of endocrine products to the maternal and fetal circulation, it also changes the feedback and control mechanisms of the HPA. One possible consequence of these alterations, depicted in FIGURE 1, is the change in the stress-response threshold during pregnancy. Our results indicate for the first time that, despite the apparent "immunization" against the effects of stress during pregnancy, stress was associated with elevated plasma ACTH and cortisol between 28 and 32 weeks.

Maternal CRH Levels in the Early Third Trimester Predict the Timing of Human Delivery

In a sample of 63 women, early third trimester levels of maternal CRH were inversely and significantly correlated with gestational age at delivery ($p < 0.001$)

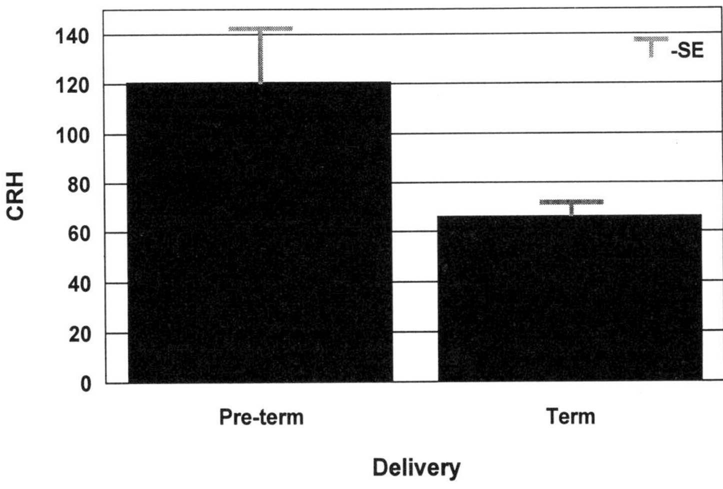

FIGURE 3. Third trimester levels of maternal CRH predict preterm birth.

after adjusting for biomedical correlates of outcome, including parity and antepartum risk. Gestational age at delivery was further dichotomized to differentiate term (after 37 weeks) and preterm (before 37 weeks) deliveries. Subjects who delivered preterm had significantly higher levels of CRH in the early third trimester of gestation ($p < 0.01$) than those who delivered at term (FIG. 3).

These results are consistent with several reports from different laboratories.[19,21,36–38] In each of these studies, plasma CRH concentrations of women in preterm labor were significantly higher than those of gestational age-matched controls. Four of these studies included additional assessments of CRH levels at least one time during gestation *before* the initiation of preterm labor.[21,36,38,39] Each study found that compared to gestational age-matched controls, CRH levels were significantly elevated in women who developed preterm labor before the clinical signs of preterm labor were detected, and that in some instances, elevated CRH preceded the signs of preterm labor by several weeks. Two of these studies[21,37] included delivery as an end point. In the Kurki *et al.* study,[37] CRH levels were measured in 23 women admitted for treatment of preterm labor and followed until delivery. Among these women in preterm labor, the 12 subjects who subsequently delivered preterm had even higher CRH levels than the other 11 who went on to deliver at term.

The most convincing evidence of the role of CRH in the timing of human delivery was the prospective, longitudinal study of 485 women.[21] In this study, CRH levels were assessed up to four times during gestation, beginning between 16 and 20 weeks. Plasma CRH levels at 18–20 weeks gestation were significantly higher in women delivering preterm ($n = 24$) than at term ($n = 308$), and were significantly lower in women delivering post-term ($n = 29$). Regression curves representing serial assessments of CRH over the course of gestation indicated that subjects delivering preterm had a precocious elevation of CRH and a steeper slope of increase as pregnancy continued. Together with results from our program, these studies strongly argue that CRH plays a pivotal role in the timing of delivery and may be responsible for mediating the effects of stress on birth outcome.

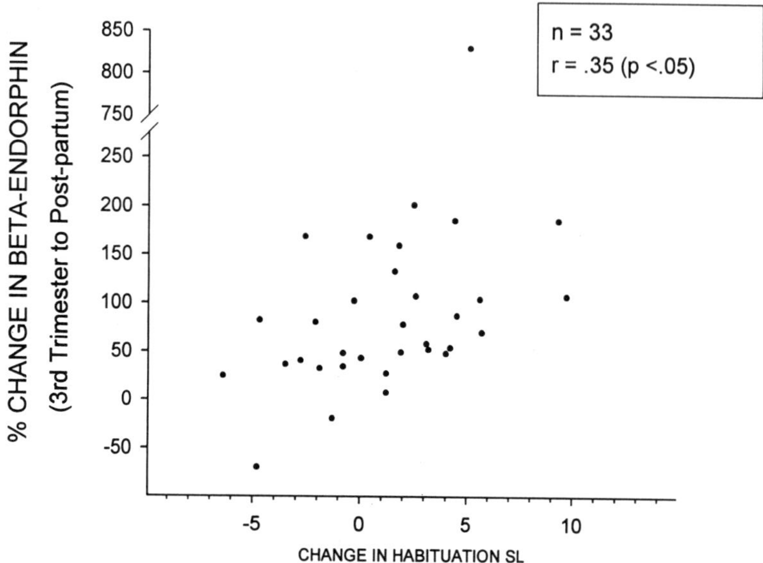

FIGURE 4. Scatter plot of the relationship between change in maternal βE during the third trimester and fetal responsivity to a novel stimulus.

Fetal Heart Rate Habituation

Previous research indicated that the procedures used in this study elicited reliable patterns of habituation in the fetus.[5] As presented in FIGURE 4, increased maternal βE during the third trimester was significantly associated with change in the fetal habituation slope (index of dishabituation) after a novel stimulus in a subsample of 33 women. Dishabituation is the *reemergence* of a previously habituated response.[40,41] The reemergence of an habituated response is typically elicited by the sensitizing or arousing effects of a novel stimulus or an altered context. Novelty is presumed to increase the level of arousal and potentiate the weak, habituated response.[41] Because dishabituation is evaluated by assessment of the reemergence of a response after a novel event, it eliminates the contribution of receptor fatigue or sensory adaptation to the response decrement.

No relationship existed between absolute concentrations of third trimester βE (or any other peptide) and FHR during challenge. The results suggest that fetal responses to the novel stimulus were attenuated in women exhibiting minimal changes (below the mean) in βE during the third trimester. In contrast, fetuses of women with very large changes of βE during the third trimester were most aroused by external novel stimuli. This is new evidence that changes in the level of maternal plasma βE during the third trimester (probably of pituitary origin) exert a significant influence on human fetal learning and, by inference, on the fetal nervous system. As reviewed above, these findings are consistent with animal studies indicating that early exposure either to stress that stimulates βE or to βE directly influences the central nervous system.[9,10,12,13] Because evidence from animal studies suggests that early exposure to endogenous opiates may have enduring effects, long-term consequences of elevated maternal βE in human infants should be evaluated.

CONCLUSIONS

Preterm birth is a major cause of infant mortality and morbidity. Despite the significance of preterm birth as a health problem in the United States, the causes remain unknown. Findings from our project indicate that stress and the HPA axis account for a significant fraction of risk for preterm birth. Our findings and those of others[21] indicate that CRH may be critically linked to the timing of birth. Psychosocial stress is associated with adverse birth outcomes, including preterm birth and elevated ACTH and cortisol levels during the third trimester. This linkage early in gestation is important because it provides a credible mechanism for the effects of prenatal stress on pregnancy outcome. The effects of βE, although closely associated with CRH and ACTH, were primarily observed on fetal behavior and not birth outcome. Because the HPA axis and stress are associated, it is a reasonable conjecture that they exert both unique and common influences on the fetus and on birth outcome.

SUMMARY

Preliminary conclusions from our research include the possibility that each of the HPA products evaluated, even though correlated (e.g., ACTH and βE), may be linked to unique and specific outcomes.

- Maternal stress during the 28–30 weeks of gestation is associated with birth outcome. Increased levels of psychosocial stress were significantly related to gestational age at birth and infant birth weight.
- Maternal stress during the third trimester was associated with increased maternal plasma levels of ACTH and cortisol. This finding is consistent with possible mechanisms whereby psychosocial stress influences birth outcome.
- CRH controls the timing of labor and delivery. Precocious elevation of CRH is related to the risk of preterm delivery. This system may be "stress-sensitive." Even though pregnant women may be *immunized* from stress, the stress signal that is transmitted (release of ACTH and cortisol) is amplified by the placental release of CRH. This possibility has at least two consequences: (1) influencing the timing of delivery and (2) desensitization of hypophyseal corticotrophs and further "protection" of the pregnant women from the results of stress (i.e., release of ACTH and βE).
- βE appears to influence fetal learning and perhaps the developing nervous system.

REFERENCES

1. BREIER, A. 1989. Experimental approaches to human stress research: Assessment of neurobiological mechanisms of stress in volunteers and psychiatric patients. Biol. Psychiatry **26:** 438–462.
2. JACOBSON, B., G. EKLUND, L. HANBERGER, D. LINNARSSON, G. SEDVALL &

M. VALVERIUS. 1987. Perinatal origin of adult self-destructive behavior. Acta Psychiatr. Scand. **76:** 364–371.

3. SANDMAN, C. A., P. C. DATTA, J. L. BARRON-QUINN, F. K. HOEHLER, C. WILLIAMS & J. M. SWANSON. 1983. Naloxone attenuates self-abusive behavior in developmentally disabled clients. Appl. Res. Ment. Retard. **4:** 5–11.

4. SANDMAN, C. A., J. L. BARRON & H. COLMAN. 1990. An orally administered opiate blocker, Naltrexone, attenuates self-injurious behavior. Am. J. Ment. Retard. **95:** 93–102.

5. SANDMAN, C. A., W. P. HETRICK, D. V. TAYLOR, J. L. BARRON, P. TOUCHETTE, I. LOTT, F. CRINELLA & V. MARINAZZI. 1993. Naltrexone reduces self-injury and improves learning. Exp. Clin. Psychopharmacol. **1:** 242–258.

6. THOMPSON, T., T. HACKENBERG, D. CERUTTI, D. BAKER & S. AXTELL. 1994. Opioid antagonist effects on self-injury in adults with mental retardation: Response form and location as determinants of medication effects. Am. J. Ment. Retard. **99:** 85–102.

7. BARRON, J. L. & C. A. SANDMAN. 1983. The relationship of sedative-hypnotic response to self-injurious behavior and stereotypy in mentally retarded clients. Am. J. Ment. Retard. **88:** 177–186.

8. BARRON, J. L. & C. A. SANDMAN. 1985. Paradoxical excitement to sedative-hypnotics in mentally retarded clients. Am. J. Ment. Defic. **2:** 124–129.

9. INSEL, T. R., G. M. KUISLEY, P. E. MAM & R. S. BRIDGES. 1990. Preterm stress has long-term effects on brain opiate receptors. Brain Res. **511:** 93–97.

10. SANCHEZ, M. D., M. V. VILARES, T. FUENTE & M. LAORDEN. 1993. The β-endorphin response to prenatal stress during postnatal development in the rat. Brain Res. Dev. **74:** 142–145.

11. CHERNICK, V. & R. J. CRIAG. 1982. Naloxone reverses neonatal depression caused by fetal asphyxia. Science **212:** 252–253.

12. SANDMAN, C. A. & N. YESSAIAN. 1986. Persisting subsensitivity of the striatal dopamine system after fetal exposure to β-endorphin. Life Sci. **39:** 1755–1763.

13. SANDMAN, C. A. & A. J. KASTIN. 1981. The influence of fragments of the LPH chain on learning, memory and attention in animals and man. Pharmacol. Ther. **13:** 39–60.

14. BECKWITH, B. E., C. A. SANDMAN, D. HOTHERSALL & A. J. KASTIN. 1977. The influence of neonatal injections of α-MSH on learning, memory and attention in rats. Physiol. Behav. **18:** 63–71.

15. STRAND, F. L., A. C. SEGARRA, C. A. ZUCCARELLI, J. KUME & K. J. ROSE. 1990. Neuropeptides as neuronal growth regulating factors: Peripheral nerve regeneration and the development of sexually dimorphic and motor behavior. Ann. N.Y. Acad. Sci. **579:** 68–90.

16. ALVES, S. E., H. M. AKBARI, G. M. ANDERSON, E. C. AZMITIA, B. C. MCEWEN & F. L. STRAND. 1997. Neonatal ACTH administration elicits long-term changes in forebrain monoamine innervation: Subsequent disruptions in hypothalamic-pituitary-adrenal and gonadal function. Ann. N.Y. Acad. Sci. This volume.

17. WILLIAMS, M. T., M. B. HENNESSY & H. N. DAVIS. 1995. CRH administered to pregnant rats alters offspring behavior and morphology. Pharmacol. Biochem. Behav. **52:** 161–167.

18. JONES, S. A., A. N. BROOKS & J. R. CHALLIS. 1989. Steroids modulate corticotropin-releasing hormone production in human fetal membranes and placenta. J. Clin. Endocrinol. Metab. **68:** 825–830.

19. PETRAGLIA, F., L. AGUZZOLI, P. FLORIO, P. BAUMANN, A. D. GENAZZANI, C. DI-CARLO & R. ROMERO. 1995. Maternal plasma and placental immunoreactive corticotrophin-releasing factor concentrations in infection-associated term and pre-term delivery. Placenta **16:** 157–164.

20. LOWRY, P. J. 1993. Corticotropin-releasing factor and its binding protein in human plasma. Ciba Found. Symp. **172:** 108–115.

21. MCLEAN, M., A. BISITS, J. DAVIES, R. WOODS, P. LOWRY & R. SMITH. 1995. A placental clock controlling the length of human pregnancy. Nature Med. **1:** 460–463.

22. LINTON, E. A., A. V. PERKINS, R. J. WOODS, F. EBEN, C. D. WOLFE, D. P. BEHAN, E. POTTER, W. W. VALE & P. J. LOWRY. 1993. Corticotropin releasing hormone-

binding protein (CRH-BP): Plasma levels decrease during the third trimester of normal human pregnancy. J. Clin. Endocrinol. Metab. **76:** 260–262.

23. PETRAGLIA, F., A. VOLPE, A. R. GENAZZANI, J. RIVIER, P. E. SAWCHENKO & W. VALE. 1990. Neuroendocrinology of the human placenta. Front. Neuroendocrinol. **11:** 6–37.

24. PETRAGLIA, F., A. GALLINELLI, D. DEVITA, K. LEWIS, L. MATHEWS & W. VALE. 1994. Activin at parturition: Changes of maternal serum levels and evidence for binding sites in placenta and fetal membranes. Obstet. Gynecol. **84:** 278–282.

25. ROBINSON, B. G., E. L. EMANUEL, D. M. FRIM & J. A. MAJZOUB. 1988. Glucocorticoid stimulation expression of corticotropin-releasing hormone gene in human placenta. Proc. Natl. Acad. Sci. USA **85:** 5244–5248.

26. SANDMAN, C. A., P. D. WADHWA, W. P. HETRICK, M. PORTO & H. V. S. PEEKE. Human fetal heart rate dishabituation at 32 weeks gestation. Child Dev. In press.

27. WADHWA, P. D., C. A. SANDMAN, M. PORTO, C. DUNKEL-SCHETTER & T. J. GARITE. 1993. The association between prenatal stress and infant birthweight and gestational age at birth: A prospective investigation. Am. J. Obstet. Gynecol. **169:** 858–865.

28. LOBEL, M., C. DUNKEL-SCHETTER & S. C. M. SCRIMSHAW. 1992. Prenatal maternal stress and prematurity: A prospective study of socioeconomically disadvantaged women. Health Psychol. **11:** 32–40.

29. HEDEGAARD, M., T. B. HENRIKSEN, S. SABROE & N. J. SECHER. 1993. Psychological distress in pregnancy and preterm delivery. Br. Med. J. **307:** 234–239.

30. CLIVER, S. P., R. L. GOLDENBURG, G. R. CUTTER, H. J. HOFFMAN, R. L. COPPER, S. J. GOTLIEB & R. O. DAVIS. 1992. The relationships among psychosocial profile, maternal size, and smoking in predicting fetal growth retardation. Obstet. Gynecol. **80:** 262–267.

31. WILLIAMSON, H. A., M. LEFEVRE & M. HECTOR. 1989. Association between life stress and serious perinatal complications. J. Fam. Pract. **29:** 489–496.

32. WADHWA, P. D., C. DUNKEL-SCHETTER, A. CHICZ-DEMET, M. PORTO & C. A. SANDMAN. The association between prenatal psychosocial factors and the maternal neuroendocrine axis in human pregnancy. Psychosom. Med. In press.

33. AXELROD, J. & T. D. REISINE. 1984. Stress hormones: Their interaction and regulation. Science **224:** 452–459.

34. HERBERT, T. B. & S. COHEN. 1993. Stress and immunity in humans: A meta-analytic review. Psychosom. Med. **55:** 364–379.

35. UR, E. 1991. Psychological aspects of hypothalamo-pituitary-adrenal activity. Ballieres Clin. Endocrinol. Metab. **5:** 79–96.

36. CAMPBELL, E. A., E. A. LINTON, C. D. WOLFE, P. R. SCRAGGS, M. T. JONES & P. J. LOWRY. 1987. Plasma corticotropin-releasing hormone concentrations during pregnancy and parturition. J. Clin. Endocrinol. Metab. **64:** 1054–1959.

37. KURKI, T., T. LAATIKAINEN, K. SALMINEN-LAPPALAINEN & O. YLIKORKALA. 1991. Maternal plasma corticotrophin-releasing hormone—Elevated in preterm labour but unaffected by indomethacin or nylidrin. Br. J. Obstet. Gynaecol. **98:** 685–691.

38. WARREN, W. B., S. L. PATRICK & R. S. GOLAND. 1992. Elevated maternal and plasma corticotropin-releasing hormone levels in pregnancies complicated by preterm labor. Am. J. Obstet. Gynecol. **166:** 1198–1207.

39. WOLFE, C. D. A., S. P. PATEL, E. A. LINTON, E. A. CAMPBELL, J. ANDERSON, A. DORNHORST, P. J. LOWRY & M. T. JONES. 1988. Plasma corticotrophin-releasing factor (CRH) in abnormal pregnancy. Br. J. Obstet. Gynecol. **95:** 1003–1006.

40. THOMPSON, R. F. & P. M. GROVES. 1970. Habituation: A dual-process theory. Psychol. Rev. **77:** 419–450.

41. MACKINTOSH, N. J. 1987. Neurobiology, psychology and habituation. Behav. Res. Ther. **25:** 81–97.

Placental CRH Modulates Maternal Pituitary-Adrenal Function in Human Pregnancy[a]

PATHIK D. WADHWA,[b,d] CURT A. SANDMAN,[c]
ALEKSANDRA CHICZ-DeMET,[c] AND MANUEL PORTO[b]

*Departments of Obstetrics and Gynecology,[b] and
Psychiatry and Human Behavior[c]
University of California, Irvine
Irvine, California*

Corticotropin-releasing hormone (CRH), a 41 amino acid neuropeptide synthesized primarily in the paraventricular nucleus of the hypothalamus, has a major role in regulating pituitary-adrenal function and the physiological response to stress.[1,2] During pregnancy, the placenta, decidua, and fetal membranes are additional sites of CRH synthesis from approximately 8 to 10 weeks gestation onward.[3,4] The expression of placental CRH rises exponentially during gestation, and it is released into maternal, fetal, and amniotic compartments.[5–7] Although placental CRH is identical to hypothalamic CRH in structure, immunoreactivity, and bioactivity,[8] and CRH levels during mid- to late gestation in maternal plasma are comparable to those capable of stimulating pituitary activity *in vitro*[1,6] and *in vivo*,[9,10] there are conflicting views and findings about the possible role of placental CRH in regulating maternal pituitary-adrenal function during pregnancy. On the one hand, although levels of pituitary [adrenocorticotropic hormone (ACTH), β-endorphins (βE)] and adrenal (cortisol) products are elevated in maternal plasma during pregnancy, the magnitude of their elevation, which ranges from a two- to fourfold increase over nonpregnant levels, is much less than that of plasma CRH, which ranges from a several hundred-fold to a thousand-fold increase over nonpregnant levels.[6,11,12] Some correlational studies have failed to find associations between CRH and ACTH or cortisol levels during pregnancy,[13,14] and administration of a 100 μg CRH bolus and a 1 μg/kg dose of exogenous CRH at 38–40 weeks gestation failed to evoke a significant pituitary or adrenal response.[15,16] A CRH-binding protein (CRH-BP), apparently unique to humans, has been found in maternal plasma during pregnancy and is thought to reduce the bioactivity of peripherally circulating CRH.[17,18] On the other hand, the relatively modest increase of ACTH and cortisol during pregnancy is consistent with findings from studies of long-term administration of exogenous CRH in nonpregnant humans.[9,10] Other correlational studies have found significant associations between CRH, ACTH, βE, and cortisol during pregnancy;[15,19–22] a 100 μg CRH bolus during mid-gestation and a 2 μg/kg dose of exogenous CRH at 38–40 weeks gestation have been shown to evoke normal pituitary and adrenal responses.[15,23] Last, in contrast to the exponentially

[a] This work was supported in part by USPHS grant HD 28413.
[d] Present address of corresponding author: Pathik D. Wadhwa, M.D., Ph.D., Department of Behavioral Science, University of Kentucky College of Medicine, 109 COMOB, Lexington, KY 40536-0086. E-mail: pwadhwa@pop.uky.edu

increasing levels of circulating CRH over the course of gestation, CRH-BP levels, which are constant in the first, second, and early third trimester and not significantly different from nonpregnant levels, fall by approximately 60% at 36–38 weeks gestation.[24,25]

The aim of this study was to clarify the role of placental CRH on maternal pituitary-adrenal function *in vivo* during pregnancy by examining the associations between maternal plasma CRH, ACTH, βE, and cortisol during the early third trimester of pregnancy.

METHOD

The sample comprised 260 adult women with a singleton, intrauterine pregnancy attending prenatal care at a large, metropolitan, teaching hospital affiliated with the University of California, Irvine. A 20-mL blood sample was withdrawn from each subject for hormone assays between 30 and 31 weeks gestation. The time of day of each blood draw was noted. Samples were centrifuged at 2000 $\times$ g (10 min), and the plasma was decanted into polypropylene tubes containing 500 KIU/mL aprotinin (Sigma Chemical Company, St. Louis, MO) and stored at -70 °C until assayed. All samples were assayed in duplicate. Plasma levels of CRH were determined by radioimmunoassay (RIA) using a commercially prepared kit (Peninsula Laboratories, Belmont, CA). The CRH assay has less than 0.01% cross-reactivity with ovine and sauvagine CRH, 36% cross-reactivity with bovine CRH, and nondetectable reactivity with human ACTH. The coefficient of variation (CV) is 5% at normal physiological levels using 4 mL or 8% using 2 mL plasma, with a minimum detectable dose (MDD; 95% confidence) of 2.04 pg per sample. Tissue linearity has been evaluated up to 4.0 mL plasma with quantitative recovery. Plasma levels of ACTH were measured by a commercially available radioimmunoassay (Nichols Institute Diagnostics, San Juan Capistrano, CA). The antiserum employed has less than 0.001% cross-reactivity with βE and ACTH fragments. The ACTH assay has a MDD of 1.0 pg/mL with a CV = 3.0% (intraassay) at 35 pg/mL and a CV = 7.8% (interassay) at 36 pg/mL. Plasma levels of βE were determined by a commercially available solid-phase two-site immunoradiometric assay (IRMA; Nichols Institute Diagnostics, San Juan Capistrano, CA). The antiserum has a 1.6% cross-reactivity with β-lipotropin at 50 pg/mL and has less than 0.01% cross-reactivity with related opiates at 5 μg/mL. The Allegro β-Endorphin Immunoassay system has a MDD of 10 pg/mL with a CV = 4.1% (intraassay) and a CV = 9.0% (interassay) at the highest concentrations expected in the present study. Plasma cortisol levels were determined by immunofluorescence using an automated procedure on an Abbott TDx Analyzer (Abbott Laboratories, Abbott Park, IL). The assay has less than 5% cross-reactivity with 11-deoxycortisol, corticosterone, and less than 1% cross-reactivity with 10 other naturally occurring steroids. The interassay and intraassay coefficients of variance are less than 9% with a MDD of 0.45 μg/dL. Data reduction for the RIA and IRMA assays were done by a computer-assisted four-parameter logistics program.[26]

RESULTS

Plasma levels of maternal hormones during pregnancy conformed with expected norms; the mean concentrations ($\pm$ SEM) of maternal CRH, ACTH, βE,

TABLE 1. Intercorrelations between Maternal CRH, ACTH, βE, and Cortisol at 30–31 Weeks Gestation

	CRH	ACTH	βE
ACTH	$.18^a$	—	
βE	$.30^b$	$.85^b$	—
Cortisol	$.09^c$	$.36^b$	$.34^b$

NOTE: $n = 260$; two-tailed test. $^a p < 0.01$; $^b p < 0.001$; $^c p < 0.05$.

and cortisol were 148.7 ± 12.63 pg/mL, 42.8 ± 1.9 pg/mL, 48.75 ± 1.52 pg/mL, and 28.87 ± 0.67 μg/dL, respectively. Hormone levels were approximately normally distributed after outliers were coded in. Log transformations were performed when appropriate. Pearson product-moment correlation coefficients were computed to examine the associations between hormone levels during pregnancy. The results are depicted in TABLE 1. ACTH and βE levels were highly correlated ($p < 0.001$), indicating their origin from a common precursor—proopiomelanocortin (POMC). CRH levels were significantly and positively correlated with ACTH ($p < 0.01$) and βE ($p < 0.001$) levels (see FIG. 1). No direct association was found between CRH and cortisol in the present sample; however, both ACTH and βE were significantly and positively correlated with cortisol (p's < 0.001). Time of day of blood draw was significantly and negatively correlated with maternal cortisol levels during pregnancy ($r = -.22$, $p < 0.001$) but not with other hormones, indicating the presence of an intact circadian variation for cortisol with higher levels in the morning and lower levels in the evening. An analysis of covariance approach employed to control the effects of time of blood draw revealed no changes in the magnitude or significance levels of the above relationships.

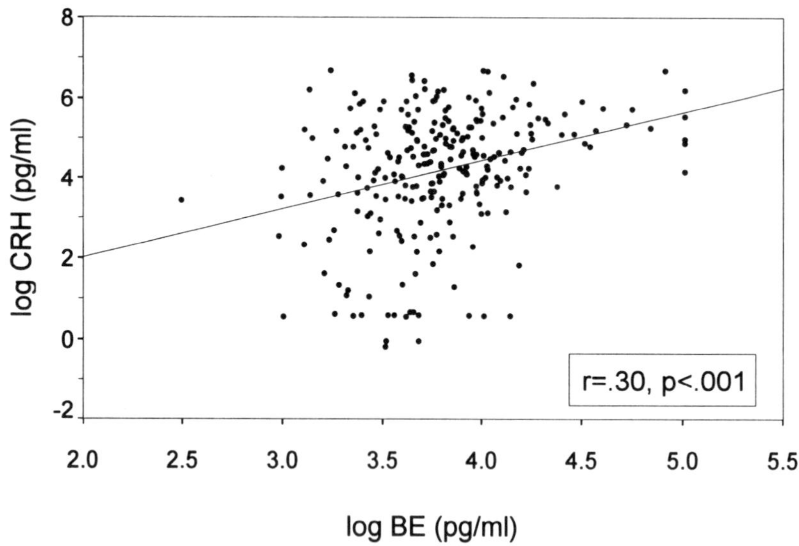

FIGURE 1. Maternal CRH and βE at 30 weeks gestation.

DISCUSSION

Placental CRH plays a major role in human parturition and fetal growth and maturation,[3,8] and deviations from the normal rate of change in CRH levels over the course of gestation have been associated with alterations in the timing of onset of spontaneous labor.[27] Various exogenous and endogenous forms of prenatal stress are also associated with adverse pregnancy outcomes, and we and others have hypothesized that placental CRH may modulate the influence of stress on outcome.[28,29] An understanding of the regulation of maternal-placental-fetal neuroendocrine function is crucial before investigating possible mechanisms that may be involved in mediating the influence of the maternal environment on fetal development and other outcomes.

Maternal plasma levels of CRH during pregnancy are presumed to reflect the expression of placental CRH because the placenta is the major source of peripheral CRH during pregnancy and because hypothalamic CRH does not cross the blood–brain barrier.[3] Although the human placenta expresses the POMC gene, maternal plasma levels of ACTH and βE are presumed to primarily reflect pituitary and not placental activity, because posttranslational processing of placental POMC preferentially favors the production of α-MSH-sized peptides rather than ACTH (1-39)-sized molecules,[30] and the placenta extensively acetylates and inactivates βE before releasing it into peripheral circulation.[31] The findings from the present study of a significant positive association of maternal CRH and ACTH and βE thereby support the premise that placental CRH modulates maternal pituitary-adrenal function in a positive manner during the early third trimester pregnancy. In contrast to the negative control on hypothalamic CRH, glucocorticoids have been shown to stimulate the expression of hCRHmRNA in the placenta, establishing a positive feedback loop that allows for the simultaneous increase of CRH, ACTH, and cortisol over the course of gestation.[32] Although there was no direct association between cortisol and CRH in the present sample, our findings supported an indirect positive effect of CRH on cortisol mediated by ACTH and βE.

Evidence from the present set of findings and from other studies[22,23] suggests that placental CRH expression does *not* exhibit a circadian rhythm; however, a diurnal variation in maternal plasma cortisol was found in the present sample, and others[11,22] have also found diurnal variations in ACTH and cortisol levels during pregnancy. These findings, taken together, support the suggestion that, in addition to placental CRH, the maternal pituitary-adrenal axis during pregnancy is also regulated by another ACTH secretagogue—probably AVP—which is secreted into the hypophyseal-portal circulation by the parvocellular neurons of the paraventricular nucleus with a circadian increase in amplitude.[22,34,35]

In conclusion, the present findings support a role for placental CRH in modulating maternal pituitary-adrenal function *in vivo* during human pregnancy, and also suggest the existence of an additional control mechanism for regulating pituitary-adrenal function during gestation.

REFERENCES

1. VALE, W., J. SPIESS, C. RIVIER & J. RIVIER. 1981. Characterization of a 41-residue ovine hypothalamic peptide that stimulates secretion of corticotropin and beta-endorphin. Science **213**: 1394–1397.
2. CHROUSOS, G. P. 1992. Regulation and dysregulation of the hypothalamic-pituitary-adrenal axis. The corticotropin-releasing hormone perspective. Endocrinol. Metab. Clin. North Am. **21(4)**: 833–858.

3. RILEY, S. C. & J. R. CHALLIS. 1991. Corticotrophin-releasing hormone production by the placenta and fetal membranes. Placenta **12(2):** 105–119.
4. WARREN, W. B. & A. J. SILVERMAN. 1995. Cellular localization of corticotrophin releasing hormone in the human placenta, fetal membranes and decidua. Placenta **16(2):** 147–156.
5. ECONOMIDES, D., E. LINTON, K. NICOLAIDES, C. H. RODECK, P. J. LOWRY & T. CHARD. 1987. Relationship between maternal and fetal corticotrophin-releasing hormone-41 and ACTH levels in human mid-trimester pregnancy. J. Endocrinol. **114(3):** 497–501.
6. GOLAND, R. S., S. L. WARDLAW, M. BLUM, P. J. TROPPER & R. I. STARK. 1988. Biologically active corticotropin-releasing hormone in maternal and fetal plasma during pregnancy. Am. J. Obstet. Gynecol. **159(4):** 884–890.
7. SASAKI, A., O. SHINKAWA, A. N. MARGIORIS, A. S. LIOTTA, S. SATO, O. MURAKAMI, M. GO, Y. SHIMIZU, K. HANEW & K. YOSHINAGA. 1987. Immunoreactive corticotropin-releasing hormone in human plasma during pregnancy, labor, and delivery. J. Clin. Endocrinol. Metab. **64(2):** 224–229.
8. PETRAGLIA, F., G. COUKOS, A. VOLPE, A. R. GENAZZANI & W. VALE. 1991. Involvement of placental neurohormones in human parturition. Ann. N. Y. Acad. Sci. **622:** 331–340.
9. SCHURMEYER, T. H., P. C. AVGERINOS, P. W. GOLD, W. T. GALLUCCI, T. P. TOMAI, G. B. CUTLER, JR., D. L. LORIAUX & G. P. CHROUSOS. 1984. Human corticotropin-releasing factor in man: Pharmacokinetic properties and dose-response of plasma adrenocorticotropin and cortisol secretion. J. Clin. Endocrinol. Metab. **59(6):** 1103–1108.
10. SCHULTE, H. M., G. P. CHROUSOS, P. W. GOLD, J. D. BOOTH, E. H. OLDFIELD, G. B. CUTLER, JR. & D. L. LORIAUX. 1985. Continuous administration of synthetic ovine corticotropin-releasing factor in man. Physiological and pathophysiological implications. J. Clin. Invest. **75(6):** 1781–1785.
11. NOLTEN, W. E., M. D. LINDHEIMER, P. A. RUECKERT, S. OPARIL & E. N. EHRLICH. 1980. Diurnal patterns and regulation of cortisol secretion in pregnancy. J. Clin. Endocrinol. Metab. **51(3):** 466–472.
12. RAISANEN, I. 1988. Plasma levels and diurnal variation of beta-endorphin, beta-lipotropin and corticotropin during pregnancy and early puerperium. Eur. J. Obstet. Gynecol. Reprod. Biol. **27(1):** 13–20.
13. ALLOLIO, B., J. HOFFMANN, E. A. LINTON, W. WINKELMANN, M. KUSCHE & H. M. SCHULTE. 1990. Diurnal salivary cortisol patterns during pregnancy and after delivery: Relationship to plasma corticotrophin-releasing-hormone. Clin. Endocrinol. **33(2):** 279–289.
14. JESKE, W., P. SOSZYNSKI, W. ROGOZINSKI, E. LUKASZEWICZ, W. LATOSZEWSKA & H. SNOCHOWSKA. 1989. Plasma GHRH, CRH, ACTH, beta-endorphin, human placental lactogen, GH and cortisol concentrations at the third trimester of pregnancy. Acta Endocrinol. **120(6):** 785–789.
15. SASAKI, A., O. SHINKAWA & K. YOSHINAGA. 1989. Placental corticotropin-releasing hormone may be a stimulator of maternal pituitary adrenocorticotropic hormone secretion in humans. J. Clin. Invest. **84(6):** 1997–2001.
16. SCHULTE, H. M., D. WEISNER & B. ALLOLIO. 1990. The corticotrophin releasing hormone test in late pregnancy: Lack of adrenocorticotrophin and cortisol response. Clin. Endocrinol. **33(1):** 99–106.
17. ORTH, D. N., & C. D. MOUNT. 1987. Specific high-affinity binding protein for human corticotropin-releasing hormone in normal human plasma. Biochem. Biophys. Res. Commun. **143(2):** 411–417.
18. SUDA, T., M. IWASHITA, F. TOZAWA, T. USHIYAMA, N. TOMORI, T. SUMITOMO, Y. NAKAGAMI, H. DEMURA & K. SHIZUME. 1988. Characterization of corticotropin-releasing hormone binding protein in human plasma by chemical cross-linking and its binding during pregnancy. J. Clin. Endocrinol. Metab. **67(6):** 1278–1283.
19. CHAN, E. C., R. SMITH, T. LEWIN, M. W. BRINSMEAD, H. P. ZHANG, J. CUBIS, K. THORNTON & D. HURT. 1993. Plasma corticotropin-releasing hormone, beta-

endorphin and cortisol inter-relationships during human pregnancy. Acta Endocrinol. **128(4):** 339–344.

20. GOLAND, R. S., I. M. CONWELL, W. B. WARREN & S. L. WARDLAW. 1992. Placental corticotropin-releasing hormone and pituitary-adrenal function during pregnancy. Neuroendocrinology **56(5):** 742–749.

21. OKAMOTO, E., T. TAKAGI, T. MAKINO, H. SATA, I. IWATA, E. NISHINO, N. MITSUDA, N. SUGITA, Y. OTSUKI & O. TANIZAWA. 1989. Immunoreactive corticotropin-releasing hormone, adrenocorticotropin and cortisol in human plasma during pregnancy and delivery and postpartum. Horm. Metab. Res. **21(10):** 566–572.

22. MAGIAKOU, M., G. MASTORAKOS, D. RABIN, A. N. MARGIORIS, B. DUBBERT, A. E. CALOGERO, C. TSIGOS, P. J. MUNSON & G. P. CHROUSOS. 1996. The maternal hypothalamic-pituitary-adrenal axis in third trimester human pregnancy. Clin. Endocrinol. **44:** 419–428.

23. SUDA, T., M. IWASHITA, T. USHIYAMA, F. TOZAWA, T. SUMITOMO, Y. NAKAGAMI, H. DEMURA & K. SHIZUME. 1989. Responses to corticotropin-releasing hormone and its bound and free forms in pregnant and nonpregnant women. J. Clin. Endocrinol. Metab. **69(1):** 38–42.

24. LINTON, E. A., A. V. PERKINS, R. J. WOODS, F. EBEN, C. D. WOLFE, D. P. BEHAN, E. POTTER, W. W. VALE & P. J. LOWRY. 1993. Corticotropin releasing hormone-binding protein (CRH-BP): Plasma levels decrease during the third trimester of normal human pregnancy. J. Clin. Endocrinol. Metab. **76(1):** 260–262.

25. PERKINS, A. V., F. EBEN, C. D. WOLFE, H. M. SCHULTE & E. A. LINTON. 1993. Plasma measurements of corticotrophin-releasing hormone-binding protein in normal and abnormal human pregnancy. J. Endocrinol. **138(1):** 149–157.

26. RODBARD, D. & Y. FELDMAN. 1978. Kinetics of two-site immunoradiometric ("sandwich") assays. I. Mathematical models for simulation, optimization, and curve fitting. Immunochemistry **15(2):** 71–76.

27. MCLEAN, M., A. BISITS, J. DAVIES, R. WOODS, P. LOWRY & R. SMITH. 1995. A placental clock controlling the length of human pregnancy. Nature Med. **1(5):** 460–463.

28. LOCKWOOD, C. J. 1994. Recent advances in elucidating the pathogenesis of preterm delivery, the detection of patients at risk, and preventative therapies. Curr. Opin. Obstet. Gynecol. **6(1):** 7–18.

29. WADHWA, P. D., C. DUNKEL-SCHETTER, A. CHICZ-DEMET, M. PORTO & C. A. SANDMAN. 1996. The association between prenatal psychosocial factors and the neuroendocrine axis in human pregnancy. Psychosom. Med. **58(5):** 432–446.

30. MARGIORIS, A. N., M. GRINO, P. PROTOS, P. W. GOLD & G. P. CHROUSOS. 1988. Corticotropin-releasing hormone and oxytocin stimulate the release of placental proopiomelanocortin peptides. J. Clin. Endocrinol. Metab. **66(5):** 922–926.

31. CHAN, E. C. & R. SMITH. 1992. Beta-endorphin immunoreactivity during human pregnancy. J. Clin. Endocrinol. Metab. **75(6):** 1453–1458.

32. ROBINSON, B. G., R. L. EMANUEL, D. M. FRIM & J. A. MAJZOUB. 1988. Glucocorticoid stimulates expression of corticotropin-releasing hormone gene in human placenta. Proc. Natl. Acad. Sci. USA **85(14):** 5244–5248.

33. PETRAGLIA, F., A. D. GENAZZANI, L. AGUZZOLI, A. GALLINELLI, D. DE VITA, A. CARUSO & A. R. GENAZZANI. 1994. Pulsatile fluctuations of plasma-gonadotropin-releasing hormone and corticotropin-releasing factor levels in healthy pregnant women. Acta Obstet. Gynecol. Scand. **73(4):** 284–289.

34. ENGLER, D., T. PHAM, M. J. FULLERTON, G. OOI, J. W. FUNDER & I. J. CLARKE. 1989. Studies of the secretion of corticotropin-releasing factor and arginine vasopressin into the hypophysial-portal circulation of the conscious sheep. I. Effect of an audiovisual stimulus and insulin-induced hypoglycemia. Neuroendocrinology **49(4):** 367–381.

35. GOMEZ, M. T., M. A. MAGIAKOU, G. MASTORAKOS & G. P. CHROUSOS. 1993. The pituitary corticotroph is not the rate limiting step in the postoperative recovery of the hypothalamic-pituitary-adrenal axis in patients with Cushing syndrome. J. Clin. Endocrinol. Metab. **77(1):** 173–177.

Steroid Effects at the Membrane Level on Oxytocin Systems

J. D. CALDWELL,[a,d] S. T. O'ROURKE,[a] M. MORRIS,[b]
C. H. WALKER,[c] R. B. CARR,[c] B. M. FAGGIN,[e]
AND G. A. MASON[c]

[a]*Department of Pharmaceutical Sciences*
North Dakota State University
Fargo, North Dakota 58105-5055

[b]*Bowman Gray School of Medicine*
Winston-Salem, North Carolina

[c]*School of Medicine*
University of North Carolina
Chapel Hill, North Carolina

The ovarian steroids estradiol and progesterone have numerous effects in the preoptic area (POA) and hypothalamus. Cytoplasmic receptors for estradiol and progesterone in neurons in the medial POA (MPOA) and hypothalamus that translocate into the cellular nucleus to act genomically have been demonstrated. Recently, however, cytoplasmic progesterone receptors have been found in neuronal processes.[1] Blaustein *et al.*[2] used electron microscopy to show that progesterone receptors in axon terminals are associated with round and clear vesicles, suggesting that steroid receptors are not only outside the perikaryon but are also associated with neuronal elements involved in transmitter release. Recent characterization of estradiol receptors (ER) in plasma membrane, structurally similar to cytoplasmic ER,[3] suggests the membrane is another site of estradiol action.

Caldwell *et al.*[4] demonstrated that estradiol conjugated to bovine serum albumin (E-6-BSA) released oxytocin (OT) from anuclear homogenates of the MPOA-medial hypothalamus within minutes suggesting that steroids act at the membrane on release of this neuropeptide. Experiments were performed to test the effect of both estradiol and progesterone treatment on *in vitro* secretion of OT. An anuclear MPOA-hypothalamic preparation was used to determine whether steroids have direct effects on secretion, independent of the nucleus. Further, OT antisera were used which are specific for the amidated peptide as compared to antisera which cross-reacts with C-terminal extended forms.[5] As shown in TABLE 1, after the vehicle buffer preincubation, the release buffer with E-6-BSA significantly increased OT release, whereas superfusion with progesterone conjugated to BSA (P-3-BSA) was without effect. In homogenates that were preincubated in E-6-BSA, E-6-BSA again significantly released OT whereas P-3-BSA greatly decreased OT levels, indicating that P-3-BSA does not have the same effect on OT release as E-6-BSA and may, under certain circumstances, have the opposite effect. It is also apparent from TABLE 1, as well as from our previous data,[4,6] that release of OT by E-6-BSA is affected by previous exposure to estradiol. It may be that

[d] Address correspondence to Jack D. Caldwell, Ph.D., 235 Sudro Hall, P.O. Box 5055, College of Pharmacy, North Dakota State University, Fargo, ND 58105-5055.

TABLE 1. Changes in Peptide Levels from Superfusion Baseline to Release Conditions Using Three Separate Antibodies

Preincubation Trt	Release Trt	OT (Pen. or OT-MM) (pg/mL)	OT-VA$_{17}$ (pg/mL)
Wash buffer	Wash buffer		0.44
Wash buffer	E-6-BSA in wash	32.0**	10.43**
Wash buffer	BSA in wash		0.25
Wash buffer	P-3-BSA in wash		0.73
Wash buffer	KCl		−2.5
Wash buffer	KCl	−21.1	
Wash buffer	E-6-BSA and KCl		248.17**
Wash buffer	BSA and KCl		0.36
E-6-BSA	BSA in wash		−3.41*
E-6-BSA	E-6-BSA in wash		−5.08
E-6-BSA	E-6-BSA in wash	5.16*	3.22
E-6-BSA	KCl	−7.07*	−6.93
E-6-BSA	KCl	−54.1	
E-6-BSA	E-6-BSA and KCl	11.8**	9.2*
E-6-BSA	BSA & KCl		−1.53
E-6-BSA	BSA and KCl	−3.9	
E-6-BSA	P-3-BSA and KCl	−212.7	
P-3-BSA	KCl		−73.47*
P-3-BSA	KCl	−6.15*	−6.255

NOTE: Synaptosome-containing homogenates from the MPOA-MH of ovariectomized rats injected daily with 5 μg E-benzoate for three days were superfused with buffers containing an electrolytically neutral balance of ions (Wash buffer) or depolarizing levels of K$^+$ (KCl). Where indicated, 100 ng/μL of E-6-BSA or P-3-BSA was added to buffers. Where only oxytocin (OT) data are indicated, the Peninsula (Pen) antibody was used for RIA. Where both columns contain data the OT-MM and OT-VA$_{17}$ antisera were used, respectively, on aliquots from the same samples. Where only OT-VA$_{17}$ is indicated, only the OT-VA$_{17}$ antiserum was used. Peptide/milliliter levels were compared between four baseline samples and six release samples using Student's t tests with $p < 0.05$ indicated by an asterisk (*) and $p < 0.01$ indicated by double asterisks (**). Data are shown as change in pg/mL from mean baseline levels.

previous exposure to estradiol either *in vivo* or *in vitro* desensitizes the membrane receptors or may affect G-protein coupling. Further, when not in combination with E-6-BSA, depolarizing concentrations of KCl have no effect or an inhibitory effect on OT secretion; this differs from the results seen with posterior pituitary, but is similar to that shown *in vivo* in nuclear regions.[7]

With regard to the comparison of the results with the different antisera. The direction of the changes is similar; that is, estradiol stimulates release whereas progesterone has little effect. However, the change in response to estradiol appears to be greater when measured with the antisera which cross-reacts with the extended OT forms. Indeed other results suggest that estradiol may increase the secretion of the incompletely processed form, perhaps affecting the secretory pathway (unpublished data).

Although both estradiol and progesterone receptors have been shown in cytoplasm, there is little evidence of whether separate membrane receptors exist for individual steroids. There is a very high density of binding sites for radiolabeled P-3-BSA (P-3-[^{125}I-BSA]) in membrane preparations from the MPOA.[8] This population of receptors shows both a low- and a high-affinity subpopulation of sites. Inhibition of these sites by the nonhydrolyzable G-protein analogue GTPγ suggests that P-3-BSA binding sites are associated with G proteins. We further demonstrated that the ribosyltransferase cholera toxin (CTX), when allowed to act on MPOA homogenates for 30 min at 37 °C, converted low-affinity P-3-BSA binding sites to a high-affinity state indicating that P-3-BSA binding sites are coupled to G proteins. Interestingly, we found that OT receptors (OTR) also are converted from a low- to a high-affinity site after preincubation with CTX,[6] suggesting that G proteins associated with both the steroid binding site and the OTR in the MPOA are CTX sensitive. Therefore, the CTX-sensitive G protein may be a link between steroid receptors and OTR in the plasma membrane.

We recently discovered that the binding characteristics for radiolabeled E-6-BSA differ from that of P-3-BSA. The binding affinity and density for E-6-[^{125}I-BSA] is different than that for P-3-[^{125}I-BSA]. FIGURE 1 shows binding of a range of concentrations of these conjugated steroids after preincubaton with CTX. CTX had no effect on binding of P-3-[^{125}I-BSA], with the possible exception of slightly increasing the binding of low concentrations, whereas CTX inhibited the binding of the lowest concentration of E-6-[^{125}I-BSA] while increasing the binding of 1 to 1.25 nM concentrations. It appears, therefore, that CTX has different effects on estradiol versus progesterone binding to MPOA tissue, suggesting a different relationship between the binding sites of these two steroids and their associated G protein. This may allow for differential responses in specific cells to estradiol versus progesterone, much as we see with steroid release of OT. The roles of estradiol versus progesterone membrane receptors need to be elucidated further, but it appears important to make a distinction between the two.

The high density of steroid membrane receptors in brain areas such as the MPOA suggests that they act on neurons of many neurotransmitter types. In fact, steroidal effects at the membrane level on neurons with adrenergic, dopaminergic, and GABAergic receptors have all been demonstrated,[9-11] suggesting that steroids

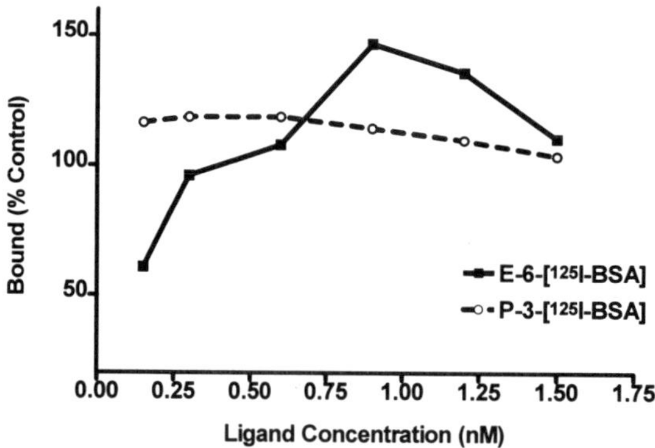

FIGURE 1. The effects of pretreatment of MPOA-AH membranes from ovariectomized rats with CTX on the binding of different concentrations (0.15–1.5 nM) of E-6-[^{125}I-BSA] or P-3-[^{125}I-BSA]. Membranes were preincubated for 30 min at 37 °C with 10^{-5} M CTX or buffer and washed twice by centrifugation (48,000 × g) to remove excess toxin. Binding is expressed as percentage of binding in tissues incubated in buffer that contained no CTX.

are capable of acting on multiple neurotransmitter systems in brain via their membrane receptors. The great density of these receptors and their association with numerous neurotransmitter systems including monoamines as well as neuropeptides suggests that they are capable of radically influencing brain activity. Such receptors may be important in altering brain function under conditions such as late pregnancy or at ovulation when levels of one steroid is elevated for a period of time and then suddenly another steroid predominates.

Much work has to be done to fully characterize the nature and function of steroid membrane receptors. The recent characterization of estradiol binding sites in membrane by Karthikeyan and Thampan[3] that suggests their primary structure is the same as cytoplasmic ERs in uterine cells is a step in this direction. The nature of estradiol interaction with G proteins in the membrane, the relationship of estradiol and progesterone binding sites associated with the membrane, analogies between such membrane receptors in uterus and brain, the importance of membrane receptors for steroid uptake, steroid actions on receptors while associated with the plasma membrane versus once they are in the cytoplasm, and actions of steroids on ion channels and synaptic vesicles are just a few of the pantheon of important questions to be answered about steroid effects in brain. Here we have demonstrated that estradiol rapidly releases OT, and possibly an extended form of OT, from anuclear brain homogenates. We also present evidence of a reciprocal relationship between membrane estradiol and progesterone binding sites in brain and present further evidence of G-protein control of this relationship.

REFERENCES

1. BLAUSTEIN, J. D. & D. H. OLSTER. 1993. Colchicine-induced accumulation of estrogen receptor and progestin receptor immunoreactivity in atypical areas in guinea-pig brain. J. Neuroendocrinol. **5(1):** 63–70.
2. BLAUSTEIN, J. D., M. N. LEHMAN, J. C. TURCOTTE & C. GREENE. 1992. Estrogen receptors in dendrites and axon terminals in the guinea pig hypothalamus. Endocrinology **131(1):** 281–290.
3. KARTHIKEYAN, N. & R. V. THAMPAN. 1996. Plasma membrane is the primary site of localization of the nonactivated estrogen receptor in the goat uterus: Hormone binding causes receptor internalization. Arch. Biochem. Biophys. **325:** 47–57.
4. CALDWELL, J. D., M. A. MORRIS, C. H. WALKER, R. B. CARR, B. M. FAGGIN & G. A. MASON. 1996. Estradiol conjugated to BSA releases oxytocin from synaptosome-containing homogenates from the medial preoptic area-medial hypothalamus. Horm. Metal. Res. **29:** 1–5.
5. GAINER, H., M. O. LIVELY & M. MORRIS. 1994. Immunological and related techniques for studying neurohypophyseal peptide processing pathways. *In* Peptidases and Neuropeptide Processing. Methods in Neurosciences. A. I. Smith, Ed. Vol. 23: 195–207. Academic Press, Inc. Orlando, FL.
6. CALDWELL, J. D., C. H. WALKER, S. T. O'ROURKE, B. M. FAGGIN, M. A. MORRIS & G. A. MASON. 1996. Analogies between oxytocin systems of the uterus and brain. Horm. Metab. Res. **28:** 65–74.
7. HATTORI, T., M. MORRIS, N. ALEXANDER & D. K. SUNDBERG. 1990. Extracellular oxytocin in the paraventricular nucleus: Hyperosmotic stimulation by in vivo microdialysis. Brain Res. **506:** 169–171.
8. CALDWELL, J. D., C. H. WALKER, B. M. FAGGIN, C. A. PEDERSEN & G. A. MASON. 1995. Characterization of ^{125}I-P-3-BSA binding sites in the medial preoptic area and hypothalamus. Brain Res. **693:** 225–232.
9. DIPAOLO, T. & P. FALARDEAU. 1985. Modulation of brain and pituitary dopamine receptors by estrogens and prolactin. Prog. Neuro-psychopharmacol. & Biol. Psychiatry **9:** 473–480.

10. UNGAR, S., M. H. MAKMAN, S. A. MORRIS & A. M. ETGEN. 1993. Estrogen uncouples β-adrenergic receptor from the stimulatory guanine nucleotide-binding protein in female rat hypothalamus. Endocrinology **133:** 2818–2826.
11. MORROW, A. L., J. R. PACE, R. H. PURDY & S. M. PAUL. 1990. Characterization of steroid interactions with GABA receptor-gated chloride ion channels: Evidence for multiple steroid recognition sites. Mol. Pharmacol. **37:** 263–270.

Vasopressin Deficiency and Phase Advancement Shifts in Circadian Rhythms with Nocturnal or Diurnal Feeding[a]

CYRILLA H. WIDEMAN, HELEN M. MURPHY,
AND GEORGE R. NADZAM

Departments of Biology and Psychology
John Carroll University
Cleveland, Ohio 44118

Two circadian systems have been identified in rats: (1) photic, which utilizes the light/dark cycle as the primary zeitgeber, and (2) nonphotic, which utilizes stimuli other than light and dark. One important nonphotic system responds differentially to restricted feeding cycles as the primary zeitgeber. The suprachiasmatic nucleus (SCN) of the hypothalamus is the primary locus for the generation of the circadian rhythm of the photic system and possesses vasopressin (VP)-containing neurons. It has been postulated that vasopressin may function as a neurotransmitter within the SCN, either as part of the actual circadian clock mechanism or as part of the output pathway to other neural structures.

In order to discover the role of VP in circadian rhythms, we used the Brattleboro rat, a genetic mutant of the Long-Evans (LE) strain, which lacks the ability to produce hypothalamic VP, exhibits hereditary diabetes insipidus (DI), and is referred to as a DI rat. VP-containing neurons that are observed in normal animals are not found in the SCN of DI rats.

Controversy exists concerning the role of VP in circadian rhythms. Some researchers have concluded that the maintenance of circadian rhythmicity is not dependent upon the presence of VP. Our laboratory, however, has shown an influence of VP on circadian rhythms by comparing DI and LE animals under ad libitum and restricted-feeding conditions in both the dark and light cycles.[1] With nocturnal restricted-feeding (when the photic and nonphotic oscillators are coupled), the overall circadian rhythms of body temperature (BT), heart rate (HR), and activity (AC) were similar. With diurnal restricted-feeding (when the oscillators are uncoupled), BT, HR, and AC of DI animals were predominantly under the influence of the nonphotic zeitgeber, whereas in LE animals, these physiological variables were influenced by both photic and nonphotic zeitgebers, leading to a confused rhythmicity.[1] The purpose of the present study was to ascertain the effects of a 6-h phase advancement (eastward shift) of the circadian cycle in DI and LE rats when the oscillators were coupled and uncoupled.

[a] This research was supported by a Focused Giving Grant from Johnson & Johnson.

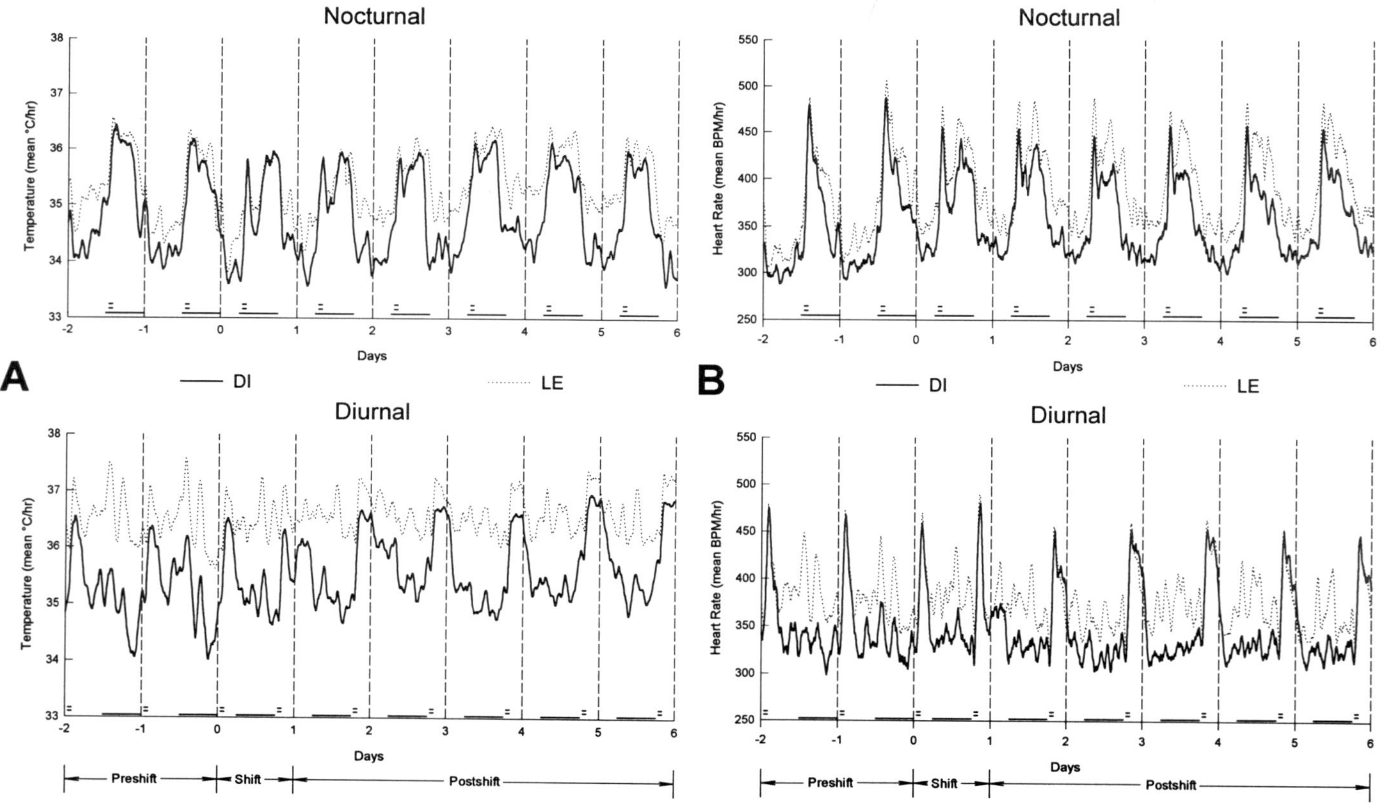
A
Nocturnal
Temperature (mean °C/hr)
38
37
36
35
34
33
-2
-1
0
1
2
3
4
5
6
Days
DI
LE
Diurnal
Temperature (mean °C/hr)
38
37
36
35
34
33
-2
-1
0
1
2
3
4
5
6
Days
Preshift
Shift
Postshift
B
Nocturnal
Heart Rate (mean BPM/hr)
550
500
450
400
350
300
250
-2
-1
0
1
2
3
4
5
6
Days
DI
LE
Diurnal
Heart Rate (mean BPM/hr)
550
500
450
400
350
300
250
-2
-1
0
1
2
3
4
5
6
Days
Preshift
Shift
Postshift

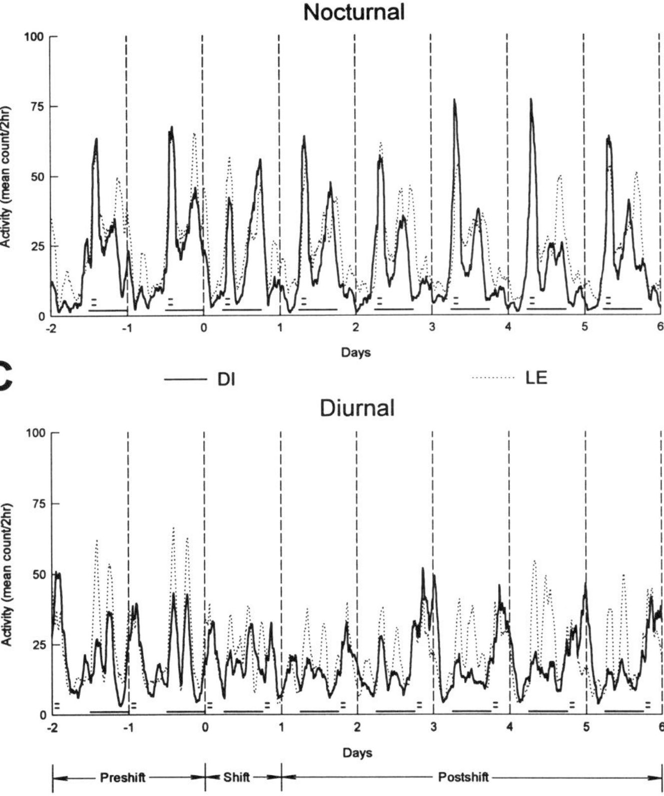

FIGURE 1. (**A**, **B**, and **C**). A summary of the circadian body temperature, heart rate, and activity, respectively, for diabetes insipidus (DI) (———) and Long-Evans (LE) (. . . . .) rats for two days preshift (last two days of the habituation period), followed by the day of shift, and five days postshift under the nocturnal feeding paradigm (*top panels*) and diurnal feeding paradigm (*bottom panels*). The figure depicts a 12-point moving average of the 5-min data points for body temperature and heart rate, and a 24-point moving average for activity. The long darkened lines on the x-axis indicate the 12-h dark phase of the light/dark cycle. The two short lines on the x-axis indicate times of feeding (one hour after the beginning of the dark or light phase) and the care and handling of animals.

METHODS

The subjects comprised 12 male LE rats and 12 male DI rats (Harlan Sprague-Dawley, Inc., Indianapolis, IN) weighing 90 to 100 g. Rats were housed in individual cages with a 12/12-h light/dark cycle. Animals were anesthetized and a biotelemetry transmitter (Model TA11CTA, Data Sciences, Inc., St. Paul, MN) capable of measuring BT, HR, and AC, was implanted subcutaneously using sterile techniques. The implant signals were transmitted to a personal computer equipped with receivers and a multiplexer (Mini-Mitter, Inc., Sunriver, OR). The individual receivers were connected to the multiplexer where BT, HR, and AC information were sequentially passed to the data acquisition system. Data were collected every five minutes for the duration of the experiment and stored on the hard disk drive of the personal computer for later analysis.

Animals were subdivided into two groups of 6 DI and 6 LE rats each. The first group was fed at the second hour of the dark cycle, and the second group was fed at the second hour of the light cycle. The amount of food presented during each feeding was 10% of the body weight of the animal calculated from the previous day. After a postimplant and preshift habituation period of 14 days, the rats were subjected to a light period shortened by six hours (simulating a 6-h time shift to the east). Feeding was shifted in synchrony with the light/dark cycle.

RESULTS

FIGURE 1A, B, and C summarizes the circadian BT, HR, and AC, respectively, for DI and LE animals for two days preshift (last two days of the habituation period), followed by the day of shift, and five days postshift under the nocturnal feeding paradigm (top panels) and diurnal feeding paradigm (bottom panels). FIGURE 1 depicts a 12-point moving average of the 5-min data points for BT and HR, and a 24-point moving average for AC.

With nocturnal feeding, both groups exhibited typical circadian rhythm patterns for nocturnal animals (i.e., higher BT, HR, and AC during the dark phase than during the light phase). The lower BT and HR observed in DI animals were likely due to the absence of VP in these animals. Overall, when these animals were subjected to a phase advancement (eastward) time shift, no significant interruptions occurred in normal circadian rhythms of BT, HR, or AC.

With diurnal feeding, however, significant differences in circadian rhythms were noted in the two strains. In the preshift period, the DI animals had a reversal of peaks and troughs in BT and HR from those observed in nocturnal feeding (i.e., higher BT and HR during the light phase than during the dark phase). On the other hand, LE animals had a confused rhythmicity with major peaks in the dark cycle, as well as a major peak in the light cycle associated with food presentation. In the postshift period, DI animal peaks of BT, HR, and AC immediately synchronized to food presentation, whereas LE animals maintained nocturnal peaks in addition to a major diurnal peak at food presentation.

CONCLUSIONS

With nocturnal feeding, when the photic and nonphotic oscillators were coupled, both groups shifted in a similar fashion and synchronized to the new dark cycle, although some significant differences were apparent within the light and dark cycles between the two strains. With diurnal feeding, when the photic and nonphotic oscillators were uncoupled, the groups shifted in a dissimilar fashion. The rhythms of BT, HR, and AC of DI animals were predominantly under the influence of the nonphotic zeitgeber, which produced a definite rhythmicity with obvious peaks and troughs. In LE animals, however, the physiological variables were influenced by both photic and nonphotic zeitgebers, leading to a confused rhythmicity. Thus, we conclude that VP has a major influence on the synchronization of the photic oscillator in circadian rhythms and causes adherence to the photic oscillator, even when the cycle is phase advanced (eastward). This influence is most apparent when the photic and nonphotic oscillators are uncoupled.

REFERENCE

1. MURPHY, H. M., C. H. WIDEMAN & G. R. NADZAM. 1996. Peptides **17:** 467–475.

c-Fos Induction in the Nucleus of the Solitary Tract of Sodium-depleted Rats by Salt Intake, Peripheral Bombesin, and the Combination

S. P. FRANKMANN,[a,c] J. HENNINGER,[a] M. J. KRUSE,[a]
K. WESTERMAN,[a] AND T. A. HOUPT[b]

[a]Department of Psychology
University of Southern Colorado
Pueblo, Colorado

[b]E. W. Bourne Behavioral Research Laboratory
Department of Psychiatry
Cornell University Medical College
White Plains, New York

INTRODUCTION

The sodium-depleted rat rapidly initiates intake of NaCl (salt) solutions when offered and satiates within 30–60 minutes. Exogenous administration of the peptide bombesin (BN) does not delay the onset of salt intake but does hasten the onset of satiety.[1,2] Salt intake in the sodium-depleted rat has been shown to induce c-Fos immunoreactivity (c-Fos) in the nucleus of the solitary tract (NTS),[3] and BN has been shown to increase c-Fos in the NTS of the nonsodium-depleted rat.[4] In this study we examined c-Fos positive cells in the NTS and area postrema (AP) of sodium-depleted rats 90 minutes after (1) access to 0.3 M NaCl, (2) BN administration or (3) combined BN and 0.3 M NaCl intake.

METHODS

Male, Sprague-Dawley rats were sodium-depleted by injection of the natriuretic/diuretic drug, furosemide (Lasix, 10 mg, sc), and overnight sodium-deficient diet and water. Twenty-four hours later, 0.3 M NaCl and water intakes were recorded in a 30-minute salt appetite test for half of the rats. For each intake group (salt appetite test or no salt appetite test), half of the rats were injected with isotonic saline vehicle (1 mL/kg), and the other half with BN (8 μg/kg) 5 minutes before the start of the appetite test. Thus, there were four sodium-depleted groups: (1) vehicle injection, no salt access, (2) vehicle injection, salt access, (3) BN injection, no salt access, and (4) BN injection, salt access. All intake data were analyzed by repeated measures ANOVA with $p < 0.05$. Ninety minutes

[c] Address correspondence to S. P. Frankmann, Department of Psychology, University of Southern Colorado, 200 Bonforte Boulevard, Pueblo, CO 81001. E-mail: spf3@cornell.edu

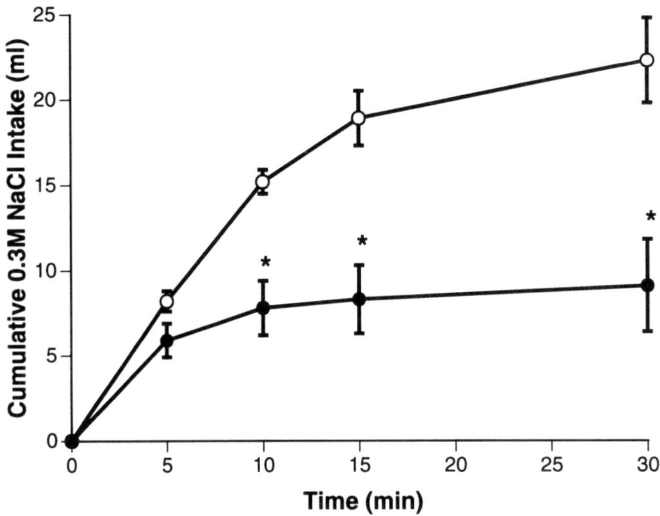

FIGURE 1. Cumulative intakes of 0.3 M NaCl following ip injection of vehicle (*white circles*) or 8 μg/kg bombesin (*black circles*). All data are the mean $\pm$ SEM. *$p < 0.05$ vs. vehicle-treated rats.

after injections, rats were transcardially perfused and the brainstems processed for c-Fos immunohistochemistry.

For immunohistochemistry, 40-μm coronal sections through the NTS and AP were taken. All treatment groups were run in each assay. After overnight incubation in sheep anti-c-Fos peptide antibody (Genosys, Woodlands, TX) followed by 1-hour incubation with biotinylated anti-sheep rabbit antibody (Vector, Burlingame, CA) and amplification of the bound secondary antibody (Vector Elite ABC kit), the antibody complex was visualized by 0.5% DAB. Images of alternate sections were digitized and the individual c-Fos positive cells plotted and counted on a computer.

RESULTS

BN significantly attenuated the salt intakes of rats compared to vehicle-treated rats [$F(1, 7) = 14.55$, $p < 0.05$; see FIG. 1], although no significant differences were found until after the 5-minute intake reading [$F(1, 7) = 3.12$, n.s. for the 5-minute comparison]. Thus, the BN had the previously reported satiety effect on salt intake.

In looking at the pattern of c-Fos induction of the four groups, several interesting differences were noted. As expected, sodium depletion alone did not induce c-Fos in the NTS/AP beyond background levels; however, salt intake following sodium depletion did induce c-Fos positive cells in the NTS but not the AP. Salt intake induced maximal c-Fos in the intermediate NTS (abutting the fourth ventricle). BN induced c-Fos positive cells in the caudal (subpostremal) NTS, intermediate NTS, and in the AP. Salt intake in combination with BN induced a

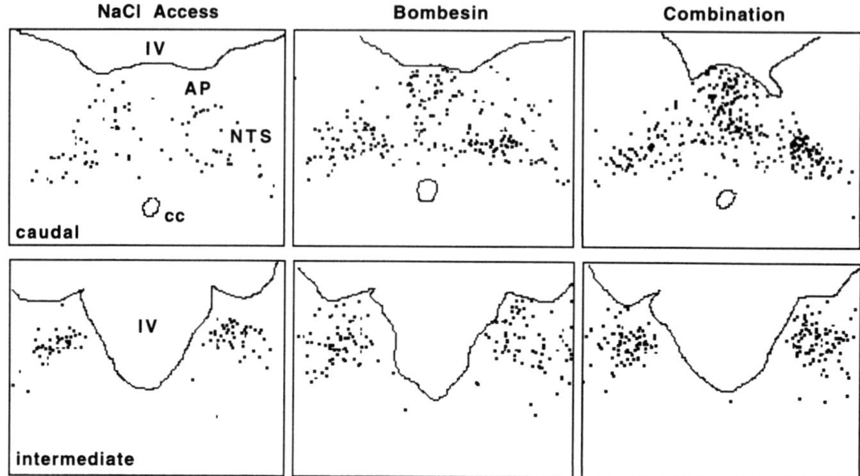

FIGURE 2. Computer-generated camera lucida examples of c-Fos-positive cells in the NTS and AP of sodium-depleted rats after 30 minute salt access, BN injection, or the combination. **Top row:** Coronal sections through the caudal NTS at the level of the AP. **Bottom row:** Coronal sections through the intermediate NTS. AP, area postrema; NTS, nucleus of the solitary tract; cc, central canal; IV, fourth ventricle.

greater number of c-Fos positive cells in both the NTS and AP than either treatment alone (see FIG. 2).

DISCUSSION

In sodium-depleted rats, salt intake alone induces c-Fos in the NTS but very little in the AP, whereas BN induces c-Fos in the NTS and the AP. The patterns of neuronal activation in the NTS after salt intake or exogenous BN overlap but are not identical; this is consistent with a role of endogenous BN-like peptides in the satiation of salt appetite. The combination of salt intake and BN induces a greater increase in the c-Fos positive cells in both the NTS and in the AP. Thus, salt intake potentiates the effect of BN in inducing c-Fos in the AP, while having little or no effect on this brain region on its own. These data suggest that not only is the behavior of salt intake altered by BN administration, but the pattern of c-Fos induction in the NTS and AP is also altered.

REFERENCES

1. FRANKMANN, S. P., J. H. DOKKO & J. GIBBS. 1993. Soc. Neurosci. Abstr. **19:** 1263.
2. FLYNN, F. W. & R. RAMOS. 1994. Behav. Neurosci. **108:** 780–788.
3. FRANKMANN, S. P., G. P. SMITH, T. H. JOH & T. A. HOUPT. 1994. Soc. Neurosci. Abstr. **20:** 586.
4. BONAZ, B., R. DE GIORGIO & Y. TACHÉ. 1993. Brain Res. **600:** 353–357.

Salt-Loading Induces Decreased POMC mRNA Levels, Increased α-MSH Immunoreactivity, and Sustained Elevated Fos Expression in Rat Pituitary Intermediate Lobe Melanotropes[a]

P. SHARMA,[b] K. E. HAGLER,[b] N. O. DYBDAL,[c]
AND B. M. CHRONWALL[b,d]

*[b]School of Biological Sciences
University of Missouri-Kansas City
Kansas City, Missouri 64108*

*[c]Genentech Inc.
South San Francisco, California 94080*

The melanotropes of the pituitary intermediate lobe are endocrine cells that release mainly α-MSH and β-endorphin, peptide hormones derived from pro-opiomelanocortin (POMC). Individual melanotropes maintain differing levels of metabolic activity.[1–5] Dopaminergic axons emanating from the hypothalamic periventricular nucleus innervate the intermediate lobe,[6] and dopamine inhibits melanotrope biosynthetic activity by means of dopamine D_2 receptors.[7,8] Salt-loading has long been known to affect melanotrope cytology,[1–3,9] which, at least in part, could be related to decreased POMC mRNA levels.[10,11] However, the physiological role of the intermediate lobe in salt-loading is not well understood, and the mechanisms underlying the responses appear complex. To further investigate this phenomenon, we determined relative POMC mRNA levels and α-MSH and Fos immunoreactivities in individual melanotropes following salt-loading.

Male Sprague-Dawley rats (150–200 g) were housed two per cage and fed Purina Autolab chow ad libitum. Experimental rats received 2% NaCl for 5 or 10 days ($n = 5$ per time point), while the control rats had free access to water ($n = 5$ per time point). Rats were sacrificed by a lethal injection of pentobarbital; pituitaries were fixed with periodate-lysine-paraformaldehyde (PLP; 10 mM $NaIO_4$, 75 mM lysine, and 2% paraformaldehyde in 38 mM $NaPO_4$ buffer, pH 7.4), cryoprotected in sucrose, and frozen. Cryostat-sections (5 μm) were thaw mounted onto Probe-on Plus slides (Fisher, Pittsburgh, PA) and used for histochemical experiments.

For *in situ* hybridization histochemistry, a 30-base oligonucleotide probe (5'-CTT GCC CCA GCG GAA GTG CTC CAT GGA GTA-3') corresponding to a

[a] This project was supported by NIH grant NS 28019 to B.M.C. and a Marion Merrell Dow Foundation postdoctoral fellowship to P.S.

[d] Corresponding author; e-mail: chronwall@cctr.umkc.edu

portion of the ACTH coding region of the POMC gene[12] was used. The probes were 3′-end labeled with either [^{35}S]-dATP (DuPont NEN Products, Wilmington, DE) or digoxigenin-11-dUTP (Boehringer Mannheim, Indianapolis, IN) using a labeling kit (DNA Tailing Kit, Boehringer Mannheim). Radioactive hybrids were detected by autoradiography and digoxigenin-labeled hybrids by immunohisto-chemistry. Hybridization and detection were performed using previously described methods.[4] Tissues were hybridized in parallel for all experimental groups to be compared. Specificity of probe hybridization to POMC mRNA was determined by several control experiments. No staining was observed on sections incubated with unlabeled probe, labeled probe without antibody-conjugate, or without probe. On pituitary sections alternately prepared for α-MSH immunohistochemistry and *in situ* hybridization the resultant staining patterns correlated well.

For immunohistochemistry, sections were incubated in primary antiserum di-

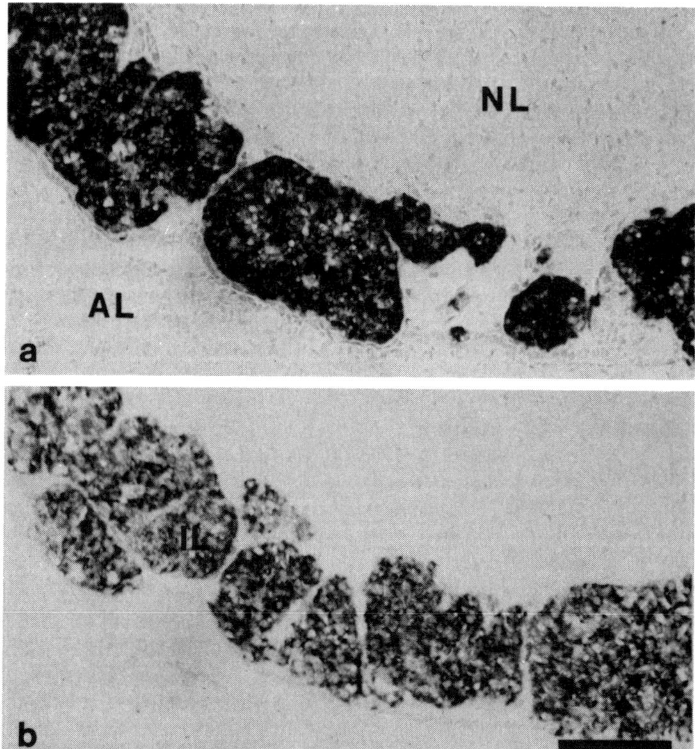

FIGURE 1. *In situ* hybridization with an oligonucleotide probe specific for POMC mRNA visualized by immunohistochemical detection of digoxigenin-labeled hybrids in the pituitary intermediate lobe of rat. Sections from a control animal **(a)** and an animal with substantial decrease of POMC mRNA after 5 days of salt-loading **(b)**. Our results show considerable inter-animal variation in response to salt-loading. Heterogeneity in POMC mRNA among individual melanotropes can be found in all sections; this phenomenon is clearly visible in **b**. NL, neural lobe; AL, anterior lobe; IL, intermediate lobe. (Magnification 350×; bar = 50 μm.)

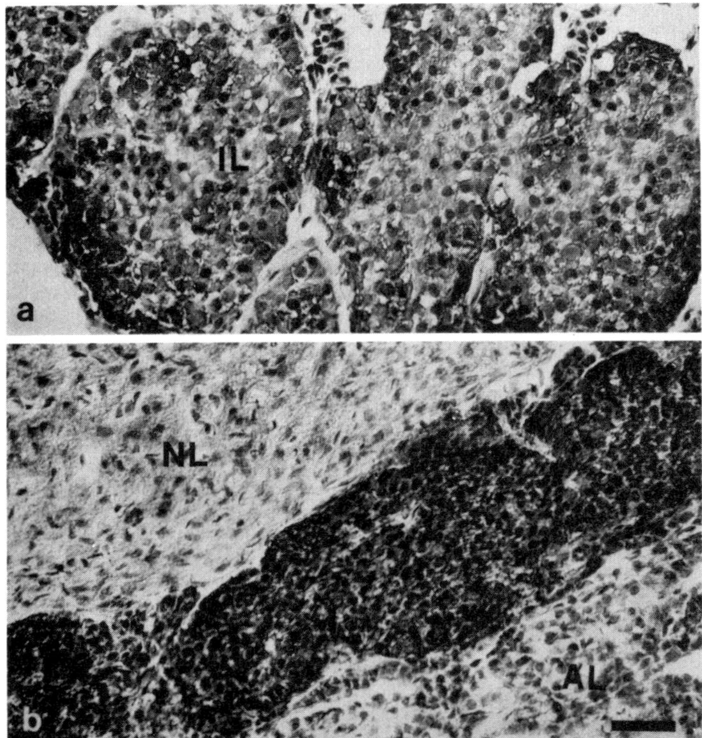

FIGURE 2. Immunohistochemistry demonstrating α-MSH immunoreactivity in melano-tropes of the intermediate lobe of the rat pituitary. Sections from a control animal **(a)** and an animal with increased α-MSH immunoreactivity after 10 days of salt-loading **(b)**. Illustrated in **b** is the shrinking of the intermediate lobe after salt-loading. (Magnification 200×; bar = 50 μm.)

luted with PBS/BSA (rabbit anti-Fos 1 : 50, Oncogene Science, Inc., Cambridge, MA; or rabbit anti-α-MSH 1 : 1500, Chemicon, Temecula, CA). Sections were rinsed and then incubated for 30 min in biotinylated goat-anti-rabbit secondary antiserum (1 : 200, Organon Technica, Durham, NC); rinsed and incubated with the ABC reagent of a Vectastain Elite ABC Kit (Vector, Burlingame, CA) for 30 min; and rinsed and treated with Pierce DAB Enhancement Solution (Pierce, Rockford, IL).

Quantification of histochemical data was made using an automated image analysis system consisting of a Dage MTI model 72 CCD camera mounted on a Zeiss Axioplan microscope connected to a PC with a hard disk drive running IM 3000B image and processing software (Micro Measure, Georgia Instruments, Roswell, GA). POMC mRNA levels were evaluated by counts of autoradiographic silver grains or by optical density (OD) measurements of immunohistochemical reaction product. Fos and α-MSH immunoreactivities were measured by densitometry. The methods for these measurements have been presented in detail.[4] Data were evaluated by ANOVA and GraphPAD InStat.

POMC mRNA levels decreased to 85% ($p < 0.001$) of control values in rats salt-loaded for 5 days as measured by OD of digoxigenin-labeled hybrids (FIG. 1); levels did not decrease further and were 89% ($p < 0.001$) of control on day 10. By grain counting, the decrease in POMC mRNA was not significant. We repeated the hybridization experiments on sections from the five animals of each treatment group simultaneously, on three different days. Each hybridization experiment yielded overall values showing the same trend of a slight to moderate decrease because of considerable inter-animal variation in control as well as in experimental groups (FIG. 1a, b). The literature[10,11] describes substantial decreases. Within the pituitary of each animal, we also observed a striking heterogeneity in the level of POMC mRNA among individual melanotropes, consistent with previous reports.[4,5] The heterogeneity was enhanced after salt-loading (FIG. 1b).

The immunoreactivity of α-MSH did not change significantly after 5 days of salt-loading (103% of controls). However, after 10 days, there was a significant increase to 129% ($p < 0.01$) of control intensities (FIG. 2). Our α-MSH data were very consistent, with less inter-animal variation than we saw in the POMC mRNA data. Using the methyl green-pyronin Y histochemical technique for RNA demonstration and the frog skin bioassay, Howe and Thody[3] report an increase in MSH content with no change in RNA.

In pituitaries from control rats, Fos protein immunoreactivity reached an arbitrary OD set at 150 in 21.5% of the area of the intermediate lobe. This area increased significantly to 158% of control values after salt-loading for 5 days with a further increase to 214% after 10 days. Because melanotropes constitute 95% of the lobe, these data indicate that an increased number of melanotropes showed a sustained high Fos level after salt-loading. In the hypothalamus, Fos-like immunoreactivity and Fos-related antigen were also detected for at least one week. In contrast, c-*fos* was only transiently induced.[13]

Fos regulates the transcriptional expression of several genes, possibly including those required for enhanced translation. If this is the case in melanotropes, it would provide one explanation for the increased α-MSH immunoreactivity in our experiments. Enhanced translation, together with decreased secretion, would result in increased melanotrope α-MSH content, even with decreased POMC mRNA expression.

REFERENCES

1. ZIEGLER, B. 1963. Licht-und elektronenmikroskopische Untersuchungen an Pars Intermedia und Neurohypophyse der Ratte. Z. Zellforsch. **59:** 486–506.
2. DUCHEN, L. W. 1962. The effects of ingestion of hypertonic saline on the pituitary gland in the rat: A morphological study of the pars intermedia and posterior lobe. Endocrinology **25:** 161–168.
3. HOWE, A. & A. J. THODY. 1970. The effect of ingestion of hypertonic saline on the melanocyte-stimulating hormone content and histology of the pars intermedia of the rat pituitary gland. J. Endocrinol. **46:** 201–208.
4. DICKERSON, D. S., B. S. HUERTER, S. J. MORRIS & B. M. CHRONWALL. 1994. POMC mRNA levels in individual melanotropes and GFAP in glial cells in rat pituitary. Peptides **15:** 247–256.
5. CHRONWALL, B. M., W. R. MILLINGTON, S. T. W. GRIFFIN, J. R. UNNERSTALL & T. L. O'DONOHUE. 1987. Histological evaluation of dopaminergic regulation of rat intermediate pituitary pro-opiomelanocortin gene expression using *in situ* hybridization and ^{3}H-thymidine uptake. Endocrinology **120:** 1201–1211.
6. GOUDREAU, J. L., S. E. LINDLEY, K. J. LOOKINGLAND & K. E. MOORE. 1992. Evidence

that hypothalamic periventricular dopamine neurons innervate the intermediate lobe of the rat pituitary. Neuroendocrinology **56:** 100–105.

7. BEAULIEU, M., R. FELDER & J. W. KEBABIAN. 1986. D-2 dopaminergic agonists and cyclic adenosine 3′,5′-monophosphate directly regulate the synthesis of α-melano-cyte-stimulating hormone-like peptides by cultured rat melanotrophs. Endocrinology **118:** 1032–1039.

8. CHRONWALL, B. M., G. R. HOOK & W. R. MILLINGTON. 1988. Dopaminergic regulation of the biosynthetic activity of individual melanotropes in the rat pituitary intermediate lobe: A morphometric analysis by light and electron microscopy and *in situ* hybridiza-tion. Endocrinology **123:** 1992–2002.

9. SCHMITT, G., M. E. STOECKEL, M. J. KLEIN & A. PORTE. 1982. Effects of experimental hypo- or hypernatremia on the fine structure of the pars intermedia of the murine pituitary. Cell Tissue Res. **223:** 641–657.

10. ELKABES, S. & Y. P. LOH. 1988. Effect of salt loading on proopiomelanocortin (POMC) messenger ribonucleic acid levels, POMC biosynthesis and secretion of POMC prod-ucts in the mouse pituitary gland. Endocrinology **123:** 1754–1760.

11. PARDY, K., D. CARTER & D. MURPHY. 1990. Dopaminergic mediation of physiological changes in proopiomelanocortin messenger ribonucleic acid expression in the neuro-intermediate lobe of the rat pituitary. Endocrinology **126:** 2960–2964.

12. DROUIN, J., M. CHAMBERLAND, J. CHARRON, L. JEANNOTTE & M. NEMER. 1985. Structure of the rat pro-opiomelanocortin (POMC) gene. FEBS Lett. **193:** 54–58.

13. SHARP, F. R., S. M. SAGAR, K. HICKS, D. LOWENSTEIN & K. HISANAGA. 1991. c-*fos* mRNA, Fos, and Fos-related antigen induction by hypertonic saline and stress. J. Neurosci. **11:** 2321–2331.

Peptidergic Mechanisms of Action in the Suprachiasmatic Nucleus[a]

KIM L. HUHMAN,[b] CHARLES F. GILLESPIE,
CHERIE L. MARVEL, AND H. ELLIOTT ALBERS

Laboratory of Neuroendocrinology and Behavior
Departments of Biology and Psychology
Georgia State University
Atlanta, Georgia 30303

The suprachiasmatic nucleus (SCN) functions as a circadian clock that generates behavioral and physiological rhythms and synchronizes, or entrains, those rhythms to the 24-h light-dark cycle.[1-4] This small, bilateral nucleus at the base of the hypothalamus contains at least 25 different neurotransmitters and neuropeptides[5] and, thus, is well suited for the investigation of the mechanisms of action of neuropeptides in the central nervous system. The purpose of this paper is to briefly illustrate some of the ways in which SCN neuropeptides interact with one another and with other SCN neurotransmitters to control circadian rhythmicity.

Information needed for the entrainment of circadian rhythms to the environment is communicated to the SCN via several afferent pathways. The retinohypothalamic tract is a direct projection to the SCN from the retina[6-8] that contains glutamate and, perhaps, substance P.[9-12] The geniculohypothalamic tract is a secondary photic projection that originates from the intergeniculate leaflet of the thalamus[13,14] and contains neuropeptide Y (NPY) and γ-aminobutyric acid (GABA).[15-18] A third major SCN afferent is a serotonergic projection that arises from the raphe nuclei of the midbrain.[19,20] Most SCN afferents terminate on neurons that contain vasoactive intestinal peptide (VIP).[21-23] These neurons also contain peptide histidine isoleucine (PHI), which is derived from the same precursor molecule as is VIP.[24] Some of these neurons also contain gastrin-releasing peptide (GRP).[25,26] Exposure to light reduces levels of VIP/PHI mRNA and immunoreactivity (IR) in the SCN, whereas light increases GRP mRNA and IR.[27-30] Endogenous rhythms of somatostatin and vasopressin mRNA in the SCN have also been demonstrated.[31,32] This rhythmicity of peptide mRNA is interesting in light of the possible role of the transcriptional process in the generation of circadian rhythms.[31]

Data from our laboratory indicate that microinjection of a cocktail of peptides containing equimolar amounts of VIP, PHI, and GRP into the SCN mimics the phase-delaying effects of light on activity rhythms, whereas microinjection of the same overall amount of any of these peptides alone or in combination with one other peptide causes significantly smaller phase delays.[26,33] Additionally, *in vitro* application of VIP, PHI, and GRP in the hypothalamic slice preparation activates SCN single units,[26] and co-administration of VIP and GRP to SCN neurons results in phase delays of SCN neuronal activity that are greater than those obtained

[a] This work was supported by National Institutes of Health grants NS34896 to K.L.H. and NS30022 and NS34586 to H.E.A.

[b] Address correspondence to Kim L. Huhman, Department of Psychology, Georgia State University, Atlanta, Georgia 30303-3083. E-mail: biokkh@panther.gsu.edu

by the administration of VIP or GRP alone.[34] These results suggest that SCN neuropeptides may interact synergistically with one another to affect SCN function (however, see also ref. 35).

A considerable body of evidence exists which suggests that an interaction occurs between peptidergic and classical neurotransmitters such as glutamate, GABA, and serotonin (5-HT) at the level of the SCN. *In vitro* application of the NMDA receptor agonist, MK-801, increases the release of VIP from SCN neurons.[34] Conversely, serotonin depletion decreases VIP mRNA in the SCN,[36] and VIP has been shown to alter binding of 5-HT to 5-HT$_1$ receptors.[37] In addition, we have shown that microinjection of the GABA$_A$ antagonist, bicuculline, can completely block the ability of NPY to phase advance circadian activity rhythms *in vivo*[38] (see FIG. 1), indicating that GABA$_A$ activity may mediate the phase-shifting activity of this neuropeptide. In the SCN, the GABA-synthesizing enzyme glutamate decarboxylase (GAD)-IR is co-localized with VIP-IR, suggesting that GABA may also be co-released with VIP.[18] GAD-IR terminals have also been shown to synapse on both VIP-IR and GRP-IR terminals[18,39] as well as on the same target neurons as do these peptide terminals.[39] Recent data from our laboratory indicate that microinjection of VIP, PHI, GRP with the GABA$_A$ agonist, muscimol, or the GABA$_A$ antagonist, bicuculline, reduces or increases, respectively, the phase-delaying effects of VIP, PHI, and GRP administration[40] (see FIG. 2). Interestingly, we found that these results parallel the effects of centrally administered muscimol and bicuculline on light-induced phase delays.[40] Microinjection of muscimol or bicuculline into the SCN immediately prior to the administration of a light pulse reduces or increases, respectively, the phase-delaying effects of light (see FIG. 2). These data support the hypothesis that neuropeptides

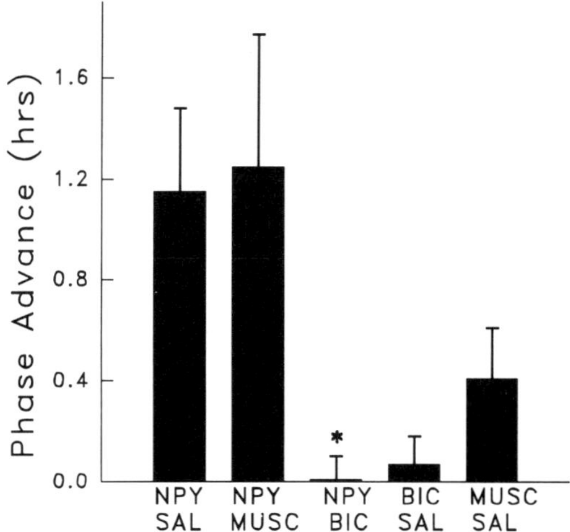

FIGURE 1. Mean phase advance (± SEM) in hours induced by microinjection into the suprachiasmatic nucleus region of either neuropeptide Y (NPY, 100 ng in 100 nL saline, $n = 8$) with saline (SAL, 50 nL), NPY with the GABA$_A$ agonist, muscimol (MUS, 4.4 nmol in 50 nL saline, $n = 4$), NPY with the GABA$_A$ antagonist, bicuculline (BIC, 0.18 nmol in 50 nL saline, $n = 10$), bicuculline with saline ($n = 5$), or muscimol with saline ($n = 5$). *Indicates less than NPY/SAL and NPY/MUSC.

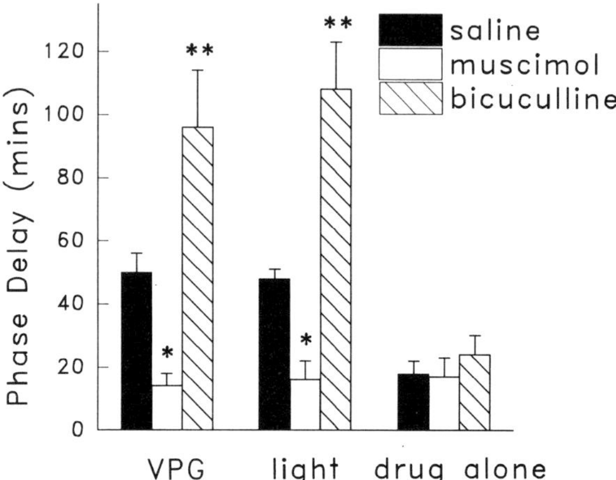

FIGURE 2. Mean phase delay ($\pm$ SEM) in minutes induced by microinjection into the suprachiasmatic nucleus region of muscimol, bicuculline or saline (for dosages, see FIG. 1) in combination with either a "peptide cocktail" containing VIP/PHI/GRP (VPG, 100 ng of each peptide in 150 nL saline, $n = 8$), a 15-min, 120-lux light pulse ($n = 8$) or given alone ($n = 5$). *Indicates less than VPG/saline or light/saline; **Indicates greater than VPG/saline or light/saline.

are involved in transducing phase-shifting information and indicate that the SCN may be an excellent neuroanatomical site within which to probe the possible synergistic effects of neuropeptides, as well as to study the interaction of neuropeptides with classical neurotransmitters.

REFERENCES

1. MEIJER, J. H. & W. J. RIETVELD. 1989. Neurophysiology of the suprachiasmatic circadian pacemaker in rodents. Physiol. Rev. **69:** 671–707.
2. MOORE, R. Y. & J. P. CARD. 1985. Visual pathways and the entrainment of circadian rhythms. Ann. N. Y. Acad. Sci. **453:** 123–133.
3. MOORE, R. Y. & V. B. EICHLER. 1972. Loss of circadian adrenal corticosterone rhythm following suprachiasmatic nuclear lesions in the rat. Brain Res. **42:** 201–206.
4. STEPHAN, F. K. & I. ZUCKER. 1972. Circadian rhythms in drinking behavior and locomotor activity of rats are eliminated by hypothalamic lesions. Proc. Natl. Acad. Sci. USA **69:** 1583–1586.
5. VAN DEN POL, A. N. & K. L. TSUJIMOTO. 1985. Neurotransmitters of the hypothalamic suprachiasmatic nucleus: Immunocytochemical analysis of 25 neuronal antigens. Neuroscience **15:** 1049–1086.
6. HENDRICKSON, A. E., N. WAGONER & W. M. COWAN. 1972. An autoradiographic and electron microscopic study of retino-hypothalamic connections. Z. Zellforsch. **135:** 1–26.
7. MOORE, R. Y. & N. J. LENN. 1972. A retinohypothalamic projection in the rat. J. Comp. Neurol. **146:** 1–14.
8. OKAMURA, H., A. BEROD, J. F. JULIEN, M. GEFFARD, K. KITAHAMA, J. MALLET & P. BOBILLIER. 1989. Demonstration of GABAergic cell bodies in the suprachiasmatic

nucleus: in situ hybridization of glutamic acid decarboxylase (GAD) mRNA and immunocytochemistry of GAD and GABA. Neurosci. Lett. **102:** 131–136.

9. CAHILL, G. M. & M. MENAKER. 1989. Effects of excitatory amino acid receptor antagonists and agonists on suprachiasmatic nucleus responses to retinohypothalamic tract volleys. Brain Res. **479:** 76–82.

10. KAGEYAMA, G. H. & R. L. MEYER. 1989. Glutamate-immunoreactivity in the retina and optic tectum of goldfish. Brain Res. **503:** 118–127.

11. BIELLO, S. 1995. Enhanced photic phase shifting after treatment with antiserum to neuropeptide Y. Brain Res. **673:** 25–29.

12. TAKATSUJI, K., J.-J. MIGUEL-HIDALGO & M. TOHYAMA. 1991. Substance P-immunoreactive innervation from the retina to the suprachiasmatic nucleus in the rat. Brain Res. **568:** 223–229.

13. RIBAK, C. E. & A. PETERS. 1975. An autoradiographic study of the projections from the lateral geniculate body of the rat. Brain Res. **92:** 261–294.

14. SWANSON, L. W., W. M. COWAN & E. G. JONES. 1974. An autoradiographic study of the efferent connections of the ventral lateral geniculate nucleus in the albino rat and cat. J. Comp. Neurol. **156:** 143–163.

15. CARD, J. P. & R. Y. MOORE. 1982. Ventral lateral geniculate nucleus efferents to the rat suprachiasmatic nucleus exhibit avian pancreatic polypeptide-like immunoreactivity. J. Comp. Neurol. **206:** 390–396.

16. ALBERS, H. E. & C. F. FERRIS. 1984. Neuropeptide Y: Role in light-dark cycle entrainment of hamster circadian rhythm. Neurosci. Lett. **50:** 163–168.

17. HARRINGTON, M. E. & B. RUSAK. 1987. Ablation of the geniculo-hypothalamic tract alters circadian activity rhythms of hamster housed under constant light. Physiol. Behav. **42:** 183–189.

18. FRANÇOIS-BELLAN, A. M., P. KACHIDIAN, G. DUSTICIER, M. C. TONON, H. VAUDRY & O. BOSLER. 1990. GABA neurons in the rat suprachiasmatic nucleus: Involvement in chemospecific synaptic circuitry and evidence for GAD-peptide colocalization. J. Neurocytol. **19:** 937–947.

19. AGHAJANIAN, G. K., F. E. BLOOM & M. H. SHEARD. 1969. Electron microscopy of degeneration within the serotonergic pathway of the rat. Brain Res. **13:** 266–273.

20. SAAVEDRA, J. M., M. PALKOVITS, M. J. BROWNSTEIN & J. AXELROD. 1974. Serotonin distribution in the nuclei of the rat hypothalamus and preoptic region. Brain Res. **77:** 157–165.

21. HISANO, S., M. CHIKAMORI-AOYAMA, S. KATOH, Y. KAGOTANI, S. DAIKOKU & K. CHIHARA. 1988. Suprachiasmatic nucleus neurons immunoreactive for vasoactive peptide have synaptic contacts with axons immunoreactive for neuropeptide Y: An immunoelectronmicroscopic study in the rat. Neurosci. Lett. **88:** 145–150.

22. IBATA, Y., Y. TAKAHASHI, H. OKAMURA, K. FUMIO, H. TERUBAYASHI, T. KUBO & N. YANAIHARA. 1989. Vasoactive intestinal peptide (VIP)-like immunoreactive neurons located in the rat suprachiasmatic nucleus receive a direct retinal projection. Neurosci. Lett. **97:** 1–5.

23. WYATT, L. M., R. B. NORGREN & M. N. LEHMAN. 1988. Retinal and neuropeptide Y innervation of the hamster suprachiasmatic nucleus: Light and electron microscopic observations. Soc. Neurosci. Abstr. **14:** 50.

24. NISHIZAWA, M., Y. HAYAKAWA, N. YANAIHARA & H. OKAMOTO. 1985. Nucleotide sequence divergence and functional constraint in VIP precursor mRNA evolution between human and rat. FEBS Lett. **183:** 55–59.

25. OKAMURA, H., S. MURAKAMI, K. UDA, T. SUGANO, Y. TAKAHASHI, C. YANAIHARA, N. YANAIHARA & Y. IBATA. 1986. Coexistence of vasoactive intestinal peptide (VIP)-, peptide histidine isoleucine amide (PHI)-, and gastrin releasing peptide (GRP)-like immunoreactivity in neurons of the rat suprachiasmatic nucleus. Biomed. Res. **7:** 295–299.

26. ALBERS, H. E., S. Y. LIOU, E. G. STOPA & R. T. ZOELLER. 1991. Interaction of co-localized neuropeptides: Functional significance in the circadian timing system. J. Neurosci. **11:** 846–851.

27. ALBERS, H. E., E. G. STOPA, R. T. ZOELLER, J. S. KAUER, J. C. KING. J. S. FINK,

H. MOBTAKER & H. WOLFE. 1990. Day-night variation in prepro vasoactive intestinal peptide/peptide histidine isoleucine mRNA with the rat suprachiasmatic nucleus. Mol. Brain Res. **7:** 85–89.

28. GOZES, I., Y. SHANI, B. LIU & J. P. H. BURBACH. 1989. Diurnal variation in vasoactive intestinal peptide messenger RNA in the suprachiasmatic nucleus of the rat. Neurosci. Res. Commun. **5:** 83–86.

29. SHINOHARA, K., K. TOMINAGA, Y. ISOBE & S. I. INOUYE. 1993. Photic regulation of peptides located in the ventrolateral subdivision of the suprachiasmatic nucleus of the rat: Daily variations in vasoactive intestinal polypeptide, gastrin releasing peptide and neuropeptide Y. J. Neurosci. **13:** 793–800.

30. ZOELLER, R. T., B. BROYLES, J. EARLEY, E. R. ANDERSON & H. E. ALBERS. 1991. Cellular levels of messenger ribonucleic acids encoding vasoactive intestinal peptide and gastrin-releasing peptide in neurons of the suprachiasmatic nucleus exhibit distinct 24-hour rhythms. J. Neuroendocrinol. **4:** 119–124.

31. TAKEUCHI, J., H. NAGASAKI, K. SHINOHARA & S.-I. INOUYE. 1992. A circadian rhythm of somatostatin messenger RNA levels, but not of vasoactive intestinal polypeptide/peptide isoleucine messenger RNA levels in rat suprachiasmatic nucleus. Mol. Cell. Neurosci. **3:** 29–35.

32. UHL, G. R. & S. M. REPPERT. 1986. Suprachiasmatic nucleus vasopressin messenger RNA: Circadian variation in normal and Brattleboro rats. Science **232:** 390–393.

33. ALBERS, H. E., C. F. GILLESPIE, T. O. BABAGBEMI & K. L. HUHMAN. 1995. Analysis of the phase shifting effects of gastrin releasing peptide when microinjected into the suprachiasmatic region. Neurosci. Lett. **191:** 63–66.

34. SHIBATA, S., M. ONO, K. TOMINAGA, T. HAMADA, A. WATANABE & S. WATANABE. 1994. Involvement of vasoactive intestinal polypeptide in NMDA-induced phase delay of firing activity rhythm in the suprachiasmatic nucleus in vitro. Neurosci. Biobehav. Rev. **18:** 591–595.

35. PIGGINS, H. D., M. C. ANTLE & B. RUSAK. 1995. Neuropeptides phase shift the mammalian circadian pacemaker. J. Neurosci. **15:** 5612–5662.

36. KAWAKAMI, F., H. OKAMURA, K. FUKUI, C. YANAIHARA, N. YANAIHARA, T. NAKAJIMA & Y. IBATA. 1985. The influence of serotonergic inputs on peptide neurons in the rat suprachiasmatic nucleus: An immunocytochemical study. Neurosci. Lett. **61:** 273–277.

37. ROSTENE, W. H., C. T. FISCHETTE & B. S. MCEWEN. 1984. Modulation by vasoactive intestinal peptide (VIP) of serotonin-1 receptors in membranes from rat hippocampus. J. Neurosci. **3:** 2414–2419.

38. HUHMAN, K. L., T. O. BABAGBEMI & H. E. ALBERS. 1995. Bicuculline blocks neuropeptide Y-induced phase advances when microinjected in the suprachiasmatic nucleus of Syrian hamsters. Brain Res. **675:** 333–336.

39. VAN DEN POL, A. N. 1986. Gamma-aminobutyrate, gastrin releasing peptide, serotonin, somatostatin, and vasopressin: Ultrastructural immunocytochemical localization in presynaptic axons in the suprachiasmatic nucleus. Neuroscience **17:** 643–659.

40. GILLESPIE, C. F., T. O. BABAGBEMI, K. L. HUHMAN & H. E. ALBERS. 1996. Bicuculline increases and muscimol reduces the phase-delaying effects of light and VIP/PHI/GRP in the suprachiasmatic region. J. Biol. Rhythms **11:** 137–144.

Structure-Activity Studies of SchistoFLRFamide-like Peptides

ANGELA B. LANGE,[a,d] I. ORCHARD,[a] Z. WANG,[a]
A. N. STARRATT,[b] AND R. J. NACHMAN[c]

[a]Department of Zoology
University of Toronto
Erindale College
Mississauga, Ontario, Canada L5L 1C6

[b]Pest Management Research Centre
Agriculture and Agri-Food Canada
London, Ontario, Canada N5V 4T3

[c]Veterinary Entomology Research Unit
United States Department of Agriculture
2881 F&B Road
College Station, Texas 77845

The sequences of two FMRFamide-related peptides from *Locusta* central nervous system have recently been determined. One peptide is identical to the previously described SchistoFLRFamide (PDVDHVFLRFamide), while the second peptide is novel and differs from SchistoFLRFamide in positions 1 and 4 (ADVGHVFLRF-amide).[1] Both are inhibitory peptides when assayed on locust oviduct, inhibiting myogenic contractions, lowering basal tonus, and inhibiting proctolin-induced contractions. These peptides are members of a subfamily of insect FMRFamide-related peptides that share the sequence XDVXHXFLRFamide and which are referred to as myosuppressins. Myosuppressins have been identified in a diverse number of insect species including the cockroach, locust, fruit fly, and tobacco hornworm.[2] Members of this subfamily are potent inhibitors of insect cardiac and visceral muscle, and in addition have effects on skeletal muscle, longitudinal flight muscle, and salivary glands.[3]

Bioassays and binding assays have shown that the His residue in the truncated analogue HVFLRFamide is critical for the retention of inhibitory biological activity, whereas VFLRFamide, in which inhibitory biological activity is lost, is the minimum sequence for binding of comparable affinity to the parent compound.[4] Histidine has previously been shown to be a critical residue for determining the biological activity of a number of peptides, including gonadotropin-releasing hormone, angiotensin II, glucagon, and luteinizing hormone-releasing hormone. Analogues in which the His imidazole group has been altered have revealed details of the important features of the His residue.[3] We have investigated the properties of the His residue in determining the inhibitory biological activity of HVFLRF-amide on locust oviducts. A number of modifications have been made in the N-terminal His residue to examine the importance of the imidazole group of histidine

---

[d] Address correspondence to Angela B. Lange, Department of Zoology, University of Toronto, Erindale College, 3359 Mississauga Road, Mississauga, Ontario, Canada L5L 1C6.
E-mail: alange@credit.erin.utoronto.ca

for biological activity, and a comparison made with the influence of these modifications on binding.[3] Substitution of His by the D-isomer or Phe produced analogues with stimulatory rather than inhibitory activity, confirming the importance of the His moiety and indicating that inhibition is not simply due to the presence of an aromatic residue. In addition, inhibitory activity was retained when the His moiety was methylated at the N-3 position of the imidazole ring, but methylation of N-1 yielded a peptide which stimulated contractions. Inhibitory activity was further retained when N^{α}-methyl-L-His and D,L-1',2',4'-triazole-3-Ala were substituted for His. The activity of this latter analogue is perhaps not surprising because two of its heterocyclic ring nitrogens are at the same positions as the imidazole nitrogens of His. Interestingly, analogues which showed activity reversal and resulted in stimulatory rather than inhibitory activity also had weaker binding, as revealed by K_i values. Thus, although the His residue in HVFLRFamide may not participate in binding directly, alterations in this His residue obviously influence binding, possibly by altering the conformation of the remaining VFLRFamide sequence. Future research on the structural and conformational requirements of these myosuppressins should aid in the development of pharmacological tools that may lead to novel insect control methods.

With the notion of insect control methods in mind, it must be acknowledged that the use of peptides per se as pesticides would appear impractical, because their chemistry renders them susceptible to degradation under field conditions and to digestion after feeding, and their polarity makes uptake through the cuticle difficult. Over the last few years, some progress has been made in the development of stable peptide analogues, although totally nonpeptide compounds may be required to overcome all of the intrinsic problems associated with targeting the peptidergic system. We have had some success in defining a nonpeptide mimetic analogue of the SchistoFLRFamide receptors on locust oviduct.[2] Benzethonium chloride (Bztc) is an agonist of the PDVDHVFLRFamide receptors found on locust oviducts. Bztc mimics the physiological effects of PDVDHVFLRFamide in being able to reversibly inhibit proctolin-induced contractions of locust oviduct. Similarly, Bztc reversibly inhibits neurally evoked contractions and is also able to inhibit spontaneous contractions of locust oviducts. The thresholds for Bztc are higher than those for the peptide, and the dose-response curves are also shifted to the right. In addition, Bztc competitively displaces [^{125}I-labeled Y^1]DVDHVFL-RFamide binding to both high- and low-affinity receptors of membrane preparations of locust oviduct, although, as with biological activity, higher doses are required and the displacement curves are shifted to the right. Bztc is therefore recognized by the binding and activation regions of these receptors. This discovery provides a unique opportunity within insects to finally target a peptide receptor for the development of future pest management strategies; current research revolves around modifications of the Bztc molecule with a view toward optimization and development of more potent analogues.

REFERENCES

1. PEEFF, N. M., I. ORCHARD & A. B. LANGE. 1994. Peptides **15:** 387–392.
2. LANGE, A. B., I. ORCHARD, Z. WANG & R. J. NACHMAN. 1995. PNAS **92:** 9250–9253.
3. LANGE, A. B., Z. WANG, I. ORCHARD & A. N. STARRATT. 1996. Peptides **17:** 375–380.
4. WANG, Z., I. ORCHARD, A. B. LANGE & X. CHEN. 1995. Neuropeptides **28:** 261–266.

FMRFamide-related Peptides in Insects, with Emphasis on the Myosuppressins

IAN ORCHARD,[a,d] B. C. DONLY,[b] M. FUSE,
A. B. LANGE,[a] S. S. TOBE,[a] AND W. G. BENDENA[c]

[a]Department of Zoology
University of Toronto
Toronto, Ontario, Canada M5S 3G5

[b]Pest Management Research Centre
Agriculture and Agri-Food Canada
London, Ontario, Canada N5V 4T3

[c]Department of Biology
Queen's University
Kingston, Ontario, Canada K7L 3N6

In insects, a vast array of FMRFamide-related peptides (FaRPs) have been isolated and shown to be distributed extensively throughout the nervous system, occurring both centrally and peripherally.[1] FaRPs appear to act as neurohormones, as well as neurotransmitters/neuromodulators with direct actions on peripheral tissues such as salivary glands, accessory glands, and muscle. These FaRPs have a variety of modulatory actions upon contraction of a variety of muscle types, including skeletal, heart, and visceral muscle.[2] In particular, a subfamily of decapeptide FaRPs, also known as myosuppressins, with the sequence XDVXHXFLRFamide has been found in several insect species and shown to inhibit spontaneous contractions of visceral muscles.[3] A member of this subfamily, PDVDHVFLRFamide (SchistoFLRFamide), is present in the nervous system of the locusts, *Schistocerca gregaria* and *Locusta migratoria*.[4,5] In addition, a second peptide ADVGHVFLRF-amide has also been sequenced from *Locusta*.[5]

In the oviduct of *Locusta migratoria* the two *Locusta* myosuppressins act as neuromodulators, inhibiting both spontaneous and induced muscle contractions.[5] Thus, they are able to inhibit proctolin-induced, glutamate-induced, and high potassium-induced contractions. Recent evidence indicates that PDVDHVFLRF-amide closes or blocks voltage-gated and some ligand-gated calcium channels in the plasma membrane, or buffers intracellular calcium.[6] These actions appear to represent true physiological roles for these myosuppressins in locust oviducts because evidence indicates (1) the presence of both peptides within the innervation of locust oviducts, (2) the presence of FMRFamide-like immunoreactivity in elec-tron-dense round granules contained in nerve endings on locust oviduct, and (3) the presence of binding sites that probably represent two receptors for PDVDHV-FLRFamide, a high-affinity and a low-affinity receptor.[7,8] Structure-activity studies using locust oviducts have identified several structural features of the neuropep-tides which are important for biological activity, and competitive binding studies on the same tissue have identified features important for binding.[9] Thus, inhibitory biological activity is retained by the N-terminal truncated peptide HVFLRFamide,

[d] E-mail: orchard@zoo.utoronto.ca

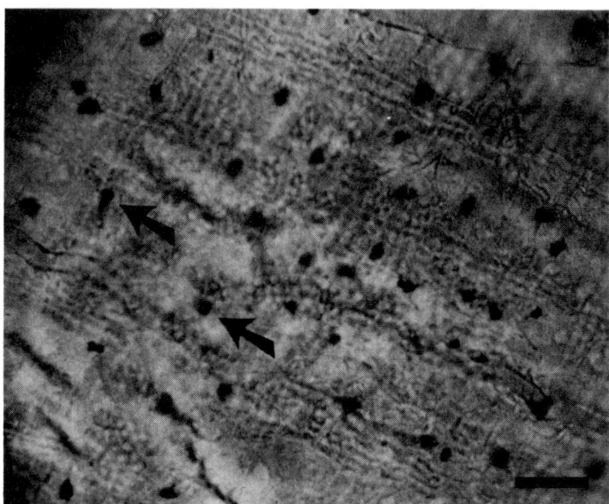

FIGURE 1. *In situ* hybridization of whole-mount preparation of midgut from *Diploptera punctata*. Positive cell bodies (*arrows*) after hybridization with DIG-labeled *Diploptera* LMS cDNA. Scale bar: 50 μm.

but totally lost in VFLRFamide. Indeed, activity reversal (minor stimulatory activity) is seen with this latter peptide. In contrast, VFLRFamide is the minimum sequence required for binding to the receptors with an affinity comparable to that of the parent compound. The His residue, therefore, is critical for activation of the receptors resulting in inhibition of muscle contraction, but is not, in itself, critical for binding. We are currently modeling the key features of the His residue that results in the inhibitory bioactivity.

More recently, we have detected the presence of an active peptide similar to another member of the subfamily of myosuppressins, leucomyosuppressin (pQDV-DHVFLRFamide, LMS), in the brain of the cockroach *Diploptera punctata*.[10] Using HPLC separation, radioimmunoassay, and subsequent bioassay, we have detected a peptide from *Diploptera* brain which co-elutes with authentic LMS. In addition, we have isolated a cDNA encoding the precursor for this peptide from cDNA libraries representing *D. punctata* brain RNA. The cDNA sequence contains an open reading frame that upon translation would result in a prepropolypeptide of 96 amino acids. Proteolytic cleavage of the predicted precursor could result in several peptides, including a 10 amino acid C-terminal peptide which, upon modification of the NH_2 and COOH-terminal amino acids, would result in a peptide identical to LMS. No other RFamide products are predicted to be processed from the 96 amino acid precursor. Southern blot analysis indicates that the gene is present in the *D. punctata* genome in a single copy. Northern blot analysis shows that the gene is predominantly expressed as a 3.8 kb mRNA in cockroach brain. Study of the expression of the *Diploptera* LMS gene in the brain, using *in situ* hybridization, indicates that expression occurs primarily in the pars intercerebralis of the protocerebrum, a region showing abundant FMRFamide-like immunoreactive neurosecretory cells. Elsewhere, expression occurs in the subesophageal ganglion and cells of the midgut (see FIG. 1). These apparent endocrine cells of the midgut are distributed unequally throughout the midgut and constitute only a fraction of the total number of FMRFamide-like immunoreactive endocrine cells. FMRFamide-like immunoreactive nerve processes are also seen extending over the midgut, although the identity of the FaRPs involved is unknown. Recent studies have indicated that leucomyosuppressin can inhibit contractions of the circular muscles of the midgut of *Diploptera* and *Locusta* (unpublished observations).

We are continuing an examination into the physiological roles of these myosuppressins in both locust and cockroaches, and furthermore are examining their relationships with the other extended FLRFamides which appear to act as myostimulants.

REFERENCES

1. DUVE, H., A. H. JOHNSEN, J. C. SEWELL, A. G. SCOTT, I. ORCHARD, J. F. REHFELD & A. THORPE. 1993. PNAS **89:** 2326–2330.
2. LANGE, A. B., N. M. PEEFF & I. ORCHARD. 1994. Peptides **15:** 1089–1094.
3. LANGE, A. B., I. ORCHARD & V. A. TE BRUGGE. 1991. J. Comp. Physiol. A **168:** 383–391.
4. ROBB, S., L. C. PACKMAN & P. D. EVANS. 1989. Biochem. Biophys. Res. Commun. **160:** 850–856.
5. PEEFF, N. M., I. ORCHARD & A. B. LANGE. 1994. Peptides **15:** 387–392.
6. WANG, Z., I. ORCHARD, A. B. LANGE & X. CHEN. 1995. Neuropeptides **28:** 147–155.
7. WANG, Z. & I. ORCHARD. 1995. Cell Tissue Res. **279:** 591–599.
8. WANG, Z., I. ORCHARD, A. B. LANGE & X. CHEN. 1995. Neuropeptides **28:** 261–266.
9. LANGE, A. B., Z. WANG, I. ORCHARD & A. N. STARRATT. 1996. Peptides **17:** 375–380.
10. DONLY, B. C., M. FUSE, I. ORCHARD, S. S. TOBE & W. G. BENDENA. 1996. Insect Biochem. Mol. Biol. **26:** 627–637.

Cloning of a cDNA from Stable Fly which Encodes a Protein with Homology to a *Drosophila* Receptor for Tachykinin-like Peptides

FELIX D. GUERRERO

USDA-ARS Knipling-Bushland
U.S. Livestock Insects Research Lab
2700 Fredericksburg Road
Kerrville, Texas 78028

Tachykinins are members of a family of peptides which act as neurohormones in mammals. These peptide hormones have been implicated in a number of mammalian motor and sensory functions, particularly related to central and peripheral neuron activity and the contraction of smooth muscle.[1] There is evidence for three classes of tachykinin receptors in mammals with variations in amino acid (aa) sequence occurring between species.[2] The classification of receptors is based on interactions with specific natural tachykinins and synthetic agonists which selectively act on only one of the three receptor classes. Tachykinin-like peptides have been purified from invertebrates,[3] and several cDNAs from *Drosophila* have been cloned which code for proteins with structural and biochemical properties similar to mammalian tachykinin receptors.[4,5]

The stable fly, *Stomoxys calcitrans*, is a major hematophagous pest of livestock causing an estimated $400 million annual loss to the United States livestock industry.[6] Current control methodologies include good animal husbandry and insecticide spraying. Ongoing studies seek alternative ways of controlling infestations as concerns increase over the use of insecticides. As part of these studies, we sought to clone the receptor for tachykinin peptides in the stable fly to study the molecular interaction between the receptor and the peptide.

A stable fly cDNA library was synthesized in a λgt22 cloning vector using polyA$^+$ RNA from whole adult flies and screened with a *Drosophila melanogaster* DTKR tachykinin receptor DNA probe.[4] One 4118 bp clone, designated STKR, was purified and sequenced and found to have an open reading frame which encoded a 678 aa protein with significant homology to the *D. melanogaster* DTKR open reading frame. The DTKR cDNA is reported to encode a 519 aa receptor-like protein with homology to vertebrae tachykinin receptors and whose mRNA is expressed in the adult insect's central nervous system and specific subsets of neurons in the developing embryo.[4] In expression studies, DTKR is activated by vertebrate substance P-like peptides. Homology between the STKR open reading frame and reported tachykinin receptors is highest in the seven putative membrane spanning regions of this receptor family. Comparison of the DTKR and STKR open reading frame sequence spanning transmembrane regions I–VII shows 227 of 283 aligned aa are identical (80%), whereas 266 of 283 (94%) are either identical or conservative substitutions as determined by the pam250S scoring matrix analysis of MacVector Sequence Analysis Software (International Biotechnologies Inc.,

New Haven, CT). Overall, the complete alignment of DTKR with a 526 aa region of STKR resulted in 70% identical aa and 77% identical or conservative aa substitutions. Characterization and expression studies of the *Stomoxys* clone SFT6 are underway to determine whether this clone codes for a functional tachykinin receptor. Preliminary experiments with RT-PCR have indicated that the receptor mRNA is expressed in all life stages of the stable fly, with lesser amounts seen in developing embryos. The gut region of adults appears to contain significant amounts of STKR mRNA, although all of the results from the RT-PCR await verification by Northern blotting and whole-mount RNA *in situ* hybridization experiments.

REFERENCES

1. MAGGI, C. 1995. The mammalian tachykinin receptors. Gen. Pharmacol. **26(5):** 911–944.
2. WATLING, K., S. GUARD, S. BOYLE, A. McKNIGHT & G. WOODRUFF. 1994. Species variants of tachykinin receptor types. Biochem. Soc. Trans. **22:** 118–122.
3. SCHOOFS, L., J. VANDEN BROECK & A. DE LOOF. 1993. The myotropic peptides of *Locusta migratoria*: Structures, distribution, functions and receptors. Insect Biochem. Mol. Biol. **23(8):** 859–881.
4. LI, X., W. WOLFGANG, Y. WU, R. NORTH & M. FORTE. 1991. Cloning, heterologous expression and developmental regulation of a *Drosophila* receptor for tachykinin-like peptides. EMBO J. **10(11):** 3221–3229.
5. MONNIER, D., J. COLAS, P. ROSAY, R. HEN, E. BORRELLI & L. MAROTEAUX. 1992. NKD, a developmentally regulated tachykinin receptor in *Drosophila*. J. Biol. Chem. **267(2):** 1298–1302.
6. KUNZ, S., K. MURRELL, G. LAMBERT, L. JAMES & C. TERRILL. 1991. Estimated losses of livestock to pests. *In* CRC Handbook of Pest Management in Agriculture. 2nd edit. D. Pimentel, Ed. Vol. 1: 69–98. CRC Press. Boca Raton, FL.

Novel Tachykinin-related Peptides in the Cockroach Nervous System and Intestine

Structure, Distribution, and Actions[a]

J. ERIC MUREN AND DICK R. NÄSSEL

Department of Zoology
Stockholm University
S-10691 Stockholm, Sweden

Peptides related to the vertebrate tachykinins have been isolated from extracts of the locust brain,[1] whole mosquitos,[2] and whole blowflies.[3] The so-called locusta-tachykinins (LomTK I–IV), culetachykinins (CusTK I–II), and callitachykinins (CavTK I–II) are known to increase the frequency and amplitude of spontaneous contractions of hindgut muscle in the cockroach *Leucophaea maderae*. The endogenous cockroach tachykinins, however, have not been identified until now. Because immunocytochemistry (ICC) with an antiserum raised against LomTK I has indicated an abundance of neurons in the *L. maderae* nervous system and intestine,[4] we decided to isolate the native tachykinin-related peptides.

Using a newly developed radioimmunoassay (RIA) for detection of LomTK immunoreactive material and the *L. maderae* hindgut contraction bioassay[5] to monitor HPLC fractions, we initiated the isolation of tachykinin-related peptides from an acidic extract of 600 midguts of *L. maderae*. Four consecutive HPLC column systems were required for the purification. Five tachykinin-related peptides were isolated and their amino acid sequences determined by Edman degradation as APSGFLGVR, APEESPKRAPSGFLGVR, NGERAPGSKKAPSGFL-GTR, APSGFMGMR, and APAMGFQVGVR. Matrix-assisted laser desorption ionization mass spectrometry (MALDI-MS) and chemical synthesis confirmed these structures and that the peptides were C-terminally amidated (TABLE 1). The five novel peptides were designated *Leucophaea* tachykinin-related peptides I–V (LemTRP I–V). All five LemTRPs are myotropic and induce dose-dependent increases in the frequency of spontaneous contractions and increase the tonus of hindgut muscle *in vitro*. LemTRP I and V, for instance, gave half of maximal responses (ED_{50}) in the hindgut contraction assay at about 5×10^{-10} M concentrations (FIG. 1).

As seen in the TABLE 1, the LemTRPs are structurally related to peptides isolated from the locust, mosquito, and blowfly, but showed a somewhat larger variability in their amino acid sequence (including the carboxy terminus). Surprisingly, one peptide, LemTRP IV, has a sequence that is identical to that of the mosquito peptide CusTK I.[2] An interesting feature of the extended peptides LemTRP II and III is that they contain doublets of basic amino acids (Lys-Arg or Lys-Lys) indicative of endoproteolytic cleavage sites.[6] If cleaved at these sites, the extended LemTRPs could each give rise to a nonapeptide, one corresponding

[a] This work was supported by the Swedish Natural Science Research Council (NFR).

TABLE 1. Amino Acid Sequences of Insect Tachykinins

LemTRP I	APSGFLGVRamide
LemTRP II	APEESP**KR**APSGFLGVRamide
LemTRP III	NGERAPGS**KK**APSGFLGTRamide
LemTRP IV*	APSGFMGMRamide
LemTRP V	APAMGFQGVRamide
LomTK I	GPSGFYGVRamide
LomTK II	APLSGFYGVRamide
LomTK III	APQAGFYGVRamide
LomTK IV	APSLGFHGVRamide
CavTK I	APTAFYGVRamide
CavTK II	GLGNNAFVGVRamide
CusTK I*	APSGFMGMRamide
CusTK II	APYGFTGMRamide

NOTE: Underlined amino acid residues are shared by LemTRP I and other tachykinin-related peptides. Amino acids shown in bold indicate putative endoproteolytic cleavage sites. Peptides marked with asterisk(*) are identical. (From refs. 1–3 and the present paper.)

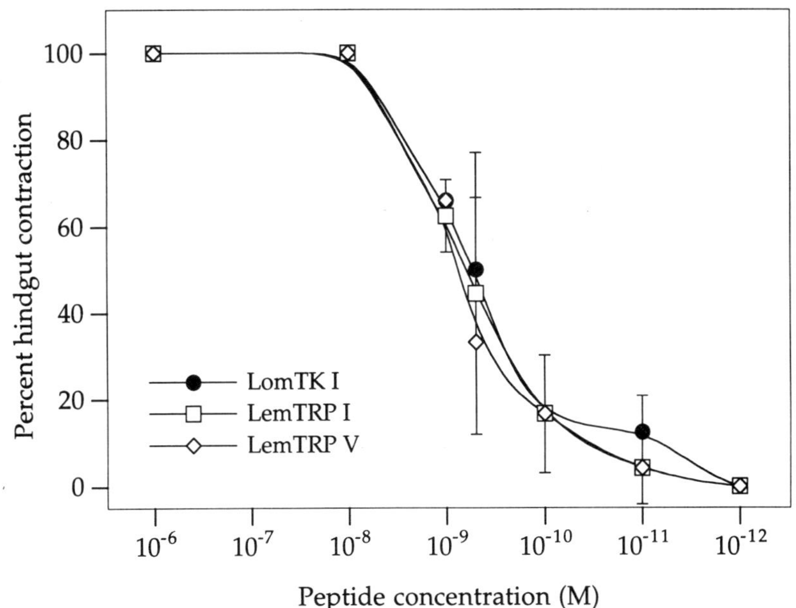

FIGURE 1. Effect of increasing doses of locustatachykinin I (LomTK I) and *Leucophaea* tachykinin-related peptides I and V (LemTRP I and V) on the isolated hindgut of *L. maderae*. Maximum response is shown as 100%. Each point represents mean ± SE of three to four preparations.

to LemTRP I and the other a LemTRP I with a single amino acid substitution. Two N-terminally extended tachykinins, neuropeptide K and neuropeptide γ, with similar Lys-Arg cleavage sites, have been demonstrated in mammals.[7]

In the RIA we showed that the hindgut contains tachykinin-like material and with ICC it was revealed that LomTK-like immunoreactive fibers derived from the terminal abdominal ganglion innervate the muscle layers of the hindgut via the proctodeal nerves. These findings indicate that the LemTRPs may have a physiological role in the regulation of the hindgut contractility. The action(s) of the TRPs in the midgut are not yet known, but with the presence of both endocrine cells and a nerve plexus containing LomTK immunoreactivity one may suggest roles both in control of enzyme release and contractility.

With the same technique as outlined above, we also isolated five tachykinin-related peptides from 1,000 dissected brains of *L. maderae*. Two of these were identical to LemTRP I and V, respectively; the other three (LemTRP VI–VIII) were found to have different sequences and may represent brain-specific isoforms (Muren and Nässel, in preparation). Thus, there are three LemTRP forms found only in midgut, another three forms found only in the brain, and finally two forms common to the two issues.

A combination of RIA and immunocytochemistry (ref. 4; Muren and Nässel, in preparation) revealed an abundance of LemTRP-like peptide in the central nervous system. In the brain all immunoreactivity appeared to be in interneurons, suggesting important roles of LemTRPs in the modulation of neurotransmission. From the ICC it appears that the LemTRPs may have distributed functions; they may be released in many different types of circuits both in sensory, integrative, and motor neuropil.

In summary, the LemTRPs exist in at least eight isoforms in the cockroach *L. maderae*. These peptides may play roles in the regulation of the activity of the different regions of the intestine and function as cotransmitters in the central nervous system.

ACKNOWLEDGMENT

Anne Karlsson is gratefully acknowledged for collection of cockroach tissues.

REFERENCES

1. SCHOOFS, L., J. VANDEN BROEK & A. DE LOOF. 1993. Insect Biochem. Mol. Biol. **23:** 859–881.
2. CLOTTENS, F. L., S. M. MEOLA, G. M. COAST, T. K. HAYES, M. S. WRIGHT, R. J. NACHMAN & G. M. HOLMAN. 1993. Regul. Pept. **49:** 145–157.
3. LUNDQUIST, C. T., F. L. CLOTTENS, G. M. HOLMAN, R. NICHOLS, R. J. NACHMAN & D. R. NÄSSEL. 1994. Peptides **15:** 761–768.
4. MUREN, J. E., C. T. LUNDQUIST & D. R. NÄSSEL. 1995. Phil. Trans. R. Soc. Lond. B **348:** 423–444.
5. HOLMAN, G. M., R. J. NACHMAN, L. SCHOOFS, T. K. HAYES, M. S. WRIGHT & A. DE LOOF. 1991. Insect Biochem. **21:** 107–112.
6. ANDREWS, P. C., K. BRAYTON & J. E. DIXON. 1987. Experientia **43:** 784–790.
7. MacDONALD, M. R., J. TAKEDA, C. M. RICE & J. E. KRAUSE. 1989. J. Biol. Chem. **264:** 15578–15592.

Dromyosuppressin and Drosulfakinin, Two Structurally Related *Drosophila* Neuropeptides, Are Uniquely Expressed in the Adult Central Nervous System[a]

R. NICHOLS,[b–d] J. McCORMICK,[c] AND I. LIM[c]

Departments of Biological Chemistry[b] and Biology[c]
University of Michigan
Ann Arbor, Michigan 48109-1048

INTRODUCTION

Peptides present in the nervous system can act as hormones, transmitters, or modulators of various physiological functions. It is well established that more than one peptide messenger can exist in a single neuron and that neuropeptides can often be grouped into a family based on structure similarity.[1] Although immunochemistry is a valuable technique in visualizing molecules present in a neuron, it is important to take into consideration the presence of structurally related peptides when designing antigens and interpreting immunocytochemical data. Antisera that distinguish between structurally related peptides are important experimental tools to determine expression and suggest function, as well as to analyze mutants and elucidate polypeptide precursor processing.

One family of neuropeptides is structurally related to the molluscan cardioexcitatory peptide FMRFamide[2] via the C-terminus -XRFamide, where X = L or M.[3,4] *Drosophila* myosuppressin (TDVDHVFLRFamide; DMS) and sulfakinin (FDDYGHMRFamide;[e] DSK) have structure similarity to FMRFamide.[5,6] To study DMS and DSK expression, we have generated and characterized DMS- and DSK-specific antisera and performed double-label immunocytochemistry. Our results indicate that DMS and DSK are expressed in different neurons in the adult central nervous system.

MATERIALS AND METHODS

The antigens, TDVDHV-MAP and FDDYGH-MAP, where MAP represents multiple antigenic peptide,[7] were designed to the variant N-terminal sequences of DMS and DSK, respectively. Antisera were raised in New Zealand white rabbits

[a] This work was supported by an NSF grant (IBN No. 9409623) and an AHA/MI grant to R.N.

[d] Address correspondence to R. Nichols, 830 N. University Street, University of Michigan, Ann Arbor, MI 48109-1048. E-mail: nicholsr@umich.edu

[e] In FDDYGHMRFamide, the bold type **Y** represents a sulfated tyrosyl residue.

and analyzed by indirect immunofluorescent analysis of whole-mount third-instar larval central nervous system tissue as previously described.[8] Antisera were purified on peptide affinity columns made by coupling the antigen to Affi-gel 10 (Bio-Rad Labs) according to the manufacturer's specifications.[8] Affinity-purified antisera were characterized by preincubation with FDDYGHMRFamide or TDVDHV-FLRFamide prior to immunocytochemistry.

Double-label immunochemistry was performed using a modified single-label method.[8] Whole-mount tissue preparations were incubated with the first primary antisera for 4–6 h, rinsed in 0.5 M sodium phosphate, pH 7.2, with 0.2% Triton X-100 and 1% sodium azide (PTN), incubated in CY3-conjugated Fab fragment goat anti-rabbit secondary antibody (Jackson ImmunoResearch Labs, West Grove, PA) for 4–6 h, rinsed in PTN for 2 h, incubated in the second primary antisera for 4–6 h, rinsed in PTN for 2 h, and incubated in FITC-conjugated goat anti-rabbit secondary antibody (Sigma, St. Louis, MO) for 4–6 h. Tissue was then extensively washed and prepared for microscopy as previously described.[8] Data, collected with a Bio-Rad MRC600 laser scanning confocal microscope equipped with a Kr-Ar laser attached to a Nikon inverted microscope, were processed with Adobe Photoshop and transferred to Kodak slide film using a Macintosh Quadra 800 and Lasergraphics LFR-X.

RESULTS AND DISCUSSION

Nomenclature used in describing the neuronal staining of DMS and DSK antisera is based on previous publications identifying FMRFamide-like immunoreactive materials.[5,6,9–11] Immunoreactivity was observed bilaterally symmetric to the midline such that reference to one neuron indicates that there were two neurons positioned bilaterally symmetric to one another. Signal intensity was strong and consistent, and no fewer than eight preparations were analyzed.

Antisera to FMRFamide have been used to stain *Drosophila* neural tissue;[5–11] however, given the number of FMRFamide-like peptides and that antisera to FMRFamide recognize the common C-terminal structure, the data are ambiguous and the expression patterns of individual FMRFamide-like peptides cannot be identified. To study the expression of peptides structurally related to FMRFamide, we generated antisera to multiple antigenic peptides designed to the variant N-terminal sequences. We chose to use MAP antigens because of the advantages that they offer including the fact that no carrier molecule is required and that multiple copies of the antigen are present in a single molecule.[7]

To study whether FMRFamide-like peptides coexist we established a double-label immunofluorescence protocol using antisera raised in different animals of the same species host animal. Generation of antisera in the same species can be advantageous in terms of technical aspects such as animal housing and blood collection.

We have reported the individual expression patterns of both DMS and DSK immunoreactive materials during development using single-label immunocytochemistry.[8,12] The neurons stained by each antisera were identified based on position, comparing the staining patterns with those of FMRFamide antisera.[9–11] Although the single-label immunochemical data are informative, they cannot unequivocally answer the question of whether these peptides are expressed in the same neuron or in neurons in close proximity; it would be more definitive to compare DMS and DSK staining patterns within the same preparation.

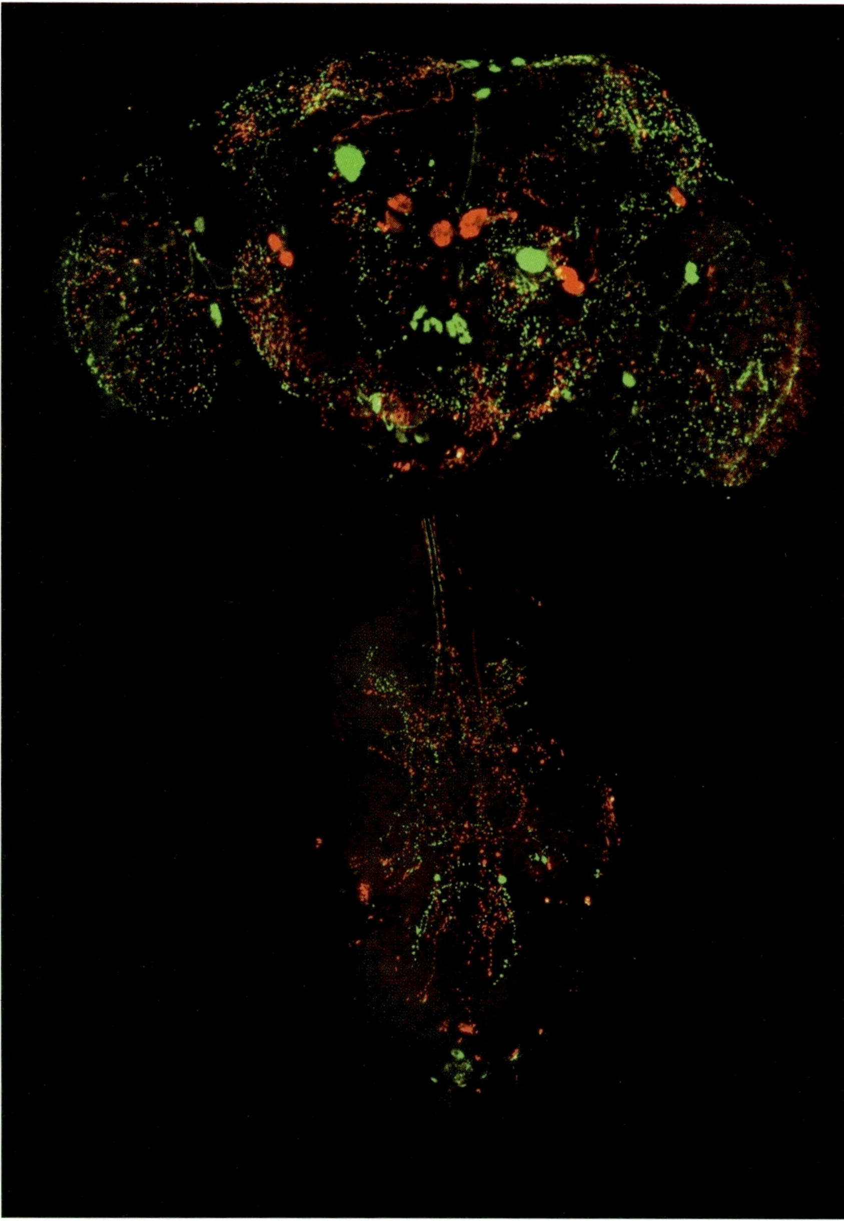

FIGURE 1. Double-label immunofluorescence of DMS and DSK in an adult central nervous system. DMS antisera are recognized by FITC-labeled (*green*) secondary antibody, whereas DSK antisera are recognized by CY3-labeled (*red*) secondary antibody. No neuron is stained by both antisera (*yellow*); some fibers appear yellow because the method of data collection results in overlaying neuronal projections that exist at different focal planes throughout the tissue.

We have previously presented double-label immunofluorescence data describing the expression patterns of DMS and DSK in larval neural tissue.[13] In this paper we describe a more versatile double-label immunofluorescence protocol than used to study expression in larva and determine the DMS and DSK staining patterns in adult neural tissue. Although DMS- and DSK-specific antisera stain several neurons in the adult *Drosophila* central nervous system (FIG. 1), the expression patterns in the adult central nervous system are unique and nonoverlapping. Numerous neurons in close proximity to one another are stained by either DMS- or DSK-specific antisera, for example, DMS-specific antisera stain a MP2 neuron, whereas DSK-specific antisera stain two MP1 neurons; however, no neuron is stained by both antisera, indicating that these structurally related neuropeptides do not coexist. The nonoverlapping expression patterns for DMS and DSK suggest that these peptides have different roles in the central nervous system.

SUMMARY

Drosophila myosuppressin (TDVDHVFLRFamide; DMS) and sulfakinin (FDDYGHMRFamide; DSK) have similar C-terminal structures. To determine the neuronal expression patterns of these structurally related peptides, we have generated DMS- and DSK-specific antisera to multiple antigenic peptides and performed double-label immunochemistry with antisera raised in different animals of the same species host animal. Our data indicate that DMS and DSK staining patterns in the adult central nervous system are unique and nonoverlapping.

REFERENCES

1. HÖKFELT, T., V. R. HOLETS, W. STAINES, B. MEISTER, T. MELANDER, M. SCHALLING, M. SCHULTZBERG, J. FREEDMAN, H. BJÖRKLUND, L. OLSON, B. LINDH, L.-G. ELFVIN, J. M. LUNDBERG, J. A. LINDGREN, B. SAMUELSSON, B. PERNOW, L. TERENIUS, C. POST, B. EVERITT & M. GOLDSTEIN. 1986. *In* Progress in Brain Research. T. Hökfelt, K. Fuxe & B. Pernow, Eds. Vol. 68: 33–70. Elsevier Science Ltd., Cambridge, UK.
2. PRICE, D. A. & M. J. GREENBERG. 1977. Science **197:** 670–672.
3. RAFFA, R. B. 1988. Peptides **9:** 915–922.
4. PRICE, D. A. & M. J. GREENBERG. 1989. Biol. Bull. **177:** 198–205.
5. NICHOLS, R. 1992. Mol. Cell. Neurosci. **3:** 342–347.
6. NICHOLS, R. 1992. J. Mol. Neurosci. **3:** 213–218.
7. POSNETT, D. N. & J. P. TAM. 1989. *In* Methods in Enzymology. J. J. Langone, Ed. Vol. 179: 739–746. Academic Press. New York.
8. MCCORMICK, J. & R. NICHOLS. 1993. J. Comp. Neurol. **338:** 272–288.
9. WHITE, K., P. HURTEAU & P. PUSNAL. 1986. J. Comp. Neurol. **247:** 430–438.
10. CHIN, A. C., E. R. REYNOLDS & R. SCHELLER. 1990. DNA Cell Biol. **9:** 263–271.
11. SCHNEIDER, L. E., M. A. O'BRIEN & P. H. TAGHERT. 1991. J. Comp. Neurol. **304:** 608–622.
12. NICHOLS, R. & I. LIM. 1996. Cell Tissue Res. **283:** 107–116.
13. TIBBETTS, M. F. & R. NICHOLS. 1993. Neuropeptides **24:** 321–325.

Occurrence and Diversity
of Neuropeptides from the Crustacean
Hyperglycemic Hormone Family
in Arthropods

A Short Review

D. SOYEZ[a]

Laboratoire de Biochimie
Equipe "Signaux et Régulations Endocrines"
EP119 CNRS
Ecole Normale Supérieure
46 rue d'Ulm
75230 Paris cedex 05, France

CRUSTACEAN HYPERGLYCEMIC HORMONE
FAMILY IN CRUSTACEANS

In crustaceans, it is well known that major physiological processes such as homeostasis, growth, and reproduction are regulated by neurohormones originating from the major neuroendocrine center, the X-organ-sinus gland complex, which is located, in decapods, within the eyestalks (see ref. 1 for review). These neuropeptides have recently been shown to be structurally related, thus defining a new peptide family designated as the crustacean hyperglycemic hormone (cHH) family, with cHH as the first described and the major neurohormone elaborated by the X-organ-sinus gland complex.[2] Until now, hyperglycemic neuropeptides had been isolated and sequenced from a number of different crustacean species (FIG. 1). cHHs are 72–73 aa residue peptides with six cysteyl residues forming three disulfide bridges in conserved positions. In addition, the primary structure of the cHHs is highly conserved among the different species with a percentage of homology above 60%. Other members of the cHH family are 75–78 aa residue inhibitory neuropeptides involved in the regulation of molting (molt-inhibiting hormone, MIH) and reproduction (gonad-inhibiting hormone, GIH; also called vitellogenesis-inhibiting hormone, VIH) (FIG. 1). MIH has been fully characterized in several species of crabs, in a crayfish, and in a shrimp, whereas VIH has been sequenced and cloned in the American lobster only. MIHs and VIH share more than 50% homology, but this value is lower than 30% when both are compared to cHHs. Major differences occur mainly at both ends of the neuropeptide molecules because the percentage of homology between cHHs and MIHs/VIH reaches 50%, with a conserved location of five cysteyl residues, when the central part (positions 23–52) of the molecules is considered. It was recently demonstrated that the mandibular organs of the crab *Cancer pagurus*, which elaborate juvenile hormone-like compounds and are therefore considered as functional equivalents of insect

[a] E-mail: Soyez@wotan.ens.fr

FIGURE 1. Alignment of amino acid sequences of the peptides from the cHH family. Identical residues (compared to Hoa cHHA) are shaded, with darker boxes for cysteyl residues. Hoa: *Homarus americanus* (cHH;[5] VIH[20,21]); Orl: *Orconectes limosus*;[22] Prb: *Procambarus bouvieri*;[23] Prc: *Procambarus clarkii* (cHH;[11] MIH[24]); Cam: *Carcinus maenas* (cHH;[25] MIH[26]); Arv: *Armadillidium vulgare*;[27] Pej: *Penaeus japonicus*;[7] Pev: *Penaeus vannamei*;[28] Cap: *Cancer pagurus*;[3] Scg: *Schistocerca gregaria*;[17] LMWP: low molecular weight protein.[18]

corpora allata, are negatively regulated by a neurohormone (mandibular organ-inhibiting hormone, MOIH) closely related to MIH and VIH.[3]

Thus, the cHH family appears to include two structural subgroups: the 72–73 aa cHHs on the one hand and the 77–78 aa MIH/VIH/MOIH on the other. The two subgroups present striking differences at the precursor level: in the cHH precursor, the hormone is separated from the signal peptide by a 33–38 aa peptide (cHH precursor-related peptide, CPRP), whereas this CPRP or equivalent is absent in the precursor of Cam MIH or Hoa VIH. In this case, the signal peptide directly flanks the hormone, without a classical dibasic aa cleavage site (see ref. 4 for review).

Peptides belonging to the cHH family are often polymorphic. This polymorphism is well documented for cHHs and may result from changes in the amino acid sequence. This phenomenon was first evidenced in the American lobster where two cHH variants differing by six amino acid residues were characterized.[5] In the crayfish *Procambarus bouvieri*, three cHH variants with different biological activities (cHH, VIH, MIH) have been isolated. Partial amino acid sequences demonstrate a few amino acid changes between the different variants.[6] A greater number of cHH variants seems to be present in penaeid shrimps, classically considered as ancestral decapods. In *Penaeus japonicus* five cHH-related peptides have been purified. Their partial amino acid sequence reveals 60% homology.[7] Similar observations were made in a closely related species, *P. duorarum* (T. K. Hayes and M. Tom, personal communication).

The functional significance of such a peptide diversity is unknown. Considering the large spectra of physiological roles devoted to the cHH-family peptides, one may speculate that each peptide displays a specific function. However, increasing evidence indicates that most of these peptides are multifunctional, as exemplified by recent results obtained in penaeid shrimps where the same peptide is both inhibitor of the steroidogenesis in molting gland (MIH activity) and hyperglycemic (cHH activity).[8,9]

Another type of cHH polymorphism results from the isomerization of one amino acid residue (Phe^3 in lobster or crayfish cHHs) from the L- to the D-configuration.[10–12] This phenomenon had been described previously in only few groups of eucaryotic organisms, but is now considered as an authentic post-translational event.[13] Such a modification in the peptide structure cannot be detected by classical analytical biochemistry or molecular biology methods, and it is very likely that its occurrence is largely underestimated. In lobster or crayfish, both peptide isomers are present in the neurohemal organ (conformational polymorphism). This structural change is correlated with modifications in the biological activities of the neuropeptides: in addition to their hyperglycemic effect which is evoked with a retarded kinetics, cHH isoforms containing a D-amino acid have been demonstrated in different crayfish species to possess MIH,[11,14] or osmoregulatory[15] bioactivity. In conclusion, peptides from the cHH family in crustaceans are structurally and functionally diversified. This diversification may occur at the gene level, by a classical duplication/mutation process or/and later at the precursor (or mature peptide) level by the nonconventional post-translational isomerization mechanism, which in fact represents an efficient mode of hormonal diversification.

cHH-RELATED PEPTIDES IN OTHER ARTHROPOD GROUPS

Evidence for the presence of cHH-related peptides besides crustaceans has been obtained in different arthropod groups including insects, arachnids, and myriapods. It has been shown in the locust *Schistocerca gregaria* that ion and fluid transport in the ileum are regulated by a neuropeptide originating from the corpora cardiaca.[16] The amino acid sequence of this Scg ITP was determined[17] (FIG. 1), and demonstrates 42% homology with the cHH from the lobster *H. americanus*. A peptide with similar secondary structure and the same disulfide pairing as the crab cHH has been purified and sequenced from the venom of the black widow spider *Latrodectus mactans* (FIG. 1).[18] However, no biological activities could be attributed to this peptide which in addition does not show any hyperglycemic effect when injected into crab, crayfish or penaeid shrimps (unpublished observations).

With the use of specific antisera developed against lobster *H. americanus* cHH, the presence of cHH-like molecules has been demonstrated by immunohisto-chemistry in neurohemal organs of the myriapod centipede *Lithobius forficatus* and in the scorpion *Euscorpius carpathicus*.[19] In this last species, a strong immuno-noreaction is present in the neurosecretory endings of the coxal plexus, associated with the coxal gland (R. Stockmann, personal communication). It is noteworthy that this gland is responsible for ion and fluid excretion. Therefore, cHH-related peptides in locust and in scorpion could be primarily involved in the regulation of ion and fluid transport. Although a major target of cHHs in crustaceans is the regulation of blood sugar level, an action of cHH isoforms on osmoregulation has been reported, as mentioned earlier.

The presence of structurally related peptides in crustaceans and insects has already been described for the RPCH/AKH[b] and PDH[b] families (see ref. 2 for review). Structure conservation does not necessarily imply function conservation. The function of a molecule relies more on tissues where specific receptors are expressed than on the structure of the molecule by itself. Considering the wide range of biological effects of the peptides belonging to the cHH family, it is expected that other members will emerge as key physiological regulators in the different arthropod groups, and possibly also outside arthropods.

ACKNOWLEDGMENT

Many thanks to Prof. René Lafont for his helpful comments on this manuscript.

REFERENCES

1. KLEINHOLZ, L. & R. KELLER. 1979. *In* Hormones and Evolution. E. J. W. Barrington, Ed. Vol. **1:** 159–213. Academic Press. New York.
2. KELLER, R. 1992. Experientia **48:** 439–448.
3. WAINWRIGHT, G., S. WEBSTER, M. C. WILKINSON, J. S. CHUNG & H. H. REES. 1996. J. Biol. Chem. **271:** 12749–12754.
4. DE KLEIJN, D. & F. VAN HERP. 1995. Comp. Biochem. Physiol. **112B:** 573–579.
5. TENSEN, C. P., D. P. V. DE KLEIJN & F. VAN HERP. 1991. Eur. J. Biochem. **200:** 103–106.
6. HUBERMAN, A., M. AGUILAR & L. S. QUACKENBUSH. 1995. Aquaculture **135:** 149–160.
7. YANG, W.-J., K. AIDA & H. NAGASAWA. 1995. Aquaculture **135:** 205–212.

[b] RPCH, red pigment concentrating hormone; AKH, adipokinetic hormone; PDH, pigment dispersing hormone.

8. SEFIANI, M., J. P. LE CAER & D. SOYEZ. 1996. Gen. Comp. Endocrinol. **103:** 41–53.
9. YANG, W. J., K. AIDA, A. TERAUCHI, H. SONOBE & H. NAGASAWA. 1996. Peptides **17:** 197–202.
10. SOYEZ, D., F. VAN HERP, J. ROSSIER, J. P. LE CAER, C. P. TENSEN & R. LAFONT. 1994. J. Biol. Chem. **269:** 18295–18298.
11. YASUDA, A., Y. YASUDA, T. FUJITA & Y. NAYA. 1994. Gen. Comp. Endocrinol. **95:** 387–398.
12. AGUILAR, M., D. SOYEZ, R. FALCHETTO, D. ARNOTT, J. SHABANOWITZ, D. HUNT & A. HUBERMAN. 1995. Peptides **16:** 1375–1383.
13. KREIL, G. 1994. Science **266:** 996–997.
14. VON GLISCYNSKI, U. 1994. PhD thesis. Rheinische Friedrich-Wilhelms-Universität Bonn. Germany.
15. CHARMANTIER-DAURES, M., G. CHARMANTIER, K. P. C. JANSSEN, D. E. AIKEN & F. VAN HERP. 1994. Gen. Comp. Endocrinol. **94:** 281–293.
16. AUDSLEY, N., C. MCINTOSH & J. E. PHILLIPS. 1992. J. Exp. Biol. **173:** 261–274.
17. MEREDITH, J., M. RING, A. MACINS, J. MARSCHALL, D. CHENG, D. THEILMAN, H. W. BROCK & J. E. PHILLIPS. 1996. J. Exp. Biol. **199:** 1053–1061.
18. GASPARINI, S., N. KIYATKIN, P. DREVET, J. P. BOULAIN, F. TACNET, P. RIPOCHE, E. FOREST, E. GRISHIN & A. MÉNEZ. 1994. J. Biol. Chem. **31:** 19803–19809.
19. LAVERDURE, A. M., C. CARETTE-DESMOUCELLES, M. BREUZET & M. DESCAMPS. 1994. Neuroscience **60:** 569–579.
20. SOYEZ, D., J. P. LE CAER, P. Y. NOEL & J. ROSSIER. 1991. Neuropeptides **20:** 25–32.
21. DE KLEIJN, D. P. V., F. SLEUTELS, G. MARTENS & F. VAN HERP. 1994. FEBS Lett. **353:** 255–258.
22. KEGEL, G., B. REICHWEIN, C. P. TENSEN & R. KELLER. 1991. Peptides **12:** 909–913.
23. HUBERMAN, A., M. B. AGUILAR, K. BREW, J. SHABANOWITZ & D. HUNT. 1993. Peptides **14:** 7–16.
24. NAGASAWA, H., W.-J. YANG, H. SHIMIZU, K. AIDA, H. TSUTSUMI, A. TERAUCHI & H. SONOBE. 1996. Biosci. Biotechnol. Biochem. **60:** 554–556.
25. KEGEL, G., B. REICHWEIN, S. WEESE, G. GAUS, J. PETER-KATALINIC & R. KELLER. 1989. FEBS Lett. **255:** 10–14.
26. WEBSTER, S. 1991. Proc. R. Soc. Lond. Biol. **244:** 247–252.
27. MARTIN, G., O. SOROKINE & A. VAN DORSSELAER. 1993. Eur. J. Biochem. **211:** 601–607.
28. SUN, P. S. 1994. Mol. Marine Biol. Biotechnol. **3:** 1–6.

The Regulation of Postfeeding Diuresis in the Migratory Locust, *Locusta migratoria*[a]

G. M. COAST,[b,d] N. AUDSLEY,[c]
AND G. J. GOLDSWORTHY[b]

[b]*Department of Biology*
Birkbeck College
Malet Street
London WC1E 7HX, United Kingdom

[c]*School of Biological Sciences*
University of Manchester
Oxford Road
Manchester M13 9PT, United Kingdom

Previous studies have shown that a diuretic hormone is synthesized by neurosecretory cells in the pars intercerebralis region of the locust brain, and is released from the storage lobe of the corpora cardiaca in response to feeding.[1] The hormone stimulates primary urine secretion by the Malpighian tubules,[1] causing an increase in fecal water loss.[2] The diuretic hormone is most likely a 46 residue corticotropin-releasing factor (CRF)-related peptide (*Locusta* diuretic hormone; *Locusta*-DH) identified in the migratory locust, *Locusta migratoria*.[3] The peptide has been immunolocalized in neurosecretory cells of the pars intercerebralis, axons from which project into the storage lobe of the corpora cardiaca.[4] It is a powerful stimulant of Malpighian tubule fluid secretion *in vitro*, and increases the rate of amaranth clearance from the hemolymph of intact starved locusts, thereby mimicking events associated with postfeeding diuresis.[1,5] Crucially, immunization of locusts with the anti-peptide antibodies blocks postfeeding diuresis.[5]

To investigate further the role of *Locusta*-DH in the control of postfeeding diuresis, a radioimmunoassay (RIA) was developed using a [^{125}I]-labeled tyrosine extended analog, [Tyr$_o$]-*Locusta*-DH. The RIA had a detection limit of 5 fmol/100 μL (50 pM), and was sensitive enough to measure circulating levels of diuretic hormone in hemolymph samples (25 μL) taken from individual insects. The assay showed no cross-reactivity with unrelated peptides, and reversed-phase HPLC analysis established that approximately 60% of the total immunoreactive material in locust hemolymph co-eluted with the synthetic peptide and was attributable to biologically active *Locusta*-DH.

Locusta-DH-like immunoreactive material was detected throughout the central nervous system, confirming the results of a previous study,[4] with the highest amounts being in the corpora cardiaca. A detailed study was made of the *Locusta*-DH content of corpora cardiaca from fifth instar and adult insects. The amount

[a] This work was taken in part from a paper presented at a satellite meeting on Insect Neuropeptides at the Seventeenth Annual Neuropeptide Conference, February 1–6, 1996, Breckenridge, Colorado.
[d] E-mail: g.coast@biol.bbk.ac.uk

of immunoreactive material in the corpora cardiaca was low (318 ± 56 fmol/gland pair) at the beginning of the fifth instar, but increased to 400–500 fmol/gland pair from days 3 through 7, when food intake was highest.[6] A further increase in the amount of *Locusta*-DH occurred before the end of the instar, reaching a maximum (1317 ± 83 fmol/gland pair) on day 8. This appears to coincide with the cessation of feeding[6] and could result from the continued transport of peptide into the corpora cardiaca at a time when no diuretic hormone is required to be released. In the adult, the amount of *Locusta*-DH-like immunoreactive material stored in the corpora cardiaca increased from 1063 ± 234 fmol/gland pair at the beginning of the instar to reach a maximum of 6378 ± 711 fmol/gland pair at 3 weeks. No consistent difference was found in the *Locusta*-DH content of corpora cardiaca from males and females, nor between fed and starved insects.

Corpora cardiaca dissected from 12-day-old adult locusts were used to investigate release of *Locusta*-DH immunoreactive material *in vitro*. Between 1 and 2% of the stored peptide, approximately 60 fmol, was released in 5 min from glands incubated in a high potassium (110 mM) saline containing 2 mM calcium. No peptide was released in the absence of external calcium. The amount of peptide released plateaus after about 30 s in high-potassium saline but, after a brief (1 min) repolarization in low potassium (9 mM) saline, a subsequent depolarization induced the release of more peptide.

After making allowances for peptide recovery and for the presence of immunoreactive material that did not co-elute with the synthetic peptide on reversed-phase HPLC, the concentration of *Locusta*-DH in hemolymph from locusts that had been starved overnight was 215 ± 30 pM. This was increased fivefold (1.49 ± 0.12 nM) in locusts that had been fed for 30 min. To follow the time course for the release of diuretic hormone, locusts that had previously been starved overnight were given access to fresh wheat grass, and the duration of feeding noted. Only after 5 min of continuous feeding was the hemolymph titer of *Locusta*-DH consistently elevated, and the highest concentrations (about 2.5 nM) were found in insects that fed for 15–16 min, approximately the maximum duration of continuous feeding in locusts.[6]

Circulating levels of *Locusta*-DH in fed insects would cause only partial stimulation of Malpighian tubule fluid secretion. Maximal stimulation may require a second peptide, locustakinin, that acts synergistically with *Locusta*-DH.[7] Nothing is known of the circulating levels of locustakinin in fed or starved locusts, but in the presence of just 0.05 nM locustakinin the response of 1 nM *Locusta*-DH is maximal.[71] The peptides are co-localized in abdominal ganglion neurosecretory cells and may be co-released from associated neurohemal structures.[8] However, the immediate stimulus for the increase in primary urine production by fed insects is the release of *Locusta*-DH, because anti-peptide antibodies completely block postfeeding diuresis.[5]

REFERENCES

1. Mordue, W. 1969. Hormonal control of Malpighian tube and rectal function in the desert locust, *Schistocerca gregaria*. J. Insect Physiol. **15:** 273–285.
2. Norris, M. J. 1961. Group effects on feeding in adult males of the desert locust, *Schistocerca gregaria* (Forsk.), in relation to sexual maturation. Bull. Entomol. Res. **51:** 731–753.
3. Kay, I., C. H. Wheeler, G. M. Coast, O. Cusinato, M. Patel & G. J. Goldsworthy. 1991. Characterization of a diuretic peptide from *Locusta migratoria*. Biol. Chem. Hoppe-Seyler **372:** 929–934.

4. PATEL, M. J-S. CHUNG, I. KAY, A. I. MALLET, C. R. GIBBON, K. S. J. THOMPSON, J. P. BACON & G. M. COAST. 1994. Localization of *Locusta*-DP in locust central nervous system and haemolymph satisfies initial hormonal criteria. Peptides **15:** 591–602.
5. PATEL, M., T. K. HAYES & G. M. COAST. 1995. Evidence for the hormonal function of a CRF-related diuretic peptide (*Locusta*-DP) in *Locusta migratoria*. J. Exp. Biol. **198:** 793–804.
6. BERNAYS, E. A. 1985. Regulation of feeding behaviour. *In* Insect Physiology, Biochemistry and Pharmacology. G. A. Kerkut & L. I. Gilbert, Eds. Vol. 4: 1–32. Pergamon Press. Oxford, UK.
7. COAST, G. M. 1995. Synergism between diuretic peptides controlling ion and fluid transport in insect Malpighian tubules. Regul. Pept. **57:** 283–296.
8. THOMPSON, K. S. J., R. C. RAYNE, C. R. GIBBON, S. T. MAY, M. PATEL, G. M. COAST & J. P. BACON. 1995. Cellular co-localization of diuretic peptides in locusts: A potent control mechanism. Peptides **16:** 95–104.

Schistostatins

L. SCHOOFS,[a,c] D. VEELAERT,[a] J. VANDEN BROECK,[a]
S. S. TOBE,[b] AND A. DE LOOF[a]

[a]Zoological Institute
Naamsestraat 59
B-3000 Leuven, Belgium

[b]Department of Zoology
University of Toronto
Toronto, Ontario, M5S 3G5, Canada

In adult locusts, juvenile hormone is required for normal development as shown by experiments in which the corpora allata were surgically removed or destroyed by precocene treatment.[1] When animals lacking corpora allata are treated with synthetic juvenile hormone, reproduction becomes normal again. Although evidence indicates that neurosecretory cells are involved in the control of JH biosynthesis in locusts, the nature of the factors involved has as yet not been determined. In the desert locust, *Schistocerca gregaria*, the juvenile hormone biosynthesis is variable and the rate of synthesis is very low. Therefore, the locust itself is not an ideal animal to use in a bioassay to identify the factors that inhibit juvenile hormone biosynthesis by the corpora allata, the allatostatins. Allatostatins have been isolated from the cockroaches *Diploptera punctata*, *Periplaneta americana*, *Blattella germanica*, from the blowfly *Calliphora vomitoria*, from the moth *Manduca sexta*, and from the cricket *Gryllus bimaculatus*.[2–4] To isolate the allatostatins in cockroaches, a very reliable bioassay has been used, which was developed by Tobe and Clark.[5] Using an antiserum directed against *D. punctata* allatostatin-2 (anti-Dip-AST-2), we demonstrated the presence of allatostatin-like peptides in the central nervous system of *S. gregaria*. Strongly immunoreactive cells were stained in the pars lateralis of the brain with axons extending to and arborizing in the corpus cardiacum and the corpora allata.[6] By means of HPLC in combination with a Dip-AST-2 radioimmunoassay, nine immunoreactive peptides were isolated from a methanolic extract of 7000 *S. gregaria* brains.[7] From some of the peptides, only a few pmoles could be isolated. A combined approach of Edman degradation sequencing and Maldi-TOF Post-source Decay mass spectrometry analyses was used to identify the peptides. The nine schistostatins (Scg-ASTs) all have the Tyr-Xaa-Phe-Gly-Leu-amide C-terminal pentamer in common, which is the core sequence for binding to the receptor on the corpora allata in *D. punctata*. One schistostatin, Scg-AST-2,[11–18] is a degradation product of Scg-AST-2 (the most potent schistostatin), which contains a dibasic cleavage site. In contrast to the other allatostatins, this peptide does not inhibit juvenile hormone biosynthesis by corpora allata in *D. punctata*, although it has the same C-terminal pentamer sequence. Schistostatin-2[12–18] may function as an antagonist by binding to the receptor without displaying allatostatic activity. In *Schistocerca*, the schistostatins inhibit the spontaneous contractions of the lateral oviducts, but have no effect on juvenile hormone biosynthesis by the corpora allata. No allatostatin-like immuno-

[c] E-mail: liliane.schoofs@bio.kuleuven.ac.be

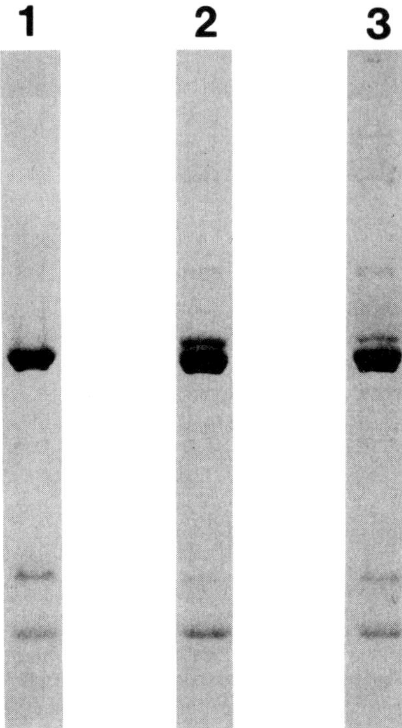

FIGURE 1. SDS-PAGE of hemolymph samples of *S. gregaria*. One microliter of hemolymph of *S. gregaria* injected every 4 h for 24 h was analyzed by SDS-PAGE. **Lane 1,** 1 μL of hemolymph from animals injected with 2 μL 0.9% NaCl (control); **lane 2,** 1 μL of hemolymph from animals injected with an equimolar cocktail of schistostatin-1 to -10 (containing 10^{-4} M of each schistostatin); **lane 3,** 1 μL of hemolymph from animals injected with an equimolar cocktail of schistostatin-1 to -10 (containing 10^{-7} M of each schistostatin).

reactivity could be detected in nerves innervating the oviduct muscle, indicating that if allatostatins act on the oviduct *in vivo*, they do so by acting as neurohormones rather than as neurotransmitters.

The pre-pro-allatostatin precursor encodes 10 schistostatins and differs from that of the cockroach pre-pro-allatostatins[8] in size, sequence, and organization.[9] It contains a lower number of peptides (10 versus 13 or 14 in cockroaches), which

TABLE 1. *In vitro* Half-Life ± SEM of the Different Schistostatins

Name	$t_{1/2}$ ± SEM (min)
Scg-AST-1	—
Scg-AST-2	120.6 ± 9.4
Scg-AST-2[11–18]	69.1 ± 4.7
Scg-AST-3	78.6 ± 5.5
Scg-AST-4	68.7 ± 6.5
Scg-AST-5	19.8 ± 0.6
Scg-AST-6	98.4 ± 5.3
Scg-AST-7	68.0 ± 2.5
Scg-AST-8	62.5 ± 3.8
Scg-AST-9	313.4 ± 37.2
Scg-AST-10	60.0 ± 4.1

are interrupted only once by an acidic spacer region (versus four times in *D. punctata* and *P. americana*). This suggests a significant evolutionary difference between cockroaches and locusts. Two peptides encoded in the precursor cDNA were not purified from the locust brain extract.[9] It is possible that these were not recognized to the same extent as the other peptides in the RIA. Scg-AST-1 indeed has a C-terminal Val residue instead of Leu. Scg-AST-9 is more hydrophilic than the other schistostatins and has an internal dibasic cleavage site.

The schistostatin gene is expressed in the central nervous system and in the gut, but not in gonads, Malpighian tubules, and skeletal muscle as was shown by Northern blot analysis. The occurrence of schistostatins in intrinsic endocrine cells of the locust midgut was confirmed by means of immunocytochemistry. Like the cockroach and blowfly ASTs,[10,11] the schistostatins have to be considered as an interesting invertebrate example of brain/gut peptides.

Schistostatins were also detected in the hemolymph. The schistostatin titer is high in the fifth instar larvae and 3-day-old adult locusts. Preliminary experiments indicated that some schistostatins show more resistance against degradation *in vitro* by hemolymph than others (TABLE 1). Scg-AST-9 seems to be the most stable one. The enzymes that degrade the allatostatins are as yet unknown. Injection of a cocktail of schistostatins in the hemolymph induces the presence of an 88-kDa hemolymph protein (FIG. 1). The partial N-terminal sequence (V?A?MVAKEA?Q-VQLG) of this protein displays no similarities with any other known peptide or protein.

REFERENCES

1. COUILLAUD, F., J. GIRARDIE, S. S. TOBE & A. GIRARDIE. 1984. Activity of disconnected corpora allata in *Locusta migratoria*: Juvenile hormone biosynthesis *in vitro* and physiological effects *in vivo*. J. Insect Physiol. **7:** 551–556.
2. STAY, B., S. S. TOBE & W. G. BENDENA. 1994. Allatostatin: Identification, primary structures, functions and distribution. Adv. Insect Physiol. **25:** 267–337.
3. BELLÉS, X., J. L. MAESTRO, M. D. PIULACHS, A. H. JOHNSON, H. DUVE & A. THORPE. 1994. Allatostatic neuropeptides from the cockroach *Blatella germanica* (L.) (Dictyoptera, Blattellidae). Identification, immunolocalization and activity. Regul. Pept. **53:** 237–247.
4. LORENZ, M. W., R. KELLNER & K. H. HOFFMANN. 1995. Identification of two allatostatins from the cricket *Gryllus bimaculatus* de Geer (Ensifera, Gryllidae): Novel members of a family of neuropeptides inhibiting juvenile hormone biosynthesis. Regul. Pept. **57:** 227–236.
5. TOBE, S. S. & N. CLARK. 1985. The effect of L-methionine concentration on juvenile hormone biosynthesis by corpora allata of the desert locust *in vitro*. Biochem J. **144:** 107–113.
6. VEELAERT, D., L. SCHOOFS, S. S. TOBE, C. G. YU, H. G. B. VULLINGS, F. COUILLAUD & A. DE LOOF. 1995. Immunological evidence for an allatostatin-like neuropeptide in the central nervous system of *Schistocerca gregaria*, *Locusta migratoria* and *Neobellieria bullata*. Cell Tissue Res. **279:** 601–611.
7. VEELAERT, D., B. DEVREESE, L. SCHOOFS, J. VAN BEEUMEN, J. VANDEN BROECK, S. S. TOBE & A. DE LOOF. 1996. Isolation and characterization of 8 myoinhibiting peptides from the desert locust, *Schistocerca gregaria*: New members of the cockroach allatostatin family. Mol. Cell. Endocrinol. **122:** 183–190.
8. DING, Q., B. C. DONLY, S. S. TOBE & W. G. BENDENA. 1995. Comparison of the allatostatin neuropeptide precursors in the distantly related cockroaches, *Periplaneta americana* and *Diploptera punctata*. Eur. J. Biochem. **234:** 737–746.

9. VANDEN BROECK, J., D. VEELAERT, S. S. TOBE & A. DE LOOF. 1996. Molecular cloning of the precursor cDNA for schistostatins, locust allatostatin-like peptides with myoinhibiting properties. Mol. Cell. Endocrinol. **122:** 191–198.
10. REICHWALD, K., G. C. UNNITHAN, N. T. DAVIS, H. AGRICOLA, & R. FEYEREISEN. 1994. Expression of the allatostatin gene in endocrine cells of the cockroach midgut. Proc. Natl. Acad. Sci. USA **91:** 11894–11898.
11. EAST, P. D., A. THORPE & H. DUVE. 1995. Leu-callatostatin gene expression the blowflies *Calliphora vomitoria* and *Lucia cuprina* studies by *in situ* hybridization: Comparison with Leu-callatostatin confocal laser scanning immunocytochemistry. Cell Tissue Res. **280:** 355–364.

Potent, AnCE Endopeptidase-resistant, Aib-containing Analogues of the Diuretic Insect Kinin Neuropeptides[a]

RONALD J. NACHMAN,[b,e] R. ELWYN ISAAC,[c]
GEOFFREY M. COAST,[d] AND G. MARK HOLMAN[b]

[b]Veterinary Entomology Research Unit
FAPRL, USDA-ARS
2881 F&B Road
College Station, Texas 77845

[c]Department of Pure and Applied Biology
University of Leeds
Leeds, United Kingdom

[d]Department of Biology, Birkbeck College
University of London
London, United Kingdom

The insect kinin neuropeptide family shares the common C-terminal pentapeptide Phe-Xaa^1-Xaa^2-Trp-Gly-NH$_2$ (Xaa^1 = His, Asn, Phe, Ser, or Tyr; Xaa^2 = Ser, Pro, or Ala) and has been isolated from such diverse sources as the cockroach *Leucophaea maderae*, cricket *Acheta domesticus*, locust *Locusta migratoria*, corn earworm *Helicoverpa zea* as well as mosquitoes *Culex salinarius* and *Aedes aegypti*.[1,2] The leucokinins influence transepithelial membrane potential and rate of fluid secretion in isolated Malpighian tubules from the mosquito, *Aedes aegypti*.[1] Coast *et al.*[3] have shown that the achetakinins at 10^{-9} M double the rate of fluid secretion by isolated Malpighian tubules of the cricket, *Acheta domesticus*, and demonstrate EC$_{50}$ values between 10^{-10} and 10^{-11} M. Therefore, this family of peptides may regulate water and ion balance in insects.

Recent experiments demonstrate that several members of the insect kinin family are hydrolyzed, and thereby inactivated, by angiotensin-converting enzyme (AnCE) from the housefly via removal of the C-terminal dipeptide amide fragment (Fig. 1).[4] Inactivation results because the hydrolysis site is located within the insect kinin C-terminal pentapeptide active core. In mammals, the Zn^{2+} metallopeptidase AnCE is responsible for the conversion of angiotensin I to the active form angiotensin II, involved in the control of blood pressure. In addition, AnCE inactivates a wide range of mammalian peptide hormones such as bradykinin, cholecystokinin, [Leu^5], and [Met^5]enkephalinamides, substance P, and LH-RH. The broad substrate specificity and widespread distribution of AnCE in mammalian tissues suggest that it plays multiple roles in addition to controlling blood pressure, although these roles have yet to be defined. Similarly, the precise role of AnCE has not been

[a] This study was supported in part by NATO Grant No. 90248 (R.J.N., G.H.C., and G.M.H.).
[e] Corresponding author.

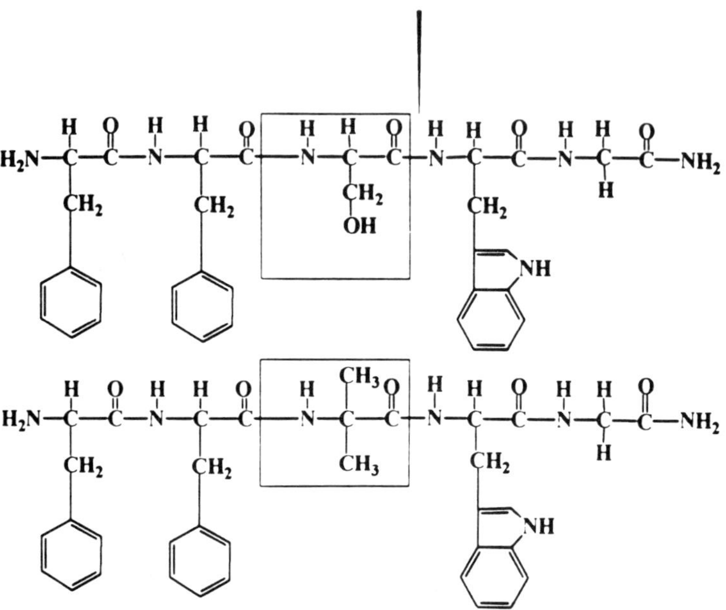

FIGURE 1. The insect kinin C-terminal pentapeptide analogue Phe-Phe-Ser-Trp-Gly-NH$_2$ is pictured (*top panel*). The arrow indicates the point at which the insect kinins are hydrolyzed, and thereby inactivated, by the endopeptidase angiotensin-converting enzyme (AnCE) from the housefly *Musca domestica*. The insect kinin analogue Phe-Phe-**Aib**-Trp-Gly-NH$_2$ (*bottom panel*) features a sterically hindered aminoisobutyrl (Aib) residue and demonstrates both complete resistance to AnCE and potent diuretic activity.

delineated in insects. However, the fact that a number of different C-terminally amidated insect neuropeptides are substrates of AnCE and the presence of the endopeptidase in hemolymph (blood) and other tissues, suggest that endopeptidase may play a role in the degradation of regulatory peptides in insects.[4]

We describe here the synthesis, diuretic activity, and AnCE susceptibility of analogues of the insect kinin C-terminal pentapeptide core, in which the Xaa^2 residue is replaced with the sterically hindered aminoisobutyric acid (Aib) residue. The analogues Phe-Phe-**Aib**-Trp-Gly-NH$_2$ and pGlu-Lys-Phe-Phe-**Aib**-Trp-Gly-NH$_2$ demonstrate complete resistance to hydrolysis by AnCE from the housefly *Musca domestica* over 120 minutes, an incubation period sufficient to hydrolyze much of the natural insect kinin leukokinin I (LK-I) (FIG. 2). Notably, the latter analogue is blocked at the N-terminus with a pGlu residue, which confers resistance to another class of endopeptidases, the aminopeptidases. The sequence of LK-I is Asp-Pro-Ala-Phe-Asn-Ser-Trp-Gly-NH$_2$, which shares with achetakinin IV (AK-IV), an Asn residue in the variable Xaa^1 position of the C-terminal pentapeptide core region.

The two analogues demonstrated potent stimulation of fluid secretion on the isolated Malpighian tubules of the cricket *Acheta domesticus*, with EC$_{50}$ values of 5.6 pM (95% CL = 3.0–10.5 pM) and 2.8 pM (95% CL = 2.1–3.8 pM), respectively. Both produce a maximal diuretic response that was not significantly differ-

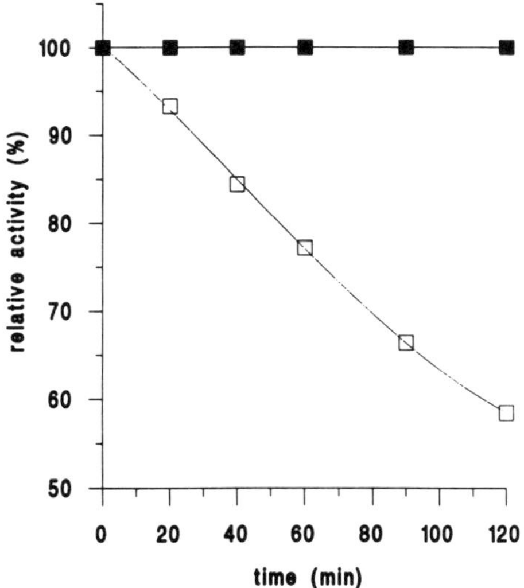

FIGURE 2. Hydrolysis trials with angiotensin-converting enzyme (AnCE) from the housefly *Musca domestica* on the naturally occurring insect kinin leucokinin I (LK-I: DPA-Phe-Asn-Ser-Trp-Gly-NH$_2$) (*open squares*) and the AnCE-resistant, insect kinin pentapeptide analogue containing Aib (*filled squares*). Refer to FIGURE 1.

ent from that obtained with the endogenous achetakinin peptides. The natural achetakinins elicit cricket Malpighian tubule secretion at EC$_{50}$ values ranging from about 20 to 325 pm.[3] The two analogues are therefore about four to eight times more potent in the Malpighian tubule assay than the most potent naturally occurring achetakinin peptide. The Aib residue is compatible with the formation of a turn at this position in the active core that is important for the diuretic activity of the insect kinins. Unfavorable steric interactions between the branched chain of the Aib alpha carbon and side chains of surrounding residues promote the formation of a turn or kink to alleviate strain.[5] This characteristic explains, at least in part, the potent biological activities observed for these kinin analogues.

In aggregate, the results suggest that incorporation of Aib residues in appropriate turn regions of other insect peptides can confer a measure of resistance to AnCE and/or other endopeptidases that degrade them, potentially without loss of biological activity. The Aib-containing and similar enzyme-resistant analogues of insect neuropeptides can provide useful tools to insect physiologists studying the neuroendocrine control of water and ion balance, as well as the physiological/behavioral consequences of challenging an insect with a diuretic signal it cannot degrade. If these analogues, whether in isolation or in combination with other factors, can disrupt the water and ion balance of insects, they hold promise for the future in the control of pest insect populations.

REFERENCES

1. NACHMAN, R. J., G. M. COAST, G. M. HOLMAN & R. C. BEIER. 1995. Diuretic activity

of C-terminal group analogues of the insect kinins in *Acheta domesticus*. Peptides **16:** 809–813.

2. BLACKBURN, M. B., R. M. WAGNER, J. SHABANOWITZ, J. P. KOCHANSKY, D. F. HUNT & A. D. RAINA. 1995. The isolation and identification of three diuretic kinins from the abdominal ventral nerve cord of adult *Helicoverpa zea*. J. Insect Physiol. **41:** 723–730.

3. COAST, G. M., G. M. HOLMAN & R. J. NACHMAN. 1990. The diuretic activity of a series of cephalomyotropic neuropeptides, the achetakinins, on isolated Malpighian tubules of the house cricket, *Acheta domesticus*. J. Insect Physiol. **36:** 481–488.

4. ISAAC, R. E., N. S. LAMANGO, R. J. NACHMAN, A. STREY & T. K. HAYES. 1996. Angiotensin-converting enzyme and the metabolism of regulatory peptides in insects. Ann. N.Y. Acad. Sci. This volume.

5. TONIOLO, C., G. M. BONORA, A. BAVOSO, E. BENEDETTI, B. DI BLASIO, V. PAVONE & C. PEDONE. 1983. Preferred conformations of peptide analogs containing α,α-disubstituted α-amino acids. Biopolymers **22:** 205–214.

Insect Myosuppressins and Sulfakinins Stimulate Release of the Digestive Enzyme α-Amylase in Two Invertebrates: The Scallop *Pecten maximus* and Insect *Rhynchophorus ferrugineus*

RONALD J. NACHMAN,[a,d] WILFRID GIARD,[b]
PASCAL FAVREL,[b] T. SURESH,[c] S. SREEKUMAR,[c]
AND G. MARK HOLMAN[a]

[a]*Veterinary Entomology Research Unit*
FAPRL, USDA-ARS
2881 F&B Road
College Station, Texas 77845

[b]*Laboratoire de Biologie et Biotechnologies Marines*
Université de Caen
Caen, France

[c]*Department of Zoology*
University College
Trivandrum, India

The myosuppressins represent a subfamily of the insect FMRFamide-related peptides that share the common C-terminal heptapeptide sequence Asp-His-Val-Phe-Leu-Arg-Phe-NH$_2$,[1] and have been identified in diverse insect species including the cockroach *Leucophaea maderae*, locusts *Schistocerca gregaria* and *Locusta migratoria*, the fleshfly *Neobelleria bullata*, and the fruitfly *Drosophila melanogaster*. Members of this family affect insect muscles, being potent inhibitors of cardiac and visceral muscle in addition to having effects on skeletal muscle and salivary glands. A related peptide, named ManducaFLRFamide, was isolated from the hawkmoth *Manduca sexta* and demonstrates myostimulatory activity on dorsal longitudinal flight muscles.[1]

The related sulfakinin insect peptide family shares the C-terminal octapeptide sequence Xaa-Asp-Tyr(SO$_3$H)-Gly-His-Met-Arg-Phe-NH$_2$ (Xaa = Glu or Asp) that terminates in HMRFamide, rather than the FLRFamide of the myosuppressins. The sulfakinins also share sequence similarities with the vertebrate gastrin/cholecystokinin (CCK) hormone family with a C-terminal gastrin octapeptide sequence of Glu-Ala-Tyr(SO$_3$H)-Gly-Trp-Met-Arg-Phe-NH$_2$. Unlike the myosuppressins, the sulfakinins stimulate contractions of the isolated cockroach hindgut, the original bioassay used to isolate them.[2,3]

In this study, we report that members of the myosuppressin and sulfakinin

---

[d] Corresponding author.

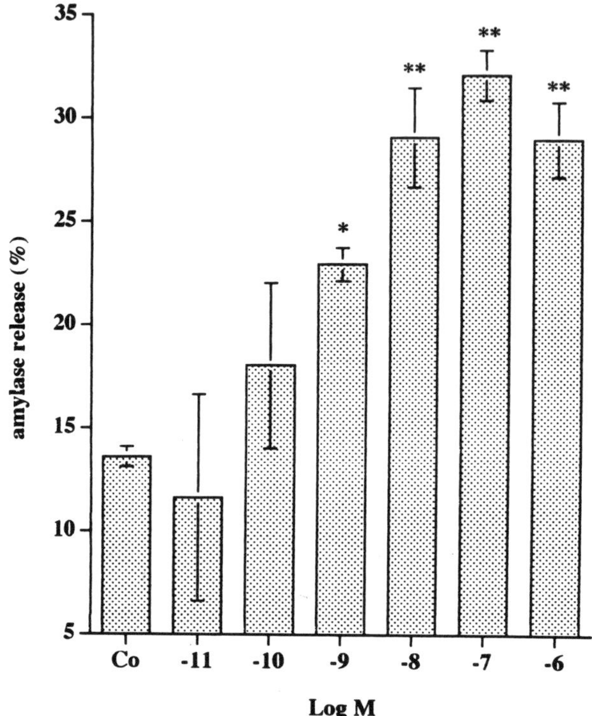

FIGURE 1. Dose-response for induction of α-amylase release from a cell suspension of the scallop (*Pecten maximus*) stomach-digestive gland complex[4] by insect leucomyosuppressin (LMS). *p <0.01; **p <0.001; $n = 6$.

peptide families stimulate release of the digestive enzyme α-amylase from digestive tract preparations of two invertebrates: the scallop *Pecten maximus* and the insect *Rhynchophorus ferrugineus*. Leucomyosuppressin (LMS: pGlu-Asp-Val-Asp-His-Val-Phe-Leu-Arg-Phe-NH_2), from the cockroach *Leucophaea maderae*, elicits over a twofold increase in α-amylase secretion from a cell suspension of the scallop stomach-digestive gland complex over controls at an EC_{50} of 0.2 nM (FIG. 1), as compared with an EC_{50} of about 20 nM for the fragments FMRFamide and FLRFamide.[4] The sulfakinin analogues [Phe[8], Nle[9]]LSK (Glu-Gln-Phe-Glu-Asp-Tyr[SO_3H]-Gly-Phe-Nle-Arg-Phe-NH_2) and (Ser[SO_3H][2])LSK-II (pGlu-Ser[SO_3H]-Asp-Asp-Tyr[SO_3H]-Gly-His-Met-Arg-Phe-NH_2) elicited amylase secretion at the reduced EC_{50} levels of 7 and 30 nM, respectively. However, both elicited the same maximal response as LMS.

In the insect *Rhynchophorus ferrugineus*, the peptides LMS, [Phe[8], Nle[9]]LSK (FLSK), FMRFamide, and the mosquito peptide culetachykinin I (CTK-I: Ala-Pro-Ser-Gly-Phe-Met-Gly-Met-Arg-NH_2) (Holman *et al.*, unpublished data) were incubated at a concentration of 1 μM with an isolated, ligated larval midgut bathed in saline. Following this, the lumen contents were assayed for levels of α-amylase.[5] As illustrated in FIGURE 2, LMS, FLSK, and FMRFamide all induced over a twofold increase in the release of α-amylase over controls, whereas the unrelated CTK-I sequence had no effect. These active, insect peptides mimicked the effect of unidentified factors in extracts of the midgut epithelial tissues of several insects.[5] Notably, expression of the LMS gene of the cockroach *Diploptera punctata* occurs in cells of the midgut, consistent with a digestive enzyme regulatory role for the myosuppressins.[6]

Replacement of a single, C-terminal–region residue (Arg for Asp) transforms the inactive mammalian hormones gastrin II and CCK into active sulfakinin analogues on the isolated cockroach hindgut bioassay.[3] However, Arg is a basic, positively charged residue, whereas Asp is an acidic, negatively charged residue. They would appear to be quite different. Evidence from structure-activity studies suggests that this difference is not as large as it would appear. For example, analogues of CCK in which the Asp residue is replaced with a neutral Pro retain a significant 10% of the potency of CCK in pancreatic binding and satiety assays in rats.[7] Similarly, replacement of the Arg residue in the analogue position of [4–11]LSK with neutral Pro also leads to retention of a significant 10% of the myostimulatory potency of the parent peptide in the insect hindgut bioassay (Nachman, unpublished data). The results suggest that the negative and positive charges of the Asp and Arg residues, respectively, are not crucial for successful receptor

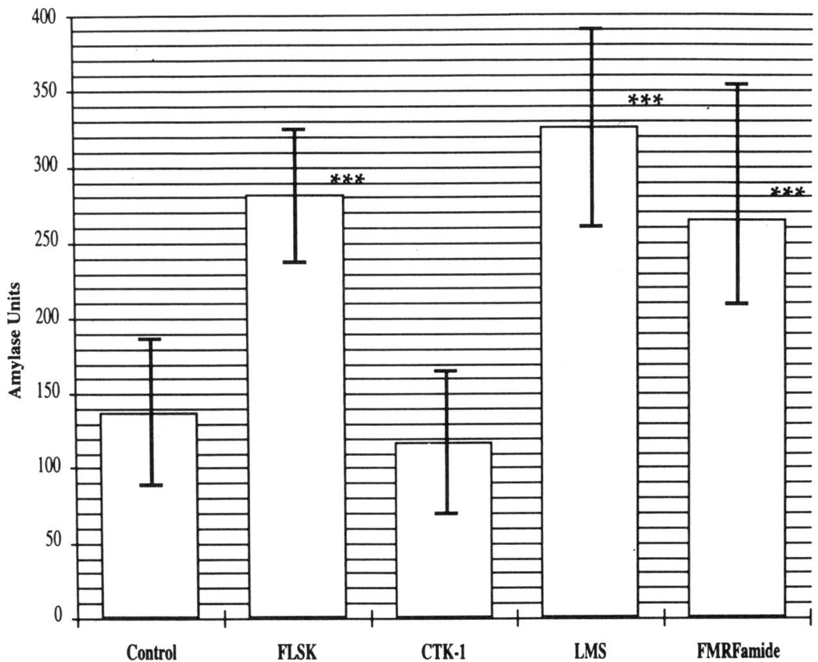

FIGURE 2. α-Amylase secretion activity from the isolated, ligated larval midgut[5] of the insect *Rhynchophorus ferrugineus* elicited by the insect peptides [Phe[8], Nle[9]]LSK (FLSK), culetachykinin I (CTK-I), leucomysosuppressin (LMS), and FMRFamide. ***p <0.001; n = 8 for each group.

interaction. Rather, they suggest that the contributions these two residues make to the conformations adopted by the respective peptide families at the receptor sites are an important factor for biological activity. Replacement of the Asp and Arg residues with Pro leads to a common sequence (Glu-Asp-Tyr[SO$_3$H]-Gly-His-Met-**Pro**-Phe-NH$_2$)[e] that would be expected to be active in both vertebrate gastrin/ CCK and invertebrate sulfakinin bioassay systems. These results raise the intriguing possibility that regulation of the release of the digestive enzyme α-amylase passed from myosuppressin/sulfakinin/FMRFamide-like neuropeptides in invertebrates to cholecystokinin in vertebrates.

REFERENCES

1. NACHMAN, R. J., E. H. OLENDER, V. A. ROBERTS, G. M. HOLMAN & D. YAMAMOTO. 1996. A nonpeptidal peptidomimetic agonist of the insect FLRFamide myosuppressin family. Peptides **17(2):** 313–320.
2. NACHMAN, R. J., G. M. HOLMAN, W. F. HADDON & N. LING. 1986. Leucosulfakinin, a sulfated insect neuropeptide with homology to gastrin and cholecystokinin. Science **234:** 71–73.
3. NACHMAN, R. J., G. M. HOLMAN & W. F. HADDON. 1988. Structural aspects of gastrin/ CCK-like insect leucosulfakinins and FMRFamide. Peptides **9(1):** 137–143.
4. FAVREL, P., W. GIARD, N. BENLIMANE, E. BOUCAUD-CAMOU & M. HENRY. 1994. A new biological activity for the neuropeptide FMRFamide: Experimental evidence for a secretagogue effect on amylase secretion in the scallop *Pecten maximus*. Experientia **50:** 1106–1110.
5. SREEKUMAR, S. 1994. Digestive enzyme secretion stimulating hormone in insects: Results of heterologous assay. Indian J. Exp. Biol. **32:** 914–915.
6. ORCHARD, I., B. C. DONLY, M. FUSE, A. B. LANGE, S. S. TOBE & W. G. BENDENA. 1996. FMRFamide-related peptides in insects, with emphasis on the myosuppressins. Ann. N. Y. Acad. Sci. This volume.
7. TILLEY, J. W., W. DANHO, V. MADISON, D. FRY, J. SWISTOK, R. MAKOFSKE, J. MICHAELEWSKY, A. SCHWARTZ, S. WEATHERFORD, J. TRISCARI & D. NELSON. 1992. Analogs of CCK incorporating conformationally constrained replacements for Asp. J. Med. Chem. **35:** 4249–4252.

[e] Replacement of the His with other aromatic residues, such as Phe and Trp (as in gastrin/ CCK) is accompanied by retention of activity in the insect myotropic assay.[3]

Angiotensin-Converting Enzyme and the Metabolism of Regulatory Peptides in Insects

R. ELWYN ISAAC,[a] NAZARIUS S. LAMANGO,[a]
RONALD J. NACHMAN,[b] A. STREY,[c]
AND TIMOTHY K. HAYES[c]

[a]Department of Biology
University of Leeds
Leeds LS2 9JT, United Kingdom

[b]Veterinary Entomology Research Unit
Food and Animal Protection Research Laboratory, USDA
2881 F&B Road
College Station, Texas 77845

[c]Laboratories for Invertebrate Neuroendocrine Research
Department of Entomology
Texas A&M University
College Station, Texas 77843

Mammalian angiotensin-converting enzyme (ACE) is a dipeptidyl carboxypeptidase best known for its role in the biosynthesis of the potent vasoconstrictor angiotensin II from angiotensin I and the inactivation of the vasodilator, bradykinin.[1] In addition to its dipeptidyl carboxypeptidase activity, ACE can also act *in vitro* as an endopeptidase hydrolyzing C-terminally amidated peptides (e.g., [Leu[5]]enkephalinamide, [Met[5]]enkephalinamide, substance P, and LH-RH). The broad substrate specificity of ACE and its widespread tissue distribution in mammals suggest that the enzyme may have a number of functionally distinct roles.[1] We have recently shown that ACE is of ancient evolutionary origin by identifying a highly homologous enzyme from the housefly, *Musca domestica*, and from *Drosophila melanogaster*.[2,3] Elucidation of the role of ACE in insects may provide clues to hitherto unknown functions of ACE in mammalian tissues.

ACE is present in many insect tissues including the hemolymph, and we have speculated that this activity might be involved in the metabolism of circulating peptide hormones. There is immunocytochemical evidence to indicate that insect bioactive peptides can function as neurotransmitters as well as hormones, and therefore the recent observations by Dr. L. Schoofs (University of Leuven, personal communication) that ACE immunoreactivity is present in neuropile regions of the insect brain suggest that the enzyme might also be responsible for inactivating peptide transmitters at synapses. Most of the insect peptide hormones that have been characterized possess an amidated C-terminus, and therefore we have investigated whether some of these insect peptides are hydrolyzed by housefly ACE. ACE was purified from bodies of *Musca domestica* by applying a soluble protein fraction to a lisinopril-sepharose affinity column (FIG. 1). This procedure yielded a single protein band on a SDS-PAGE gel and a 700-fold purification of ACE. Housefly ACE hydrolyzed angiotensin I, bradykinin, [Leu[5]]enkephalin, and [Met[5]]enkephalin by cleaving the C-terminal dipeptide. Insect ACE, like the

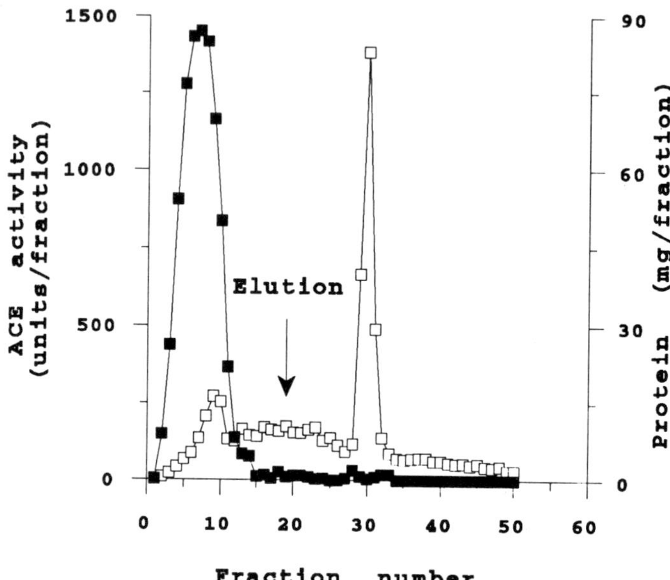

FIGURE 1. Purification of *Musca domestica* ACE using a lisinopril-sepharose affinity column. A 40,000 × *g* supernatant prepared from bodies of adult *M. domestica* was loaded onto the affinity column in 10 mm Tris/HCl, pH 8.3, containing 0.2 M $(NH_4)_2SO_4$. After extensive washing with the loading buffer, elution of the enzyme was achieved by washing the column with 10 mm Tris/HCl, pH 8.3. Protein (*filled squares*) was estimated using bovine serum albumin as a standard, and ACE activity (*open squares*) was assayed using Hip-His-Leu as the substrate.

mammalian enzyme, was also able to hydrolyze [Leu[5]]enkephalinamide, [Met[5]]enkephalinamide, substance P, and LH-RH, all of which have an amidated C-terminus. A range of insect peptide hormones were also tested as substrates (TABLE 1), and all except proctolin and crustacean cardioactive peptide (CCAP) were hydrolyzed. Peptides with a proline residue in the penultimate position are

TABLE 1. Peptides Hydrolyzed by *Musca domestica* Angiotensin-Converting Enzyme

Mammalian Peptides	Insect Peptides
Angiotensin I	Culex depolarizing peptide I
Bradykinin	Culex depolarizing peptide II
[Leu[5]]enkephalin	Leucokinin I
[Met[5]]enkephalin	Leucokinin II
[Leu[5]]enkephalinamide	Locustatachykinin I
[Met[5]]enkephalinamide	Locustatachykinin II
Substance P	Cockroach myoactive peptide I
LH-RH	Locust adipokinetic hormone I
	Stick insect hypertrehalosemic hormone
	Red pigment concentrating hormone

known to be poor substrates for mammalian ACE and because proctolin has a penultimate proline, it was not unexpected that this insect peptide would be resistant to hydrolysis. CCAP has a cystine bond which probably results in secondary structure that prevents access to the active site. The penultimate C-terminal peptide bond was identified as the site of initial hydrolysis for culex depolarizing peptide II, leucokinin I, leucokinin II, and locustatachykinin I. In conclusion, we have shown that housefly ACE, like its mammalian homologue, can hydrolyze peptides with an amidated C-terminus. Because the core structure for bioactivity often resides in the C-terminus of insect regulatory peptides, it is likely that cleavage of C-terminal dipeptides will result in a biologically inactive fragment. Thus, ACE is a candidate inactivation enzyme for a number of insect neuropeptides. However, the precise role of insect ACE in the central nervous system and in peripheral tissues must be established by studying the effects of enzyme inhibitors on the metabolism of peptides *in vivo*.

REFERENCES

1. ERDOS, E. G. 1990. Hypertension **16:** 363–370.
2. LAMANGO, N. S. & R. E. ISAAC. 1994. Biochem. J. **299:** 651–657.
3. CORNELL, M. J., T. A. WILLIAMS, N. S. LAMANGO, D. COATES, P. CORVOL, F. SOUBRIER, J. HOHEISEL, H. LEHRACH & R. E. ISAAC. 1995. J. Biol. Chem. **270:** 13613–13619.

Biochemical Characterization of Mosquito Kinin and Related Receptors[a]

TIMOTHY K. HAYES,[b,g] ALLISON STREY,[b]
STEPHANIE BELK,[c] G. MARK HOLMAN,[d]
RONALD J. NACHMAN,[d] DAVID PETZEL,[e] JAN READIO,[b]
ROGER MEOLA,[b] TOM PANNABECKER,[f]
AND KLAUS BEYENBACH[f]

*bDepartment of Entomology
Texas A&M University
College Station, Texas 77843*

*cSweetbriar College
Sweetbriar, Virginia 24595*

*dFood Animal Protection Laboratory
USDA-ARS
College Station, Texas 77843*

*eDivision of Physiology, School of Medicine
Creighton University
Omaha, Nebraska 68178*

*fSection of Physiology, Veterinary Research Tower
Cornell University
Ithaca, New York 14853*

The anatomy of the diuretic system in mosquitoes includes the Malpighian tubule, the gut, and the rectum. The potential involvement of gut movement and thus myotropic peptides in the regulation of salt and water balance was suspected because of this anatomical relationship. Thus, several families (i.e., proctolin, AKH-hypertrehalosemic hormone (HTH), pyrokinin, sulfakinins, insect tachykinins, and the insect kinins) of small insect myotropic peptides were tested in a transepithelial voltage (TEV) assay of mosquito Malpighian tubules. Of these families, only the insect kinin family stimulated a response from the Malpighian tubules (i.e., a depolarization).[1]

The insect kinin family is found in several insect species and the members share a conserved C-terminal pentamer: FX(S/P)WGamide. At least six of these kinin peptides have been isolated and characterized from two mosquito species— *Culex salinarius*[2]: NPFHSWGamide; NNANVFYPWGamide; XKYVSKQKF-FSWGamide; and *Aedes aegypti*[3]: NPFHAWGamide; NNPNVFPWGamide; NSKYVSKQKFYSWGamide. Kinins have been shown to increase the paracellular Cl⁻ conductance across the transepithelial membrane of the mosquito Malpi-

[a] We are grateful for grants from the National Science Foundation (IBN-9419990 to T.K.H.) and the National Institutes of Health (RO1 NS20137 to T.K.H.) for supporting aspects of this work.
[g] Present address: Timothy K. Hayes, Bayer Corporation, NCSU Centennial Campus, 1017 Main Campus Drive, Suite 3800, Raleigh, NC 27606.

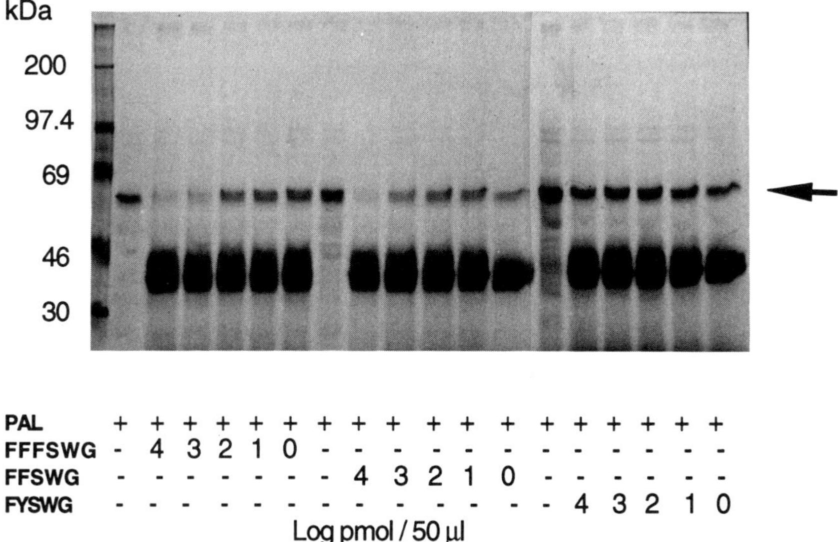

FIGURE 1. SDS-PAGE analysis of mosquito Malpighian tubule membrane proteins: kinin agonists protect against photoaffinity labeling of kinin receptors. The *arrow* highlights the two proteins (60 and 62 kDa) that are photoaffinity labeled. The sequence of each kinin agonist is listed and each peptide is used in the C-terminal amide form. The other intense bands are nonspecifically labeled and are related to the commercial preparation of BLG present at very high relative concentrations with the membranes.

ghian tubule.[4] Structure-activity studies have indicated that the C-terminal penta-peptide core is required for activity and that the conserved Phe and Trp were the most important residues within that core for depolarization of mosquito Malpighian tubules.[5]

Kinin structure-activity information for mosquito Malpighian tubules was used to design a photoaffinity label (PAL, see sequence on next page), which has been used as the primary tool to study kinin receptor-like binding proteins. The PAL is composed of a receptor interactive unit (the C-terminal pentamer), a site for photocross-linking (4-benzoyl-phenylalanine, Bpa), and a Tyr for radioiodination. The amino acids of the N-terminal dipeptide of the PAL are in the D-configuration to resist breakdown by aminopeptidases.

Photoaffinity labeling resulted in the specific (i.e., PALing can be protected against by a kinin agonist) labeling of a protein around 60 kDa and a slightly larger protein (i.e., approximately 62 kDa) that range in size to as high as 69 kDa in some preparations (FIG. 1). The visualization of either of these proteins was highly dependent on the presence of a cocktail of protease inhibitors and a high concentration of a "non-specific protein" such as β-lactoglobulin (BLG) or bovine serum albumin (BSA) present in the buffers for membrane preparation and photolysis. Likewise, the ratio of PAL/membrane and the photolysis time were important for the detection of specific labeling. The specific PALing result was dependent on the presence of the membrane, the PAL, and the photolysis (TABLE 1).

A connection between the protein band visualized on the gel by PALing and

TABLE 1. Example Agonists Used for PAL Protection

Least potent analogue	FYSWG-amide
More potent analogue	FFSWG-amide
Most potent analogue (superagonist)	FFFSWG-amide
Photoaffinity label	ay-(Bpa)-KFFSWG-amide

NOTE: Amino acids indicated by lowercase single-letter abbreviation are in the D-configuration.

the pharmacology and physiology of the kinin receptor in the Malpighian tubule was established. Nonradioactive kinin agonists protected against PALing in a rank order that reflected their potency in the TEV bioassay (FIG. 1). Likewise, other myoactive peptide families that failed to stimulate a response in the TEV assay failed to protect against PALing of the kinin receptor (FIG. 2). The example illustrated in FIGURE 2 uses (pQVNFSPGWGTamide; HTH). Note that this peptide contains the aromatic amino acids F and W yet in a different sequence position from that found in the kinins. HTH did not influence the Malpighian tubule TEV assay and did not protect against PALing by the kinin-based PAL. PALs that were designed to meet the specifications of other myoactive peptides in other

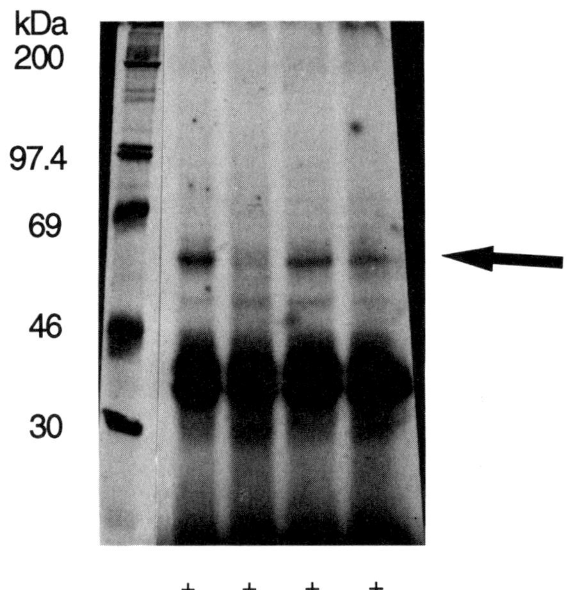

FIGURE 2. SDS-PAGE analysis of mosquito Malpighian tubule membrane proteins: HTH fails to protect against photoaffinity labeling of kinin receptors. The *arrow* highlights the protein (60 kDa) that is photoaffinity labeled. The sequence of the kinin agonist is listed, and the peptide is used in the C-terminal amide form.

insect systems failed to specifically label a Malpighian tubule receptor-like binding protein just as the peptides failed to stimulate the TEV bioassay.

The use of the kinin-based PAL has detected receptor-like binding proteins (i.e., approximately 60 and 62 kDa) in membranes from larval *Aedes albopictus* cell cultures (ATCI# CL 1660). These proteins have similar characteristics as the kinin receptors first detected in mosquito Malpighian tubule membranes.

REFERENCES

1. HAYES, T. K., T. L. PANNABECKER, D. J. HINCKLEY, G. M. HOLMAN, R. J. NACHMAN, D. H. PETZEL & K. W. BEYENBACH. 1989. Life Sci. **44:** 1259–1266.
2. HAYES, T. K., G. M. HOLMAN, T. L. PANNABECKER, M. S. WRIGHT, A. A. STREY, R. J. NACHMAN, D. F. HOEL, J. K. OLSON & K. W. BEYENBACH. 1994. Regul. Pept. **52:** 235–248.
3. VEENSTRA, J. 1994. Biochem. Biophy. Res. Commun. **202:** 715–719.
4. PANNABECKER, T. L., T. K. HAYES & K. W. BEYENBACH. 1993. J. Membr. Biol. **132:** 63–76.
5. STREY, A., T. HAYES, M. WRIGHT, R. MEOLA, J. KELLY, G. HOLMAN, R. NACHMAN, F. CLOTTENS & D. PETZEL. 1994. *In* Peptides: Chemistry, Structure and Biology. R. Hodges & J. Smith, Eds.: 704–706. ESCOM Science Publishers B. V. Leiden, the Netherlands.

Molecular Characterization of the *Manduca sexta*-type of Allatostatin in the Migratory Moth *Pseudaletia unipuncta*

W. G. BENDENA,[a,e] I. S. JANSONS,[a] M. CUSSON,[b]
J. N. McNEIL,[c] P. KOLADICH,[d] AND S. S. TOBE[d]

[a]*Department of Biology*
Queen's University
Kingston, Ontario, Canada K7L 3N6

[b]*Laurentian Forestry Centre*
Ste. Foy, Québec, Canada G1V 4C7

[c]*Département de biologie*
Université Laval
Ste. Foy, Québec, Canada G1K 7P4

[d]*Department of Zoology*
University of Toronto
Toronto, Ontario, Canada M5S 1A1

Pseudaletia unipuncta, the true armyworm, exhibits several juvenile hormone (JH)-dependent processes which include the onset of calling, pheromone synthesis, and oocyte maturation. Three forms of JH (JHI, JHII, and JHIII) or JH acids (JHAI, JHAII, and JHAIII) are released from the corpora allata (CA) of females and males, respectively. The quantities of the three JH or JHA homologues produced by CA in both males and females vary with age and environmental parameters such as photoperiod and temperature.[1] Defined ratios of the JHs may be necessary for optimal hormonal response.[1] Studies on *in vitro* JH biosynthesis by CA of *P. unipuncta* have shown that the low rates of JH release early in the life of adult females coincide with the absence of calling behavior and pheromone production whereas the subsequent increases in JH release coincide with the onset of pheromone biosynthesis and release.[2] Evidence also suggests that JH regulates migratory flight,[1] and that a relationship exists between CA activity and the maturation of the pheromonal communication system in both sexes. This includes the response of male *P. unipuncta* to female calling and sex pheromone release. JHA production by male CA correlates with male responsiveness to female pheromone. Rankin and Riddiford proposed a model to explain the control of both migratory flight and oogenesis by JH.[2] In brief, migratory flight is initiated when the JH titer exceeds a minimum level, and if the JH titer increases above a higher threshold level, sexual maturation is stimulated. The relationship between the JH titer and these events may not be as simple as suggested. The changes in the ratios of the most abundant homologues of JH in *P. unipuncta* (JHI and JHII in females; JHAI

[e] Corresponding author; e-mail: bendenaw@biology.queensu.ca

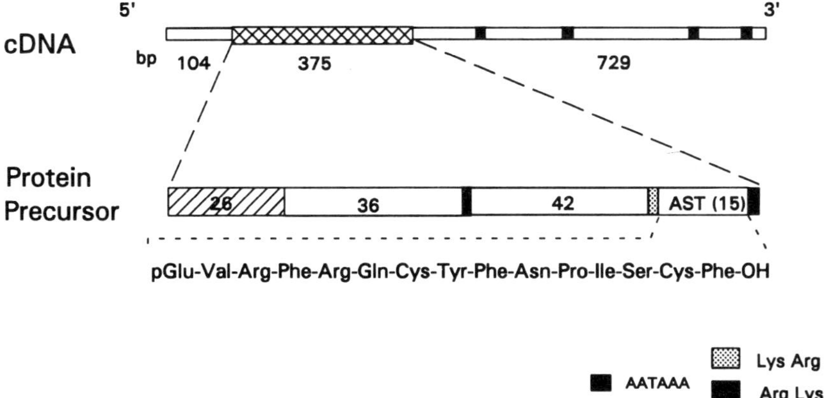

FIGURE 1. Schematic diagram of the *P. unipuncta* AST cDNA and polypeptide precursor. The 375 bp open reading frame (*cross-hatched box*) encodes the polypeptide precursor which begins with a 26 amino acid hydrophobic leader (*box with diagonal lines*). Black and shaded boxes within the polypeptide precursor indicate potential dibasic proteolytic cleavage sites as indicated. The 15 amino acid peptide corresponds to the AST polypeptide previously purified from *M. sexta*.

and JHAII in males) as brought about by changes in total JH biosynthesis may influence the reproductive state or the onset of migratory flight. These changes can be regulated by environmental conditions.[1] In response to colder temperatures, *P. unipuncta* may abstain from reproduction and initiate migratory flight to over-wintering areas. Alternatively, under summer-like conditions the moths will likely reproduce and die without ever migrating.

A major component in the regulation of JH biosynthesis under varying conditions in *P. unipuncta* may reside in the action of the neuropeptides allatotropin and allatostatin, which act in a stimulatory or inhibitory fashion, respectively. The only allatotropin isolated to date is a α-amidated tridecapeptide from late pharate adult head extracts of the tobacco hornworm, *Manduca sexta*.[3] This peptide appears to function specifically in lepidopteran adults. In contrast, allatostatins (ASTs) have been purified from several insect species, including the cockroaches *D. punctata*,[4] *P. americana*[5] and *Blattella germanica*,[6] the cricket *G. bimaculatus*,[7] and the moth *M. sexta*.[8] Allatostatins from both cockroaches and crickets are characterized by a common carboxyl terminus—Tyr/Phe-Xaa-Phe-Gly-Leu/Ile-NH$_2$ (where Xaa = Ser, Gly, Asp, Asn, Ala)—which is required for activity.[4] An additional family of nonapeptide allatostatins has also been isolated from crickets which share sequence similarly to a locust myosuppressin.[7] This finding is not surprising in light of the myotropic activities associated with the cockroach allatostatins.[8]

The allatostatin of *M. sexta* is a pentadecapeptide with the sequence pGlu-Val-Arg-Phe-Arg-Glu-Cys-Tyr-Phe-Asn-Pro-Ile-Ser-Cys-Phe (pGlu = pyrogluta-mate),[9] which is distinct from that of cockroaches and crickets. The action of this allatostatin is reversible and is specific to Lepidoptera.[9] Based on the amino acid sequence of the *M. sexta* allatostatin, a 45-nucleotide antisense oligodeoxynucleo-tide probe (5'GAA GCA IGA GAT IGG GTT GAA GTA GCA CTG ICG GAA

ICG CAC CTG-3') was used to screen a *P. unipuncta* brain cDNA library. The isolated cDNA contained an insert of 1,229 nucleotides comprising a 104-nucleotide 5' untranslated sequence, a 375 bp open reading frame that encodes a 125 amino acid protein precursor, a 729-nucleotide 3' untranslated region and a 21-nucleotide poly(A) tail (FIG. 1). In contrast to the cockroach AST precursors, the AST polypeptide precursor of *P. unipuncta* contains only one AST peptide, located at the carboxyl-terminus. The AST amino acid sequence derived from the DNA sequence is identical to the amino acid sequence of *M. sexta* AST. Endoproteolytic cleavage at dibasic cleavage sites would result in release of the AST and two additional peptides of 36 and 42 amino acids. These latter peptides share no sequence similarity to any known peptide. Southern blot analysis has confirmed that AST sequences exist as single copy in the genome of *P. unipuncta*. However, multiple allelic polymorphisms were detected in DNA samples from multiple individuals. Cross-species high-stringency hybridization with a variety of moth species revealed single or multiple hybridizing bands (FIG. 2), suggesting that the genomes of all Lepidoptera contain similar sequences. Whole-mount *in situ* hybridization of *P. unipuncta* adult brains revealed that the AST gene is expressed by a cluster of approximately 24 cells located in the deutocerebrum. Expression was also detected in a symmetrical pair of cells of the subesophageal ganglion and in the lateral regions of the protocerebrum. Antibody staining has also revealed

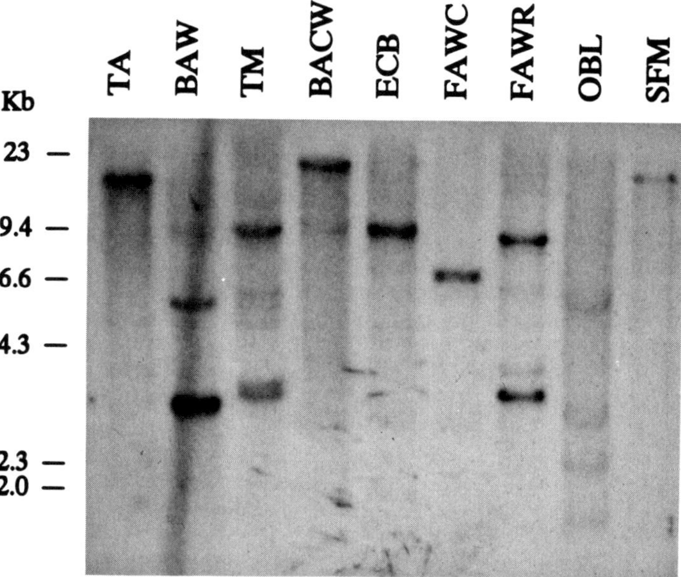

FIGURE 2. Cross-species Southern hybridization analysis of total genomic DNA (10 μg/ lane) using a [^{32}P]-labeled fragment of the *P. unipuncta* AST cDNA. Genomic DNA samples from moths were digested with Eco RI: TA, *Pseudaletia unipuncta*; BAW, *Mamestra configurata*; TM, *Lacanobia oleracia*; BACW, *Actebia fennica*; ECB, *Ostrinia nubilalis*; FAWC, *Spodoptera frugiperda* (corn strain); FAWR, *Spodoptera frugiperda* (rice strain); OBL, *Choristoneura rosaceana*; SFM, *Ostrinia nubilalis*. Relative positions of DNA size markers (λ DNA digested with Hind III) are shown on the left of the figure.

cells expressing the AST peptide in the same regions of the brain. In *P. unipuncta*, steady-state AST mRNA levels expressed from brain tissue were found to be low in sixth instar larvae, prepupae and early pupae. Relative transcript levels were maximal in late pupae and the first three days of adult life in both sexes. This was unexpected because both JH and JHA levels increase during the first 5 days of life; if AST is a primary regulator of JH production, its expression would be expected to decrease during this time. In addition, *in vitro* assay demonstrated that 1 μM AST was unable to inhibit JH biosynthesis by CA of *P. unipuncta* sixth instar larvae or newly emerged adults, but inhibited CA of 5-day-old females by 60%. This contrasts with *in vitro* assays with CA from adult *M. sexta* and *Helicoverpa zea* in which inhibition of JH biosynthesis was 100% (0.1 μM) and 77% (0.5 μM), respectively. Thus, taken together the expression and activity data suggest that the Lepidopteran AST exerts only a weak allatostatic activity in *P. unipuncta* under the summer rearing conditions and for the developmental stages tested. This within-order interspecies difference in AST activity may be analogous to the interorder variation in allatostatin activity in the cockroach *D. punctata* and the blowfly, *Calliphora vomitoria*. Both species produce a family of peptides with similar primary structures, and, although they have high allatostatic activity in the cockroach, they do not affect JH biosynthesis in the blowfly.

Many neuropeptides have also been shown to have multiple activities. Cockroach ASTs, in addition to inhibiting JH biosynthesis by CA, also inhibit both myogenic and proctolin-induced contractions of the hindgut.[8,10,11] Similarly, the *M. sexta* allatotropin that stimulates JH biosynthesis also acts as a cardioaccelerator.[12] AST in *P. unipuncta* may therefore have major physiological actions beyond the modulation of JH production.

Our understanding of regulation of JH titers in *P. unipuncta* and other adult Lepidopterans and their importance to the regulation of physiological processes is still vague. To obtain a clear picture of the role of ASTs and JH in adult reproduction and development, further work will be necessary to examine an array of Lepidoptera with different life histories and rearing conditions.

REFERENCES

1. McNeil, J. N., M. Cusson, J. Delisle, I. Orchard & S. S. Tobe. 1995. *In* Migration: Physical Factors and Physiological Mechanisms. V. A. Drake & A. G. Gatehouse, Eds.: 279–302. Cambridge University Press. Cambridge, UK.
2. Rankin, M. A. & L. M. Riddiford. 1978. J. Insect Physiol. **24:** 31–38.
3. Kataoka, H., A. Toschi, J. P. Li, R. L. Carney, D. A. Schooley & S. J. Kramer. 1989. Science **243:** 1481–1483.
4. Stay, B., S. S. Tobe & W. G. Bendena. 1994. Adv. Insect Physiol. **25:** 267–338.
5. Weaver, R. J., Z. A. Freeman, M. G. Pickering & J. P. Edwards. 1994. Comp. Biochem. Physiol. B **107C:** 119–127.
6. Belles, X., J. L. Maestro, M. D. Piulachs, A. H. Johnsen, H. Duve & A. Thorpe. 1994. Regul. Pept. **53:** 237–248.
7. Lorenz, M. W., R. Kellner & K. H. Hoffman. 1995. J. Biol. Chem. **270:** 21103–21108.
8. Lange, A. B., W. G. Bendena & S. S. Tobe. 1995. J. Insect Physiol. **41:** 581–588.
9. Kramer, S. J., A. Toschi, C. A. Miller, H. Kataoka, G. B. Quistad, J. P. Li, R. L. Carney & D. A. Schooley. 1991. Proc. Natl. Acad. Sci. USA **88:** 9458–9462.
10. Donly, B. C., Q. Ding, S. S. Tobe & W. G. Bendena. 1993. Proc. Natl. Acad. Sci. USA **90:** 8807–8811.
11. Q. Ding, B. C. Donly, S. S. Tobe & W. G. Bendena. 1995. Eur. J. Biochem. **234:** 737–746.
12. J. A. Veenstra, H. Lehman & N. T. Davis. 1994. J. Exp. Biol. **188:** 347–354.

Index of Contributors